Graduate Texts in Mathematics 260

For other titles published in this series, go to
www.springer.com/series/136

Jürgen Herzog · Takayuki Hibi

Monomial Ideals

 Springer

Jürgen Herzog
Universität Duisburg-Essen
Fachbereich Mathematik
Campus Essen
Universitätsstraße 2
D-45141 Essen
Germany
juergen.herzog@uni-due.de

Takayuki Hibi
Department of Pure
and Applied Mathematics
Graduate School of Information Science
and Technology
Osaka University
Toyonaka, Osaka 560-0043
Japan
hibi@math.sci.osaka-u.ac.jp

ISSN 0072-5285
ISBN 978-1-4471-2594-5   ISBN 978-0-85729-106-6 (eBook)
DOI 10.1007/978-0-85729-106-6
Springer London Dordrecht Heidelberg New York

British Library Cataloguing in Publication Data
A catalogue record for this book is available from the British Library

Mathematics Subject Classification (2010): 13D02, 13D40, 13F55, 13H10, 13P10

Cover design: VTEX, Vilnius

Printed on acid-free paper

Springer is part of Springer Science+Business Media (www.springer.com)

To our wives Maja and Kumiko, our children Susanne, Ulrike, Masaki and Ayako, and our grandchildren Paul, Jonathan, Vincent, Nelson, Sofia and Jesse

Preface

Commutative algebra has developed in step with algebraic geometry and has played an essential role as the foundation of algebraic geometry. On the other hand, homological aspects of modern commutative algebra became a new and important focus of research inspired by the work of Melvin Hochster. In 1975, Richard Stanley [Sta75] proved affirmatively the upper bound conjecture for spheres by using the theory of Cohen–Macaulay rings. This created another new trend of commutative algebra, as it turned out that commutative algebra supplies basic methods in the algebraic study of combinatorics on convex polytopes and simplicial complexes. Stanley was the first to use concepts and techniques from commutative algebra in a systematic way to study simplicial complexes by considering the Hilbert function of Stanley–Reisner rings, whose defining ideals are generated by squarefree monomials. Since then, the study of squarefree monomial ideals from both the algebraic and combinatorial points of view has become a very active area of research in commutative algebra.

In the late 1980s the theory of Gröbner bases came into fashion in many branches of mathematics. Gröbner bases, together with initial ideals, provided new methods. They have been used not only for computational purposes but also to deduce theoretical results in commutative algebra and combinatorics. For example, based on the fundamental work by Gel'fand, Kapranov, Zelevinsky and Sturmfels, far beyond the classical techniques in combinatorics, the study of regular triangulations of a convex polytope by using suitable initial ideals turned out to be a very successful approach, and, after the pioneering work of Sturmfels [Stu90], the algebraic properties of determinantal ideals have been explored by considering their initial ideal, which for a suitable monomial order is a squarefree monomial ideal and hence is accessible to powerful techniques.

At about the same time Galligo, Bayer and Stillman observed that generic initial ideals have particularly nice combinatorial structures and provide a basic tool for the combinatorial and computational study of the minimal free resolution of a graded ideal of the polynomial ring. Algebraic shifting, which was introduced by Kalai and which contributed to the modern development

of enumerative combinatorics on simplicial complexes, can be discussed in the
frame of generic initial ideals.

The present monograph invites the reader to become acquainted with cur-
rent trends in combinatorial commutative algebra, with the main emphasis
on basic research into monomials and monomial ideals. Apart from a few
exceptions, where we refer to the books [BH98], [Kun08] and [Mat80], only
basic knowledge of commutative algebra is required to understand most of
the monograph. Part I is a self-contained introduction to the modern theory
of Gröbner bases and initial ideals. Its highlight is a quick introduction to
the theory of Gröbner bases (Chapter 2), and it also offers a detailed description of, and information about, generic initial ideals (Chapter 4). Part II
covers Hilbert functions and resolutions and some of the combinatorics related
to monomial ideals, including the Kruskal–Katona theorem and algebraic aspects of Alexander duality. In Part III we discuss combinatorial applications of
monomial ideals. The main topics include edge ideals of finite graphs, powers
of ideals, algebraic shifting theory and an introduction to polymatroids.

We now discuss the contents of the monograph in detail together with a
brief history of commutative algebra and combinatorics on monomials and
monomial ideals.

Chapter 1 summarizes fundamental material on monomial ideals. In particular, we consider the integral closure of monomial ideals, squarefree normally torsionfree ideals, squarefree monomial ideals and simplicial complexes,
Alexander duality and polarization of monomial ideals.

In Chapter 2 a short introduction to the main features of Gröbner basis
theory is given, including the Buchberger criterion and algorithm. These basic
facts are discussed in a comprehensive but compact form.

Chapter 3 presents one of the most fundamental results on initial ideals,
which says that the graded Betti numbers of the initial ideal $\mathrm{in}_<(I)$ are greater
than or equal to the corresponding graded Betti numbers of I. This fact is
used again and again in this book, especially in shifting theory.

Chapter 4 concerns generic initial ideals. This theory plays an essential
role in the combinatorial applications considered in Part III. Therefore, for the
sake of completeness, we present in Chapter 4 the main theorems on generic
initial ideals together with their complete proofs. Generic initial ideals are
Borel-fixed. They belong to the more general class of Borel type ideals for
which various characterizations are given. Generic annihilator numbers and
extremal Betti numbers are introduced, and it is shown that extremal Betti
numbers are invariant under taking generic initial ideals.

Chapter 5 is devoted to establishing the theory of Gröbner bases in the
exterior algebra, and uses exterior techniques to give a proof of the Alexander
duality theorem which establishes isomorphisms between simplicial homology
and cohomology of a simplicial complex and its Alexander dual.

Chapter 6 offers basic material on combinatorics of monomial ideals. First
we recall the concepts of Hilbert functions and Hilbert polynomials, and their
relationship to the f-vector of a simplicial complex is explained. We study in

detail the combinatorial characterization of Hilbert functions of graded ideals due to Macaulay together with its squarefree analogue, the Kruskal–Katona theorem, which describes the possible face numbers of simplicial complexes. Lexsegment ideals as well as squarefree lexsegment ideals play the key role in the discussion.

Chapter 7 discusses minimal free resolutions of monomial ideals. We derive formulas for the graded Betti numbers of stable and squarefree stable ideals, and use these formulas to deduce the Bigatti–Hulett theorem which says that lexsegment ideals have the largest graded Betti numbers among all graded ideals with the same Hilbert function. We also present the squarefree analogue of the Bigatti–Hulett theorem, and give the comparison of Betti numbers over the symmetric and exterior algebra.

Chapter 8 begins with Hochster's formula to compute the graded Betti numbers of Stanley–Reisner ideals and Reisner's Cohen–Macaulay criterion for simplicial complexes. Then the Eagon–Reiner theorem and variations of it are discussed. In particular, ideals with linear quotients, componentwise linear ideals, sequentially Cohen–Macaulay ideals and shellable simplicial complexes are studied.

Chapter 9 deals with the algebraic aspects of Dirac's theorem on chordal graphs and the classification problem for Cohen–Macaulay graphs. First the classification of bipartite Cohen–Macaulay graphs is given. Then unmixed graphs are characterized and we present the result which says that a bipartite graph is sequentially Cohen–Macaulay if and only if it is shellable. It follows the classification of Cohen–Macaulay chordal graphs. Finally the relationship between the Hilbert–Burch theorem and Dirac's theorem on chordal graphs is explained.

Chapter 10 is devoted to the study of powers of monomial ideals. We begin with a brief introduction to toric ideals and Rees algebras, and present a Gröbner basis criterion which guarantees that all powers of an ideal have a linear resolution. As an application it is shown that all powers of monomial ideals with 2-linear resolution have a linear resolution. Then the depth of powers of monomial ideals, and Mengerian and unimodular simplicial complexes are considered.

Chapter 11 offers a self-contained and systematic presentation of modern shifting theory from the viewpoint of generic initial ideals as well as of graded Betti numbers. Combinatorial, exterior and symmetric shifting are introduced and the comparison of the graded Betti numbers for the distinct shifting operators is studied. It is shown that the extremal graded Betti numbers of a simplicial complex and its symmetric and exterior shifted complex are the same. Finally, super-extremal Betti numbers are considered to give an algebraic proof of the Björner–Kalai theorem.

In Chapter 12 we consider discrete polymatroids and polymatroidal ideals. After giving a short introduction to the combinatorics and geometry of discrete polymatroids, the algebraic properties of its base ring are studied. We close

Chapter 12 by presenting polymatroidal and weakly polymatroidal ideals, which provide large classes of ideals with linear quotients.

It becomes apparent from the above detailed description of the topics discussed in this monograph that the authors have chosen those combinatorial topics which are strongly related to monomial ideals. Binomial ideals, toric rings and convex polytopes are not the main topic of this book. We refer the reader to Sturmfels [Stu96], Miller–Sturmfels [MS04] and Bruns–Gubeladze [BG09]. We also do not discuss the pioneering work by Richard Stanley on the upper bound conjecture for spheres. For this topic we refer the reader to Bruns–Herzog [BH98], Hibi [Hib92] and Stanley [Sta95].

We have tried as much as possible to make our presentation self-contained, and we believe that combinatorialists who are familiar with only basic materials on commutative algebra will understand most of this book without having to read other textbooks or research papers. However, for the convenience of the reader who is not so familiar with commutative algebra and convex geometry we have added an appendix in which we explain some fundamental algebraic and geometric concepts which are used in this book. In addition, researchers working on applied mathematics who want to learn Gröbner basis theory quickly as a basic tool for their work need only consult Chapter 2. Since shifting theory is rather technical, the reader may skip Chapters 4–7 and 11 (which are required for the understanding of shifting theory) on a first reading.

We conclude each chapter with a list of problems. They are intended to complement and provide better understanding of the topics treated in each chapter.

We are grateful to Viviana Ene and Rahim Zaare-Nahandi for their comments and for suggesting corrections in some earlier drafts of this monograph.

Essen, Osaka   *Jürgen Herzog*
February 2010   *Takayuki Hibi*

Contents

Part III Combinatorics

Part I

Gröbner bases

1

Monomial Ideals

Monomials form a natural K-basis in the polynomial ring $S = K[x_1, \ldots, x_n]$ defined over the field K. An ideal I which is generated by monomials, a so-called monomial ideal, also has a K-basis of monomials. As a consequence, a polynomial f belongs to I if and only if all monomials in f appearing with a nonzero coefficient belong to I. This is one of the reasons why algebraic operations with monomial ideals are easy to perform and are accessible to combinatorial and convex geometric arguments. One may take advantage of this fact when studying general ideals in S by considering its initial ideal with respect to some monomial order.

1.1 Basic properties of monomial ideals

1.1.1 The K-basis of a monomial ideal

Let K be a field, and let $S = K[x_1, \ldots, x_n]$ be the polynomial ring in n variables over K. Let $\mathbb{R}^n_+$ denote the set of those vectors $\mathbf{a} = (a_1, \ldots, a_n) \in \mathbb{R}^n$ with each $a_i \geq 0$, and $\mathbb{Z}^n_+ = \mathbb{R}^n_+ \cap \mathbb{Z}^n$. In addition, we denote as usual, the set of positive integers by $\mathbb{N}$.

Any product $x_1^{a_1} \cdots x_n^{a_n}$ with $a_i \in \mathbb{Z}_+$ is called a **monomial**. If $u = x_1^{a_1} \cdots x_n^{a_n}$ is a monomial, then we write $u = \mathbf{x}^{\mathbf{a}}$ with $\mathbf{a} = (a_1, \ldots, a_n) \in \mathbb{Z}^n_+$. Thus the monomials in S correspond bijectively to the lattice points in $\mathbb{R}^n_+$, and we have

$$\mathbf{x}^{\mathbf{a}} \mathbf{x}^{\mathbf{b}} = \mathbf{x}^{\mathbf{a}+\mathbf{b}}.$$

The set $\mathrm{Mon}(S)$ of monomials of S is a K-basis of S. In other words, any polynomial $f \in S$ is a unique K-linear combination of monomials. Write

$$f = \sum_{u \in \mathrm{Mon}(S)} a_u u \quad \text{with} \quad a_u \in K.$$

Then we call the set

J. Herzog, T. Hibi, *Monomial Ideals*, Graduate Texts in Mathematics 260, DOI 10.1007/978-0-85729-106-6_1, © Springer-Verlag London Limited 2011

$$\text{supp}(f) = \{u \in \text{Mon}(S) : a_u \neq 0\}$$

the **support** of f.

Definition 1.1.1. An ideal $I \subset S$ is called a **monomial ideal** if it is generated by monomials.

An important property of monomial ideals is given in the following.

Theorem 1.1.2. *The set $\mathcal{N}$ of monomials belonging to I is a K-basis of I.*

Proof. It is clear that the elements of $\mathcal{N}$ are linearly independent, as $\mathcal{N}$ is a subset of $\text{Mon}(S)$.

Let $f \in I$ be an arbitrary polynomial. We will show that $\text{supp}(f) \subset \mathcal{N}$. This then yields that $\mathcal{N}$ is a system of generators of the K-vector space I.

Indeed, since $f \in I$, there exist monomials $u_1, \ldots, u_m \in I$ and polynomials $f_1, \ldots, f_m \in S$ such that $f = \sum_{i=1}^{m} f_i u_i$. It follows that $\text{supp}(f) \subset \bigcup_{i=1}^{m} \text{supp}(f_i u_i)$. Note that $\text{supp}(f_i u_i) \subset \mathcal{N}$ for all i, since each $v \in \text{supp}(f_i u_i)$ is of the form $w u_i$ with $w \in \text{Mon}(S)$, and hence belongs to I. It follows that $\text{supp}(f) \subset \mathcal{N}$, as desired. $\square$

Recall from basic commutative algebra that an ideal $I \subset S$ is graded if, whenever $f \in I$, all homogeneous components of f belong to I. Monomial ideals can be characterized similarly.

Corollary 1.1.3. *Let $I \subset S$ be an ideal. The following conditions are equivalent:*

(a) *I is a monomial ideal;*
(b) *for all $f \in S$ one has: $f \in I$ if and only if $\text{supp}(f) \subset I$.*

Proof. (a) $\Rightarrow$ (b) follows from Theorem 1.1.2.

(b) $\Rightarrow$ (a): Let $f_1, \ldots, f_m$ be a set of generators of I. Since $\text{supp}(f_i) \subset I$ for all i, it follows that $\bigcup_{i=1}^{m} \text{supp}(f_i)$ is a set of monomial generators of I. $\square$

Let $I \subset S$ be an ideal. We overline an element or a set to denote its image modulo I.

Corollary 1.1.4. *Let I be a monomial ideal. The residue classes of the monomials not belonging to I form a K-basis of the residue class ring S/I.*

Proof. Let $\mathcal{W}$ be the set of monomials not belonging to I. It is clear that $\overline{\mathcal{W}}$ is a set of generators of the K-vector space S/I. Suppose there is a non-trivial linear combination

$$\sum_{w \in \mathcal{W}} a_w \bar{w} = 0$$

of zero. Then $f = \sum_{w \in \mathcal{W}} a_w w \in I$. Hence Corollary 1.1.3 implies that $w \in I$ for $a_w \neq 0$, a contradiction. $\square$

1.1.2 Monomial generators

In any algebra course one learns that the polynomial ring $S = K[x_1, \ldots, x_n]$ is Noetherian. This is Hilbert's basis theorem. We will give a proof of this theorem in the next chapter. Here we only need that any monomial ideal is finitely generated. This is a direct consequence of Dickson's lemma, also proved in the next chapter.

The set of monomials which belong to I can be described as follows:

Proposition 1.1.5. *Let $\{u_1, \ldots, u_m\}$ be a monomial system of generators of the monomial ideal I. Then the monomial v belongs to I if and only if there exists a monomial w such that $v = wu_i$ for some i.*

Proof. Suppose that $v \in I$. Then there exist polynomials $f_i \in S$ such that $v = \sum_{i=1}^{m} f_i u_i$. It follows that $v \in \bigcup_{i=1}^{m} \mathrm{supp}(f_i u_i)$, and hence $v \in \mathrm{supp}(f_i u_i)$ for some i. This implies that $v = wu_i$ for some $w \in \mathrm{supp}(f_i)$. The other implication is trivial. □

For a graded ideal all minimal sets of generators have the same cardinality. For monomial ideals one even has:

Proposition 1.1.6. *Each monomial ideal has a unique minimal monomial set of generators. More precisely, let G denote the set of monomials in I which are minimal with respect to divisibility. Then G is the unique minimal set of monomial generators.*

Proof. Let $G_1 = \{u_1, \ldots, u_r\}$ and $G_2 = \{v_1, \ldots, v_s\}$ be two minimal sets of generators of the monomial ideal I. Since $u_i \in I$, there exists v_j such that $u_i = w_1 v_j$ for some monomial w_1. Similarly there exists u_k and a monomial w_2 such that $v_j = w_2 u_k$. It follows that $u_i = w_1 w_2 u_k$. Since G_1 is a minimal set of generators of I, we conclude that $k = i$ and $w_1 w_2 = 1$. In particular, $w_1 = 1$ and hence $u_i = v_j \in G_2$. This shows that $G_1 \subset G_2$. By symmetry we also have $G_2 \subset G_1$. □

It is common to denote the unique minimal set of monomial generators of the monomial ideal I by $G(I)$.

1.1.3 The $\mathbb{Z}^n$-grading

Let $\mathbf{a} \in \mathbb{Z}^n$; then $f \in S$ is called **homogeneous of degree a** if f is of the form $c\mathbf{x}^{\mathbf{a}}$ with $c \in K$. The polynomial ring S is obviously $\mathbb{Z}^n$-graded with graded components

$$S_{\mathbf{a}} = \begin{cases} K\mathbf{x}^{\mathbf{a}}, & \text{if } \mathbf{a} \in \mathbb{Z}_+^n, \\ 0, & \text{otherwise.} \end{cases}$$

An S-module M is called $\mathbb{Z}^n$-graded if $M = \bigoplus_{\mathbf{a} \in \mathbb{Z}^n} M_{\mathbf{a}}$ and $S_{\mathbf{a}} M_{\mathbf{b}} \subset M_{\mathbf{a}+\mathbf{b}}$ for all $\mathbf{a}, \mathbf{b} \in \mathbb{Z}^n$.

Let M, N be $\mathbb{Z}^n$-graded S-modules. A module homomorphism $\varphi : N \to M$ is called **homogeneous module homomorphism** if $\varphi(N_\mathbf{a}) \subset M_\mathbf{a}$ for all $\mathbf{a} \in \mathbb{Z}^n$, and N is called a $\mathbb{Z}^n$-graded submodule of M if $N \subset M$ and the inclusion map is homogeneous. In this case the factor module M/N inherits a natural $\mathbb{Z}^n$-grading with components $(M/N)_\mathbf{a} = M_\mathbf{a}/N_\mathbf{a}$ for all $\mathbf{a} \in \mathbb{Z}^n$.

Observe that an ideal $I \subset S$ is a $\mathbb{Z}^n$-graded submodule of S if and only if it is a monomial ideal, in which case S/I is also naturally $\mathbb{Z}^n$-graded.

1.2 Algebraic operations on monomial ideals

1.2.1 Standard algebraic operations

It is obvious that sums and products of monomial ideals are again monomial ideals. Moreover, if I and J are monomial ideals, then $G(I+J) \subset G(I) \cup G(J)$ and $G(IJ) \subset G(I)G(J)$.

Given two monomials u and v, we denote by $\gcd(u, v)$ the greatest common divisor and by $\operatorname{lcm}(u, v)$ the least common multiple of u and v.

For the intersection of monomial ideals we have

Proposition 1.2.1. *Let I and J be monomial ideals. Then $I \cap J$ is a monomial ideal, and $\{\operatorname{lcm}(u, v) : u \in G(I), v \in G(J)\}$ is a set of generators of $I \cap J$.*

Proof. Let $f \in I \cap J$. By Corollary 1.1.3, $\operatorname{supp}(f) \subset I \cap J$. Again applying Corollary 1.1.3 we see that $I \cap J$ is a monomial ideal.

Let $w \in \operatorname{supp}(f)$; then since $\operatorname{supp}(f) \subset I \cap J$, there exists $u \in G(I)$ and $v \in G(J)$ such that $u|w$ and $v|w$. It follows that $\operatorname{lcm}(u, v)$ divides w. Since $\operatorname{lcm}(u, v) \in I \cap J$ for all $u \in G(I)$ and $v \in G(J)$, we conclude that $\{\operatorname{lcm}(u, v) : u \in G(I), v \in G(J)\}$ is indeed a set of generators of $I \cap J$.   □

Let $I, J \subset S$ be two ideals. The set

$$I : J = \{f \in S : fg \in I \text{ for all } g \in J\}$$

is an ideal, called the **colon ideal** of I with respect to J.

Proposition 1.2.2. *Let I and J be monomial ideals. Then $I : J$ is a monomial ideal, and*

$$I : J = \bigcap_{v \in G(J)} I : (v).$$

Moreover, $\{u/\gcd(u, v) : u \in G(I)\}$ is a set of generators of $I : (v)$.

Proof. Let $f \in I : J$. Then $fv \in I$ for all $v \in G(J)$. In view of Corollary 1.1.3 we have $\operatorname{supp}(f)v = \operatorname{supp}(fv) \subset I$. This implies that $\operatorname{supp}(f) \subset I : J$. Thus Corollary 1.1.3 yields that $I : J$ is a monomial ideal.

The given presentation of $I : J$ as an intersection is obvious, and it is also clear that $\{u/\gcd(u, v) : u \in G(I)\} \subset I : (v)$. So now let $w \in I : (v)$. Then there exists $u \in G(I)$ such that u divides wv. This implies that $u/\gcd(u, v)$ divides w, as desired.   □

1.2.2 Saturation and radical

Let $I \subset S$ be a graded ideal. We denote by $\mathfrak{m} = (x_1, \ldots, x_n)$ the graded maximal ideal of S.

The **saturation** $\tilde{I}$ of I is the ideal

$$I : \mathfrak{m}^\infty = \bigcup_{k=1}^{\infty} I : \mathfrak{m}^k,$$

while the ideal $\sqrt{I} = \{f \in S : f^k \in I \text{ for some } k\}$ is called the **radical of** I.

The ideal I is called **saturated** if $I = \tilde{I}$ and is called a **radical ideal** if $I = \sqrt{I}$.

Proposition 1.2.3. *The saturation and the radical of a monomial ideal are again monomial ideals.*

Proof. By Proposition 1.2.2, $I : \mathfrak{m}^k$ is a monomial ideal for all k. Since $\tilde{I}$ is the union of these ideals, it is a monomial ideal.

Let $f = c\mathbf{x}^{\mathbf{a}_1} + \cdots \in \sqrt{I}$ with $0 \neq c \in K$. Then $f^k \in I$, and consequently $\mathrm{supp}(f^k) \subset I$, since I is a monomial ideal. Let $\mathrm{supp}(f) = \{\mathbf{x}^{\mathbf{a}_1}, \ldots, \mathbf{x}^{\mathbf{a}_r}\}$. The convex hull of the set $\{\mathbf{a}_1, \ldots, \mathbf{a}_r\} \subset \mathbb{R}^n$ is a polytope. We may assume that $\mathbf{a}_1$ is a vertex of this polytope, in other words, $\mathbf{a}_1$ does not belong to the convex hull of $\{\mathbf{a}_2, \ldots, \mathbf{a}_r\}$.

Assume $(\mathbf{x}^{\mathbf{a}_1})^k = (\mathbf{x}^{\mathbf{a}_1})^{k_1}(\mathbf{x}^{\mathbf{a}_2})^{k_2} \cdots (\mathbf{x}^{\mathbf{a}_r})^{k_r}$ with $k = k_1 + k_2 + \cdots + k_r$ and $k_1 < k$. Then

$$\mathbf{a}_1 = \sum_{i=2}^{r} (k_i/(k - k_1))\mathbf{a}_i \quad \text{with} \quad \sum_{i=2}^{r}(k_i/(k - k_1)) = 1,$$

so $\mathbf{a}_1$ is not a vertex, a contradiction. It follows that the monomial $(x^{\mathbf{a}_1})^k$ cannot cancel against other terms in f^k and hence belongs to $\mathrm{supp}(f^k)$, which is a subset of I. Therefore $\mathbf{x}^{\mathbf{a}_1} \in \sqrt{I}$ and $f - c\mathbf{x}^{\mathbf{a}_1} \in \sqrt{I}$. By induction on the cardinality of $\mathrm{supp}(f)$ we conclude that $\mathrm{supp}(f) \subset \sqrt{I}$. Thus Corollary 1.1.3 implies that $\sqrt{I}$ is a monomial ideal. $\square$

The radical of a monomial ideal I can be computed explicitly. A monomial $\mathbf{x}^{\mathbf{a}}$ is called **squarefree** if the components of $\mathbf{a}$ are 0 or 1. Let $u = \mathbf{x}^{\mathbf{a}}$ be a monomial. We set

$$\sqrt{u} = \prod_{i, a_i \neq 0} x_i.$$

One has $\sqrt{u} = u$ if and only if u is squarefree.

Proposition 1.2.4. *Let I be a monomial ideal. Then $\{\sqrt{u} : u \in G(I)\}$ is a set of generators of $\sqrt{I}$.*

Proof. Obviously $\{\sqrt{u} : u \in G(I)\} \subset \sqrt{I}$. Since $\sqrt{I}$ is a monomial ideal it suffices to show that each monomial $v \in \sqrt{I}$ is a multiple of some $\sqrt{u}$ with $u \in G(I)$. In fact, if $v \in \sqrt{I}$ then $v^k \in I$ for some integer $k \geq 0$, and therefore $v^k = wu$ for some $u \in G(I)$ and some monomial w. This yields the desired conclusion.   □

A monomial ideal I is called a **squarefree monomial ideal** if I is generated by squarefree monomials. As a consequence of Proposition 1.2.4 we have

Corollary 1.2.5. *A monomial ideal I is a radical ideal, that is, $I = \sqrt{I}$, if and only if I is a squarefree monomial ideal.*

1.3 Primary decomposition and associated prime ideals

1.3.1 Irreducible monomial ideals

A presentation of an ideal I as an intersection $I = \bigcap_{i=1}^{m} Q_i$ of ideals is called **irredundant** if none of the ideals Q_i can be omitted in this presentation.

We have the following fundamental fact.

Theorem 1.3.1. *Let $I \subset S = K[x_1, \ldots, x_n]$ be a monomial ideal. Then $I = \bigcap_{i=1}^{m} Q_i$, where each Q_i is generated by pure powers of the variables. In other words, each Q_i is of the form $(x_{i_1}^{a_1}, \ldots, x_{i_k}^{a_k})$. Moreover, an irredundant presentation of this form is unique.*

Proof. Let $G(I) = \{u_1, \ldots, u_r\}$, and suppose some u_i is not a pure power, say u_1. Then we can write $u_1 = vw$ where v and w are coprime monomials, that is, $\gcd(v, w) = 1$ and $v \neq 1 \neq w$. We claim that $I = I_1 \cap I_2$ where $I_1 = (v, u_2, \ldots, u_r)$ and $I_2 = (w, u_2, \ldots, u_r)$.

Obviously, I is contained in the intersection. Conversely, let u be a monomial in $I_1 \cap I_2$. If u is a multiple of one of the u_i, then $u \in I$. If not, then u is a multiple of v and of w, and hence of u_1, since v and w are coprime. In any case, $u \in I$.

If either $G(I_1)$ or $G(I_2)$ contains an element which is not a pure power, we proceed as before and obtain after a finite number of steps a presentation of I as an intersection of monomial ideals generated by pure powers. By omitting those ideals which contain the intersection of the others we end up with an irredundant intersection.

Let $Q_1 \cap \cdots \cap Q_r = Q'_1 \cap \cdots \cap Q'_s$ two irredundant intersections of ideals generated by pure powers. We will show that for each $i \in [r]$ there exists $j \in [s]$ such that $Q'_j \subset Q_i$. By symmetry we then also have that for each $k \in [s]$ there exists an $\ell \in [r]$ such that $Q_\ell \subset Q'_k$. This will then imply that $r = s$ and $\{Q_1, \ldots, Q_r\} = \{Q'_1, \ldots, Q'_s\}$.

In fact, let $i \in [r]$. We may assume that $Q_i = (x_1^{a_1}, \ldots, x_k^{a_k})$. Suppose that $Q_j' \not\subset Q_i$ for all $j \in [s]$. Then for each j there exists $x_{\ell_j}^{b_j} \in Q_j' \setminus Q_i$. It follows that either $\ell_j \notin [k]$ or $b_j < a_{\ell_j}$. Let

$$u = \mathrm{lcm}\{x_{\ell_1}^{b_1}, \ldots, x_{\ell_s}^{b_s}\}.$$

We have $u \in \bigcap_{j=1}^{s} Q_j' \subset Q_i$. Therefore there exists $i \in [k]$ such that $x_i^{a_i}$ divides u. But this is obviously impossible.   □

A monomial ideal is called **irreducible** if it cannot be written as proper intersection of two other monomial ideals. It is called **reducible** if it is not irreducible.

Corollary 1.3.2. *A monomial ideal is irreducible if and only if it is generated by pure powers of the variables.*

Proof. Let $Q = (x_{i_1}^{a_1}, \ldots, x_{i_k}^{a_k})$, and suppose $Q = I \cap J$ where I and J are monomial ideals properly containing Q. By Theorem 1.3.1 we have $I = \bigcap_{i=1}^{r} Q_i$ and $J = \bigcap_{j=1}^{s} Q_j'$ where the Q_i and Q_j' are generated by powers of the variables. Thus we get the presentation

$$Q = \bigcap_{i=1}^{r} Q_i \cap \bigcap_{j=1}^{s} Q_j'.$$

By omitting suitable ideals in the intersection on the right-hand side, we obtain an irredundant presentation of Q. The uniqueness statement in Theorem 1.3.1 then implies that $Q = Q_i$ or $Q = Q_j'$ for some i or j, a contradiction.

Conversely, if $G(Q)$ contains a monomial $u = vw$ with $\gcd(v, w) = 1$ and $v \neq 1 \neq w$, then, as in the proof of Theorem 1.3.1, Q can be written a proper intersection of monomial ideals.   □

Theorem 1.3.1 in combination with Corollary 1.3.2 now says that each monomial ideal has a unique presentation as an irredundant intersection of irreducible monomial ideals.

The proof of Theorem 1.3.1 shows us how we can find such a presentation. The following example illustrates the procedure.

Example 1.3.3. Let
$$I = (x_1^2 x_2, x_1^2 x_3^2, x_2^2, x_2 x_3^2).$$
Then

$$
\begin{aligned}
I &= (x_1^2, x_1^2 x_3^2, x_2^2, x_2 x_3^2) \cap (x_2, x_1^2 x_3^2, x_2^2, x_2 x_3^2) = (x_1^2, x_2^2, x_2 x_3^2) \cap (x_2, x_1^2 x_3^2) \\
&= (x_1^2, x_2^2, x_2) \cap (x_1^2, x_2^2, x_3^2) \cap (x_2, x_1^2) \cap (x_2, x_3^2) \\
&= (x_1^2, x_2^2, x_3^2) \cap (x_1^2, x_2) \cap (x_2, x_3^2).
\end{aligned}
$$

If I is a squarefree monomial ideal, the above procedure yields that the irreducible monomial ideals appearing in the intersection of I are all of the form $(x_{i_1}, \ldots, x_{i_k})$. These are obviously exactly the monomial prime ideals. Thus we have shown

Corollary 1.3.4. *A squarefree monomial ideal is an intersection of monomial prime ideals.*

Let R be a ring and $I \subset R$ an ideal. A prime ideal P is called a **minimal prime ideal** of I, if $I \subset P$ and there is no prime ideal containing I which is properly contained in P. We denote the set minimal prime ideals of I by $\mathrm{Min}(I)$.

We recall the following general fact.

Lemma 1.3.5. *Suppose I has irredundant presentation $I = P_1 \cap \cdots \cap P_m$ as an intersection of prime ideals. Then $\mathrm{Min}(I) = \{P_1, \ldots, P_m\}$*

Proof. Suppose P_i is not a minimal prime ideal of I. Then there exists a prime ideal P with $I \subset P$, and P is properly contained in P_i. Since $P_j R_{P_i} = R_{P_i}$ for $i \neq j$ and since localization commutes with intersections, it follows that $IR_{P_i} = P_i R_{P_i}$, contradicting the fact that PR_{P_i} contains IR_{P_i} and is properly contained in $P_i R_{P_i}$.

On the other hand, if P is a prime ideal containing I, then $P_1 P_2 \cdots P_m \subset P_1 \cap \cdots \cap P_p \subset P$. So one of the P_i must be contained in P. Hence if P is a minimal prime ideal of I, then $P = P_i$.

Combining Corollary 1.3.4 with Lemma 1.3.5 we obtain

Corollary 1.3.6. *Let $I \subset S$ be a squarefree monomial ideal. Then*

$$I = \bigcap_{P \in \mathrm{Min}(I)} P,$$

and each $P \in \mathrm{Min}(I)$ is a monomial prime ideal.

1.3.2 Primary decompositions

Let R be a Noetherian ring and M a finitely generated R-module. A prime ideal $P \subset R$ is called an **associated prime ideal** of M, if there exists an element $x \in M$ such that $P = \mathrm{Ann}(x)$. Here $\mathrm{Ann}(x)$ is the **annihilator** of x, that is to say, $\mathrm{Ann}(x) = \{a \in R: ax = 0\}$. The set of associated prime ideals of M is denoted $\mathrm{Ass}(M)$.

A prime ideal $P \subset R$ is called a **minimal prime ideal** of M, if $M_P \neq 0$, and for each prime ideal Q properly contained in P one has $M_Q = 0$. Observe that P is minimal prime ideal of R/I if and only if $I \subset P$, and there is no prime ideal $I \subset Q$ which is properly contained in P. It is known that $\mathrm{Ass}(M)$

is a finite set containing all minimal prime ideals of M. For this and other basic properties of associated prime ideals we refer to Matsumura [Mat86].

Recall that an ideal I in a Noetherian ring R is P-**primary**, if $\mathrm{Ass}(R/I) = \{P\}$. In an abuse of notation, one often writes $\mathrm{Ass}(I)$ instead of $\mathrm{Ass}(R/I)$.

Proposition 1.3.7. *The irreducible ideal* $(x_{i_1}^{a_1}, \ldots, x_{i_k}^{a_k})$ *is* $(x_{i_1}, \ldots, x_{i_k})$-*primary.*

Proof. Let $Q = (x_{i_1}^{a_1}, \ldots, x_{i_k}^{a_k})$ and $P = (x_{i_1}, \ldots, x_{i_k})$. Since P is a minimal prime ideal of Q, it follows that $P \in \mathrm{Ass}(Q)$.

Notice that $P^m \subset Q$ for $m = \sum_{i=1}^k a_i$. Therefore P is the only minimal prime ideal containing Q. Hence if P' is an associated prime ideal of Q, then $P \subset P'$.

We have $P' = Q : (g)$ for some polynomial g. Suppose $P' \neq P$. Then P' contains a polynomial f with the property that none of the elements $u \in \mathrm{supp}(f)$ is divisible by the variables x_{i_j}. Therefore f is regular on S/Q. Since $fg \in Q$, we conclude that $g \in Q$ and so $Q : (g) = S$, a contradiction. $\square$

A presentation of an ideal I as intersection $I = \bigcap_{i=1}^r Q_i$ where each Q_i is a primary ideal is called a **primary decomposition** of I. Let $\{P_i\} = \mathrm{Ass}(Q_i)$. The primary decomposition is called **irredundant primary decomposition** if none of the Q_i can be omitted in this intersection and if $P_i \neq P_j$ for all $i \neq j$. If $I = \bigcap_{i=1}^r Q_i$ is an irredundant primary decomposition of I, then the Q_i is called the P_i-**primary components** of I, and one has $\mathrm{Ass}(I) = \{P_1, \ldots, P_r\}$. Only the primary components belonging to the minimal prime ideals of I are uniquely determined. Indeed, if $P \in \mathrm{Ass}(I)$ is a minimal prime ideal of I, then the P-primary component of I is the kernel of the natural ring homomorphism $R \to (R/I)_P$, see [Kun08].

Proposition 1.3.7 implies that the decomposition of an ideal into irreducible ideals is a primary decomposition. But of course it may not be irredundant. However, since an intersection of P-primary ideals is again P-primary we may construct an irredundant primary decomposition of a monomial ideal I from a presentation $I = \bigcap_{i=1}^r Q_i$ as given in Theorem 1.3.1 by letting the P-primary component of I be the intersection of all Q_i with $\mathrm{Ass}(Q_i) = \{P\}$. The following example illustrates this.

Example 1.3.8. The ideal $I = (x_1^3, x_2^3, x_1^2 x_3^2, x_1 x_2 x_3^2, x_2^2 x_3^2)$ has the irredundant presentation as intersection of irreducible ideals

$$I = (x_1^3, x_2^3, x_3^2) \cap (x_1^2, x_2) \cap (x_1, x_2^2).$$

We have $\mathrm{Ass}(x_1^2, x_2) = \mathrm{Ass}(x_1, x_2^2) = \{(x_1, x_2)\}$. Intersecting (x_1^2, x_2) and (x_1, x_2^2) we obtain the (x_1, x_2)-primary ideal $(x_1^2, x_1 x_2, x_2^2)$ and hence the irredundant primary decomposition

$$I = (x_1^3, x_2^3, x_3^2) \cap (x_1^2, x_1 x_2, x_2^2).$$

Even though a primary decomposition of a monomial ideal I may not be unique, the primary decomposition, obtained from an irredundant intersection of irreducible ideals as described above, is unique. We call it the **standard primary decomposition** of I.

From the standard primary decomposition we deduce immediately

Corollary 1.3.9. *The associated prime ideals of a monomial ideal are monomial prime ideals.*

Corollary 1.3.10. *Let $I \subset S$ be a monomial ideal, and let $P \in \mathrm{Ass}(I)$. Then there exists a monomial v such that $P = I : v$.*

Proof. Since $P \in \mathrm{Ass}(I)$, there exists $f \in S$ such that $P = I : f$. Thus for each $x_i \in P$ we have that $x_i f \in I$. Since I is a monomial ideal, this implies that $x_i u \in I$ for all $u \in \mathrm{supp}(f)$. It follows that $P = I : f \subset \bigcap_{u \in \mathrm{supp}(f)} I : u$. On the other hand, if $g \in \bigcap_{u \in \mathrm{supp}(f)} I : u$, then $ug \in I$ for all $u \in \mathrm{supp}(f)$, and hence $gf \in I : f = P$. Consequently, $P = \bigcap_{u \in \mathrm{supp}(f)} I : u$. Since P is irreducible, it follows that $P = I : u$ for some $u \in \mathrm{supp}(f)$. □

1.4 Integral closure of ideals

1.4.1 Integral closure of monomial ideals

We introduce normal and normally torsionfree ideals. These concepts will be use in Chapter 10.

Definition 1.4.1. *Let R be a ring and I an ideal in R. An element $f \in R$ is **integral** over I, if there exists an equation*

$$f^k + c_1 f^{k-1} + \cdots + c_{k-1} f + c_k = 0 \quad \text{with} \quad c_i \in I^i. \tag{1.1}$$

*The set of elements $\bar{I}$ in R which are integral over I is the **integral closure** of I. The ideal I is **integrally closed**, if $I = \bar{I}$, and I is **normal** if all powers of I are integrally closed.*

Equation (1.1) is called an **equation of integral dependence** of f over I.

The integral closure of an ideal is again an ideal [SH06, Corollary 1.3.1]. For a monomial ideal it can be described as follows.

Theorem 1.4.2. *Let $I \subset S$ be a monomial ideal. Then $\bar{I}$ is a monomial ideal generated by all monomials $u \in S$ for which there exists an integer k such that $u^k \in I^k$.*

Proof. We first show that if J is monomial ideal and $u \in \bar{J}$ is monomial, then there exists an integer k such that $u^k \in J^k$. Indeed, let $u^m + c_1 u^{m-1} + \cdots + c_{m-1}u + c_m = 0$ be an equation of integral dependence of u over J, and let $a_i \in K$ be the coefficient of u^i in the polynomial c_i. Then $u^m + a_1 u^m + a_2 u^m + \cdots + a_m u^m = 0$. This is only possible if some $a_k \neq 0$. It then follows that $u^k \in \text{supp}(c_k)$. Since $c_k \in J^k$ and J^k is a monomial ideal, Corollary 1.1.3 implies that $u^k \in J^k$.

In order to prove that $\bar{I}$ is a monomial ideal, we first extend the base field K by a transcendental element t to obtain the field $L = K(t)$, and prove in a first step that $\overline{IT}$ is a monomial ideal, where $T = L[x_1, \ldots, x_n]$. Let $f \in \overline{IT}$. By Corollary 1.1.3 it is enough to show that $\text{supp}(f) \subset \overline{IT}$, and we show this by induction on the cardinality of $\text{supp}(f)$. The assertion is trivial if $|\text{supp}(f)| = 1$. Now suppose that the support of f consists of more than one monomial. Let $u = \mathbf{x}^{\mathbf{a}}$ be the lexicographical smallest monomial in f; see Subsection 2.1.2. Then, according to Lemma 3.1.1, there exists an integer vector $\omega = (\omega_1, \ldots, \omega_n) \in \mathbb{Z}_+^n$ such that $\sum_{i=1}^n b_i \omega_i > \sum_{i=1}^n a_i \omega_i$ for all $x_1^{b_1} \cdots x_n^{b_n}$ in the support of f which are different from u.

Let $\varphi : T \to T$ be the automorphism with $\varphi(x_i) = t^{\omega_i} x_i$ for $i = 1, \ldots, n$. Then $\varphi(IT) = IT$, since IT is a monomial ideal, and $\varphi(f)$ is integral over $\varphi(IT) = IT$. Hence the polynomial $g = t^{-c}\varphi(f)$ with $c = \sum_{i=1}^n a_i \omega_i$ is integral over IT as well, and so is the polynomial $h = f - g$. By the choice of ω and the integer c we get $\text{supp}(h) = \text{supp}(f) \setminus \{u\}$. Hence our induction hypothesis implies that $\text{supp}(f) \setminus \{u\} \subset \overline{IT}$. Since a nonzero scalar multiple of u can be obtained by subtracting from f a linear combination of the monomials $v \in \text{supp}(f) \setminus \{u\}$ (which all belong to $\overline{IT}$), it follows that $u \in \overline{IT}$, too.

Now in order to see that $\bar{I}$ is a monomial ideal, let $f \in \bar{I}$. Then $f \in \overline{IT}$ and so $u \in \overline{IT}$ for all $u \in \text{supp}(f)$, since $\overline{IT}$ is a monomial ideal. But then, as we have seen, for each $u \in \text{supp}(f)$ there exists an integer k such that $u^k \in (IT)^k$. Thus, since $G(IT) = G(I)$, Proposition 1.1.5 yields that u^k is a product of k monomials in I. This implies that $u \in \bar{I}$, and yields the desired result. $\square$

Let $I \subset S$ be a monomial ideal. The convex hull $\mathcal{C}(I)$ of the set of lattice points $\{\mathbf{a} : \mathbf{x}^{\mathbf{a}} \in I\}$ in $\mathbb{R}^n$ is called the **Newton polyhedron** of I.

Corollary 1.4.3. *Let $I \subset S$ be a monomial ideal. Then $\bar{I}$ is generated by the monomials $\mathbf{x}^{\mathbf{a}}$ with $\mathbf{a} \in \mathcal{C}(I)$.*

Proof. By Theorem 1.4.2, $\mathbf{x}^{\mathbf{a}} \in \bar{I}$ if and only if there exists an integer $k > 0$ such that $(\mathbf{x}^{\mathbf{a}})^k \in I^k$. It follows from Proposition 1.1.5 that this is the case if and only if there exist $\mathbf{x}^{\mathbf{a}_1}, \ldots, \mathbf{x}^{\mathbf{a}_k} \in I$ such that $(\mathbf{x}^{\mathbf{a}})^k = \mathbf{x}^{\mathbf{a}_1} \cdots \mathbf{x}^{\mathbf{a}_k}$. This is equivalent to saying that

$$\mathbf{a} = (1/k)(\mathbf{a}_1 + \cdots + \mathbf{a}_k). \tag{1.2}$$

If equation (1.2) holds, then $\mathbf{a} \in \mathcal{C}(I)$. Conversely, if $\mathbf{a} \in \mathcal{C}(I)$, then there exist $\mathbf{b}_1, \ldots, \mathbf{b}_m$ with $\mathbf{x}^{\mathbf{b}_i} \in I$, and there exist $q_i \in \mathbb{Q}_+$ such that $\mathbf{a} = q_1 \mathbf{b}_1 + \cdots +$

$q_m \mathbf{b}_m$ and $\sum_{i=1}^{m} q_i = 1$. We write $q_i = k_i/k$ as fraction of nonnegative integers. Then $\mathbf{a} = (1/k)(k_1 \mathbf{b}_1 + \cdots + k_m \mathbf{b}_m)$. Since $\sum_{i=1}^{m} k_i = k$, this presentation of $\mathbf{a}$ has the desired form (1.2).   □

1.4.2 Normally torsionfree squarefree monomial ideals

Let R be a Noetherian ring and $I \subset R$ an ideal. We define the kth symbolic power $I^{(k)}$ of I as the intersection of those primary components of I^k which belong to the minimal prime ideals of I. In other words,

$$I^{(k)} = \bigcap_{P \in \mathrm{Min}(I)} \mathrm{Ker}(R \to (R/I^k)_P).$$

Proposition 1.4.4. *Let $I \subset S$ be a squarefree monomial ideal. Then*

$$I^{(k)} = \bigcap_{P \in \mathrm{Min}(I)} P^k.$$

Proof. Because of Corollary 1.3.6 we have $IS_P = PS_P$ for $P \in \mathrm{Min}(I)$. It follows that $I^k S_P = P^k S_P$. Thus it is clear that $P^k \subset \mathrm{Ker}(S \to (S/I^k)_P)$.

Conversely, if $f \in \mathrm{Ker}(S \to (S/I^k)_P)$, then there exist $g \in I^k$ and $h \in S \backslash P$ such that $f/1 = g/h$. Therefore, $fh = g$.

The prime ideal P is a monomial prime ideal. Each element $r \in S$ has a unique presentation $r = \sum_i r_i$, where for each $\mathbf{x}^{\mathbf{a}} \in \mathrm{supp}(r_i)$ one has $\sum_{j,\, x_j \in P} a_j = i$. For $r, s \in S$ we have $(rs)_i = \sum_{j=0}^{i} r_j s_{i-j}$. The conditions on h and g imply that $h_0 \neq 0$ and that $g_i = 0$ for $i < k$. Thus the equation $fh = g$ yields $f_i = 0$ for $i < k$, which implies that $f \in P^k$.   □

Definition 1.4.5. *An ideal $I \subset R$ is called **normally torsionfree** if*

$$\mathrm{Ass}(I^k) \subset \mathrm{Ass}(I)$$

for all k.

Theorem 1.4.6. *Let $I \subset S$ be a squarefree monomial ideal. Then the following conditions are equivalent:*

(a) *I is normally torsionfree;*
(b) *$I^{(k)} = I^k$ for all k.*

If the equivalent conditions hold, then I is a normal ideal.

Proof. Let $I^k = \bigcap_{P \in \mathrm{Ass}(I^k)} Q(P)$ be an irredundant primary decomposition of I^k. Then $I^{(k)} = I^k$, if and only if $\bigcap_{P \in \mathrm{Ass}(I^k)} Q(P) = \bigcap_{P \in \mathrm{Min}(I^k)} Q(P)$, and this is the case if and only if $\mathrm{Ass}(I^k) = \mathrm{Min}(I^k) = \mathrm{Min}(I) = \mathrm{Ass}(I)$. The last equation is a consequence of Corollary 1.3.6. This proves the equivalence of (a) and (b).

In order to prove that I is a normal ideal we have to show that for each k, the ideal I^k is integrally closed. Thus for a monomial $u \in S$ for which $u^\ell \in I^{k\ell}$ for some integer $\ell > 0$, we need to show that $u \in I^k$; see Theorem 1.4.2. Since by assumption $I^j = I^{(j)}$ for all j, and since $I^{(j)} = \bigcap_{P \in \mathrm{Min}(I)} P^j$ according to Proposition 1.4.4, it amounts to proving that whenever $u^\ell \in \bigcap_{P \in \mathrm{Min}(I)} P^{k\ell}$ for some integer $\ell > 0$, then $u \in \bigcap_{P \in \mathrm{Min}(I)} P^k$. But this is easily seen, because if $u = x_1^{a_1} \cdots x_n^{a_n}$, then $u^\ell = x_1^{a_1\ell} \cdots x_n^{a_n\ell}$. To say that $u^\ell \in \bigcap_{P \in \mathrm{Min}(I)} P^{k\ell}$ is equivalent to saying that $a_i\ell \geq k\ell$ for all i for which $x_i \in P$ and all $P \in \mathrm{Min}(I)$. This then implies that $a_i \geq k$ for all i for which $x_i \in P$ and all $P \in \mathrm{Min}(I)$, which yields the desired conclusion. □

1.5 Squarefree monomial ideals and simplicial complexes

The purpose of the present section is to summarize the combinatorics on squarefree monomial ideals.

1.5.1 Simplicial complexes

Let $[n] = \{1, \ldots, n\}$ be the **vertex set** and Δ a **simplicial complex** on $[n]$. Thus Δ is a collection of subsets of $[n]$ such that if $F \in \Delta$ and $F' \subset F$, then $F' \in \Delta$. Often it is also required that $\{i\} \in \Delta$ for all $i \in [n]$; however, we will not assume this condition.

Each element $F \in \Delta$ is called a **face** of Δ. The dimension of a face F is $|F| - 1$. Let $d = \max\{|F| : F \in \Delta\}$ and define the **dimension** of Δ to be $\dim \Delta = d - 1$. An **edge** of Δ is a face of dimension 1. A **vertex** of Δ is a face of dimension 0. A **facet** is a maximal face of Δ (with respect to inclusion). Let $\mathcal{F}(\Delta)$ denote the set of facets of Δ. It is clear that $\mathcal{F}(\Delta)$ determines Δ. When $\mathcal{F}(\Delta) = \{F_1, \ldots, F_m\}$, we write $\Delta = \langle F_1, \ldots, F_m \rangle$. More generally, if we have a set $\{G_1, \ldots, G_s\}$ of faces of Δ, we denote by $\langle G_1, \ldots, G_s \rangle$ the subcomplex of Δ consisting of those faces of Δ which are contained in some G_i.

We say that a simplicial complex is **pure** if all facets have the same cardinality. A **nonface** of Δ is a subset F of $[n]$ with $F \notin \Delta$. Let $\mathcal{N}(\Delta)$ denote the set of minimal nonfaces of Δ.

Let $f_i = f_i(\Delta)$ denote the number of faces of Δ of dimension i. Thus in particular $f_0 = n$, if $\{i\} \in \Delta$ for all $i \in [n]$. The sequence $f(\Delta) = (f_0, f_1, \ldots, f_{d-1})$ is called the **f-vector** of Δ. Letting $f_{-1} = 1$, we define the **h-vector** $h(\Delta) = (h_0, h_1, \ldots, h_d)$ of Δ by the formula

$$\sum_{i=0}^{d} f_{i-1}(t-1)^{d-i} = \sum_{i=0}^{d} h_i t^{d-i}.$$

To visualize a simplicial complex we often use its **geometric realization**. For example, Figure 1.1 represents the simplicial complex Δ of dimension 2 on the vertex set $[5]$ with

$$\mathcal{F}(\Delta) = \{\{1,2,4\}, \{1,2,5\}, \{2,3\}, \{3,4\}\}$$

and with

$$\mathcal{N}(\Delta) = \{\{1,3\}, \{3,5\}, \{4,5\}, \{2,3,4\}\}.$$

One has $f(\Delta) = (5,7,2)$ and $h(\Delta) = (1,2,0,-1)$.

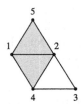

Fig. 1.1. The geometric realization of Δ

Let Δ be a simplicial complex on $[n]$ of dimension $d-1$. For each $0 \leq i \leq d-1$ the i**th skeleton** of Δ is the simplicial complex $\Delta^{(i)}$ on $[n]$ whose faces are those faces F of Δ with $|F| \leq i+1$. We say that a simplicial complex Δ is **connected** if there exists a sequence of facets $F = F_0, F_1, \ldots, F_{q-1}, F_q = G$ such that $F_i \cap F_{i+1} \neq \emptyset$. Observe that Δ is connected if and only if $\Delta^{(1)}$ is connected.

1.5.2 Stanley–Reisner ideals and facet ideals

Let, as before, $S = K[x_1, \ldots, x_n]$ be the polynomial ring in n variables over a field K and Δ a simplicial complex on $[n]$. For each subset $F \subset [n]$ we set

$$\mathbf{x}_F = \prod_{i \in F} x_i.$$

The **Stanley–Reisner ideal** of Δ is the ideal I_Δ of S which is generated by those squarefree monomials $\mathbf{x}_F$ with $F \notin \Delta$. In other words,

$$I_\Delta = (\mathbf{x}_F : F \in \mathcal{N}(\Delta)).$$

The **facet ideal** of Δ is the ideal $I(\Delta)$ of S which is generated by those squarefree monomials $\mathbf{x}_F$ with $F \in \mathcal{F}(\Delta)$. Thus if $\Delta = \langle F_1, \ldots, F_m \rangle$, then

$$I(\Delta) = (\mathbf{x}_{F_1}, \ldots, \mathbf{x}_{F_m}).$$

Proposition 1.5.1. *The set of all monomials* $x_1^{a_1} \cdots x_n^{a_n}$ *of S with $\{i \in [n] : a_i \neq 0\} \in \Delta$ is a K-basis of S/I_Δ.*

Proof. Let $u = x_1^{a_1} \cdots x_n^{a_n}$ be a monomial of S. If $\{i \in [n] : a_i \neq 0\} \notin \Delta$ then by definition $\sqrt{u} \in I_\Delta$. Thus $u \in I_\Delta$. On the other hand, if $u \in I_\Delta$, then $\sqrt{u} \in I_\Delta$, since I_Δ is a radical ideal; see Corollary 1.2.5. Thus a subset F of $\{i \in [n] : a_i \neq 0\}$ is a nonface of Δ. Since Δ is a simplicial complex, the set $\{i \in [n] : a_i \neq 0\}$ cannot be a face of Δ. Thus we have shown that $x_1^{a_1} \cdots x_n^{a_n} \notin I_\Delta$ if and only if $\{i \in [n] : a_i \neq 0\} \in \Delta$. Hence the assertion follows from Corollary 1.1.4 $\square$

1.5.3 The Alexander dual

Given a simplicial complex Δ on $[n]$, we define $\Delta^\vee$ by

$$\Delta^\vee = \{[n] \setminus F : F \notin \Delta\}.$$

Lemma 1.5.2. *The collection of sets $\Delta^\vee$ is a simplicial complex and*

$$(\Delta^\vee)^\vee = \Delta.$$

Proof. Let $F \in \Delta^\vee$ and $F' \subset F$. Then $[n] \setminus F \notin \Delta$. Since $[n] \setminus F \subset [n] \setminus F'$, it follows that $[n] \setminus F' \notin \Delta$. Thus $F' \in \Delta^\vee$. This shows that $\Delta^\vee$ is a simplicial complex. It is obvious that $(\Delta^\vee)^\vee = \Delta$. $\square$

The simplicial complex $\Delta^\vee$ is called the **Alexander dual** of Δ.
Note that

$$\mathcal{F}(\Delta^\vee) = \{[n] \setminus F : F \in \mathcal{N}(\Delta)\}.$$

For each subset $F \subset [n]$ we set $\bar{F} = [n] \setminus F$ and let

$$\bar{\Delta} = \langle \bar{F} : F \in \mathcal{F}(\Delta) \rangle.$$

Lemma 1.5.3. *One has*

$$I_{\Delta^\vee} = I(\bar{\Delta}).$$

Proof. A squarefree monomial $\mathbf{x}_F$ belongs to $G(I_{\Delta^\vee})$ if and only if F is a minimal nonface of $\Delta^\vee$. In other words, F is a nonface of $\Delta^\vee$ and all proper subsets of F are faces of $\Delta^\vee$. This is equivalent to saying that $\bar{F}$ is a face of Δ and no subset of $[n]$ which properly contains $\bar{F}$ is a face of Δ. This is the case if and only if $\bar{F}$ is a facet of Δ. Hence $I_{\Delta^\vee} = I(\bar{\Delta})$, as desired. $\square$

For each subset $F \subset [n]$ we set

$$P_F = (x_i : i \in F).$$

Lemma 1.5.4. *The standard primary decomposition of I_Δ is*

$$I_\Delta = \bigcap_{F \in \mathcal{F}(\Delta)} P_{\bar{F}}.$$

Proof. Let $u = x_1^{a_1} \cdots x_n^{a_n}$ be a monomial of S and $F_u = \{i \in [n] : a_i \neq 0\}$. If $u \in I_\Delta$, then by Proposition 1.5.1, $F_u \notin \Delta$. Thus no facet of Δ contains F_u. Hence $F_u \cap (\bar{F}) \neq \emptyset$ for all facets F of Δ. Thus $u \in \bigcap_{F \in \mathcal{F}(\Delta)} P_{\bar{F}}$.

On the other hand, if $u \notin I_\Delta$, then again by Proposition 1.5.1, $F_u \in \Delta$. Hence there is a facet F of Δ with $F_u \subset F$. Then $u \notin P_{\bar{F}}$. Hence $u \notin \bigcap_{F \in \mathcal{F}(\Delta)} P_{\bar{F}}$. □

Lemma 1.5.3 and Lemma 1.5.4 supply us with an effective method to compute $G(I_{\Delta^\vee})$.

Corollary 1.5.5. *Let $I_\Delta = P_{F_1} \cap \cdots \cap P_{F_m}$ be the standard primary decomposition of I_Δ, where each $F_j \subset [n]$. Then $G(I_{\Delta^\vee}) = \{x_{F_1}, \ldots, x_{F_m}\}$.*

Example 1.5.6. Let Δ be the simplicial complex of Figure 1.1. Since

$$I_\Delta = (x_3, x_4) \cap (x_3, x_5) \cap (x_1, x_4, x_5) \cap (x_1, x_2, x_5),$$

the ideal $I_{\Delta^\vee}$ is generated by $x_3 x_4, x_3 x_5, x_1 x_4 x_5$ and $x_1 x_2 x_5$.

Let $I \subset S$ be an arbitrary squarefree monomial ideal. Then there is a unique simplicial complex Δ such that $I = I_\Delta$. For simplicity, we often write $I^\vee$ to denote the ideal $I_{\Delta^\vee}$.

1.6 Polarization

Polarization is a process, in fact a deformation, that assigns to an arbitrary monomial ideal a squarefree monomial ideal in a new set of variables. The construction of the polarization is based on the following

Lemma 1.6.1. *Let $I \subset S = K[x_1, \ldots, x_n]$ be a monomial ideal with $G(I) = \{u_1, \ldots, u_m\}$ where $u_i = \prod_{j=1}^n x_j^{a_{ij}}$ for $i = 1, \ldots, m$. Fix an integer $j \in [n]$ and suppose that $a_{ij} > 1$ for at least one $i \in [m]$. Let $T = S[y]$ be the polynomial ring over S in the variable y and let $J \subset T$ be the monomial ideal with $G(J) = \{v_1, \ldots, v_m\}$ where $v_i = u_i$ if $a_{ij} = 0$, and $v_i = (u_i/x_j)y$ if $a_{ij} \geq 1$. Then $y - x_j$ a nonzero divisor modulo J and $(T/J)/(y - x_j)(T/J) \cong S/I$.*

Proof. Suppose $y - x_j$ is a zero divisor modulo J. Then $y - x_j \in P$ for some $P \in \text{Ass}(J)$. Since by Corollary 1.3.9, P is a monomial prime ideal, it follows that $y, x_j \in P$. Hence there exists $w \in S \setminus J$ such that $yw, x_j w \in J$. Since J is a monomial ideal, we may assume that w is a monomial. Then there exist $v_k, v_\ell \in G(J)$ and monomials w_1, w_2 such that $yw = w_1 v_k$ and $x_j w = w_2 v_\ell$.

Since $w \notin J$ it follows that x_j divides v_ℓ and this implies that y divides v_ℓ. Consequently, y divides w. The variable y does not divide w_1 since $w \notin J$. Therefore, the equation $yw = w_1 v_k$ implies that y^2 divides v_k, a contradiction. □

Repeated application of Lemma 1.6.1 leads to the following construction: let $I \subset S = K[x_1, \ldots, x_n]$ be a monomial ideal with $G(I) = \{u_1, \ldots, u_m\}$ where $u_i = \prod_{j=1}^{n} x_j^{a_{ij}}$ for $i = 1, \ldots, m$. For each j let $a_j = \max\{a_{ij}: i = 1, \ldots, m\}$ and let T be the polynomial ring over K in the variables

$$x_{11}, x_{12} \ldots, x_{1a_1}, x_{21}, x_{22} \ldots, x_{2a_2}, \ldots, x_{n1}, x_{n2}, \ldots, x_{na_n}.$$

Let $J \subset T$ be the squarefree monomial ideal with $G(J) = \{v_1, \ldots, v_m\}$ where

$$v_i = \prod_{j=1}^{n} \prod_{k=1}^{a_{ij}} x_{jk} \quad \text{for} \quad i = 1, \ldots, m.$$

The monomial v_i is called the polarization of u_i, and the ideal J the **polarization** of I. As an immediate consequence of Lemma 1.6.1 we now have

Proposition 1.6.2. *Let $I \subset S$ be a monomial ideal and $J \subset T$ its polarization. Then the sequence $\mathbf{z}$ of linear forms*

$$x_{11} - x_{12}, \ldots, x_{11} - x_{1a_1}, x_{21} - x_{22}, \ldots, x_{21} - x_{2a_2}, \ldots, x_{n1} - x_{n2}, \ldots, x_{n1} - x_{na_n}$$

is a T/J-sequence (i.e. a regular sequence on T/J), and one has the following isomorphism of graded K-algebras

$$(T/J)/(\mathbf{z})(T/J) \cong S/I.$$

A monomial ideal I and its polarization J share many homological and algebraic properties. Thus, by polarization, many questions concerning monomial ideals can be reduced to squarefree monomial ideals. Most important is that the graded Betti numbers of I and J are the same. For unexplained concepts and notation we refer the reader to the appendices and to later chapters.

Corollary 1.6.3. *Let $I \subset S$ be a monomial ideal and $J \subset T$ its polarization. Then*

(a) $\beta_{ij}(I) = \beta_{ij}(J)$ *for all i and j;*
(b) $H_{S/I}(t) = (1-t)^\delta H_{T/J}(t)$ *where $\delta = \dim T - \dim S$;*
(c) $\text{height } I = \text{height } J$;
(d) $\text{proj dim } S/I = \text{proj dim } T/J$ *and $\text{reg } S/I = \text{reg } T/J$;*
(e) *S/I is Cohen–Macaulay (resp. Gorenstein) if and only if T/J is Cohen–Macaulay (resp. Gorenstein).*

Proof. (a) Since $\mathbf{z}$ is a T/J-sequence, Corollary A.3.5 and Theorem A.3.4 imply that

$$\text{Tor}_i^T(T/(\mathbf{z}), T/J) = H_i(\mathbf{z}; T/J) = 0.$$

Hence if $\mathbb{F}$ is a graded minimal free T-resolution of T/J it follows that $\mathbb{F}/(\mathbf{z})\mathbb{F}$ is acyclic with

$$H_0(\mathbb{F}/(\mathbf{z})\mathbb{F}) \cong (T/J)/(\mathbf{z})(T/J) \cong S/I.$$

If $F_i = \bigoplus_j T(-j)^{\beta_{ij}(J)}$, then $F_i/(\mathbf{z})F_i \cong \bigoplus_j S(-j)^{\beta_{ij}(J)}$. Thus $\mathbb{F}/(\mathbf{z})\mathbb{F}$ is a free S-resolution of S/I. Obviously, it is again a minimal resolution, which then implies that $\beta_{ij}(I) = \beta_{ij}(J)$.

(b) follows from Formula (6.3) in Subsection 6.1.3.

(c) is a consequence of (b) and the fact that the Hilbert function of a module determines its dimension; see Theorem 6.1.3.

(d) is an immediate implication of (a).

(e) By the Auslander–Buchsbaum formula (see Corollary A.4.3) one has $\operatorname{proj\,dim} M + \operatorname{depth} M = n$ for any finitely generated graded S-module M. On the other hand, M is Cohen–Macaulay if and only of $\operatorname{depth} M = \dim M$. Thus (a) and (c) together imply that S/I is Cohen–Macaulay if and only if T/J is Cohen–Macaulay.

Since the Gorenstein property can be characterized by the fact (see A.6.6) that the last non-vanishing Betti number of S/I resp. T/J is equal to 1, we see that (a) implies the remaining assertion of (e) as well.   □

Problems

1.1. Let $I \subset S = K[x_1, \ldots, x_n]$ be a monomial ideal.
(a) Show that $\dim_K S/I < \infty$ if and only if there exists an integer $a \in \mathbb{Z}_+$ such that $x_i^a \in I$ for all i.
(b) Given integers $a_i \in \mathbb{Z}_+$, compute $\dim_K S/I$ for $I = (x_1^{a_1}, \ldots, x_n^{a_n})$.

1.2. We use the standard notation $[n]$ for the set $\{1, 2, \ldots, n\}$. Let $F \subset [n]$. We denote by $P_F \subset K[x_1, \ldots, x_n]$ the monomial ideal generated by the variables x_i with $i \in F$. Given an integer $d \in [n]$, compute the intersection $I = \bigcap_{F, |F|=d} P_F$; in other words, describe the elements of $G(I)$.

1.3. Let $d > 0$ be an integer, and let $I \subset K[x_1, \ldots, x_n]$ be the monomial ideal generated by all monomials $x_1^{a_1} x_2^{a_2} \cdots x_n^{a_n}$ with $\sum_{i=1}^n a_i = d$ and $a_i < d$ for all i. Compute the saturation $\tilde{I}$ and the radical $\sqrt{I}$ of I.

1.4. The smallest integer k such that $I : \mathfrak{m}^k = I : \mathfrak{m}^{k+1}$ is called the **saturation number** of I. What is the saturation number of I in Problem 1.3?

1.5. Find an example of a monomial ideal I for which $\widetilde{I^2} \neq (\tilde{I})^2$.

1.6. Show that the monomial ideal $(x_1^2, x_1 x_2)$ has infinitely many different irredundant primary decompositions.

1.7. For $n = 3$, find the standard primary decomposition of the ideals described in Problem 1.3.

1.8. Let $P = (x_1, \ldots, x_r) \subset S = K[x_1, \ldots, x_n]$. Show that a monomial ideal Q is P-primary if and only if there exists a monomial ideal $Q' \subset T = K[x_1, \ldots, x_r]$ such that $\dim_K T/Q' < \infty$ and $Q = Q'S$.

1.9. Compute the integral closure of the monomial ideal $(x^3, y^5) \subset K[x, y]$.

1.10. Show that the ideal $(xy, xz, yz) \subset K[x, y, z]$ is not normally torsionfree but normal.

1.11. Let Δ be the simplicial complex on the vertex set $[5]$ whose Stanley–Reisner ideal is $I_\Delta = (x_1 x_4, x_1 x_5, x_2 x_5, x_1 x_2 x_3, x_3 x_4 x_5)$. Compute $I_{\Delta^\vee}$.

1.12. Let $u_1, \ldots, u_m$ be a (not necessarily minimal) system of generators of the monomial ideal I. Let v_i be the polarization of u_i for $i = 1, \ldots, m$, J the polarization of I and J' the ideal generated by $v_1, \ldots, v_m$. Show that $G(J) = G(J')$.

1.13. Let $I_1, I_2 \subset S$ be monomial ideals and let J_1 be the polarization of I_1, J_2 the polarization of I_2 and J the polarization of $I_1 \cap I_2$. Show that $G(J) = G(J_1 \cap J_2)$. Prove a similar result for the sum of I_1 and I_2.

Notes

Monomial ideals are the bridge between commutative algebra and combinatorics. Another reason for their importance is the fact that monomial ideals appear as initial ideals of arbitrary ideals; see Chapter 2. Since many properties of an initial ideal are inherited by its original ideal, it is an often applied strategy to study general ideals via their initial ideal, thereby reducing a given question concerning an ideal to that of a monomial ideal.

The systematic study of squarefree monomial ideals began with the work of Stanley [Sta75] and Reisner [Rei76]. In 1983 Stanley wrote his influential book *Combinatorics and Commutative Algebra*, where he discussed the upper bound conjecture for spheres by using algebraic properties of squarefree monomial ideals.

Almost all topics explained here are already contained in several standard textbooks on combinatorics and commutative algebra, including Bruns–Herzog [BH98], Eisenbud [Eis95], Hibi [Hib92], Miller–Sturmfels [MS04], Stanley [Sta95] and Villarreal [Vil01].

In the presentation of the integral closure of monomial ideals in Subsection 1.4.1 we follow closely the book of Swanson and Hunecke [SH06]. Normally torsionfree squarefree monomial ideals are special classes of normal ideals. They also have the property that their symbolic powers coincide with the ordinary powers. The relationship of this property with Mengerian simplicial complexes is discussed in Chapter 10.

More recently the application of Alexander duality of simplicial complexes has turned out to be a powerful technique in the study of algebraic and combinatorial properties of squarefree monomial ideals. This will be discussed in Chapters 8 and 9.

The technique of polarization which allows to pass from arbitrary monomial ideals to squarefree monomial ideals was first used by Hartshorne in his paper "Connectedness of the Hilbert scheme" [Har66]. It became a popular tool in the study of monomial ideals after Hochster's article "Cohen–Macaulay rings, combinatorics and simplicial complexes", which appeared in [Hoc77].

2

A short introduction to Gröbner bases

We summarize fundamental material on Gröbner bases, including Dickson's lemma and Buchberger's criterion and algorithm. Our presentation is a quick and self-contained introduction to the theory.

2.1 Dickson's lemma and Hilbert's basis theorem

2.1.1 Dickson's lemma

Let, as before, $S = K[x_1, \ldots, x_n]$ denote the polynomial ring in n variables over a field K with each $\deg x_i = 1$, and $\mathrm{Mon}(S)$ the set of monomials of S.

For monomials $\mathbf{x}^\mathbf{a} = x_1^{a_1} x_2^{a_2} \cdots x_n^{a_n}$ and $\mathbf{x}^\mathbf{b} = x_1^{b_1} x_2^{b_2} \cdots x_n^{b_n}$ of S, we say that $\mathbf{x}^\mathbf{b}$ **divides** $\mathbf{x}^\mathbf{a}$ if each $b_i \leq a_i$. We write $\mathbf{x}^\mathbf{b} \,|\, \mathbf{x}^\mathbf{a}$ if $\mathbf{x}^\mathbf{b}$ divides $\mathbf{x}^\mathbf{a}$.

Let $\mathcal{M}$ be a nonempty subset of $\mathrm{Mon}(S)$. A monomial $\mathbf{x}^\mathbf{a} \in \mathcal{M}$ is said to be a **minimal element** of $\mathcal{M}$ with respect to divisibility if whenever $\mathbf{x}^\mathbf{b} \,|\, \mathbf{x}^\mathbf{a}$ with $\mathbf{x}^\mathbf{b} \in \mathcal{M}$, then $\mathbf{x}^\mathbf{b} = \mathbf{x}^\mathbf{a}$. Let $\mathcal{M}^{\min}$ denote the set of minimal elements of $\mathcal{M}$.

Theorem 2.1.1 (Dickson's lemma). *Let $\mathcal{M}$ be a nonempty subset of $\mathrm{Mon}(S)$. Then $\mathcal{M}^{\min}$ is a finite set.*

Proof. We prove Dickson's lemma by using induction on n, the number of variables of $S = K[x_1, x_2, \ldots, x_n]$. Let $n = 1$. If d is the smallest integer for which $x_1^d \in \mathcal{M}$, then $\mathcal{M}^{\min} = \{x_1^d\}$. Thus $\mathcal{M}^{\min}$ is a finite set.

Let $n \geq 2$ and $B = K[\mathbf{x}] = K[x_1, x_2, \ldots, x_{n-1}]$. We use the notation y instead of x_n. Thus $S = K[x_1, x_2, \ldots, x_{n-1}, y]$. Let $\mathcal{M}$ be a nonempty subset of $\mathrm{Mon}(S)$. Write $\mathcal{N}$ for the subset of $\mathrm{Mon}(B)$ which consists of those monomials $\mathbf{x}^\mathbf{a}$, where $\mathbf{a} \in \mathbb{Z}_+^{n-1}$, such that $\mathbf{x}^\mathbf{a} y^b \in \mathcal{M}$ for some $b \geq 0$. Our assumption of induction says that $\mathcal{N}^{\min}$ is a finite set. Let $\mathcal{N}^{\min} = \{u_1, u_2, \ldots, u_s\}$. By the definition of $\mathcal{N}$, for each $1 \leq i \leq s$, there is $b_i \geq 0$ with $u_i y^{b_i} \in \mathcal{M}$. Let $b = \max\{b_1, b_2, \ldots, b_s\}$. Now, for each $0 \leq \xi < b$, define the subset $\mathcal{N}_\xi$ of $\mathcal{N}$ to be

J. Herzog, T. Hibi, *Monomial Ideals*, Graduate Texts in Mathematics 260,   23
DOI 10.1007/978-0-85729-106-6_2, © Springer-Verlag London Limited 2011

$$\mathcal{N}_\xi = \{\mathbf{x}^\mathbf{a} \in \mathcal{N} : \mathbf{x}^\mathbf{a} y^\xi \in \mathcal{M}\}.$$

Again, our assumption of induction says that, for each $0 \le \xi < b$, the set $\mathcal{N}_\xi^{\min}$ is finite. Let $\mathcal{N}_\xi^{\min} = \{u_1^{(\xi)}, u_2^{(\xi)}, \ldots, u_{s_\xi}^{(\xi)}\}$. We now show that each monomial belonging to $\mathcal{M}$ is divisible by one of the monomials in the following list:

$$u_1 y^{b_1}, u_2 y^{b_2}, \ldots, u_s y^{b_s},$$
$$u_1^{(0)}, u_2^{(0)}, \ldots, u_{s_0}^{(0)},$$
$$u_1^{(1)} y, u_2^{(1)} y, \ldots, u_{s_1}^{(1)} y,$$
$$\cdots\cdots$$
$$u_1^{(b-1)} y^{b-1}, u_2^{(b-1)} y^{b-1}, \ldots, u_{s_{b-1}}^{(b-1)} y^{b-1}.$$

In fact, since for each monomial $w = \mathbf{x}^\mathbf{a} y^\gamma \in \mathcal{M}$ with $\mathbf{x}^\mathbf{a} \in \mathrm{Mon}(B)$ one has $\mathbf{x}^\mathbf{a} \in \mathcal{N}$, it follows that if $\gamma \ge b$, then w is divisible by one of the monomials $u_1 y^{b_1}, u_2 y^{b_2}, \ldots, u_s y^{b_s}$, and that if $0 \le \gamma < b$, then w is divisible by one of the monomials $u_1^{(\gamma)} y^\gamma, u_2^{(\gamma)} y^\gamma, \ldots, u_{s_\gamma}^{(\gamma)} y^\gamma$. Hence $\mathcal{M}^{\min}$ is a subset of the set of monomials listed above. Thus $\mathcal{M}^{\min}$ is finite, as desired.   □

2.1.2 Monomial orders

Recall that a **partial order** on a set P is a relation $\le$ on P such that, for all $x, y, z \in P$ one has

(i) $x \le x$ (reflexivity);
(ii) $x \le y$ and $y \le x \Rightarrow x = y$ (antisymmetry);
(iii) $x \le y$ and $y \le z \Rightarrow x \le z$ (transitivity).

A **total order** on a set P is a partial order $\le$ on P such that, for any two elements x and y belonging to P, one has either $x \le y$ or $y \le x$.

A **monomial order** on S is a total order $<$ on $\mathrm{Mon}(S)$ such that

(i) $1 < u$ for all $1 \ne u \in \mathrm{Mon}(S)$;
(ii) if $u, v \in \mathrm{Mon}(S)$ and $u < v$, then $uw < vw$ for all $w \in \mathrm{Mon}(S)$.

Example 2.1.2. (a) Let $\mathbf{a} = (a_1, a_2, \ldots, a_n)$ and $\mathbf{b} = (b_1, b_2, \ldots, b_n)$ be vectors belonging to $\mathbb{Z}_+^n$. We define the total order $<_{\mathrm{lex}}$ on $\mathrm{Mon}(S)$ by setting $\mathbf{x}^\mathbf{a} <_{\mathrm{lex}} \mathbf{x}^\mathbf{b}$ if either (i) $\sum_{i=1}^n a_i < \sum_{i=1}^n b_i$, or (ii) $\sum_{i=1}^n a_i = \sum_{i=1}^n b_i$ and the leftmost nonzero component of the vector $\mathbf{a} - \mathbf{b}$ is negative. It follows that $<_{\mathrm{lex}}$ is a monomial order on S, which is called the **lexicographic order** on S induced by the ordering $x_1 > x_2 > \cdots > x_n$.

(b) Let $\mathbf{a} = (a_1, a_2, \ldots, a_n)$ and $\mathbf{b} = (b_1, b_2, \ldots, b_n)$ be vectors belonging to $\mathbb{Z}_+^n$. We define the total order $<_{\mathrm{rev}}$ on $\mathrm{Mon}(S)$ by setting $\mathbf{x}^\mathbf{a} <_{\mathrm{rev}} \mathbf{x}^\mathbf{b}$ if either (i) $\sum_{i=1}^n a_i < \sum_{i=1}^n b_i$, or (ii) $\sum_{i=1}^n a_i = \sum_{i=1}^n b_i$ and the rightmost nonzero component of the vector $\mathbf{a} - \mathbf{b}$ is positive. It follows that $<_{\mathrm{rev}}$ is a monomial order on S, which is called the **reverse lexicographic order** on S induced by the ordering $x_1 > x_2 > \cdots > x_n$.

(c) Let $\mathbf{a} = (a_1, a_2, \ldots, a_n)$ and $\mathbf{b} = (b_1, b_2, \ldots, b_n)$ be vectors belonging to $\mathbb{Z}_+^n$. We define the total order $<_{\text{purelex}}$ on $\text{Mon}(S)$ by setting $\mathbf{x}^{\mathbf{a}} <_{\text{purelex}} \mathbf{x}^{\mathbf{b}}$ if the leftmost nonzero component of the vector $\mathbf{a} - \mathbf{b}$ is negative. It follows that $<_{\text{purelex}}$ is a monomial order on S, which is called the **pure lexicographic order** on S induced by the ordering $x_1 > x_2 > \cdots > x_n$. (Can we also define the pure reverse lexicographic order?)

Let $\pi = i_1 i_2 \cdots i_n$ be a permutation of $1, 2, \ldots, n$. How can we define the lexicographic order (or the reverse lexicographic order) induced by the ordering $x_{i_1} > x_{i_2} > \cdots > x_{i_n}$? The answer is easy. For a monomial $u = x_1^{a_1} x_2^{a_2} \cdots x_n^{a_n}$ of S, we set

$$u^{\pi} = x_1^{b_1} x_2^{b_2} \cdots x_n^{b_n}, \quad \text{where} \quad b_j = a_{i_j}.$$

We then define the total order $<_{\text{lex}}^{\pi}$ (resp. $<_{\text{rev}}^{\pi}$) on $\text{Mon}(S)$ by setting $u <_{\text{lex}}^{\pi} v$ if $u^{\pi} <_{\text{lex}} v^{\pi}$ (resp. $u^{\pi} <_{\text{rev}} v^{\pi}$), where $u, v \in \text{Mon}(S)$. It follows that $<_{\text{lex}}^{\pi}$ (resp. $<_{\text{rev}}^{\pi}$) is a monomial order on S, which is called the lexicographic order (resp. reverse lexicographic order) on S induced by the ordering $x_{i_1} > x_{i_2} > \cdots > x_{i_n}$.

Unless otherwise stated, we *only* consider monomial orders satisfying

$$x_1 > x_2 > \cdots > x_n.$$

Example 2.1.3. Fix a vector $\omega = (\omega_1, \omega_2, \ldots, \omega_n) \in \mathbb{R}^n$ with each $\omega_i \geq 0$. Given an arbitrary monomial order $<$ on S, we introduce the total order $<_{\omega}$ on $\text{Mon}(S)$ by setting $\mathbf{x}^{\mathbf{a}} <_{\omega} \mathbf{x}^{\mathbf{b}}$ if either (i) $\sum_{i=1}^n \omega_i(a_i - b_i)$ is negative or (ii) $\sum_{i=1}^n \omega_i(a_i - b_i) = 0$ and $\mathbf{x}^{\mathbf{a}} < \mathbf{x}^{\mathbf{b}}$, where $\mathbf{a} = (a_1, a_2, \ldots, a_n)$ and $\mathbf{b} = (b_1, b_2, \ldots, b_n)$. Then $<_{\omega}$ is a monomial order on S.

2.1.3 Gröbner bases

We will work with a fixed monomial order $<$ on S. Let $f = \sum_{u \in \text{Mon}(S)} a_u u$ be a nonzero polynomial of S with each $a_u \in K$. The **initial monomial** of f with respect to $<$ is the biggest monomial with respect to $<$ among the monomials belonging to $\text{supp}(f)$. We write $\text{in}_<(f)$ for the initial monomial of f with respect to $<$. The **leading coefficient** of f is the coefficient of $\text{in}_<(f)$ in f.

Lemma 2.1.4. *Let u, v be monomials of S and f, g nonzero polynomials of S. Then one has*

(i) *if u divides v, then $u \leq v$;*
(ii) $\text{in}_<(uf) = u \, \text{in}_<(f)$;
(iii) $\text{in}_<(fg) = \text{in}_<(f) \, \text{in}_<(g)$.
(iv) $\text{in}_<(f + g) \leq \max\{\text{in}_<(f), \text{in}_<(g)\}$ *with equality if $\text{in}_<(f) \neq \text{in}_<(g)$.*

Proof. (i) In fact, if u divides v and if $v = uw$ with $w \in \mathrm{Mon}(S)$, then since $1 \leq w$ one has $1 \cdot u \leq w \cdot u$. Thus $u \leq v$, as desired.

(ii) Let $w \in \mathrm{supp}(f)$ with $w < \mathrm{in}_<(f)$, then $uw < u\,\mathrm{in}_<(f)$. Hence $\mathrm{in}_<(uf) = u\,\mathrm{in}_<(f)$.

(iii) Let $w \in \mathrm{supp}(f)$ with $w < \mathrm{in}_<(f)$ and $w' \in \mathrm{supp}(g)$ with $w' < \mathrm{in}_<(g)$. Then $ww' < w\,\mathrm{in}_<(g) < \mathrm{in}_<(f)\,\mathrm{in}_<(g)$. Hence $\mathrm{in}_<(fg) = \mathrm{in}_<(f)\,\mathrm{in}_<(g)$.

(iv) is obvious. $\square$

Let $I \subset S$ be a monomial ideal. It follows that I is generated by a subset $\mathcal{N} \subset \mathrm{Mon}(S)$ if and only if $(I \cap \mathrm{Mon}(S))^{\min} \subset \mathcal{N}$. Hence $(I \cap \mathrm{Mon}(S))^{\min}$ is a **unique** minimal system of monomial generators of I (see also Proposition 1.1.6). Dickson's Lemma guarantees that $(I \cap \mathrm{Mon}(S))^{\min}$ is a finite set. Thus in particular every monomial ideal I of S is finitely generated.

Let I be a nonzero ideal of S. The **initial ideal** of I with respect to $<$ is the monomial ideal of S which is generated by $\{\mathrm{in}_<(f) : 0 \neq f \in I\}$. We write $\mathrm{in}_<(I)$ for the initial ideal of I. Thus

$$\mathrm{in}_<(I) = (\{\mathrm{in}_<(f) : 0 \neq f \in I\}).$$

Since $(\mathrm{in}_<(I) \cap \mathrm{Mon}(S))^{\min}$ is the minimal system of monomial generators of $\mathrm{in}_<(I)$ and since $(\mathrm{in}_<(I) \cap \mathrm{Mon}(S)) = \{\mathrm{in}_<(f) : 0 \neq f \in I\}$, there exist a finite number of nonzero polynomials $g_1, g_2, \ldots, g_s$ belonging to I such that $\mathrm{in}_<(I)$ is generated by their initial monomials $\mathrm{in}_<(g_1), \mathrm{in}_<(g_2), \ldots, \mathrm{in}_<(g_s)$.

Definition 2.1.5. *Let I be a nonzero ideal of S. A finite set of nonzero polynomials $\{g_1, g_2, \ldots, g_s\}$ with each $g_i \in I$ is said to be a **Gröbner basis** of I with respect to $<$ if the initial ideal $\mathrm{in}_<(I)$ of I is generated by the monomials $\mathrm{in}_<(g_1), \mathrm{in}_<(g_2), \ldots, \mathrm{in}_<(g_s)$.*

A Gröbner basis of I with respect to $<$ exists. If $\mathcal{G}$ is a Gröbner basis of I with respect to $<$, then every finite set $\mathcal{G}'$ with $\mathcal{G} \subset \mathcal{G}' \subset I$ is also a Gröbner basis of I with respect to $<$. If $\mathcal{G} = \{g_1, \ldots, g_s\}$ is a Gröbner basis of I with respect to $<$ and if $f_1, \ldots, f_s$ are nonzero polynomials belonging to I with each $\mathrm{in}_<(f_i) = \mathrm{in}_<(g_i)$, then $\{f_1, \ldots, f_s\}$ is a Gröbner basis of I with respect to $<$.

Example 2.1.6. (a) Let $S = K[x_1, x_2, \ldots, x_7]$ and $<_{\mathrm{lex}}$ the lexicographic order on S induced by $x_1 > x_2 > \cdots > x_7$. Let $f = x_1x_4 - x_2x_3$ and $g = x_4x_7 - x_5x_6$ with their initial monomials $\mathrm{in}_{<_{\mathrm{lex}}}(f) = x_1x_4$ and $\mathrm{in}_{<_{\mathrm{lex}}}(g) = x_4x_7$. Let $I = (f, g)$. Then $\{f, g\}$ is not a Gröbner basis of I with respect to $<_{\mathrm{lex}}$. In fact, the polynomial $h = x_7f - x_1g = x_1x_5x_6 - x_2x_3x_7$ belongs to I, but its initial monomial $\mathrm{in}_{<_{\mathrm{lex}}}(h) = x_1x_5x_6$ can be divided by neither $\mathrm{in}_{<_{\mathrm{lex}}}(f)$ nor $\mathrm{in}_{<_{\mathrm{lex}}}(g)$. Hence $\mathrm{in}_{<_{\mathrm{lex}}}(h) \notin (\mathrm{in}_{<_{\mathrm{lex}}}(f), \mathrm{in}_{<_{\mathrm{lex}}}(g))$. Thus $\mathrm{in}_{<_{\mathrm{lex}}}(I) \neq (\mathrm{in}_{<_{\mathrm{lex}}}(f), \mathrm{in}_{<_{\mathrm{lex}}}(g))$. In other words, $\{f, g\}$ is not a Gröbner basis of I with respect to $<_{\mathrm{lex}}$. It will be shown in Example 2.3.6 that $\{f, g, h\}$ is a Gröbner basis of I with respect to $<_{\mathrm{lex}}$.

(b) Let $S = K[x_1, x_2, \ldots, x_7]$ and $<_{\text{rev}}$ the reverse lexicographic order on S induced by $x_1 > x_2 > \cdots > x_7$. Let $f = x_2 x_3 - x_1 x_4$ and $g = x_4 x_7 - x_5 x_6$. Later by using Corollary 2.3.4 it turns out that $\{f, g\}$ is a Gröbner basis of $I = (f, g)$ with respect to $<_{\text{rev}}$.

Lemma 2.1.7. *Let $<$ be a monomial order on $S = K[x_1, \ldots, x_n]$. Then, for any monomial u of S, there is no infinite descending sequence of the form*

$$\cdots < u_2 < u_1 < u_0 = u. \tag{2.1}$$

Proof. Suppose, on the contrary, that one has an infinite descending sequence (2.1) and write $\mathcal{M}$ for the set of monomials $\{u_0, u_1, u_2, \ldots\}$. It follows from Dickson's Lemma that $\mathcal{M}^{\min}$ is a finite set, say $\mathcal{M}^{\min} = \{u_{i_1}, u_{i_2}, \ldots, u_{i_s}\}$ with $i_1 < i_2 < \cdots < i_s$. Then the monomial u_{i_s+1} is divided by u_{i_j} for some $1 \le j \le s$. Thus $u_{i_j} < u_{i_s+1}$, which contradicts $i_j < i_s + 1$. □

2.1.4 Hilbert's basis theorem

Theorem 2.1.8. *Let I be a nonzero ideal of $S = K[x_1, \ldots, x_n]$ and $\mathcal{G} = \{g_1, \ldots, g_s\}$ a Gröbner basis of I with respect to a monomial order $<$ on S. Then $I = (g_1, \ldots, g_s)$. In other words, every Gröbner basis of I is a system of generators of I.*

Proof. (Gordan) Let $0 \ne f \in I$. Since $\text{in}_<(f) \in \text{in}_<(I)$, it follows that there is g_{i_0} such that $\text{in}_<(g_{i_0})$ divides $\text{in}_<(f)$. Let $\text{in}_<(f) = w_0 \text{in}_<(g_{i_0})$ with $w_0 \in \text{Mon}(S)$. Let $h_0 = f - c_{i_0}^{-1} c_0 w_0 g_{i_0}$, where c_0 is the coefficient of $\text{in}_<(f)$ in f and where c_{i_0} is the coefficient of $\text{in}_<(g_{i_0})$ in g_{i_0}. Then $h_0 \in I$. Since $\text{in}_<(w_0 g_{i_0}) = w_0 \text{in}_<(g_{i_0})$ it follows that $\text{in}_<(h_0) < \text{in}_<(f)$. If $h_0 = 0$, then $f \in (g_1, \ldots, g_s)$.

Let $h_0 \ne 0$. Then the same technique as we used for f can be applied for h_0. Thus $h_1 = f - c_{i_1}^{-1} c_1 w_1 g_{i_1} - c_{i_0}^{-1} c_0 w_0 g_{i_0}$, where c_1 is the coefficient of $\text{in}_<(h_0)$ in h_0 and where c_{i_1} is the coefficient of $\text{in}_<(g_{i_1})$ in g_{i_1}. Then $h_1 \in I$ and $\text{in}_<(h_1) < \text{in}_<(h_0)$. If $h_1 = 0$, then $f \in (g_1, \ldots, g_s)$.

If $h_1 \ne 0$, then we proceed as before. Lemma 2.1.7 guarantees that this procedure must terminate. Thus we obtain an expression of the form $f = \sum_{q=0}^{N} c_{i_q}^{-1} c_q w_q g_{i_q}$. In particular, f belongs to $(g_1, g_2, \ldots, g_s)$. Thus $I = (g_1, g_2, \ldots, g_s)$, as desired. □

Corollary 2.1.9 (Hilbert's basis theorem). *Every ideal of the polynomial ring is finitely generated.*

It is natural to ask if the converse of Theorem 2.1.8 is true or false. That is to say, if $I = (f_1, f_2, \ldots, f_s)$ is an ideal of $S = K[x_1, \ldots, x_n]$, then does there exist a monomial order $<$ on S such that $\{f_1, f_2, \ldots, f_s\}$ is a Gröbner basis of I with respect to $<$?

Example 2.1.10. Let $S = K[x_1, x_2, \ldots, x_{10}]$ and I the ideal of S generated by

$$f_1 = x_1x_8 - x_2x_6, \quad f_2 = x_2x_9 - x_3x_7, \quad f_3 = x_3x_{10} - x_4x_8,$$
$$f_4 = x_4x_6 - x_5x_9, \quad f_5 = x_5x_7 - x_1x_{10}.$$

We claim that there exists *no* monomial order $<$ on S such that $\{f_1, \ldots, f_5\}$ is a Gröbner basis of I with respect to $<$.

Suppose, on the contrary, that there exists a monomial order $<$ on S such that $\mathcal{G} = \{f_1, \ldots, f_5\}$ is a Gröbner basis of I with respect to $<$. First, note that each of the five polynomials

$$x_1x_8x_9 - x_3x_6x_7, \; x_2x_9x_{10} - x_4x_7x_8, \; x_2x_6x_{10} - x_5x_7x_8,$$
$$x_3x_6x_{10} - x_5x_8x_9, \; x_1x_9x_{10} - x_4x_6x_7$$

belongs to I. Let, say, $x_1x_8x_9 > x_3x_6x_7$. Since $x_1x_8x_9 \in \mathrm{in}_<(I)$, there is $g \in \mathcal{G}$ such that $\mathrm{in}_<(g)$ divides $x_1x_8x_9$. Such $g \in \mathcal{G}$ must be f_1. Hence $x_1x_8 > x_2x_6$. Thus $x_2x_6 \notin \mathrm{in}_<(I)$. Hence there exists no $g \in \mathcal{G}$ such that $\mathrm{in}_<(g)$ divides $x_2x_6x_{10}$. Hence $x_2x_6x_{10} < x_5x_7x_8$. Thus $x_5x_7 > x_1x_{10}$. Continuing these arguments yields

$$x_1x_8x_9 > x_3x_6x_7, \; x_2x_9x_{10} > x_4x_7x_8, \; x_2x_6x_{10} < x_5x_7x_8,$$
$$x_3x_6x_{10} > x_5x_8x_9, \; x_1x_9x_{10} < x_4x_6x_7$$

and

$$x_1x_8 > x_2x_6, \; x_2x_9 > x_3x_7, \; x_3x_{10} > x_4x_8,$$
$$x_4x_6 > x_5x_9, \; x_5x_7 > x_1x_{10}.$$

Hence

$$(x_1x_8)(x_2x_9)(x_3x_{10})(x_4x_6)(x_5x_7) > (x_2x_6)(x_3x_7)(x_4x_8)(x_5x_9)(x_1x_{10}).$$

However, both sides of the above inequality coincide with $x_1x_2 \cdots x_{10}$. This is a contradiction.

2.2 The division algorithm

2.2.1 The division algorithm

The division algorithm generalizes the following well-known result in high school algebra: given polynomials f and $g \neq 0$ in one variable x, there exist unique polynomials q and r such that $f = gq + r$, where either $r = 0$ or $\deg r < \deg g$.

Theorem 2.2.1 (The division algorithm). *Let $S = K[x_1, \ldots, x_n]$ denote the polynomial ring in n variables over a field K and fix a monomial order $<$ on S. Let $g_1, g_2, \ldots, g_s$ be nonzero polynomials of S. Then, given a polynomial $0 \neq f \in S$, there exist polynomials $f_1, f_2, \ldots, f_s$ and f' of S with*

$$f = f_1g_1 + f_2g_2 + \cdots + f_sg_s + f', \tag{2.2}$$

such that the following conditions are satisfied:

(i) *if $f' \neq 0$ and if $u \in \mathrm{supp}(f')$, then none of the initial monomials*
$\mathrm{in}_<(g_1), \mathrm{in}_<(g_2), \ldots, \mathrm{in}_<(g_s)$ *divides u, i.e. no monomial $u \in \mathrm{supp}(f')$*
belongs to $(\mathrm{in}_<(g_1), \mathrm{in}_<(g_2), \ldots, \mathrm{in}_<(g_s))$*;*
(ii) *if $f_i \neq 0$, then*

$$\mathrm{in}_<(f) \geq \mathrm{in}_<(f_i g_i).$$

The right-hand side of equation (2.2) is said to be a **standard expression**
for f with respect to $g_1, g_2, \ldots, g_s$, and the polynomial f' is said to be a
remainder of f with respect to $g_1, g_2, \ldots, g_s$. One also says that f **reduces**
to f' with respect $g_1, \ldots, g_s$.

Proof (of Theorem 2.2.1). Let $I = (\mathrm{in}_<(g_1), \ldots, \mathrm{in}_<(g_s))$. If none of the monomials $u \in \mathrm{supp}(f)$ belongs to I, then the desired expression can be obtained
by setting $f' = f$ and $f_1 = \cdots = f_s = 0$.

Now, suppose that a monomial $u \in \mathrm{supp}(f)$ belongs to I and write u_0
for the monomial which is biggest with respect to $<$ among the monomials
$u \in \mathrm{supp}(f)$ belonging to I. Let, say, $\mathrm{in}_<(g_{i_0})$ divide u_0 and $w_0 = u_0/\mathrm{in}_<(g_{i_0})$.
We rewrite

$$f = c_0' c_{i_0}^{-1} w_0 g_{i_0} + h_1,$$

where c_0' is the coefficient of u_0 in f and c_{i_0} is that of $\mathrm{in}_<(g_{i_0})$ in g_{i_0}. One
has

$$\mathrm{in}_<(w_0 g_{i_0}) = w_0 \, \mathrm{in}_<(g_{i_0}) = u_0 \leq \mathrm{in}_<(f).$$

If either $h_1 = 0$ or, in case of $h_1 \neq 0$, none of the monomials $u \in \mathrm{supp}(h_1)$
belongs to I, then $f = c_0' c_{i_0}^{-1} w_0 g_{i_0} + h_1$ is a standard expression of f with
respect to $g_1, g_2, \ldots, g_s$ and h_1 is a remainder of f.

If a monomial of $\mathrm{supp}(h_1)$ belongs to I and if u_1 is the monomial which
is biggest with respect to $<$ among the monomials $u \in \mathrm{supp}(h_1)$ belonging to
I, then one has

$$u_0 > u_1.$$

In fact, if a monomial u with $u > u_0 (= \mathrm{in}_<(w_0 g_{i_0}))$ belongs to $\mathrm{supp}(h_1)$,
then u must belong to $\mathrm{supp}(f)$. This is impossible. Moreover, u_0 itself cannot
belong to $\mathrm{supp}(h_1)$.

Let, say, $\mathrm{in}_<(g_{i_1})$ divide u_1 and $w_1 = u_1/\mathrm{in}_<(g_{i_1})$. Again, we rewrite

$$f = c_0' c_{i_0}^{-1} w_0 g_{i_0} + c_1' c_{i_1}^{-1} w_1 g_{i_1} + h_2,$$

where c_1' is the coefficient of u_1 in h_1 and c_{i_1} is that of $\mathrm{in}_<(g_{i_1})$ in g_{i_1}. One
has

$$\mathrm{in}_<(w_1 g_{i_1}) < \mathrm{in}_<(w_0 g_{i_0}) \leq \mathrm{in}_<(f).$$

Continuing these procedures yields the descending sequence

$$u_0 > u_1 > u_2 > \cdots$$

Lemma 2.1.7 thus guarantees that these procedures will stop after a finite
number of steps, say N steps, and we obtain an expression

$$f = \sum_{q=0}^{N-1} c'_q c_{i_q}^{-1} w_q g_{i_q} + h_N,$$

where either $h_N = 0$ or, in case $h_N \neq 0$, none of the monomials $u \in \text{supp}(h_N)$ belongs to I, and where

$$\text{in}_<(w_q g_{i_q}) < \cdots < \text{in}_<(w_0 g_{i_0}) \leq \text{in}_<(f).$$

Thus, by letting $\sum_{i=1}^{s} f_i g_i = \sum_{q=0}^{N-1} c'_q c_{i_q}^{-1} w_q g_{i_q}$ and $f' = h_N$, we obtain an expression $f = \sum_{i=1}^{s} f_i g_i + f'$ satisfying the conditions (i) and (ii), as desired. $\square$

Example 2.2.2. Let $<_{\text{lex}}$ denote the lexicographic order on $S = K[x, y, z]$ induced by $x > y > z$. Let $g_1 = x^2 - z, g_2 = xy - 1$ and $f = x^3 - x^2 y - x^2 - 1$. Each of

$$\begin{aligned}
f = x^3 - x^2 y - x^2 - 1 &= x(g_1 + z) - x^2 y - x^2 - 1 \\
&= xg_1 - x^2 y - x^2 + xz - 1 = xg_1 - (g_1 + z)y - x^2 + xz - 1 \\
&= xg_1 - yg_1 - x^2 + xz - yz - 1 = xg_1 - yg_1 - (g_1 + z) + xz - yz - 1 \\
&= (x - y - 1)g_1 + (xz - yz - z - 1)
\end{aligned}$$

and

$$\begin{aligned}
f = x^3 - x^2 y - x^2 - 1 &= x(g_1 + z) - x^2 y - x^2 - 1 \\
&= xg_1 - x^2 y - x^2 + xz - 1 = xg_1 - x(g_2 + 1) - x^2 + xz - 1 \\
&= xg_1 - xg_2 - x^2 + xz - x - 1 = xg_1 - xg_2 - (g_1 + z) + xz - x - 1 \\
&= (x - 1)g_1 - xg_2 + (xz - x - z - 1)
\end{aligned}$$

is a standard expression of f with respect to g_1 and g_2, and each of $xz - yz - z - 1$ and $xz - x - z - 1$ is a remainder of f.

Until the end of the present section, we work with a fixed monomial order $<$ on $S = K[x_1, \ldots, x_n]$. Example 2.2.2 says that in the division algorithm a remainder of f is, in general, not unique. However,

Lemma 2.2.3. *If $\mathcal{G} = \{g_1, \ldots, g_s\}$ is a Gröbner basis of $I = (g_1, \ldots, g_s)$, then for any nonzero polynomial f of S, there is a unique remainder of f with respect to $g_1, \ldots, g_s$.*

Proof. Suppose there exist remainders f' and f'' with respect to $g_1, \ldots, g_s$ with $f' \neq f''$. Since $0 \neq f' - f'' \in I$, the initial monomial $w = \text{in}_<(f' - f'')$ must belong to $\text{in}_<(I)$. However, since $w \in \text{supp}(f') \cup \text{supp}(f'')$, it follows that none of the monomials $\text{in}_<(g_1), \ldots, \text{in}_<(g_s)$ divides w. Hence $\text{in}_<(I) \neq (\text{in}_<(g_1), \ldots, \text{in}_<(g_s))$: a contradiction. $\square$

Corollary 2.2.4. *If $\mathcal{G} = \{g_1, \ldots, g_s\}$ is a Gröbner basis of $I = (g_1, \ldots, g_s)$, then a nonzero polynomial f of S belongs to I if and only if the unique remainder of f with respect to $g_1, \ldots, g_s$ is 0.*

Proof. First, in general, if a remainder of a nonzero polynomial f of S with respect to $g_1, g_2, \ldots, g_s$ is 0, then f belongs to $I = (g_1, g_2, \ldots, g_s)$.

Second, suppose that a nonzero polynomial f belongs to I and $f = f_1 g_1 + f_2 g_2 + \cdots + f_s g_s + f'$ is a standard expression of f with respect to $g_1, g_2, \ldots, g_s$. Since $f \in I$, one has $f' \in I$. If $f' \neq 0$, then $\mathrm{in}_<(f') \in \mathrm{in}_<(I)$. Since $\mathcal{G}$ is a Gröbner basis of I, it follows that $\mathrm{in}_<(I) = (\mathrm{in}_<(g_1), \mathrm{in}_<(g_2), \ldots, \mathrm{in}_<(g_s))$. However, since f' is a remainder, none of the monomials $u \in \mathrm{supp}(f')$ can belong to $(\mathrm{in}_<(g_1), \mathrm{in}_<(g_2), \ldots, \mathrm{in}_<(g_s))$. $\qquad\square$

We conclude this subsection with the presentation of important properties of initial ideals.

Proposition 2.2.5. *Let I be a nonzero ideal of $S = K[x_1, \ldots, x_n]$, and $<$ a monomial order on S. Then*

(a) *the set of monomials which do not belong to $\mathrm{in}_<(I)$ form a K-basis of S/I.*
(b) *$\dim_K I_j = \dim_K \mathrm{in}_<(I)_j$ for all j, if in addition, $I \subset S$ is a graded ideal.*

Proof. (a) Let $g_1, \ldots, g_m$ be a Gröbner basis of I, let $f \in S$ and f' the remainder of f with respect to $g_1, \ldots, g_m$. Then $f + I = f' + I$ and $\mathrm{supp}(f') \cap \mathrm{in}_<(I) = \emptyset$. This shows that S/I is generated by the monomials $u \in S \setminus \mathrm{in}_<(I)$. Suppose there exist monomials $u_1 > u_2 > \ldots > u_r$ in $S \setminus \mathrm{in}_>(I)$ which are linearly dependent modulo I. Say, $f = \sum_{i=1}^r a_i u_i \in I$ for some $a_i \in K$. We may assume $a_1 \neq 0$. Then $\mathrm{in}_<(f) \in I$, a contradiction.

(b) It follows from (a) that the monomials of degree j in $S \setminus \mathrm{in}_<(I)$ form a K-basis of $(S/I)_j$. Since they also form a K-basis of $(S/\mathrm{in}_<(I))_j$ it follows that $\dim_K S_j - \dim_K I_j = \dim_K S_j - \mathrm{in}_<(I)_j$, and hence $\dim_K I_j = \dim_K \mathrm{in}_<(I)_j$. $\qquad\square$

Proposition 2.2.6. *Let $I \subset J$ be nonzero ideals of $S = K[x_1, \ldots, x_n]$ with $I \neq J$, and let $<$ and $<'$ be monomial orders on S. Then*

(a) *$\mathrm{in}_<(I) \subset \mathrm{in}_<(J)$ and $\mathrm{in}_<(I) \neq \mathrm{in}_<(J)$.*
(b) *$\mathrm{in}_<(I) = \mathrm{in}_{<'}(I)$, if $\mathrm{in}_<(I) \subset \mathrm{in}_{<'}(I)$.*

Proof. (a) $\mathrm{in}_<(I)$ is generated by all monomials $\mathrm{in}_<(f)$ with $f \in I$. Since $I \subset J$, each $f \in I$ belongs to J. Therefore $\mathrm{in}_<(I) \subset \mathrm{in}_<(J)$. If $I \neq J$, then there exists $f \in J \setminus I$. Let f' be the remainder of f with respect to a Gröbner basis of I. Then $f' \neq 0$, $f' \in J$ and $\mathrm{supp}(f') \not\subset \mathrm{in}_<(I)$. It follows that $\mathrm{in}_<(f') \in \mathrm{in}_<(J) \setminus \mathrm{in}_<(I)$.

(b) By Proposition 2.2.5, the set of monomials in $S \setminus \mathrm{in}_<(I)$ as well as the set of monomials in $S \setminus \mathrm{in}_{<'}(I)$ form a K-basis of S/I. Suppose that $\mathrm{in}_<(I) \subset \mathrm{in}_{<'}(I)$. Then $S \setminus \mathrm{in}_{<'}(I)$ is a proper subset of $S \setminus \mathrm{in}_<(I)$, a contradiction. $\qquad\square$

2.2.2 Reduced Gröbner bases

A Gröbner basis $\mathcal{G} = \{g_1, g_2, \ldots, g_s\}$ is called **reduced** if the following conditions are satisfied:

(i) The coefficient of $\mathrm{in}_<(g_i)$ in g_i is 1 for all $1 \leq i \leq s$;
(ii) If $i \neq j$, then none of the monomials of $\mathrm{supp}(g_j)$ is divisible by $\mathrm{in}_<(g_i)$.

Theorem 2.2.7. *A reduced Gröbner basis exists and is uniquely determined.*

Proof. (Existence) Let I be a nonzero ideal of S and $\{u_1, \ldots, u_s\}$ the unique minimal system of monomial generators of $\mathrm{in}_<(I)$. Thus, for $i \neq j$, the monomial u_i cannot be divided by u_j. For each $1 \leq i \leq s$, we choose a polynomial $g_i \in I$ with $\mathrm{in}_<(g_i) = u_i$.

Let $g_1 = f_2 g_2 + f_3 g_3 + \cdots + f_s g_s + h_1$ be a standard expression of g_1 with respect to $g_2, g_3, \ldots, g_s$, where h_1 a remainder. It follows from the property (ii) required in the division algorithm that $\mathrm{in}_<(g_1)$ coincides with one of the monomials $\mathrm{in}_<(f_2)\,\mathrm{in}_<(g_2), \cdots, \mathrm{in}_<(f_s)\,\mathrm{in}_<(g_s), \mathrm{in}_<(h_1)$. Since $u_1 = \mathrm{in}_<(g_1)$ can be divided by none of the monomials $\mathrm{in}_<(g_2), \ldots, \mathrm{in}_<(g_s)$, one has $\mathrm{in}_<(h_1) = \mathrm{in}_<(g_1)$. Hence $\{h_1, g_2, g_3, \ldots, g_s\}$ is a Gröbner basis of I. Since the monomial h_1 is a remainder of a standard expression of g_1 with respect to $g_2, g_3, \ldots, g_s$, each monomial of $\mathrm{supp}(h_1)$ is divided by none of the monomials $\mathrm{in}_<(g_2), \mathrm{in}_<(g_3), \ldots, \mathrm{in}_<(g_s)$.

Similarly, if h_2 is a remainder of a standard expression of g_2 with respect to $h_1, g_3, g_4, \ldots, g_s$, then one has $\mathrm{in}_<(h_2) = \mathrm{in}_<(g_2)$ and each monomial of $\mathrm{supp}(h_2)$ is divided by none of the monomials $\mathrm{in}_<(h_1), \mathrm{in}_<(g_3), \ldots, \mathrm{in}_<(g_s)$. Moreover, $\{h_1, h_2, g_3, g_4, \ldots, g_s\}$ is a Gröbner basis of I. Since $\mathrm{in}_<(h_2) = \mathrm{in}_<(g_2)$, each monomial of $\mathrm{supp}(h_1)$ is divided by none of the monomials $\mathrm{in}_<(h_2), \mathrm{in}_<(g_3), \ldots, \mathrm{in}_<(g_s)$.

Continuing these procedures yields the polynomials $h_3, h_4, \ldots, h_s$ we obtain a Gröbner basis $\{h_1, h_2, \ldots, h_s\}$ which satisfies condition (ii). Dividing h_i by the coefficient of $\mathrm{in}_<(h_i)$ for all i, we obtain a reduced Gröbner basis of I.

(Uniqueness) Let $\{g_1, \ldots, g_s\}$ and $\{h_1, \ldots, h_t\}$ be reduced Gröbner bases of I. Since $\{\mathrm{in}_<(g_1), \ldots, \mathrm{in}_<(g_s)\}$ and $\{\mathrm{in}_<(h_1), \ldots, \mathrm{in}_<(h_t)\}$ are the minimal system of monomial generators of the initial ideal $\mathrm{in}_<(I)$ of I, we may assume that $s = t$ and $\mathrm{in}_<(g_i) = \mathrm{in}_<(h_i)$ for all $1 \leq i \leq s\,(= t)$. If $g_i \neq h_i$, then $0 \neq g_i - h_i \in I$ and $\mathrm{in}_<(g_i - h_i) < \mathrm{in}_<(g_i)$. In particular $\mathrm{in}_<(g_i)$ cannot divide $\mathrm{in}_<(g_i - h_i)$. Since the monomial $\mathrm{in}_<(g_i - h_i)$ must appear in either $\mathrm{supp}(g_i)$ or $\mathrm{supp}(h_i)$, it follows that $\mathrm{in}_<(g_i - h_i)$ cannot be divided by $\mathrm{in}_<(g_j)$ with $j \neq i$. Hence $\mathrm{in}_<(g_i - h_i) \notin \mathrm{in}_<(I)$. This contradicts $g_i - h_i \in I$. $\qquad\square$

We write $\mathcal{G}_{\mathrm{red}}(I; <)$ for *the* reduced Gröbner basis of I with respect to $<$.

Corollary 2.2.8. *Let I and J be nonzero ideals of S. Then $I = J$ if and only if $\mathcal{G}_{\mathrm{red}}(I; <) = \mathcal{G}_{\mathrm{red}}(J; <)$.*

2.3 Buchberger's criterion

Let, as before, $S = K[x_1, \ldots, x_n]$ denote the polynomial ring over a field K. We work with a fixed monomial order $<$ on S and will omit the phrase "with respect to $<$", if there is no danger of confusion.

2.3.1 S-polynomials

Given nonzero polynomials f and g of S. Recall that $\text{lcm}(\text{in}_<(f), \text{in}_<(g))$ stands for the least common multiple of $\text{in}_<(f)$ and $\text{in}_<(g)$. Let c_f denote the coefficient of $\text{in}_<(f)$ in f and c_g the coefficient of $\text{in}_<(g)$ in g. The polynomial

$$S(f,g) = \frac{\text{lcm}(\text{in}_<(f), \text{in}_<(g))}{c_f \, \text{in}_<(f)} f - \frac{\text{lcm}(\text{in}_<(f), \text{in}_<(g))}{c_g \, \text{in}_<(g)} g$$

is called the **S-polynomial** of f and g.

We say that f **reduces to** 0 with respect to $g_1, g_2, \ldots, g_s$ if, in the division algorithm, there is a standard expression (2.2) of f with respect to $g_1, g_2, \ldots, g_s$ with $f' = 0$.

Lemma 2.3.1. *Let f and g be nonzero polynomials and suppose that $\text{in}_<(f)$ and $\text{in}_<(g)$ are relatively prime, i.e. $\text{lcm}(\text{in}_<(f), \text{in}_<(g)) = \text{in}_<(f)\text{in}_<(g)$. Then $S(f,g)$ reduces to 0 with respect to f, g.*

Proof. To simplify notation we will assume that each of the coefficients of $\text{in}_<(f)$ in f and $\text{in}_<(g)$ in g is equal to 1. Let $f = \text{in}_<(f) + f_1$ and $g = \text{in}_<(g) + g_1$. Since $\text{in}_<(f)$ and $\text{in}_<(g)$ are relatively prime, it follows that

$$\begin{aligned}
S(f,g) &= \text{in}_<(g)f - \text{in}_<(f)g \\
&= (g - g_1)f - (f - f_1)g \\
&= f_1 g - g_1 f.
\end{aligned}$$

We claim $(\text{in}_<(f_1)\text{in}_<(g) =) \, \text{in}_<(f_1 g) \neq \text{in}_<(g_1 f) \, (= \text{in}_<(g_1)\text{in}_<(f))$. In fact, if $\text{in}_<(f_1)\text{in}_<(g) = \text{in}_<(g_1)\text{in}_<(f)$, then, since $\text{in}_<(f)$ and $\text{in}_<(g)$ are relatively prime, it follows that $\text{in}_<(f)$ must divide $\text{in}_<(f_1)$. However, since $\text{in}_<(f_1) < \text{in}_<(f)$, this is impossible. Let, say, $\text{in}_<(f_1)\text{in}_<(g) < \text{in}_<(g_1)\text{in}_<(f)$. Then $\text{in}_<(S(f,g)) = \text{in}_<(g_1 f)$ and $S(f,g) = f_1 g - g_1 f$ turns out to be a standard expression of $S(f,g)$ in terms of f and g. Hence $S(f,g)$ has remainder 0 with respect to f, g. $\square$

2.3.2 Buchberger's criterion

We now come to the most important theorem in the theory of Gröbner bases.

Theorem 2.3.2 (Buchberger's criterion). *Let I be a nonzero ideal of S and $\mathcal{G} = \{g_1, \ldots, g_s\}$ a system of generators of I. Then $\mathcal{G}$ is a Gröbner basis of I if and only if the following condition is satisfied:*

(∗) *For all $i \neq j$, $S(g_i, g_j)$ reduces to 0 with respect to $g_1, \ldots, g_s$.*

Proof. "Only if": Since $S(g_i, g_j) \in I$, it follows from Corollary 2.2.4 that if $\mathcal{G} = \{g_1, \ldots, g_s\}$ is a Gröbner basis of I, then $S(g_i, g_j)$ reduces to 0 with respect to $g_1, \ldots, g_s$.

"If": Let $\mathcal{G} = \{g_1, g_2, \ldots, g_s\}$ be a system of generators of I which satisfies the condition (∗). If a nonzero polynomial f belongs to I, then we write $\mathcal{H}_f$ for the set of sequences $\mathbf{h} = (h_1, h_2, \ldots, h_s)$ with each $h_i \in S$ such that $f = \sum_{i=1}^{s} h_i g_i$. We associate each sequence $\mathbf{h} \in \mathcal{H}_f$ with the monomial $\delta_{\mathbf{h}} = \max\{\text{in}_<(h_i g_i) : h_i g_i \neq 0\}$. Among such monomials $\delta_{\mathbf{h}}$ with $\mathbf{h} \in \mathcal{H}_f$, we are interested in the monomial

$$\delta_f = \min\{\delta_{\mathbf{h}} : \mathbf{h} \in \mathcal{H}_f\}.$$

One has $\text{in}_<(f) \leq \delta_f$. It then follows that $\mathcal{G}$ is a Gröbner basis of I if $\text{in}_<(f) = \delta_f$ for all nonzero polynomials f belonging to I. In fact, if $\text{in}_<(f) = \delta_f$ and if $\delta_f = \delta_{\mathbf{h}}$ with $\mathbf{h} = (h_1, h_2, \ldots, h_s) \in \mathcal{H}_f$, then $\text{in}_<(f) = \text{in}_<(h_i g_i)$ for some $1 \leq i \leq s$. Hence $\text{in}_<(f) \in (\text{in}_<(g_1), \text{in}_<(g_2), \ldots, \text{in}_<(g_s))$.

Our goal is to show that $\text{in}_<(f) = \delta_f$ for all nonzero polynomials $f \in I$. Suppose that there exists $f \in I$ with $f \neq 0$ such that $\text{in}_<(f) < \delta_f$ and choose a sequence $\mathbf{h} = (h_1, h_2, \ldots, h_s) \in \mathcal{H}_f$ with $\delta_f = \delta_{\mathbf{h}}$. Then

$$
\begin{aligned}
f &= \sum_{\text{in}_<(h_i g_i)=\delta_f} h_i g_i + \sum_{\text{in}_<(h_i g_i)<\delta_f} h_i g_i \\
&= \sum_{\text{in}_<(h_i g_i)=\delta_f} c_i \, \text{in}_<(h_i) g_i + \sum_{\text{in}_<(h_i g_i)=\delta_f} (h_i - c_i \, \text{in}_<(h_i)) g_i \\
&\quad + \sum_{\text{in}_<(h_i g_i)<\delta_f} h_i g_i,
\end{aligned}
$$

where c_i is the coefficient of $\text{in}_<(h_i)$ in h_i. Since $\text{in}_<(f) < \delta_f$, it follows that

$$\text{in}_<\Big(\sum_{\text{in}_<(h_i g_i)=\delta_f} c_i \, \text{in}_<(h_i) g_i \Big) < \delta_f.$$

By virtue of Lemma 2.3.3 below, it turns out that $\sum_{\text{in}_<(h_i g_i)=\delta_f} c_i \, \text{in}_<(h_i) g_i$ is a linear combination of those S-polynomials $S(\text{in}_<(h_j)g_j, \text{in}_<(h_k)g_k)$ with $\text{in}_<(h_j g_j) = \text{in}_<(h_k g_k) = \delta_f$. In case of $\text{in}_<(h_j g_j) = \text{in}_<(h_k g_k) = \delta_f$, one can easily compute that

$$S(\text{in}_<(h_j)g_j, \text{in}_<(h_k)g_k) = (\delta_f / \text{lcm}(\text{in}_<(g_j), \text{in}_<(g_k)) S(g_j, g_k).$$

Let $u_{jk} = \delta_f / \text{lcm}(\text{in}_<(g_j), \text{in}_<(g_k))$. It then follows that there exists an expression of the form

$$\sum_{\text{in}_<(h_i g_i)=\delta_f} c_i \, \text{in}_<(h_i) g_i = \sum_{j,k} c_{jk} u_{jk} S(g_j, g_k), \quad c_{jk} \in K, \qquad (2.3)$$

with
$$\text{in}_<(u_{jk}S(g_j,g_k)) < \delta_f.$$

By using the condition $(*)$, there exists an expression of the form

$$S(g_j,g_k) = \sum_{i=1}^{s} p_i^{jk} g_i, \quad p_i^{jk} \in S, \tag{2.4}$$

with
$$\text{in}_<(p_i^{jk} g_i) \le \text{in}_<(S(g_j,g_k)).$$

Combining the equalities (2.4) with (2.3) yields the equality

$$\sum_{\text{in}_<(h_i g_i)=\delta_f} c_i \, \text{in}_<(h_i)g_i = \sum_{j,k} c_{jk} u_{jk} \Big(\sum_{i=1}^{s} p_i^{jk} g_i \Big). \tag{2.5}$$

If we write the right-hand side of (2.5) as $\sum_{i=1}^{s} h_i' g_i$, then

$$\text{in}_<(h_i' g_i) < \delta_f.$$

Consequently, the polynomial f finally can be expressed as

$$f = \sum_{i=1}^{s} h_i'' g_i, \quad \text{in}_<(h_i'' g_i) < \delta_f. \tag{2.6}$$

The existence of such an expression (2.6) contradicts the definition of δ_f. □

Lemma 2.3.3. *Let w be a monomial of S and $f_1, f_2, \ldots, f_s$ polynomials of S with $\text{in}_<(f_i) = w$ for all $1 \le i \le s$. Let $g = \sum b_i f_i \ (\ne 0)$ be a linear combination of $f_1, f_2, \ldots, f_s$ with each $b_i \in K$ and suppose that $\text{in}_<(g) < w$. Then g is a linear combination of the S-polynomials $S(f_j, f_k)$, $1 \le j, k \le s$.*

Proof. Let c_i denote the coefficient of $w = \text{in}_<(f_i)$ in f_i. Then

$$\sum_{i=1}^{s} b_i c_i = 0.$$

Let $g_i = (1/c_i)f_i$. Then

$$S(f_j, f_k) = g_j - g_k, \quad 1 \le j, k \le s.$$

Now, we compute that

$$\sum_{i=1}^{s} b_i f_i = \sum_{i=1}^{s} b_i c_i g_i$$
$$= b_1 c_1(g_1 - g_2) + (b_1 c_1 + b_2 c_2)(g_2 - g_3)$$
$$+ (b_1 c_1 + b_2 c_2 + b_3 c_3)(g_3 - g_4)$$
$$+ \cdots + (b_1 c_1 + \cdots + b_{s-1} c_{s-1})(g_{s-1} - g_s)$$
$$+ (b_1 c_1 + \cdots + b_s c_s)g_s.$$

Since $\sum_{i=1}^{s} b_i c_i = 0$, it follows that

$$\sum_{i=1}^{s} b_i f_i = \sum_{i=2}^{s} (b_1 c_1 + \cdots + b_{i-1} c_{i-1}) S(f_{i-1}, f_i),$$

as desired.   □

In applying Buchberger's criterion it is not always necessary to check whether *all* S-polynomials $S(g_i, g_j)$ with $i \neq j$ reduce to 0 with respect to $g_1, \ldots, g_s$. This may substantially save time in Buchberger's algorithm which is described in the next subsection.

The first result in this direction is the following

Corollary 2.3.4. *If $g_1, \ldots, g_s$ are nonzero polynomials of S such that $\mathrm{in}_<(g_i)$ and $\mathrm{in}_<(g_j)$ are relatively prime for all $i \neq j$, then $\{g_1, \ldots, g_s\}$ is a Gröbner basis of $I = (g_1, \ldots, g_s)$.*

Secondly we have

Proposition 2.3.5. *Let I be a nonzero ideal of S and $\mathcal{G} = \{g_1, \ldots, g_s\}$ a system of generators of I. Consider the S-module epimorphism $\epsilon \colon S^s \to (\mathrm{in}(g_1), \ldots, \mathrm{in}(g_s))$ which for $i = 1, \ldots, s$ maps the canonical basis element e_i to $\mathrm{in}(g_i)$. Then*

(a) *for all $i, j = 1, \ldots, s$ with $i < j$ the elements*

$$r_{ij} = \frac{\mathrm{lcm}(\mathrm{in}_<(g_i), \mathrm{in}_<(g_j))}{\mathrm{in}_<(g_i)} e_j - \frac{\mathrm{lcm}(\mathrm{in}_<(g_j), \mathrm{in}_<(g_i))}{\mathrm{in}_<(g_i)} e_i$$

generate $\mathrm{Ker}(\epsilon)$.

(b) *Let $\mathcal{R}$ be any subset of the relations r_{ij} with the property that $\mathcal{R}$ generates $\mathrm{Ker}(\epsilon)$. Then $\mathcal{G}$ is a Gröbner basis of I if and only if $S(g_i, g_j)$ reduces to 0 with respect to $g_1, \ldots, g_s$ for all i, j such that $r_{ij} \in \mathcal{R}$.*

Proof. (a) Set $u_i = \mathrm{in}(g_i)$ and $\deg e_i = \deg u_i = \mathbf{a}_i$ for $i = 1, \ldots, s$. Then ϵ is a $\mathbb{Z}^n$-graded S-module homomorphism, and hence $\mathrm{Ker}(\epsilon)$ is generated by $\mathbb{Z}^n$-graded elements. Let $r = \sum_{i=1}^{s} r_i e_i$ be a nonzero element in $\mathrm{Ker}(\epsilon)$ of $\mathbb{Z}^n$-degree $\mathbf{a}$. Then each $r_i \in S$ is homogeneous of degree $\mathbf{a} - \mathbf{a}_i$. Hence each $r_i \neq 0$ is of the form $c_i v_i$, where $c_i \in K$ and where v_i is a monomial of degree $\mathbf{a} - \mathbf{a}_i$. Since $r \in \mathrm{Ker}(\epsilon)$, it follows that $0 = \epsilon(r) = \sum_i c_i v_i u_i = (\sum_i c_i) \mathbf{x}^{\mathbf{a}}$, where each of the sums is taken over those i for which $r_i \neq 0$. For simplicity let us assume that $c_i \neq 0$ for $i = 1, \ldots, k$ and $c_i = 0$ for $i = k+1, \ldots, s$. Then $\sum_{i=1}^{k} c_i = 0$ and $r = \sum_{i=2}^{k} c_i (v_i e_i - v_1 e_1)$. Each of the summands $c_i (v_i e_i - v_1 e_1)$ in r belongs to $\mathrm{Ker}(\epsilon)$ and hence is a multiple of $(\mathrm{lcm}(u_1, u_i)/u_i) e_i - (\mathrm{lcm}(u_1, u_i)/u_1) e_1$, as desired.

(b) We go back to the proof of Theorem 2.3.2. In equation (2.3) we express the sum $(*) \sum_{\text{in}_<(h_i g_i) = \delta_f} c_i \text{in}_<(h_i) g_i$ as a linear combination of S-polynomials. We only need to show: if $\mathcal{R}$ is a set of relations of type r_{ij} generating $\text{Ker}(\epsilon)$, then $(*)$ can be written as a linear combination of those S-polynomials $S(g_i, g_j)$ for which $r_{ij} \in \mathcal{R}$. Indeed, since it is assumed that $\text{in}_<(f) < \delta_f$, it follows from $(*)$ that $\sum_{\text{in}_<(h_i g_i) = \delta_f} c_i \text{in}_<(h_i) \text{in}_<(g_i) = 0$. Therefore part (a) implies that $\sum_{\text{in}_<(h_i g_i) = \delta_f} c_i \text{in}_<(h_i) e_i$ is a linear combination of the relations $r_{ij} \in \mathcal{R}$. For simplicity we may assume that the coefficient of $\text{in}_<(g_i)$ in g_i is 1 for all i. Then, replacing the basis elements e_i by the polynomials g_i the relation r_{ij} becomes the S-polynomial $S(g_i, g_j)$, and hence it follows that $(*)$ is a linear combination of the S-polynomials $S(g_i, g_j)$ with $r_{ij} \in \mathcal{R}$, as desired. □

2.3.3 Buchberger's algorithm

The Buchberger criterion supplies an algorithm to compute a Gröbner basis starting from a system of generators of an ideal.

Let $\{g_1, g_2, \ldots, g_s\}$ be a system of generators of a nonzero ideal I of S. Compute the S-polynomials $S(g_i, g_j)$. If all $S(g_i, g_j)$ reduce to 0 with respect to $g_1, \ldots, g_s$, then, by the Buchberger criterion, $\{g_1, g_2, \ldots, g_s\}$ is a Gröbner basis. Otherwise one of the $S(g_i, g_j)$ has a nonzero remainder g_{s+1}. Then none of the monomials $\text{in}_<(g_1), \text{in}_<(g_2), \ldots, \text{in}_<(g_s)$ divides $\text{in}_<(g_{s+1})$. In other words, the inclusion

$$(\text{in}_<(g_1), \text{in}_<(g_2), \ldots, \text{in}_<(g_s)) \subset (\text{in}_<(g_1), \text{in}_<(g_2), \ldots, \text{in}_<(g_s), \text{in}_<(g_{s+1}))$$

is strict.

Notice that $g_{s+1} \in I$. Now we replace $\{g_1, g_2, \ldots, g_s\}$ by $\{g_1, \ldots, g_s, g_{s+1}\}$ and compute all the S-polynomials for this new system of generators.

If all S-polynomials reduce to 0 with respect to $g_1, g_2, \ldots, g_s, g_{s+1}$, then $\{g_1, g_2, \ldots, g_s, g_{s+1}\}$ is a Gröbner basis. Otherwise there is a nonzero remainder g_{s+2} and we obtain the new system of generators $\{g_1, g_2, \ldots, g_{s+1}, g_{s+2}\}$, and the inclusion

$$(\text{in}_<(g_1), \text{in}_<(g_2), \ldots, \text{in}_<(g_s), \text{in}_<(g_{s+1}))$$
$$\subset (\text{in}_<(g_1), \text{in}_<(g_2), \ldots, \text{in}_<(g_s), \text{in}_<(g_{s+1}), \text{in}_<(g_{s+2}))$$

is strict.

By virtue of Dickson's lemma, it follows that these procedures will terminate after a finite number of steps, and a Gröbner basis can be obtained.

In fact, if this were not the case, then a strictly increasing infinite sequence of monomial ideals

$$(\text{in}_<(g_1), \text{in}_<(g_2), \ldots, \text{in}_<(g_s)) \subset (\text{in}_<(g_1), \ldots, \text{in}_<(g_s), \text{in}_<(g_{s+1}))$$
$$\subset \cdots \subset (\text{in}_<(g_1), \ldots, \text{in}_<(g_s), \text{in}_<(g_{s+1}), \ldots, \text{in}_<(g_j)) \subset \cdots$$

would arise. However, if $\mathcal{M} = \{\mathrm{in}_<(g_1), \ldots, \mathrm{in}_<(g_s), \mathrm{in}_<(g_{s+1}), \ldots\}$ and if

$$\mathcal{M}^{\min} = \{\mathrm{in}_<(g_{i_1}), \mathrm{in}_<(g_{i_2}), \ldots, \mathrm{in}_<(g_{i_q})\}, \qquad i_1 < i_2 < \cdots < i_q,$$

then, for all $j > i_q$, one would have

$$(\mathrm{in}_<(g_{i_1}), \mathrm{in}_<(g_{i_2}), \ldots, \mathrm{in}_<(g_{i_q}))$$
$$= (\mathrm{in}_<(g_1), \mathrm{in}_<(g_2), \ldots, \mathrm{in}_<(g_{i_q}), \mathrm{in}_<(g_{i_q+1}), \ldots, \mathrm{in}_<(g_j)),$$

which is a contradiction.

The above algorithm to find a Gröbner basis starting from a system of generators of I is said to be **Buchberger's algorithm**.

Example 2.3.6. We continue Example 2.1.6. Let $S = K[x_1, x_2, \ldots, x_7]$ and $<_{\mathrm{lex}}$ the lexicographic order on S induced by $x_1 > x_2 > \cdots > x_7$. Let $f = x_1 x_4 - x_2 x_3$ and $g = x_4 x_7 - x_5 x_6$ with their initial monomials $\mathrm{in}_{<_{\mathrm{lex}}}(f) = x_1 x_4$ and $\mathrm{in}_{<_{\mathrm{lex}}}(g) = x_4 x_7$. Let $I = (f, g)$. Then $\{f, g\}$ is not a Gröbner basis of I with respect to $<_{\mathrm{lex}}$. Now, as a remainder of $S(f, g) = x_7 f - x_1 g = x_1 x_5 x_6 - x_2 x_3 x_7$ with respect to f and g, we choose $S(f, g)$ itself. Let $h = x_1 x_5 x_6 - x_2 x_3 x_7$ with $\mathrm{in}_{<_{\mathrm{lex}}}(h) = x_1 x_5 x_6$. Then $\mathrm{in}_{<_{\mathrm{lex}}}(g)$ and $\mathrm{in}_{<_{\mathrm{lex}}}(h)$ are relatively prime. On the other hand, $S(f, h) = x_2 x_3(x_4 x_7 - x_5 x_6)$ reduces to 0 with respect to f, g, h. It follows from the Buchberger criterion that $\{f, g, h\}$ is a Gröbner basis of I with respect to $<_{\mathrm{lex}}$.

A **binomial** is a polynomial of the form $u - v$ where u and v are monomials. A **binomial ideal** is an ideal generated by binomials.

Proposition 2.3.7. *Let $I \subset S$ be an ideal and $<$ a monomial order on S.*

(a) *If I is graded, then the reduced Gröbner basis of I with respect to $<$ consists of homogeneous polynomials.*

(b) *If I is a binomial ideal, then the reduced Gröbner basis of I consists of binomials.*

Proof. (a) If f and g are homogeneous polynomials, then the S-polynomial $S(f, g)$ is again homogeneous. In the division algorithm, if $g_1, \ldots, g_s$ and f are homogeneous polynomials, then a remainder f' of f with respect to $g_1, \ldots, g_s$ is again homogeneous. The above two facts, together with Buchberger's algorithm, guarantee that a homogeneous ideal I possesses a Gröbner basis consisting of homogeneous polynomials. Thus by using the algorithm to compute the reduced Gröbner basis of I, which is discussed in the proof of Theorem 2.2.7, it turns out that the reduced Gröbner basis of I consists of homogeneous polynomials.

(b) Replacing "homogeneous polynomial" with "binomial" in the proof of (a) yields a proof of (b). $\square$

Problems

2.1. Give a direct proof of Dickson's Lemma for $n = 2$.

2.2. (a) Show that there is a unique monomial order on $K[x_1]$.
(b) Let $n \geq 2$. Show that there are infinite many monomial orders on $S = K[x_1, \ldots, x_n]$.

2.3. Let $S = K[x_1, \ldots, x_6]$. Let $f = x_1 x_5 - x_2 x_4$ and $g = x_2 x_4 - x_3 x_6$.
(a) Find a monomial order $<$ on S such that $\{f, g\}$ is a Gröbner basis of $I = (f, g)$ with respect to $<$.
(b) Find a K-basis of S/I consisting of monomials.

2.4. Let $S = K[x_1, \ldots, x_6]$. Let $f = x_1 x_5 - x_2 x_4$, $g = x_1 x_6 - x_3 x_4$ and $h = x_2 x_6 - x_3 x_5$.
(a) Find a monomial order $<$ on S such that $\{f, g, h\}$ is a Gröbner basis of $I = (f, g, h)$ with respect to $<$.
(b) Find a K-basis of S/I consisting of monomials.

2.5. Work with the same situation as in Example 2.2.2. Compute a remainder of the following polynomials:
(i) $x^5 - x^2 y^3 - x^3 - 1$;
(ii) $y^4 - x^2 y^2 + y^3 - 1$.

2.6. Show that $\mathcal{G} = \{g_1, \ldots, g_s\}$ is a Gröbner basis of $I = (g_1, \ldots, g_s)$ if and only if each nonzero polynomial $f \in I$ reduces to 0 with respect to $g_1, \ldots, g_s$.

2.7. By using the Buchberger algorithm compute the reduced Gröbner basis of the ideal discussed in Example 2.1.10 with respect to the lexicographic order $<_{\text{lex}}$.

2.8. Let $S = K[x_1, x_2, \ldots, x_8]$ and I the ideal of S generated by

$$f_1 = x_2 x_8 - x_4 x_7, \qquad f_2 = x_1 x_6 - x_3 x_5, \qquad f_3 = x_1 x_3 - x_2 x_4.$$

(a) Show that there exists no monomial order $<$ on S such that $\{f_1, f_2, f_3\}$ is a Gröbner basis of I with respect to $<$.
(b) By using the Buchberger algorithm compute the reduced Gröbner basis of I with respect to the lexicographic order $<_{\text{lex}}$.

2.9. Let $S = K[x_1, \ldots, x_n]$ and $B = K[x_m, \ldots, x_n]$, where $1 \leq m \leq n$. Let $<_S$ be a monomial order on S.
(a) For $u, v \in \text{Mon}(B)$, we define $u <_B v$ if $u <_S v$. Show that $<_B$ is a monomial order on B.
(b) Let I be an ideal of S. Show that $I \cap B$ is an ideal of B.
(c) Let $\mathcal{G}$ be a Gröbner basis of a nonzero ideal I of S with respect to $<_S$ and suppose that, for each $g \in \mathcal{G}$, one has $g \in B$ if $\text{in}_{<_S}(g) \in B$. Show that $\mathcal{G} \cap B$ is a Gröbner basis of $I \cap B$ with respect to $<_B$. Thus in particular $I \cap B = 0$ if $\mathcal{G} \cap B = \emptyset$.
(d) Let $<_{\text{purelex}}$ be the pure lexicographic order on S. Show that, for a nonzero polynomial f of S, one has $f \in B$ if $\text{in}_{<_{\text{purelex}}}(f) \in B$.

2.10. Let $f_1, \ldots, f_s$ be binomials and m a monomial. Show that every remainder of m with respect to $f_1, \ldots, f_s$ is again a monomial.

Notes

Nowadays, we can easily find well-written textbooks that introduce Gröbner bases, for example Adams–Loustaunau [AL94], Becker–Weispfenning [BW93], Cox–Little–O'Shea [CLO92] and Kreutzer–Robbiano ([KR00] and [KR05]). The computational aspects of commutative algebra are highlighted in the book [GP08] by Greuel and Pfister and the book of Vasconcelos [Vas98]. A short but rather comprehensive introduction to Gröbner bases is given in the book of Eisenbud [Eis95, Chapter 15].

In history, the lexicographic order was first used by C. F. Gauss in his proof (e.g. [CLO92, pp. 312–314]) of the fundamental theorem of symmetric polynomials, i.e. every symmetric polynomial can be written uniquely as a polynomial of the elementary symmetric functions.

Gordan's proof ([Gor00]) of Theorem 2.1.8 might be the earliest use of the technique of Gröbner bases. Later, in 1927, initial ideals with respect to lexicographic order essentially appeared in Macaulay [Mac27] to characterize the possible Hilbert function of homogeneous ideals of the polynomial ring.

In the mid-1960s, Buchberger introduced the notion of Gröbner basis in his thesis, where a Gröbner basis criterion and an algorithm to compute a Gröbner basis was presented.

At the same time, in 1964, Hironaka independently introduced "standard bases" in his major paper on resolution of singularities of an algebraic variety. Standard bases are analogous to Gröbner bases in the formal power series ring.

Sturmfels [Stu96] discusses the Gröbner basis technique in the theory of convex polytopes. The article of Bruns and Conca [BC03] is a compact presentation of how to apply Gröbner basis theory to the study of determinantal ideals. For information about computer algebra systems we refer the reader to [CLO92, Appendix C], [CLO98] and Eisenbud [Eis95, Chapter 15].

3

Monomial orders and weights

For a given ideal I and a given monomial order $<$, the initial ideal $\mathrm{in}_<(I)$ of I can also be obtained as the initial ideal with respect to a suitable integral weight. This point of view allows us to consider $\mathrm{in}_<(I)$ as the special fibre of a flat family which is parameterized by the elements of the base field and whose general fibre is I. From this fact we deduce that the graded Betti numbers of $\mathrm{in}_<(I)$ are greater than or equal to the corresponding graded Betti numbers of I. This fundamental observation has many applications.

3.1 Initial terms with respect to weights

3.1.1 Gradings defined by weights

Let $\mathbf{w} = (w_1, \ldots, w_n) \in \mathbb{N}^n$ be an integer vector. We call this vector a **weight** and define a new grading on $S - K[x_1, \ldots, x_n]$, different from the standard grading, by setting $\deg_{\mathbf{w}} x_i = w_i$ for $i = 1, \ldots, n$. Then

$$\deg_{\mathbf{w}} \mathbf{x}^{\mathbf{a}} = \langle \mathbf{a}, \mathbf{w} \rangle = \sum_{i=1}^{n} a_i w_i.$$

A polynomial $f \in S$ is called **homogeneous of degree** j with respect to the weight $\mathbf{w}$, if $\deg_{\mathbf{w}} u = j$ for all $u \in \mathrm{supp}(f)$.

For example, if we let $\mathbf{w} = (1, 2, 3)$, then $f = 3x_1^6 - x_1 x_2 x_3$ is homogeneous of degree 6 with respect to this weight.

Now we fix a weight $\mathbf{w}$, and let S_j be the K-vector space spanned by all homogeneous polynomials of degree j. The vector space S_j is finite-dimensional, and the monomials u with $\deg_{\mathbf{w}} u = j$ form a K-basis of this vector space. Moreover,

$$S = \bigoplus_{j} S_j.$$

J. Herzog, T. Hibi, *Monomial Ideals*, Graduate Texts in Mathematics 260, 41
DOI 10.1007/978-0-85729-106-6_3, © Springer-Verlag London Limited 2011

Therefore, each polynomial $f \in S$ can be uniquely written as $f = \sum_j f_j$ with $f_j \in S_j$. The summands f_j are called the **homogeneous components** of f.

The degree of f is defined to be $\deg_{\mathbf{w}} f = \max\{j : f_j \neq 0\}$, and if $i = \deg_{\mathbf{w}} f$, then f_i is called the **initial term** of f, and is denoted by $\mathrm{in}_{\mathbf{w}}(f)$. Of course, $\mathrm{in}_{\mathbf{w}}(f)$ is in general not a monomial but a polynomial.

Let $I \subset S$ be an ideal. Similarly as for monomial orders we define the **initial ideal** of I with respect to $\mathbf{w}$ as

$$\mathrm{in}_{\mathbf{w}}(I) = (\{\mathrm{in}_{\mathbf{w}}(f) : f \in I\}).$$

By its definition, $\mathrm{in}_{\mathbf{w}}(I)$ is a homogeneous ideal with respect to the grading given by $\mathbf{w}$, and of course is finitely generated. A set of polynomials $f_1, \ldots, f_m \in I$ such that $\mathrm{in}_{\mathbf{w}}(I) = (\mathrm{in}_{\mathbf{w}}(f_1), \ldots, \mathrm{in}_{\mathbf{w}}(f_m))$ is called a **standard basis** of I with respect to $\mathbf{w}$.

3.1.2 Initial ideals given by weights

The similarities between Gröbner bases and standard bases are apparent. We shall now see that the Gröbner basis of an ideal may be viewed as the standard basis with respect to a suitable weight. To see this we shall need

Lemma 3.1.1. *Given a monomial order $<$ and a finite number of pairs of monomials $(u_1, v_1), \ldots, (u_m, v_m)$ such that $u_i > v_i$ for all i. Then there exists a weight $\mathbf{w}$ such that $\deg_{\mathbf{w}} u_i > \deg_{\mathbf{w}} v_i$ for all i.*

Proof. Let $u_i = \mathbf{x}^{\mathbf{a}_i}$ and $v_i = \mathbf{x}^{\mathbf{b}_i}$ for $i = 1, \ldots, m$. We are looking for an integral vector $\mathbf{w} \in \mathbb{N}^n$ such that $\langle \mathbf{a}_i - \mathbf{b}_i, \mathbf{w} \rangle > 0$ for all i. Suppose no such $\mathbf{w}$ exists. Then by the Farkas Lemma (see [Sch98, Section 7.3]) there exist $c_i \in \mathbb{Z}_+$ with $c_i > 0$ for at least one i such that the vector $\mathbf{g} = \sum_{i=1}^{m} c_i(\mathbf{a}_i - \mathbf{b}_i)$ has entries ≤ 0. Then this implies that $\prod_{i=1}^{m}(\mathbf{x}^{\mathbf{b}_i})^{c_i} = \prod_{i=1}^{m}(\mathbf{x}^{\mathbf{a}_i})^{c_i}\mathbf{x}^{-\mathbf{g}}$, contradicting the fact that $u_i > v_i$ for all i, see Lemma 2.1.4. □

Now we have

Theorem 3.1.2. *Given an ideal $I \subset S$ and a monomial order $<$, there exists a weight $\mathbf{w}$ such that*

$$\mathrm{in}_{<}(I) = \mathrm{in}_{\mathbf{w}}(I).$$

Proof. Let $g_1, \ldots, g_m$ be a Gröbner basis of I. We consider all pairs $(\mathrm{in}_{<}(g_i), u)$ where $u \in \mathrm{supp}(g_i)$ and $u \neq \mathrm{in}_{<}(g_i)$. There are finitely many such pairs. Hence by Lemma 3.1.1 there exists a weight $\mathbf{w}$ such that $\deg_{\mathbf{w}} \mathrm{in}_{<}(g_i) > \deg_{\mathbf{w}} u$ for all $u \in \mathrm{supp}(g_i)$ with $u \neq \mathrm{in}_{<}(g_i)$ and for all i. It follows that $\mathrm{in}_{\mathbf{w}}(g_i) = c_i \mathrm{in}_{<}(g_i)$, where c_i is the coefficient of $\mathrm{in}_{<}(g_i)$ in g_i. In particular we see that

$$\mathrm{in}_{<}(I) = (\mathrm{in}_{\mathbf{w}}(g_1), \ldots, \mathrm{in}_{\mathbf{w}}(g_m)) \subset \mathrm{in}_{\mathbf{w}}(I).$$

Consequently, we obtain

$$\mathrm{in}_{<}(I) = \mathrm{in}_{<}(\mathrm{in}_{<}(I)) \subset \mathrm{in}_{<}(\mathrm{in}_{\mathbf{w}}(I)) = \mathrm{in}_{<_{\mathbf{w}}}(I),$$

where $<_{\mathbf{w}}$ is the monomial order defined in Example 2.1.3. Thus the assertion follows from statement (b) in Proposition 2.2.6. □

3.2 The initial ideal as the special fibre of a flat family

3.2.1 Homogenization

Similarly to the standard gradings there is a process of homogenizing for the grading defined by a weight $\mathbf{w}$.

Let $f \in S$ be a nonzero polynomial with homogeneous components f_j (with respect to the weight $\mathbf{w}$). We introduce a new variable t, and define the **homogenization** of f with respect to $\mathbf{w}$ as the polynomial

$$f^h = \sum_j f_j t^{\deg_{\mathbf{w}} f - j} \in S[t]$$

Note that f^h is homogeneous in $S[t]$ with respect to the **extended weight** $\mathbf{w'} = (w_1, \ldots, w_n, 1) \in \mathbb{N}^{n+1}$ which assigns to the new variable t the degree 1.

In case of the standard grading (where $\mathbf{w} = (1, 1, \ldots, 1)$) this is the usual homogenization.

One easily verifies that for any two polynomials $f, g \in S$ one has

$$(fg)^h = f^h g^h, \tag{3.1}$$

and

$$(f + g)^h = t^{\deg_{\mathbf{w}}(f+g) - \deg_{\mathbf{w}} f} f^h + t^{\deg_{\mathbf{w}}(f+g) - \deg_{\mathbf{w}} g} g^h. \tag{3.2}$$

The second equation is only valid in $S[t, t^{-1}]$. Indeed, the exponents of t may be negative.

Let $I \subset S$ be an ideal. The **homogenization** of I is defined to be the ideal

$$I^h = (\{f^h : f \in I\}) \subset S[t].$$

Let $f_1, \ldots, f_m$ be a system of generators of I. In general $I^h \neq (f_1^h, \ldots, f_m^h)$. However, a system of generators of I^h can be computed by using Gröbner bases.

We first show

Lemma 3.2.1. *Let* $f \in S[t]$ *be homogeneous. Then* $f \in I^h$ *if and only if* $f = t^m g^h$ *for some* $g \in I$ *and some* $m \in \mathbb{Z}_+$.

Proof. The "if" part of the assertion is obvious. Conversely, assume that $f \in I^h$. Then $f = \sum_{i=1}^r g_i f_i^h$ with $f_i \in I$ and $g_i \in S[t]$ homogeneous.

For any homogeneous polynomial $p \in S[t]$, let $\bar{p} \in S$ denote the dehomogenization of p which is obtained from p by substituting t by 1. Then

$$\bar{f} = \sum_{i=1}^r \bar{g}_i \bar{f}_i^h = \sum_{i=1}^r \bar{g}_i f_i \in I.$$

Since $f = t^m \bar{f}^h$ for some $m \in \mathbb{Z}_+$, we may take $g = \bar{f}$. $\qquad\square$

Given a weight $\mathbf{w}$ on S. A monomial order $<$ is said to be **graded** with respect $\mathbf{w}$, if whenever $\deg_\omega(u) < \deg_\omega(v)$ for $u, v \in \mathrm{Mon}(S)$, then $u < v$. For example, the lexicographic order and reverse lexicographic order introduced in Example 2.1.2 are graded with respect to the standard grading. More generally, if $<$ is any monomial order, then the monomial order $<_\omega$ introduced in Example 2.1.3 is graded with respect to $\mathbf{w}$.

For a monomial order $<$ which is graded with respect to $\mathbf{w}$ we define a natural extension $<'$ to $S[t]$ as follows:

$$\mathbf{x}^\mathbf{a}t^c <' \mathbf{x}^\mathbf{b}t^d \quad \text{if and only if} \quad \text{(i) } \mathbf{x}^\mathbf{a} < \mathbf{x}^\mathbf{b}, \text{ or (ii) } \mathbf{x}^\mathbf{a} = \mathbf{x}^\mathbf{b} \text{ and } c < d.$$

This monomial order has the property that $\mathrm{in}_<(g) = \mathrm{in}_{<'}(g^h)$ for all nonzero $g \in S$.

Proposition 3.2.2. *Let $I \subset S$ be an ideal and $\mathcal{G} = \{g_1, \ldots, g_s\}$ a Gröbner basis of I with respect to a monomial order $<$ which is graded with respect to $\mathbf{w}$. Then $\mathcal{G}^h = \{g_1^h, \ldots, g_s^h\}$ is a Gröbner basis of I^h with respect to $<'$.*

Proof. Since I^h is homogeneous it suffices to show that for any homogeneous element $f \in I^h$ one has $\mathrm{in}_{<'}(f) \in (\mathrm{in}_{<'}(g_1^h), \ldots, \mathrm{in}_{<'}(g_s^h))$. In fact, $\mathrm{in}_{<'}(f) = \mathrm{in}_{<'}(f_j)$ for some homogeneous component f_j of f, and since I^h is homogeneous, all homogeneous components of f belong to I^h.

By virtue of Lemma 3.2.1 we have $f = t^m g^h$ for some $g \in I$ and some $m \in \mathbb{Z}_+$. By the choice of our monomial order,

$$\mathrm{in}_{<'}(f) = t^m \, \mathrm{in}_{<'}(g^h) = t^m \, \mathrm{in}_<(g).$$

Since $\mathcal{G}$ is a Gröbner basis of I, there exists $u \in \mathrm{Mon}(S)$ such that $\mathrm{in}_<(g) = u \, \mathrm{in}_<(g_i)$ for some i, and since $\mathrm{in}_<(g_i) = \mathrm{in}_{<'}(g_i^h)$ we obtain

$$\mathrm{in}_{<'}(f) = t^m u \, \mathrm{in}_{<'}(g_i^h),$$

as desired. $\square$

Example 3.2.3. Let $I = (x_1 x_2 - 1, x_1^2 - x_2)$. Then I has the Gröbner basis $\{x_1 x_2 - 1, x_1^2 - x_2, x_2^2 - x_1\}$ with respect to the reverse lexicographic order. Thus the homogenization of I with respect to the standard grading is given by

$$I^h = (x_1 x_2 - t^2, x_1^2 - tx_2, x_2^2 - tx_1).$$

3.2.2 A one parameter flat family

Let $I \subset S$ be an ideal. The inclusion $K[t] \subset S[t]$ induces a natural K-algebra homomorphism $K[t] \rightarrow S[t]/I^h$. This gives $S[t]/I^h$ a natural $K[t]$-module structure. We will show that $S[t]/I^h$ is a flat $K[t]$-module. One even has

Proposition 3.2.4. *$S[t]/I^h$ is a free $K[t]$-module.*

Proof. Let $<$ be monomial order which is graded with respect to $\mathbf{w}$. According to Proposition 3.2.2, $\{g_1^h, \ldots, g_s^h\}$ is Gröbner basis of I^h provided $\{g_1, \ldots, g_s\}$ is Gröbner basis of I. Moreover, one has $\mathrm{in}_{<'}(g_i^h) = \mathrm{in}_{<}(g_i)$ for all i. Therefore, by Proposition 2.2.5, the residue classes modulo I^h of the monomials in $S[t]$ which do not belong to $(\mathrm{in}_{<}(g_1), \ldots, \mathrm{in}_{<}(g_s))S[t]$ establish a K-basis of $S[t]/I^h$. This implies at once that $S[t]/I^h$ is a free $K[t]$-module whose basis consists of the residue classes modulo I^h of monomials in S which do not belong to $(\mathrm{in}_{<}(g_1), \ldots, \mathrm{in}_{<}(g_s))$. $\qquad\square$

Corollary 3.2.5. *For all $a \in K$, the element $t - a$ is a nonzero divisor of $S[t]/I^h$.*

Proof. We know from Proposition 3.2.4 that $S[t]/I^h$ is a free $K[t]$-module. Let $(e_j)_{j \in J}$ be a $K[t]$-basis, and suppose $(t-a)f = 0$ for some $f \in S[t]/I^h$. Write $f = \sum_j f_j e_j$ with $f_j \in K[t]$. Then $0 = \sum_j (t-a)f_j e_j$, and so $(t-a)f_j = 0$ for all j. This implies that all $f_j = 0$. Consequently, $f = 0$. $\qquad\square$

A **one parameter flat family** of K-algebras is a family of K-algebras R_a, $a \in K$, for which there exists a K-algebra R and a flat K-algebra homomorphism $K[t] \to R$ whose fibres $R/(t-a)R$ are isomorphic to R_a for all $a \in K$. The K-algebra R_0 is called the **special fibre**, and R_a for $a \neq 0$ a **general fibre** of the family.

Corollary 3.2.6. *Let $I \subset S$ be an ideal, and let $\mathbf{w}$ be a weight. Then there exists a one parameter flat family of K-algebras whose special fibre is isomorphic to $S/\mathrm{in}_{\mathbf{w}}(I)$ and whose general fibres are all isomorphic to S/I.*

Proof. The one parameter flat family is defined by the graded flat K-algebra homomorphism $K[t] \to S[t]/I^h$. It is clear that the substitution $t \mapsto 0$ maps I^h to $\mathrm{in}_{\mathbf{w}}(I)$. Thus the special fibre of the family is the one announced.

On the other hand, by Proposition 3.2.2 there exists a system of generators $g_1, \ldots, g_s$ of I (in fact a Gröbner basis) such that $g_1^h, \ldots, g_s^h$ is a system of generators of I^h. Say, $g_i = \sum_u c_u^i u$; then the substitution $t \mapsto a$ for $a \in K$ with $a \neq 0$, maps g_i^h to $g_{i,a} = \sum_u c_u^i a^{\deg_{\mathbf{w}} g_i - \deg_{\mathbf{w}} u} u$.

The automorphism $\varphi : S \to S$ with $\varphi(x_i) = a^{w_i} x_i$ for all i, maps $g_{i,a}$ to $a^{\deg_{\mathbf{w}} g_i} g_i$. This shows that the general fibre is isomorphic to S/I. $\qquad\square$

3.3 Comparison of I and in(I)

Throughout this section $I \subset S$ will always be a graded ideal (with respect to the standard grading of S). We fix a weight $\mathbf{w}$. Then $I^h \subset S[t]$ has a system of generators of the form $g_1^h, \ldots, g_s^h$, where $g_1, \ldots, g_s$ is a suitable homogeneous system of generators of I, see Proposition 3.2.2.

If we assign to each x_i the bidegree $(w_i, 1)$ and to t the bidegree $(1, 0)$, then all the generators g_i^h are bihomogeneous, and hence I^h is a bigraded ideal. Therefore I^h has a minimal bigraded resolution $\mathbb{F}$. The minimality of the resolution is equivalent to the condition that all entries in the matrices describing the differentials of the resolution $\mathbb{F}$ belong to $\mathfrak{n} = (x_1, \ldots, x_n, t)$. The reader who is not so familiar with resolutions is referred to Appendix A2.

We set $T = S[t]$, and let

$$\mathbb{F} : 0 \longrightarrow F_p \longrightarrow F_{p-1} \longrightarrow \cdots \longrightarrow F_1 \longrightarrow F_0 \longrightarrow T/I^h \longrightarrow 0$$

with $F_i = \bigoplus_{k,j} T(-k, -j)^{\beta_{i,(k,j)}}$.

Since t is a nonzero divisor on T with $t \in \mathfrak{n}$, the complex $\bar{\mathbb{F}} = \mathbb{F}/t\mathbb{F}$ is a bigraded minimal free S-resolution of $(T/I^h)/t(T/I^h) = S/\operatorname{in}_{\mathbf{w}}(I)$, and it has the same bigraded shifts as $\mathbb{F}$. Note that the second component of the shifts in the resolution are the ordinary shifts of the standard graded ideal $\operatorname{in}_{\mathbf{w}}(I)$. Thus we have shown that

$$\beta_{ij}(S/\operatorname{in}_{\mathbf{w}}(I)) = \sum_k \beta_{i,(k,j)} \quad \text{for all } i \text{ and } j \tag{3.3}$$

On the other hand, let $\mathbb{G} = \mathbb{F}/(t-1)\mathbb{F}$. Since $t-1$ is a nonzero divisor on T, the complex $\mathbb{G}$ is a free S-resolution of $(T/I^h)/(t-1)(T/I^h) = S/I$. However, $t-1$ is only homogeneous with respect to the second component of the bidegree. Therefore, $\mathbb{G}$ is no longer a bigraded resolution of S/I. However, the second components of the shifts in the resolution of $\mathbb{F}$ are preserved. Hence $\mathbb{G}$ is a graded free resolution of the standard graded ring S/I. But in general $\mathbb{G}$ may not be a minimal free resolution of S/I, because $t - 1 \notin \mathfrak{n}$.

A comparison with the minimal graded free resolution of S/I then implies that

$$\beta_{ij}(S/I) \leq \sum_k \beta_{i,(k,j)} \quad \text{for all } i \text{ and } j. \tag{3.4}$$

Thus combining (3.3) with (3.4) we obtain the important

Theorem 3.3.1. *Let I be a graded ideal $I \subset S$ and $\mathbf{w}$ a weight. Then*

$$\beta_{ij}(I) \leq \beta_{ij}(\operatorname{in}_{\mathbf{w}}(I)) \quad \text{for all } i \text{ and } j.$$

The following example illuminates what happens in this deformation process.

Example 3.3.2. Let $I = (x_1 x_2 - x_3^2, -x_1 x_3 + x_2^2, x_1^2 - x_2 x_3) \subset S = K[x_1, x_2, x_3]$. Then with respect to the lexicographic order we obtain the Gröbner basis

$$\{-x_2^3 + x_3^3, x_1 x_2 - x_3^2, -x_1 x_3 + x_2^2, x_1^2 - x_2 x_3\}.$$

Thus the homogenization of I with respect to the weight $(2, 1, 1)$ gives the ideal

$$I^h = (-x_2^3 + x_3^3, x_1 x_2 - x_3^2 t, -x_1 x_3 + x_2^2 t, x_1^2 - x_2 x_3 t^2).$$

The bigraded T-resolution of I^h is then given by

$$\mathbb{F} : 0 \to T(-6,-4) \xrightarrow{\ \alpha\ } T(-4,-3) \oplus T(-5,-3)^2 \oplus T(-5,-4)$$
$$\xrightarrow{\ \beta\ } T(-3,-3) \oplus T(-3,-2)^2 \oplus T(-4,-2) \to I^h \to 0$$

Here the maps are described by the matrices

$$\alpha = \begin{pmatrix} x_1 \\ x_2 \\ -x_3 \\ t \end{pmatrix} \quad \text{and} \quad \beta = \begin{pmatrix} t & 0 & 0 & -x_1 \\ x_3 & x_2 t & x_1 & -x_2^2 \\ x_2 & -x_1 & -x_3 t & -x_3^2 \\ 0 & -x_3 & -x_2 & 0 \end{pmatrix}.$$

If we now specialize t to 0, then we obtain from $\mathbb{F}$ the minimal graded free S-resolution $\bar{\mathbb{F}}$ of $\operatorname{in}_{\mathbf{w}}(I) = (x_1 x_3, x_1 x_2, x_1^2, -x_2^3 + x_3^3)$, namely

$$\bar{\mathbb{F}} : 0 \to S(-4) \longrightarrow S(-3)^3 \oplus S(-4) \longrightarrow S(-3) \oplus S(-2)^3 \longrightarrow \operatorname{in}_{\mathbf{w}}(I) \to 0,$$

with the maps described by the matrices

$$\begin{pmatrix} x_1 \\ x_2 \\ -x_3 \\ 0 \end{pmatrix} \quad \text{and} \quad \begin{pmatrix} 0 & 0 & 0 & -x_1 \\ x_3 & 0 & x_1 & -x_2^2 \\ x_2 & -x_1 & 0 & -x_3^2 \\ 0 & -x_3 & -x_2 & 0 \end{pmatrix}.$$

On the other hand, if we specialize t to 1, then we obtain a graded free S-resolution $\mathbb{G}$ of I with the same shifts as in the resolution of $\bar{\mathbb{F}}$. But now this resolution is not minimal, because the entries of the matrices describing the maps in the resolution contain units. Indeed, the matrices are

$$\begin{pmatrix} x_1 \\ x_2 \\ -x_3 \\ 1 \end{pmatrix} \quad \text{and} \quad \begin{pmatrix} 1 & 0 & 0 & -x_1 \\ x_3 & x_2 & x_1 & -x_2^2 \\ x_2 & -x_1 & -x_3 & -x_3^2 \\ 0 & -x_3 & -x_2 & 0 \end{pmatrix}.$$

However by cancelling isomorphic summands in $\mathbb{G}$ one obtains the graded minimal free S-resolution of I:

$$0 \to S(-3)^2 \longrightarrow S(-2)^3 \longrightarrow I \to 0,$$

where

$$\begin{pmatrix} -x_2 & x_1 & x_3 \\ -x_1 & x_3 & x_2 \end{pmatrix}$$

is the relation matrix of I.

Theorem 3.3.1 together with Theorem 3.1.2 now yields

Corollary 3.3.3. Let $<$ be a monomial order on S. Then for any graded ideal $I \subset S$ one has

$$\beta_{ij}(I) \leq \beta_{ij}(\operatorname{in}_<(I)) \quad \text{for all } i \text{ and } j.$$

There are two important invariants attached to a graded ideal $I \subset S$ defined in terms of the minimal graded free resolution of S/I:

(1) the **projective dimension**of S/I,

$$\operatorname{proj\,dim} S/I = \max\{i : \beta_{ij}(S/I) \neq 0 \quad \text{for some } j\}, \quad \text{and}$$

(2) the **regularity** of I,

$$\operatorname{reg} I = \max\{j : \beta_{i,i+j}(I) \neq 0 \quad \text{for some } i\}.$$

The following theorem summarizes the comparison between $\operatorname{in}_<(I)$ and I.

Theorem 3.3.4. *Let $I \subset S$ be a graded ideal and $<$ a monomial order on S. Then*

(a) $\dim S/I = \dim S/\operatorname{in}_<(I)$;
(b) $\operatorname{proj\,dim} S/I \leq \operatorname{proj\,dim} S/\operatorname{in}_<(I)$;
(c) $\operatorname{reg} S/I \leq \operatorname{reg} S/\operatorname{in}_<(I)$;
(d) $\operatorname{depth} S/I \geq \operatorname{depth} S/\operatorname{in}_<(I)$;

Proof. (a) follows from the fact that the residue classes of the monomials which do not belong to $\operatorname{in}_<(I)$ form a K-basis of S/I. Indeed, this implies that S/I and $S/\operatorname{in}_<(I)$ have the same Hilbert function and hence the same dimension; see Theorem 6.1.3.

(b) and (c) are immediate consequences of Corollary 3.3.3, while (d) follows from (b) and the Auslander–Buchsbaum formula which says that

$$\operatorname{depth} M + \operatorname{proj\,dim} M = \dim S$$

for any finitely generated graded S-module M; see Corollary A.4.3.   □

Corollary 3.3.5. *S/I is Cohen–Macaulay (resp. Gorenstein) if $S/\operatorname{in}_<(I)$ has the corresponding property.*

Proof. Since, by definition, a finitely generated graded S-module M is Cohen–Macaulay if and only if $\dim M = \operatorname{depth} M$, Theorem 3.3.4 implies the statement about Cohen–Macaulayness.

Concerning the Gorenstein property we use the fact (see A.6.6) that for a graded ideal $J \subset S$ we have: S/J is Gorenstein if and only if S/J is Cohen–Macaulay and the last non-vanishing Betti number of S/J is equal to 1. Therefore, using again Theorem 3.3.4 and Corollary 3.3.3, the assertion follows.   □

Example 3.3.6. Of course it may happen, and in indeed in most cases it does, that S/I is Cohen–Macaulay but $S/\operatorname{in}_<(I)$ is not. For example, consider the ideal $I = (x_1^2 - x_2 x_3, x_1 x_2 - x_3^2, x_2^2 - x_1 x_3) \subset S = K[x_1, x_2, x_3]$. Then $\operatorname{in}_{<_{\mathrm{lex}}}(I) = (x_1 x_3, x_1 x_2, x_1^2, x_2^3)$, and the resolutions are

$$0 \longrightarrow S(-3)^2 \longrightarrow S(-2)^3 \longrightarrow I \longrightarrow 0,$$

and

$$0 \longrightarrow S(-4) \longrightarrow S(-3)^3 \oplus S(-4) \longrightarrow S(-2)^3 \oplus S(-3) \longrightarrow \mathrm{in}_{<_{\mathrm{lex}}}(I) \longrightarrow 0.$$

We have $\dim S/I = \dim S/\mathrm{in}_{<_{\mathrm{lex}}}(I) = 1$. From the resolutions we deduce that $\mathrm{depth}\, S/I = 1$ and that $\mathrm{depth}\, S/\mathrm{in}_{<_{\mathrm{lex}}}(I) = 0$. Thus S/I is Cohen–Macaulay and $S/\mathrm{in}_{<_{\mathrm{lex}}}(I)$ is not Cohen–Macaulay. We also see that $2 = \mathrm{reg}\, I < \mathrm{reg}\, \mathrm{in}_{<_{\mathrm{lex}}}(I) = 3$.

Theorem 3.3.4 remains true more generally if we replace $\mathrm{in}_<(I)$ by $\mathrm{in}_{\mathbf{w}}(I)$ everywhere in the statements.

As a final useful result concerning the comparison of I with $\mathrm{in}(I)$ we show

Proposition 3.3.7. *Let $I \subset S$ be a graded ideal and suppose that $\mathrm{in}_{\mathbf{w}}(I)$ is a prime (resp. a radical) ideal. Then I is a prime (resp. a radical) ideal. In particular, if $\mathrm{in}_<(I)$ is a squarefree monomial ideal, then I is a radical ideal.*

Proof. Let $I^h \in S[t]$ be the homogenization of I with respect to the weight $\mathbf{w}$. Then I^h is a graded ideal in $S[t]$, if we set $\deg x_i = w_i$ and $\deg t = 1$. We claim that I^h is a prime ideal (resp. a radical) ideal, if $\mathrm{in}_{\mathbf{w}}(I)$ has this property. Once this claim is shown, the desired conclusion follows since $IS[t, t^{-1}] \cong I^h S[t, t^{-1}]$. This isomorphism is induced by the automorphism $\varphi \colon S[t, t^{-1}] \to S[t, t^{-1}]$ which is defined by $x_i \mapsto t^{w_i} x_i$ for $i = 1, \dots, n$.

In order to prove the claim we use Corollary 3.2.5 together with the following fact: let R be a finitely generated positively graded K-algebra, and let $t \in R$ be a homogeneous nonzero divisor of R such that R/tR is a domain or a reduced ring. Then R has this property, too.

Indeed, set $\bar{R} = R/tR$ and denote by $\bar{a} \in \bar{R}$ the residue class of an element $a \subset R$. Suppose $\bar{R}$ is reduced and that $a^n = 0$ for some n, then $\bar{a}^n = 0$, and so $\bar{a} = 0$. Therefore $a = bt$ for some $b \in R$, and so $b^n t^n = 0$. However, since t is a nonzero divisor on R, it follows that $b^n = 0$. Again, since $\bar{R}$ is reduced it follows that $\bar{b} = 0$ which implies that $b \in (t)$, so that $a \in (t)^2$. By induction we have that $a \in (t)^k$ for all k. By Krull's intersection theorem, $\bigcap_{k \geq 0}(t)^k = (0)$, and so $a = 0$, as desired. In the same way one shows that the property of $\bar{R}$ being a domain can be lifted to R. $\qquad\square$

Problems

3.1. What is the homogenization of the ideal $I = (x_1 x_2 - x_3, x_2 x_3 - x_1, x_1 x_3 - x_2)$ with respect to the standard grading?

3.2. Give an example of a graded ideal $I \subset S$ for which $\mathrm{depth}\, I > \mathrm{depth}\, \mathrm{in}_<(I)$ and $\mathrm{reg}\, I < \mathrm{reg}\, \mathrm{in}_<(I)$.

3.3. A graded ideal $I \subset S$ is called a complete intersection if I is generated by a regular sequence $f_1, \ldots, f_m$ of homogeneous polynomials. Let I be graded ideal such that $\mathrm{in}_<(I)$ is a complete intersection.
(a) Show that I is a complete intersection.
(b) Give an example of a complete intersection I for which $\mathrm{in}_<(I)$ is not even Cohen–Macaulay.

3.4. Let $n \geq 2$ and $I \subset S = K[x_1, \ldots, x_n]$ be the ideal generated by the elements $x_1^2 - x_i^2$, $i = 2, \ldots, n$ and by the monomials $x_i x_j$ with $1 \leq i < j \leq n$. Show that S/I is Gorenstein but $\mathrm{in}_<(I)$ is never Gorenstein, no matter which monomial order $<$ on S we choose.

3.5. (a) Show that I is a complete intersection if $\mathrm{in}_<(I)$ is a complete intersection for some monomial order on S.
(b) Show that $I = (x_1^2 - x_2^2, x_1 x_2) \subset K[x_1, x_2]$ is a complete intersection, but $\mathrm{in}_<(I)$ is never a complete intersection, no matter which monomial order $<$ on $K[x_1, x_2]$ we choose.

3.6. Let $f \in S = K[x_1, \ldots, x_n]$ be a nonzero polynomial. We denote by f^* the highest nonzero homogeneous component of f. For an ideal $I \subset S$ we let I^* be the ideal generated by all f^* with $f \in I$ and $f \neq 0$. Show that I is a prime (resp. radical) ideal, if I^* has this property.

3.7. Show that the ideal $(x_1 y_2 - x_2 y_1, x_1 y_3 - x_3 y_1) \subset K[x_1, x_2, x_3, y_1, y_2, y_3]$ is a radical ideal, but not a prime ideal.

Notes

The fact that for any graded ideal in the polynomial ring and for any monomial order the graded Betti numbers of the initial ideal are greater than or equal to the graded Betti numbers of the original ideal has many theoretical and practical applications. Indeed, this inequality of graded Betti numbers implies that if the initial ideal has nice algebraic properties, then so does the original ideal. For example, an elegant proof of the fact that determinantal ideals are Cohen–Macaulay was given by Sturmfels [Stu90] by showing that the initial of a determinantal ideal is the Stanley–Reisner ideal of a shellable simplicial complex. Another interesting example of how to use the comparison with the initial ideal is given in the theory of Koszul algebras, namely if the initial ideal of a graded ideal is generated by quadratic monomials, then the quotient algebra of the original ideal is Koszul. Here it is used that any quadratic monomial ideal defines a Koszul algebra, as shown in [BF85]. A simple proof of this fact can also be found in [HHR00]. More detailed information about the comparison of I and $\mathrm{in}(I)$ can be found in the survey article [BC04] by Bruns and Conca. There one can also find more references to this topic.

4

Generic initial ideals

Generic initial ideals play an essential role in geometry as well as in shifting theory, which will be studied in Chapter 11. Let $S = K[x_1, \ldots, x_n]$ be a polynomial ring over an infinite field K, and $<$ a monomial order on S which satisfies $x_1 > x_2 > \cdots > x_n$. For a graded ideal $I \subset S$ it will be shown that there exists a nonempty open set U of linear automorphisms of S such that $\mathrm{in}_<(\alpha I)$ does not depend on $\alpha \in U$. The resulting initial ideal is called the generic initial ideal of I with respect to $<$. It turns out that generic initial ideals are Borel-fixed, and are even strongly stable if the base field is of characteristic 0. Generic annihilator numbers will be introduced and it will be shown that the extremal Betti numbers of an ideal and its generic initial coincide.

4.1 Existence

Let K be a field. Recall that a subset of the affine space K^m is called **Zariski closed** if it is the set of common zeroes of a set of polynomials in m variables. A **Zariski open** subset of K^m is by definition the complement of a Zariski closed subset. The topology so defined on K^m is called the **Zariski topology**.

Note that if K is a finite field, then any subset of K^m is Zariski closed, and consequently any subset of K^m is Zariski open as well.

Throughout this chapter we will assume that K is an infinite field, because otherwise the statements which refer to Zariski open sets would be meaningless.

An important property of Zariski open sets is given in

Lemma 4.1.1. *Let* $U_1, \ldots, U_r \subset K^m$ *be nonempty Zariski open sets. Then* $U_1 \cap \ldots \cap U_r \neq \emptyset$.

Proof. It is enough to show that $U \cap U' \neq \emptyset$, if U and U' are nonempty Zariski open sets of K^m. Let $A = K^m \setminus U$ and $A' = K^m \setminus U'$, and assume that A is

J. Herzog, T. Hibi, *Monomial Ideals*, Graduate Texts in Mathematics 260, DOI 10.1007/978-0-85729-106-6_4, © Springer-Verlag London Limited 2011

the common set of zeroes of the polynomials $f_1, \ldots, f_r$ and A' is the common set of zeroes of the polynomials $g_1, \ldots, g_s$. Let $\mathbf{x} \in U$ and $\mathbf{x}' \in U'$. Then there exist f_i and g_j with $f_i(\mathbf{x}) \neq 0$ and $g_j(\mathbf{x}') \neq 0$. It follows that $f_i g_j \neq 0$. Since K is infinite, there exists $\mathbf{x}'' \in K^m$ such that $f_i g_j(\mathbf{x}'') \neq 0$. This implies $f_i(\mathbf{x}'') \neq 0$ and $g_j(\mathbf{x}'') \neq 0$. Hence $\mathbf{x}'' \in U \cap U'$. $\qquad\square$

Lemma 4.1.1 says that any nonempty Zariski open set is a dense subset of K^m.

Let $S = K[x_1, \ldots, x_n]$ be the polynomial ring in n variables and let $\mathrm{GL}_n(K)$ denote the general linear group, that is, the group of all invertible $n \times n$-matrices with entries in K. Any $\alpha \in \mathrm{GL}_n(K)$, $\alpha = (a_{ij})$ induces an automorphism

$$\alpha : S \to S, \quad f(x_1, \ldots, x_n) \mapsto f\left(\sum_{i=1}^{n} a_{i1} x_i, \ldots, \sum_{i=1}^{n} a_{in} x_i\right).$$

This type of automorphism of S is called a **linear automorphism**.

The set $\mathrm{M}_n(K)$ of all $n \times n$ matrices may be identified with the points in $K^{n \times n}$, the coordinates of the points being the entries of the corresponding matrices. It is then clear that $\mathrm{GL}_n(K)$ is a Zariski open subset of $\mathrm{M}_n(K)$, because $\alpha \in \mathrm{M}_n(K)$ belongs to $\mathrm{GL}_n(K)$ if and only if $\det \alpha \neq 0$. This is the case if and only if α does not belong to the Zariski closed set which is defined as the set of zeroes of the polynomial $\det(x_{ij}) \in K[\{x_{ij}\}_{i,j=1,\ldots,n}]$.

Since $\mathrm{GL}_n(K)$ itself is open, a subset of $\mathrm{GL}_n(K)$ is open if and only if it is a Zariski open subset of $K^{n \times n}$.

Theorem 4.1.2. *Let $I \subset S$ be a graded ideal and $<$ a monomial order on S. Then there exists a nonempty open subset $U \subset \mathrm{GL}_n(K)$ such that $\mathrm{in}_<(\alpha I) = \mathrm{in}_<(\alpha' I)$ for all $\alpha, \alpha' \in U$.*

Definition 4.1.3. The ideal $\mathrm{in}_<(\alpha I)$ with $\alpha \in U$ and $U \subset \mathrm{GL}_n(K)$ as given in Theorem 4.1.2 is called the **generic initial ideal** of I with respect to the monomial order $<$. It is denoted $\mathrm{gin}_<(I)$.

In preparation of the proof of Theorem 4.1.2 we introduce some concepts and notation: Let $d, t \in \mathbb{N}$ with $t \leq \dim_K S_d$. We consider the tth exterior power $\bigwedge^t S_d$ of the K-vector space S_d, cf. Chapter 5.

Given a monomial order $<$ on S, an element $u_1 \wedge u_2 \wedge \cdots \wedge u_t$ where each u_i is a monomial of degree d and where $u_1 > u_2 > \cdots > u_t$, will be called a **standard exterior monomial** of $\bigwedge^t S_d$. It is clear that the standard exterior monomials form a K-basis of $\bigwedge^t S_d$. In particular, any element $f \in \bigwedge^t S_d$ is a unique linear combination of standard exterior monomials. The **support** of f is the set $\mathrm{supp}(f)$ of standard exterior monomials which appear in f with a nonzero coefficient.

We order the standard exterior monomials lexicographically by setting

$$u_1 \wedge u_2 \wedge \cdots \wedge u_t > v_1 \wedge v_2 \wedge \cdots \wedge v_t,$$

if $u_i > v_i$ for the smallest index i with $u_i \neq v_i$. This allows us to define the **initial monomial** $\text{in}_<(f)$ of a nonzero element $f \in \bigwedge^t S_d$ as the largest standard exterior monomial in the support of f.

Now let $V \subset S_d$ be a t-dimensional linear subspace of S_d and $f_1, \ldots, f_t$ a K-basis of V. Then the 1-dimensional K-vector space $\bigwedge^t V$ is generated by $f_1 \wedge f_2 \wedge \cdots \wedge f_t$.

We let $\text{in}_<(V)$ be the K-vector space generated by all the monomials $\text{in}_<(f)$ with $0 \neq f \in V$. Then we have

Lemma 4.1.4. *Let $w_1, \ldots, w_t$ be monomials in S_t with $w_1 > w_2 > \cdots > w_t$. The following conditions are equivalent:*

(a) *the monomials $w_1, \ldots, w_t$ form a K-basis of $\text{in}_<(V)$;*
(b) *if $w_i = \text{in}_<(g_i)$ with $g_i \in V$, then $g_1, \ldots, g_t$ is a K-basis of V and*

$$\text{in}_<(g_1 \wedge \cdots \wedge g_t) = \text{in}_<(g_1) \wedge \cdots \wedge \text{in}_<(g_t);$$

(c) *if $f_1, \ldots, f_t$ is a K-basis of V, then $\text{in}_<(f_1 \wedge \cdots \wedge f_t) = w_1 \wedge \cdots \wedge w_t$.*

Proof. (a) $\Rightarrow$ (b): Let $0 \neq f \in V$; then $\text{in}_<(f) = w_i$ for some i. Hence there exists $a \in K$ such that $f - ag_i = 0$ or $\text{in}_<(f - ag_i) < w_i$. Arguing by induction on i we see that $f - ag_i$ is a K-linear combination of $g_{i+1}, \ldots, g_t$. This shows that $g_1, \ldots, g_t$ is a system of generators of V.

Suppose $\sum_{i=1}^t a_i g_i = 0$ with $a_i \in K$ and not all $a_i = 0$. Let j be the smallest integer such that $a_j \neq 0$. Then $\text{in}_<(\sum_{i=1}^t a_i g_i) = w_j \neq 0$: a contradiction. Thus the elements $g_1, \ldots, g_t$ are linearly independent.

Since $w_1 > w_2 > \cdots > w_t$ it follows that $\text{in}_<(g_1) \wedge \cdots \wedge \text{in}_<(g_t)$ is a standard exterior monomial. Let $g = g_1 \wedge \cdots \wedge g_t$; then, up to a sign, a standard exterior monomial of $\text{supp}(g)$ is of the form $u_1 \wedge u_2 \wedge \cdots \wedge u_t$ with $u_i \in \text{supp}(g_i)$. Since $w_i \geq u_i$ it follows that any standard exterior monomial of $\text{supp}(g)$ is less than or equal to $w_1 \wedge w_2 \wedge \cdots \wedge w_t$. On the other hand, since $w_1 \wedge w_2 \wedge \cdots \wedge w_t \in \text{supp}(g)$, the desired conclusion follows.

(b) $\Rightarrow$ (c): Since $f_1 \wedge \cdots \wedge f_t$ and $g_1 \wedge \cdots \wedge g_t$ differ only by a nonzero scalar, we have

$$\text{in}_<(f_1 \wedge \cdots \wedge f_t) = \text{in}_<(g_1 \wedge \cdots \wedge g_t) = w_1 \wedge \cdots \wedge w_t.$$

(c) $\Rightarrow$ (a): We claim that each $w_i \in \text{in}_<(V)$. Then, since $\dim_K V = \dim_K \text{in}_<(V)$, this implies that $w_1, \ldots, w_t$ is a K-basis of $\text{in}_<(V)$, as we want to show. Indeed, since $\text{in}_<(f_1 \wedge \cdots \wedge f_t) = \text{in}_<(h_1 \wedge \cdots \wedge h_t)$ for any other basis $h_1, \ldots, h_t$ of V, we may assume that $\text{in}_<(f_1) > \text{in}_<(f_2) > \cdots > \text{in}_<(f_t)$. But then $w_i = \text{in}_<(f_i)$. $\square$

Now let $\alpha \in \text{GL}_n(K)$ be a linear automorphism of S, and $f_1, f_2, \ldots, f_t$ a K-basis of V. Then $\alpha(f_1), \alpha(f_2), \ldots, \alpha(f_t)$ is a K-basis of the vector subspace $\alpha V \subset S_d$, and if $\text{in}_<(\alpha(f_1) \wedge \cdots \wedge \alpha(f_t)) = w_1 \wedge w_2 \wedge \cdots \wedge w_t$, then $\text{in}_<(\alpha V)$ has the K-basis $w_1, \ldots, w_t$; see Lemma 4.1.4.

Lemma 4.1.5. *Let $w_1 \wedge \cdots \wedge w_t$ be the largest standard exterior monomial of $\bigwedge^t S_d$ with the property that there exists $\alpha \in \mathrm{GL}_n(K)$ with*

$$\mathrm{in}_<(\alpha(f_1) \wedge \cdots \wedge \alpha(f_t)) = w_1 \wedge \cdots \wedge w_t.$$

Then the set $U = \{\alpha \in \mathrm{GL}_n(K) : \mathrm{in}_<(\alpha(f_1) \wedge \cdots \wedge \alpha(f_t)) = w_1 \wedge \cdots \wedge w_t\}$ is a nonempty Zariski open subset of $\mathrm{GL}_n(K)$.

Proof. By its definition, $U \neq \emptyset$. Let $p(\alpha)$ be the coefficient of $w_1 \wedge \cdots \wedge w_t$ in the presentation of $\alpha(f_1) \wedge \cdots \wedge \alpha(f_t)$ as a linear combination of standard exterior monomials. Then α belongs to U if and only if $p(\alpha) \neq 0$.

Basic linear algebra implies that $p(\alpha)$ is a polynomial function of the entries α whose coefficients are determined by $f_1, \ldots, f_t$; see Example 4.1.6. This yields the desired conclusion. $\quad\square$

Example 4.1.6. Let $S = K[x_1, x_2]$, and $<$ the lexicographic monomial order on S. Then the standard exterior monomials in $\bigwedge^2 S_2$ are:

$$x_1^2 \wedge x_1 x_2 > x_1^2 \wedge x_2^2 > x_1 x_2 \wedge x_2^2.$$

Let $f_1 = x_1^2$, $f_2 = x_2^2$ and $\alpha \in \mathrm{GL}_2(K)$. Then $\alpha(f_1) = \alpha_{11}^2 x_1^2 + 2\alpha_{11}\alpha_{21} x_1 x_2 + \alpha_{21}^2 x_2^2$ and $\alpha(f_2) = \alpha_{12}^2 x_1^2 + 2\alpha_{12}\alpha_{22} x_1 x_2 + \alpha_{22}^2 x_2^2$. Therefore,

$$\alpha(f_1) \wedge \alpha(f_2) = (2\alpha_{11}^2 \alpha_{12}\alpha_{22} - 2\alpha_{12}^2 \alpha_1\alpha_{21}) x_1^2 \wedge x_1 x_2 + \cdots,$$

and so $p(\alpha) = 2(\alpha_{11}^2 \alpha_{12}\alpha_{22} - \alpha_{12}^2 \alpha_1\alpha_{21})$.

Now we are ready to give the proof of Theorem 4.1.2.

Proof. Let $d \in \mathbb{Z}_+$ with $I_d \neq 0$. We define the nonempty Zariski open subset $U_d \subset \mathrm{GL}_n(K)$ for the linear subspace $I_d \subset S_d$ similarly to how we defined in Lemma 4.1.5 the Zariski open subset $U \subset \mathrm{GL}_n(K)$ for $V \subset S_d$. For those $d \in \mathbb{Z}_+$ with $I_d = 0$, we set $U_d = \mathrm{GL}_n(K)$.

Let $\alpha \in U_d$ and set $J_d = \mathrm{in}_<(\alpha I_d)$. By the definition of U_d and by Lemma 4.1.4, J_d does not depend on the particular choice of $\alpha \in U_d$. We claim that $J = \bigoplus_d J_d$ is an ideal. In fact, for a given $d \in \mathbb{Z}_+$, we have $U_d \cap U_{d+1} \neq \emptyset$. Then for any $\alpha \in U_d \cap U_{d+1}$ it follows that

$$S_1 J_d = S_1 \mathrm{in}_<(\alpha I_d) \subset \mathrm{in}_<(\alpha I_{d+1}) = J_{d+1},$$

which shows that J is indeed an ideal.

Let c be the highest degree of a generator of J, and let $U = U_1 \cap U_2 \cap \cdots \cap U_c$. For any $\alpha \in U$ we will show that $J_d = \mathrm{in}_<(\alpha I_d)$ for all d. This is obviously the case for $d \leq c$, because $\alpha \in U_d$ for all $d \leq c$. Now let $d \geq c$. We show by induction on d, that $J_d = \mathrm{in}_<(\alpha I_d)$. For $d = c$, there is nothing to prove. Now let $d > c$. Applying the induction hypothesis we get

$$J_d = S_1 J_{d-1} = S_1 \mathrm{in}_<(\alpha I_{d-1}) \subset \mathrm{in}_<(\alpha I_d).$$

Since $\dim_K J_d = \dim_K \mathrm{in}_<(\alpha I_d)$ we conclude that $J_d = \mathrm{in}_<(\alpha I_d)$.

The (nonempty) Zariski open set U just defined, has the desired property. $\quad\square$

The proof of Theorem 4.1.2 gives us the following additional information about $\mathrm{gin}_<(I)$.

Proposition 4.1.7. *Let* $t = \dim_K I_d$ *and let* $w_1 \wedge w_2 \wedge \cdots \wedge w_t$ *be the standard exterior monomial generating* $\bigwedge^t \mathrm{gin}_<(I)_d$. *Then*

$$w_1 \wedge w_2 \wedge \cdots \wedge w_t = \max\{\mathrm{in}_<(f) : 0 \neq f \in \overset{t}{\bigwedge} \alpha(I)_d, \ \alpha \in \mathrm{GL}_n(K)\}.$$

How can $\mathrm{gin}_<(I)$ be computed? The nonempty Zariski open set U with $\mathrm{gin}_<(I) = \mathrm{in}_<(\alpha I)$ for all $\alpha \in U$ is a dense subset of K^m with respect to the Zariski topology. (Here $m = n^2$.) Suppose K is a subfield of the real numbers, for example $K = \mathbb{Q}$, then U is also a dense subset of K^m with respect to the standard topology on K^m. Indeed, suppose U is the complement of the Zariski closed set $A \subset K^m$, which is defined as the common set of zeroes of the polynomials f_j, $j \in J$. Then a point $\mathbf{x} \in K^m$ belongs to U if and only if $f_j(\mathbf{x}) \neq 0$ for at least one j. To simplify notation we set $f = f_j$. Let $\mathbf{y} \in K^m$ and $\epsilon > 0$. We set $U_\epsilon = \{\mathbf{z} \in K^m : |\mathbf{z} - \mathbf{y}| < \epsilon\}$ and show that $U_\epsilon \cap U \neq \emptyset$. Suppose this is not the case, then $f(\mathbf{z}) = 0$ for all $\mathbf{z} \in U_\epsilon$. But then also

$$\frac{\partial f}{\partial x_j}(\mathbf{a}) = \lim_{\substack{t \to 0, \\ \mathbf{a}+t\mathbf{e}_j \in U_\epsilon}} \frac{f(\mathbf{a} + t\mathbf{e}_j) - f(\mathbf{a})}{t} = 0$$

for all $\mathbf{a} \in U_\epsilon$, where $\mathbf{e}_j$ denotes the jth standard basis vector of K^m. By induction it follows that all higher partial derivatives of f vanish. This implies that $f = 0$, a contradiction.

Now as we know that $U \subset K^m$ is a dense subset of K^m in the standard topology, it is hard to avoid U when choosing a point $\mathbf{x} \in K^m$. In other words, if we pick $\mathbf{x} \in K^m$ randomly, i.e. "generic enough", then most likely $\mathbf{x}$ will belong to U. This is how most computer algebra systems compute $\mathrm{gin}_<(I)$. An uncertainty remains. Therefore it is advisable to make several random choices of coordinates. If the result is always the same, then there is a good reason to believe that this is $\mathrm{gin}_<(I)$.

4.2 Stability properties of generic initial ideals

4.2.1 The theorem of Galligo and Bayer–Stillman

The subgroup $\mathcal{B} \subset \mathrm{GL}_n(K)$ of all nonsingular upper triangular matrices is called the **Borel subgroup** of $\mathrm{GL}_n(K)$. A matrix $\alpha = (a_{ij}) \in \mathcal{B}$ is called an **upper elementary** matrix, if $a_{ii} = 1$ for all i and if there exist integers $1 \leq k < l \leq n$ such that $a_{kl} \neq 0$ while $a_{ij} = 0$ for all $i \neq j$ with $\{i, j\} \neq \{k, l\}$. Recall from linear algebra that the subgroup $\mathcal{D} \subset \mathcal{B}$ of all nonsingular diagonal matrices together with the set of all upper elementary matrices generate $\mathcal{B}$.

The main goal of this section is to show that $\mathrm{gin}_<(I)$ is fixed under the action of $\mathcal{B}$. To be precise, we have

Theorem 4.2.1 (Galligo, Bayer–Stillman). *Let $I \subset S$ be a graded ideal and $<$ a monomial order on S. Then $\mathrm{gin}_<(I)$ is a Borel-fixed ideal, that is, $\alpha(\mathrm{gin}_<(I)) = \mathrm{gin}_<(I)$ for all $\alpha \in \mathcal{B}$.*

Proof. We first notice that an invertible diagonal matrix δ keeps monomial ideals fixed, because if $d_1, \ldots, d_n$ is the diagonal of δ and u is a monomial, then $\delta(u) = u(d_1, \ldots, d_n)u$ and $u(d_1, \ldots, d_n) \in K \setminus \{0\}$. Here $u(d_1, \ldots, d_n)$ denotes the evaluation of the monomial u at the point $(d_1, \ldots, d_n)$, that is, if $u = x_1^{a_1} x_2^{a_2} \cdots x_n^{a_n}$, then $u(d_1, \ldots, d_n) = d_1^{a_1} d_2^{a_2} \cdots d_n^{a_n}$.

Suppose now there is an element $\alpha \in \mathcal{B}$ with $\alpha(\mathrm{gin}_<(I)) \neq \mathrm{gin}_<(I)$. Then, since $\mathcal{B}$ is generated by invertible diagonal matrices (which fix $\mathrm{gin}_<(I)$) and by upper elementary matrices, we may assume that α is an upper elementary matrix.

Now since $\alpha(\mathrm{gin}_<(I)) \neq \mathrm{gin}_<(I)$, there exists $d \in \mathbb{Z}_+$ with $\alpha(\mathrm{gin}_<(I)_d) \neq \mathrm{gin}_<(I)_d$. Let $t = \dim_K I_d = \dim_K \mathrm{gin}_<(I)_d$. We let α act as a K-linear map on $\bigwedge^t S_d$ by setting

$$\alpha(w_1 \wedge w_2 \wedge \cdots \wedge w_t) = \alpha(w_1) \wedge \alpha(w_2) \wedge \cdots \wedge \alpha(w_t)$$

for all standard exterior monomials $w_1 \wedge w_2 \wedge \cdots \wedge w_t$ in $\bigwedge^t S_d$.

If $w = w_1 \wedge w_2 \wedge \cdots \wedge w_t$ is a standard exterior monomial generating $\bigwedge^t \mathrm{gin}_<(I)_d$, then, since α is an upper triangular matrix and since $\alpha(\mathrm{gin}_<(I)) \neq \mathrm{gin}_<(I)$, it follows that there exists a standard exterior monomial $u \in \mathrm{supp}(\alpha(w))$ with $u > w$.

Let $\beta \in \mathrm{GL}_n(K)$ with $\mathrm{gin}_<(I) = \mathrm{in}_<(\beta I)$, and let $f_1, \ldots, f_t$ be a K-basis of βI_d with $\mathrm{in}_<(f_i) = w_i$ and set $f = f_1 \wedge f_2 \wedge \cdots \wedge f_t$. We will show that $u \in \mathrm{supp}(\gamma(f))$ for a suitable $\gamma \in \mathcal{B}$, contradicting Proposition 4.1.7.

The element $\gamma \in \mathcal{B}$ that we are going to construct will be of the form $\alpha\delta$ for a suitable nonsingular diagonal matrix δ. If we apply such δ with diagonal $d_1, \cdots, d_n$ to f, then all standard exterior monomials $v = v_1 \wedge \cdots \wedge v_t \in \mathrm{supp}(f)$ for which $v_1 v_2 \cdots v_t$ is the same monomial $m \in S_{td}$ are mapped to the scalar multiple $m(d_1, \ldots, d_n)$ of themselves. Thus if we take the sum of all terms $a_v v_1 \wedge \cdots \wedge v_t$ in f with $v_1 v_2 \cdots v_t = m$ and call this sum f_m, then we have

$$f = f_{m_0} + \sum_{m \neq m_0} f_m,$$

where $f_{m_0} = a w_1 \wedge w_2 \wedge \cdots \wedge w_t = a \, \mathrm{in}_<(f)$ and $a \in K \setminus \{0\}$, and get

$$\delta(f) = a m_0(d_1, \ldots, d_n) \, \mathrm{in}_<(f) + \sum_{m \neq m_0} m(d_1, \ldots, d_t) f_m.$$

Hence after applying α we obtain

$$(\alpha\delta)(f) = a m_0(d_1, \ldots, d_n) \alpha(\mathrm{in}_<(f)) + \sum_{m \neq m_0} m(d_1, \ldots, d_t) \alpha(f_m).$$

Since u appears in $\alpha(\mathrm{in}_<(f))$, say with coefficient $c \in K \neq \{0\}$, we see that the coefficient of u in $(\alpha\delta)(f)$ equals

$$acm_0(d_1,\ldots,d_n) + \sum_{m\in W} c_m m(d_1,\ldots,d_t)$$

with $c_m \in K$ and where W is the set of all m for which $u \in \mathrm{supp}(\alpha(f_m))$.

Since $p = acm_0 + \sum_{m\in W} c_m m$ is a nonzero polynomial in S and since K is infinite, there exists $d_1,\ldots,d_n$ with $p(d_1,\ldots,d_n) \neq 0$. Thus if we choose δ with this diagonal and let γ be $\alpha\delta$, then $u \in \mathrm{supp}(\gamma(f))$, as desired.   □

4.2.2 Borel-fixed monomial ideals

We say that a graded monomial ideal $I \subset S$ is **Borel-fixed** if it is fixed under the action of $\mathcal{B}$. Theorem 4.2.1 theorem tells us that the generic initial ideal of a graded ideal is Borel-fixed.

If char $K = 0$, then the Borel-fixed ideals can be easily characterized, as we shall see in a moment.

Definition 4.2.2. Let $I \subset S$ be a monomial ideal. Then I is called **strongly stable** if one has $x_i(u/x_j) \subset I$ for all monomials $u \in I$ and all $i < j$ such that x_j divides u.

The defining property of a strongly stable ideal needs to be checked only for the set of monomial generators of a monomial ideal.

Lemma 4.2.3. *Let I be a monomial ideal. Suppose of all $u \in G(I)$, and for all integers $1 \le i < j \le n$ such that x_j divides u one has $x_i(u/x_j) \in I$. Then I is strongly stable.*

Proof. Let $v \in I$ be a monomial and $1 \le i < j \le n$ integers such that x_j divides v. There exists $u \in G(I)$ and a monomial w such that $v = uw$. If $x_j|u$, then $x_i(u/x_j) \in I$ by assumption, and so $x_i(v/x_j) = x_i(u/x_j)w \in I$. If $x_j|w$, then $x_i(v/x_j) = x_iu(w/x_j) \in I$.   □

Proposition 4.2.4. (a) *Let $I \subset S$ be a graded ideal. Then I is a monomial ideal, if I is Borel-fixed.*
(b) *Let I be a Borel-fixed ideal and a the largest exponent appearing among the monomial generators of I. If char $K = 0$ or char $K > a$, then I is strongly stable.*
(c) *If I is strongly stable, then I is Borel-fixed.*

Proof. (a) We show that if $f \in I$ is a nonzero homogeneous polynomial, and $u \in \mathrm{supp}(f)$, then there exists a homogeneous polynomial $g \in I$ with $\mathrm{supp}(g) = \mathrm{supp}(f) \setminus \{u\}$.

Suppose $f = a_u u + \sum_{v\neq u} a_v v$, and $\alpha \in \mathcal{B}$ is a diagonal matrix with diagonal $c_1, c_2, \ldots, c_n$. Then $\alpha(f) = a_u u(c_1,\ldots,c_n)u + \sum_{v\neq u} a_v v(c_1,\ldots,c_n)v$. Since K

is infinite, we may choose $c_1, \ldots, c_n$ such that $u(c_1, \ldots, c_n) \neq v(c_1, \ldots, c_n)$ for all $v \neq u$. Let $g = u(c_1, \ldots, c_n)f - \alpha(f)$. Then, indeed, we have $\operatorname{supp}(g) = \operatorname{supp}(f) \setminus \{u\}$.

(b) Let $u \in I$ be a monomial, x_j a variable which divides u and $1 \leq i < j$ a number. Let $\alpha \in \mathcal{B}$ be the upper elementary matrix which induces the linear automorphism on S with $x_k \mapsto x_k$ for $k \neq j$ and $x_j \mapsto x_i + x_j$.

Suppose that $u = x_1^{a_1} x_2^{a_2} \cdots x_n^{a_n}$; then

$$\alpha(u) = x_1^{a_1} x_2^{a_2} \cdots (x_i + x_j)^{a_j} \cdots x_n^{a_n} = u + a_j x_i (u/x_j) + \cdots.$$

Since I is Borel-fixed, it follows that $\alpha(u) \in I$, and since I is a monomial ideal, we have $\operatorname{supp}(\alpha(u)) \subset I$. The assumption on the characteristic on K and the above calculation then shows that $x_i(u/x_j) \in I$.

(c) Let $u \in I$ be a monomial. Since the upper elementary matrices and the diagonal matrices generate $\mathcal{B}$ and since $\delta I = I$ for any diagonal matrix δ, it is enough to show that for $1 \leq i < j$ and $\alpha \in \mathrm{GL}_n(K)$ with $\alpha(x_k) = x_k$ for $j \neq k$ and $\alpha(x_j) = cx_i + x_j$ with $c \in K \setminus \{0\}$, one has $\alpha(u) \in I$.

If $u = x_1^{a_1} x_2^{a_2} \cdots x_n^{a_n}$, then

$$\alpha(u) = \sum_{k=0}^{a_j} \binom{a_j}{k} c^k x_i^k (u/x_j^k).$$

Since I is strongly stable, we see that all monomials in the support of $\alpha(u)$ belong to I. □

There is another interesting characterization of strongly stable ideals. To describe it, we introduce the **Borel order**. This is the partial order on $\operatorname{Mon}(S_d)$ defined as follows: let $u, v \in \operatorname{Mon}(S_d)$, where $u = x_{i_1} x_{i_2} \cdots x_{i_d}$ with $i_1 \leq i_2 \leq \cdots \leq i_d$ and $v = x_{j_1} x_{j_2} \cdots x_{j_d}$ with $j_1 \leq j_2 \leq \cdots \leq j_d$. We say that $u \geq_{\text{Borel}} v$ if $i_k \leq j_k$ for $k = 1, \ldots, d$.

Lemma 4.2.5. (a) *Let $u, v \in \operatorname{Mon}(S_d)$ with $u >_{\text{Borel}} v$. Then*

(i) *there exist monomials $w_1, \ldots, w_r \in \operatorname{Mon}(S_d)$ with $w_1 = u$ and $w_r = v$, and such that for each k there exist integers $i > j$ such that x_j divides w_k and $w_{k+1} = x_i(w_k/x_j)$. In particular we have*

$$u = w_1 >_{\text{Borel}} w_2 >_{\text{Borel}} \cdots >_{\text{Borel}} w_r = v.$$

(ii) *$u > v$ with respect to any monomial order $>$ on S with $x_1 > x_2 > \cdots > x_n$.*

(b) *A monomial ideal $I \subset S$ is strongly stable, if and only if for all d, $\operatorname{Mon}(I_d)$ is an order filter in $\operatorname{Mon}(S_d)$ with respect to the Borel order. In other words, if $v \in \operatorname{Mon}(I_d)$ and $u \in \operatorname{Mon}(S_d)$ with $u >_{\text{Borel}} v$, then $u \in I_d$.*

Proof. (a)(i) Let $u = x_{i_1} x_{i_2} \cdots x_{i_d}$ with $i_1 \leq i_2 \leq \cdots \leq i_d$ and $v = x_{j_1} x_{j_2} \cdots x_{j_d}$ with $j_1 \leq j_2 \leq \cdots \leq j_d$. Since $u >_{\text{Borel}} v$, there exists an integer k such that $i_d = j_d, \ldots, i_{k+1} = j_{k+1}$ and $i_k < j_k$. It follows that $i_k + 1 \leq$

$j_k \leq j_{k+1} = i_{k+1}$, and we set $w_1 = u$ and $w_2 = x_{i_1} \ldots x_{i_{k-1}} x_{i_k+1} x_{i_{k+1}} \cdots x_{i_d}$. Then $w_2 = x_{i_k+1}(w_1/x_{i_k})$. If $w_2 = v$, we are done. Otherwise we apply the same argument to w_2 and obtain the next monomial w_3. In a finite number of steps we arrive at v and the construction of the sequence $w_1, w_2, \ldots, w_r$ is completed.

(a)(ii) By (i) it suffices to consider the case that $v = x_i(u/x_j)$ with $i > j$. Then $x_i u = x_j v > x_i v$, and so $u > v$.

(b) is an immediate consequence of (a)(i). $\square$

Now we have

Proposition 4.2.6. *Let $I \subset S$ be a graded ideal and $<$ a monomial order on S. Then the following holds:*

(a) $\text{gin}_<(I)$ *is strongly stable, if* $\text{char } K = 0$.
(b) **(Conca)** $\text{gin}_<(I) = I$ *if and only if I is Borel-fixed.*

Proof. (a) By Theorem 4.2.1 the ideal $\text{gin}_<(I)$ is Borel-fixed. Thus the statement follows from Proposition 4.2.4.

(b) One direction of the assertion is a consequence of Theorem 4.2.1. For the other direction we use the fact that a matrix α whose principal minors are all nonzero, can be written as a product $\beta\gamma$ where β is an invertible lower triangular matrix and γ an invertible upper triangular matrix. This is an open condition. Thus we may choose $\alpha \in \text{GL}_n(K)$ with $\text{gin}_<(I) = \text{in}_<(\alpha I)$ and which has a product presentation $\alpha = \beta\gamma$, as described above.

For the invertible lower triangular matrix β and any monomial u one has $\beta(u) = au + \cdots$ with $a \in K \setminus \{0\}$ and with $u >_{\text{Borel}} v$ for all $v \in \text{supp}(\beta(u))$ such that $v \neq u$. It follows therefore from Lemma 4.2.5(a)(ii) that for every homogeneous polynomial f, one has $\text{in}_<(\beta(f)) = \text{in}_<(f)$. This implies that $\text{in}_<(\beta I) = \text{in}_<(I)$.

Therefore,

$$\text{gin}_<(I) = \text{in}_<(\beta\gamma I) = \text{in}_<(\gamma I) = \text{in}_<(I) = I.$$

Here we used that $\gamma I = I$, since by assumption, I is Borel-fixed. The last equation holds, since I is a monomial ideal. $\square$

Theorem 4.2.1 together with Proposition 4.2.6(b) implies

Corollary 4.2.7. *Let $I \subset S$ be a graded ideal and $<$ a monomial order on S. Then $\text{gin}_<(\text{gin}_<(I)) = \text{gin}_<(I)$.*

Example 4.2.8. Proposition 4.2.6 is false in positive characteristic. For example, assume $\text{char } K = p > 0$, and consider the ideal $I = (x_1^p, x_2^p) \subset K[x_1, x_2]$. Let $\alpha \in \text{GL}_n(K)$ be any element, say, $\alpha(x_i) = a_{i1}x_1 + a_{i2}x_2$. Then

$$\alpha(x_i^p) = \alpha(x_i)^p = (a_{i1}x_1 + a_{i2}x_2)^p = a_{i1}^p x_1^p + a_{i2}^p x_2^p.$$

Since the matrix with entries a_{ij}^p is also nonsingular, we see that $\alpha I = I$. Hence I is Borel-fixed but not strongly stable.

We say that a monomial ideal $I \subset S = K[x_1, \ldots, x_n]$ is of **Borel type** if

$$I : x_i^\infty = I : (x_1, \ldots, x_i)^\infty \quad \text{for} \quad i = 1, \ldots, n.$$

Let u be a monomial. We denote by $\nu_i(u)$ be highest power of x_i which divides u. Borel type ideals can be characterized as follows:

Proposition 4.2.9. *Let $I \subset S$ be a monomial ideal. The following conditions are equivalent:*

(a) *I is of Borel type.*
(b) *For each monomial $u \in I$ and all integers i, j, s with $1 \leq j < i \leq n$ and $s > 0$ such that $x_i^s | u$ there exists an integer $t \geq 0$ such that $x_j^t(u/x_i^s) \in I$.*
(c) *For each monomial $u \in I$ and all integers i, j with $1 \leq j < i \leq n$ there exists an integer $t \geq 0$ such that $x_j^t(u/x_i^{\nu_i(u)}) \in I$.*
(d) *If $P \in \text{Ass}(S/I)$, then $P = (x_1, \ldots, x_j)$ for some j.*

Observe that condition (b) or (c) is satisfied for all monomials $u \in I$ if and only if it is satisfied for all $u \in G(I)$. Thus one has to check only finitely many conditions.

Proof (of Proposition 4.2.9). (a) $\Rightarrow$ (b): Let $u \in I$ be a monomial such that $x_i^s | u$ for some $s > 0$, and let $j < i$. Then $u = x_i^s v$ with $v \in I : x_i^\infty$. Condition (a) implies that $I : x_i^\infty \subset I : x_j^\infty$. Therefore, there exists an integer $t \geq 0$ such that $x_j^t(u/x_i^s) = x_j^t v \in I$.

(b) $\Rightarrow$ (a): We will show that $I : x_i^\infty \subset I : x_j^\infty$ for $j < i$. This will imply (a). Let $u \in I : x_i^\infty$ be a monomial. Then $x_i^s u \in I$ for some $s > 0$, and so (b) implies that $x_j^t u \in I$ for some t, that is, $u \in I : x_j^\infty$.

The implication (b) $\Rightarrow$ (c) is trivial. For the converse, let $u \in I$ be a monomial such that $x_i^s | u$ for some $s > 0$, and let $j < i$. By (c) there exists an integer $t \geq 0$ such that $x_j^t(u/x_i^{\nu_i(u)}) \in I$. Therefore, $x_j^t(u/x_i^s) = x_i^{\nu_i(u)-s} x_j^t(u/x_i^{\nu_i(u)}) \in I$.

(b) $\Rightarrow$ (d): Let $P \in \text{Ass}(S/I)$; then, according to Proposition 1.3.10, there exists a monomial v such that $P = I : v$. Notice that $v \notin I$, since $P \neq S$. By Corollary 1.3.9 the ideal P is generated by a subset of the variables. Let $x_i \in P$ and let $j < i$. We show that $x_j \in P$. Indeed, let $u = x_i v$. Then $u \in I$, and hence by condition (b) there exists an integer $t \geq 0$ such that $x_j^t v = x_j^t(u/x_i) \in I$. Therefore, $x_j^t \in I : v = P$. Since $v \notin I$ it follows that $t > 0$ and since P is a prime ideal we conclude then that $x_j \in P$.

(d) $\Rightarrow$ (b): Consider the (unique) standard primary decomposition of $I = \bigcap_{i=1}^m Q_i$, cf. Section 1.3. Then each Q_i is of the form $(x_{i_1}^{a_1}, \ldots, x_{i_j}^{a_j})$ and $P_i = \sqrt{P_i} = (x_{i_1}, \ldots, x_{i_j})$ is an associated prime ideal of S/I. Thus (d) implies that each Q_i is of the form $(x_1^{a_1}, \ldots, x_j^{a_j})$. Write $I = J \cap Q_m$. Obviously, Q_m satisfies condition (b), and proceeding by induction on m, we may as well assume that J satisfies condition (b). Thus if $u \in I$ and $x_i^s | u$, then for any $j < i$ there exists t_1 and t_2 such that $x^{t_1}(u/x_i^s) \in J$ and $x^{t_2}(u/x_i^s) \in Q_m$. It follows that $x^t(u/x_i^s) \in I$ for $t = \max\{t_1, t_2\}$. $\square$

Ideals of Borel type include strongly stable ideals. But they are also important because of the following result.

Theorem 4.2.10 (Bayer–Stillman). *Borel-fixed ideals are of Borel type.*

Proof. We know from Proposition 4.2.4(a) that I is a monomial ideal. We will show that I satisfies condition (c) of Proposition 4.2.9. Let $u \in I$ with $a = \nu_i(u)$, and let $1 \leq j < i$. We want to find an integer t such that $x_j^t(u/x_i^a) \in I$. If $a = 0$, there is nothing to show. Suppose now that $a > 0$. Since I is Borel-fixed, the polynomial $\sum_{k=0}^a \binom{a}{k} x_j^k(u/x_i^k)$ belongs to I (cf. the proof of Proposition 4.2.4(c)). Thus, since I is a monomial ideal, it follows that $x_j^k(u/x_i^k) \in I$ for all k with $\binom{a}{k} \neq 0$ in K. Hence if char $K = 0$, then $x_j^k(u/x_i^k) \in I$ for all $k = 0, \ldots, a$. Now assume that char $K = p > 0$, and let $a = \sum_i a_i p^i$ be the p-adic expansion of a. Let j be an index such that $a_j \neq 0$, and let $k = p^j$. Then $\binom{a}{k} = a_j \neq 0$ in K. This follows from the following identity

$$\binom{a}{k} = \prod_i \binom{a_i}{k_i} \bmod p,$$

of Lucas, where $k = \sum_i k_i p^i$ is the p-adic expansion of k.

Therefore in all cases there exists an integer k with $1 \leq k \leq a$ such that $x_j^k(u/x_i^k) \in I$. Set $a' = a - k$ and $u' = x_j^k(u/x_i^k)$. Then $\nu_i(u') = a' < a$. Arguing by induction on a, we may assume that there exists an integer t' such that $x_j^{t'}(u'/x_i^{a'}) \in I$. Thus if set $t = t' + k$, then $x_j^t(u/x_i^a) \in I$, as desired. $\square$

4.3 Extremal Betti numbers

We introduce the generic annihilator numbers of a graded K-algebra. These numbers are closely related to Betti numbers. We will use this approach to prove Theorem 4.3.17 on extremal Betti numbers by Bayer, Charalambous and Popescu. From this we will deduce the classical results on generic initial ideals by Bayer and Stillman.

4.3.1 Almost regular sequences and generic annihilator numbers

Let M be a finitely generated graded S-module, where $S = K[x_1, \ldots, x_n]$ is the polynomial ring and K is an infinite field (which is the standard assumption in this chapter).

Let $y \in S_1$ be a linear form. Then multiplication with y induces the homogeneous homomorphism $M(-1) \to M$, $m \mapsto ym$. Let

$$0 :_M y = \{m \in M : ym = 0\}.$$

be the submodule of M consisting of all elements of M which are annihilated by y. Note that $0:_M y$ is a graded submodule of M, and that

$$\mathrm{Ker}(M(-1) \to M) = (0 :_M y)(-1).$$

If y is a nonzero divisor on M, then $0 :_M y = 0$. Instead if we require that $0 :_M y$ is a module of finite length, then only finitely many graded components of $0 :_M y$ are nonzero. This is equivalent to saying that the multiplication map $y: M_{i-1} \to M_i$ is injective for all $i \gg 0$. We call an element $y \in S_1$ with this property an **almost regular** element on M.

For any finitely generated graded S-module there exists an almost regular element. Indeed we have

Lemma 4.3.1. *Let M be a finitely generated graded S-module. Then the set*

$$U = \{y \in S_1 \colon y \text{ is almost regular on } M\}$$

is a nonempty Zariski open subset of S_1.

Proof. Let N be the graded submodule of M consisting of those elements of M which are annihilated by some power of $\mathfrak{m} = (x_1, \ldots, x_n)$. (Note that N is just the 0th local cohomology module of M. But we will not use this fact.) Obviously, $N = \bigcup_k (0:_M \mathfrak{m}^k)$. Since $0:_M \mathfrak{m} \subset 0:_M \mathfrak{m}^2 \subset \cdots$ is an ascending sequence of submodules of M and since M is Noetherian, there exists a number k_0 such that $0:_M \mathfrak{m}^{k_0} = 0:_M \mathfrak{m}^{k_0+1} = \cdots$. In other words, we have $N = 0:_M \mathfrak{m}^{k_0}$. In particular, $\mathfrak{m}^{k_0} N = 0$, so that N has finite length.

Let $P_1, \ldots, P_r$ be the associated prime ideals of M/N. Observe that $\mathfrak{m} \notin \mathrm{Ass}(M/N)$, because otherwise there would exist an element $0 \neq m+N \in M/N$ with $\mathfrak{m}(m + N) = 0$. This would imply that $\mathfrak{m}m \in N$. But then we would have that $\mathfrak{m}^{k_0+1}m = 0$, and hence $m \in N$, a contradiction.

Now since all $P_i \neq \mathfrak{m}$, we have that $P_i \cap S_1$ is a proper linear subspace of S_1. Since K is an infinite field, S_1 cannot be the union of finitely many proper linear subspaces. Indeed, if the union of the linear subspaces would be S_1, then the intersection of the complements of these linear subspaces in S_1 would be the empty set, contradicting Lemma 4.1.1 since these complements are nonempty Zariski open subsets of S_1.

It follows that the set $U = S_1 \setminus \bigcup_{i=1}^r P_i$ is nonempty. It is also a Zariski open subset of S_1, since a finite union of linear subspaces of S_1 is a Zariski closed subset of S_1.

Let $y \in U$. Then y is a nonzero divisor on M/N since y belongs to no associated prime ideal of M/N.

Choose any element $m \in 0:_M y$. Then $y(m + N) = 0$, and so $m + N = 0$, since y is a nonzero divisor on M/N. Hence we have shown that $0:_M y \subset N$. In particular, $0:_M y$ has finite length and y is almost regular on M.   □

We call a sequence $\mathbf{y} = y_1, \ldots, y_r$ with $y_i \in S_1$ an **almost regular** sequence on M, if y_i is an almost regular element on $M/(y_1, \ldots, y_{i-1})M$ for all $i = 1, \ldots, r$.

As an immediate consequence of Lemma 4.3.1 we obtain

Corollary 4.3.2. *Let M be a finitely generated graded S-module. Then there exists K-basis of S_1 which is an almost regular sequence on M.*

An explicit example of an almost regular sequence is given in

Proposition 4.3.3. *Let $I \subset S$ be a monomial ideal of Borel type. Then $x_n, x_{n-1}, \ldots, x_1$ is an almost regular sequence on S/I.*

Proof. By using an induction argument it suffices to show that x_n is almost regular. But this is obvious since by Proposition 4.2.9(d) the element x_n does not belong to any associated prime ideal of S/I which is different from $\mathfrak{m} = (x_1, \ldots, x_n)$. $\square$

We denote by $A_{i-1}(\mathbf{y}; M)$ the graded module $0 :_{M/(y_1,\ldots,y_{i-1})M} y_i$ and call the numbers

$$\alpha_{ij}(\mathbf{y}; M) = \begin{cases} \dim_K A_i(\mathbf{y}; M)_j, & \text{if } i < n, \\ \beta_{0j}(M), & \text{if } i = n. \end{cases}$$

the **annihilator numbers** of M with respect to the sequence $\mathbf{y}$. For each i one has $\alpha_{ij}(\mathbf{y}; M) = 0$ for almost all j, in case $\mathbf{y}$ is an almost regular sequence.

We have the following vanishing and non-vanishing property for the annihilator numbers.

Proposition 4.3.4. *Let $\mathbf{y}$ a K-basis of S_1 which is almost regular on M and set $\alpha_i(\mathbf{y}; M) = \sum_j \alpha_{ij}(\mathbf{y}; M)$ for all i. Then $\alpha_i(\mathbf{y}; M) = 0$ if and only $i < \operatorname{depth} M$.*

Proof. Set $\alpha_i = \alpha_i(\mathbf{y}; M)$ for all i. Then $\alpha_{i-1} = \dim_K 0 :_{M/(y_1,\ldots,y_{i-1})M} y_i$ for $i = 1, \ldots, n$ and $\alpha_n = \beta_0(M)$.

Assume first that $\operatorname{depth} M > 0$. Since $0 :_M y_1$ is a submodule of M of finite length, and since $\operatorname{depth} M > 0$, we see that $\alpha_0 = 0$ and that y_1 is a nonzero divisor on M. Therefore $\operatorname{depth} M/y_1 M = \operatorname{depth} M - 1$. Arguing by induction on the depth of M we may assume $\alpha_i(y_2, \ldots, y_n; M/y_1 M) = 0$ for $i < \operatorname{depth} M/y_1 M = \operatorname{depth} M - 1$. Since $\alpha_i = \alpha_{i-1}(y_2, \ldots, y_n; M/y_1 M)$ for all i, we see that $\alpha_i = 0$ for $i < \operatorname{depth} M$.

Next we want to show that $\alpha_i \neq 0$ for $i \geq t = \operatorname{depth} M$. Replacing M by $M/(y_1, \ldots, y_t)M$, we may assume that $\operatorname{depth} M = 0$, and show (i) $\alpha_0 \neq 0$ and (ii) $\operatorname{depth} M/y_1 M = 0$. The desired assertion follows then from (i) and (ii) and by induction on n.

(i) follows from the fact that $0 :_M \mathfrak{m} \subset 0 :_M y_1$ and that $0 :_M \mathfrak{m} \neq 0$ since $\operatorname{depth} M > 0$.

For the proof of (ii) assume to the contrary that $\operatorname{depth} M/y_1 M > 0$. For each integer $i > 0$ consider the map $\varphi_i : y_1^{i-1} M/y_1^i M \to y_1^i M/y_1^{i+1} M$ induced by multiplication with y_1. Clearly φ_i is surjective. Let $a + y_1^i M \in \operatorname{Ker} \varphi_i$; then $y_1 a \in y_1^{i+1} M$. Hence there exists $b \in y_1^i M$ such that $y_1 a = y_1 b$. Let $c = a - b$; then $c \in (0 :_M y_1)$ and $a + y_1^i M = c + y_1^i M$. This shows that $\operatorname{Ker} \varphi_i$ is a module of finite length for all i.

We will show by induction on i that φ_i is indeed an isomorphism for all i. To this end it suffices to show that $\operatorname{Ker}\varphi_i = 0$ for all i. For $i = 1$, we have $\operatorname{Ker}\varphi_1 \subset M/y_1 M$. Since $\operatorname{Ker}\varphi_1$ has finite length and since $\operatorname{depth} M/y_1 M > 0$ by assumption, it follows that $\operatorname{Ker}\varphi_1 = 0$. Assume now that φ_j is an isomorphism for all $j < i$. Then $y_1^{i-1}M/y_1^i M \cong M/y_1 M$, so that $\operatorname{depth} y_1^{i-1}M/y_1^i M > 0$. Since $\operatorname{Ker}\varphi_i \subset y_1^{i-1}M/y_1^i M$ and $\operatorname{Ker}\varphi_i$ has finite length, it follows that $\operatorname{Ker}\varphi_i = 0$.

The proof of (i) shows that y_2 is a nonzero divisor on $M/y_1 M$ and since $y_1^{i-1}M/y_1^i M \cong M/y_1 M$ for all i, it follows that y_2 is a nonzero divisor on $y_1^{i-1}M/y_1^i M$ for all i, which by virtue of the exact sequence

$$0 \to y_1^i M \longrightarrow y_1^{i-1}M \to y_1^{i-1}M/y_1^i M \longrightarrow 0$$

yields that $(0 :_M y_2) \cap y_1^{i-1}M \subset y_1^i M$ for all i. This implies that $(0 :_M y_2) \subset \bigcap_{i \geq 0} y_1^i M = 0$. Hence y_2 is a nonzero divisor on M. This is a contradiction, since $\operatorname{depth} M = 0$ by assumption. $\qquad\square$

Next we study Koszul homology of almost regular sequences. For some basic facts on Koszul homology we refer the reader to Appendix A.3.

A sequence $\mathbf{y} \subset S_1$ is a regular sequence on M if and only if $H_i(\mathbf{y}; M) = 0$ for all $i > 0$, see Theorem A.3.4. Thus the following result is not surprising.

Proposition 4.3.5. *Let M be a finitely generated graded S-module, and $\mathbf{y} = y_1, \ldots, y_r$ a sequence of elements in S_1. The following conditions are equivalent:*

(a) $\mathbf{y}$ *is an almost regular sequence on M.*
(b) $H_j(y_1, \ldots, y_i; M)$ *has finite length for all $j > 0$ and all $i = 1, \ldots, r$.*
(c) $H_1(y_1, \ldots, y_i; M)$ *has finite length for all $i = 1, \ldots, r$.*

Proof. (a) $\Rightarrow$ (b): We prove the assertion by induction on i. We have $H_j(y_1; M) = 0$ for $j > 1$ and $H_1(y_1; M) \cong 0 :_M y_1$. This module is of finite length by assumption.

Now let $i > 1$. Then there is the long exact sequence of graded Koszul homology

$$\to H_j(y_1, \ldots, y_{i-1}; M) \to H_{j+1}(y_1, \ldots, y_i; M) \to H_j(y_1, \ldots, y_{i-1}; M)(-1) \to$$
$$\cdots \to H_1(y_1, \ldots, y_{i-1}; M) \to H_1(y_1, \ldots, y_i; M) \to A_{i-1}(\mathbf{y}; M)(-1) \to 0.$$

Since $A_{i-1}(\mathbf{y}; M)$ has finite length and since by induction hypothesis the modules $H_j(y_1, \ldots, y_{i-1}; M)$ have finite length for all $j > 0$, the exact sequence implies that also $H_j(y_1, \ldots, y_i; M)$ has finite length for all $j > 0$.

(b) $\Rightarrow$ (c) is trivial.

(c) $\Rightarrow$ (a): the beginning of the above exact sequence

$$H_1(y_1, \ldots, y_{i-1}; M) \to H_1(y_1, \ldots, y_i; M) \to A_{i-1}(\mathbf{y}; M)(-1) \to 0$$

shows that all the annihilators $A_{i-1}(\mathbf{y}; M)$ have finite length. $\qquad\square$

The length of the annihilators $A_{i-1}(\mathbf{y}; M)$ may depend on the almost regular sequence $\mathbf{y}$. However, we have

Theorem 4.3.6. *Let $I \subset S$ be a graded ideal. With each $\gamma = (g_{ij}) \in \mathrm{GL}_n(K)$ we associate the sequence $\mathbf{y} = \gamma(\mathbf{x})$ with $y_j = \sum_{i=1}^{n} g_{ij} x_i$ for $j = 1, \ldots, n$. Then there exists a nonempty Zariski open subset $U \subset \mathrm{GL}_n(K)$ such that $\gamma(\mathbf{x})$ is almost regular for all $\gamma \in U$. Moreover, the open set U has the property that $\dim_K A_{i-1}(\gamma(\mathbf{x}); S/I)_j = \dim_K A_{i-1}(x_n, x_{n-1}, \ldots, x_1; S/\mathrm{gin}_{<_{\mathrm{rev}}}(I))_j$ for all i and j and all $\gamma \in U$.*

The following lemma is crucial for the proof of Theorem 4.3.6, and it is the step in the chain of arguments where the reverse lexicographic order is indispensable.

Lemma 4.3.7. *Let $I \subset S$ be a graded ideal. Then for all i one has*

$$\mathrm{in}_{<_{\mathrm{rev}}}(I, x_{i+1}, \ldots, x_n) = (\mathrm{in}_{<_{\mathrm{rev}}}(I), x_{i+1}, \ldots, x_n), \quad \text{and}$$

$$\mathrm{in}_{<_{\mathrm{rev}}}((I, x_{i+1}, \ldots, x_n) :_S x_i) = (\mathrm{in}_{<_{\mathrm{rev}}}(I), x_{i+1}, \ldots, x_n) :_S x_i.$$

Proof. We only prove the statements for the colon ideals. That the left-hand side is contained in the right-hand side is easy to see and true not only for the reverse lexicographic order but for any other monomial order as well.

For the converse inclusion it suffices to show that each monomial u in $(\mathrm{in}_{<_{\mathrm{rev}}}(I), x_{i+1}, \ldots, x_n) :_S x_i$ belongs to $\mathrm{in}_{<_{\mathrm{rev}}}((I, x_{i+1}, \ldots, x_n) :_S x_i)$. We may assume that no x_j with $j > i$ divides u. Then there exists a homogeneous element $f \in I$ with $ux_i = \mathrm{in}_{<_{\mathrm{rev}}}(f)$. Because we use the reverse lexicographic order it follows that $f = cux_i + h$ with $h \in (x_i, \ldots, x_n)$ and $c \in K \setminus \{0\}$. Write $h = g_i x_i + \cdots + g_n x_n$, and set $f_1 = cu + g_i$. Then $f_1 x_i \in (I, x_{i+1}, \ldots, x_n)$ and $\mathrm{in}_{<_{\mathrm{rev}}}(f_1) = u$. This shows that $u \in \mathrm{in}_{<_{\mathrm{rev}}}((I, x_{i+1}, \ldots, x_n) :_S x_i)$. $\square$

Before giving the proof of Theorem 4.3.6 we need the following technical result.

Lemma 4.3.8. *Let $U \subset \mathrm{GL}(n; K)$ be a Zariski open subset, and $\sigma \in \mathrm{GL}(n; K)$. We set $U^{-1} = \{\varphi^{-1} : \varphi \in U\}$ and $U\sigma = \{\varphi\sigma : \varphi \in U\}$. Then U^{-1} and $U\sigma$ are Zariski open sets in $\mathrm{GL}(n; K)$.*

Proof. Let $\xi = (x_{ij})$ be an $n \times n$ matrix of indeterminates. We write $U = \mathrm{GL}_n(K) \setminus A$ where A is the common set of zeroes of the polynomials $f_1(\xi), \ldots, f_m(\xi)$ in the variables x_{ij}. Let $D(f_i) = \{\varphi \in \mathrm{GL}_n(K) : f_i(\varphi) \neq 0\}$. Then U is the union of the Zariski open sets $D(f_i)$. We let $g_i(\xi) = f_i(\xi\sigma^{-1})$; then $g_i(\varphi\sigma) = f_i(\varphi)$ for all $\varphi \in \mathrm{GL}_n(K)$. Therefore, $\varphi\sigma \in D(g_i)$ if and only if $\varphi \in D(f_i)$, and so $D(g_i) = D(f_i)\sigma$. Thus $U\sigma = \bigcup_{i=1}^{m} D(f_i)\sigma = \bigcup_{i=1}^{m} D(g_i)$ is Zariski open.

Since $U = \bigcup_{i=1}^{m} D(f_i)$, it follows that $U^{-1} = \bigcup_{i}^{m} D(f_i)^{-1}$. It suffices therefore to show that each $D(f_i)^{-1}$ is Zariski open. Let $\eta = ((-1)^{i+j}\delta_{ji}/\delta)$, where $\delta = \det(\xi)$ and δ_{ji} is the minor of ξ which is obtained by skipping the jth column and ith row. Suppose $d = \deg f_i$. Then $g_i(\xi) = \delta^d f_i(\eta)$ is a polynomial in the variables x_{ij}, and for each $\varphi \in \mathrm{GL}_n(K)$ we have $g_i(\varphi^{-1}) = cf_i(\varphi)$ with $c = \det(\varphi)^{-d} \neq 0$. Therefore, $\varphi^{-1} \in D(g_i)$ if and only if $\varphi \in D(f_i)$. In other words, $D(f_i)^{-1} = D(g_i)$ is Zariski open. $\qquad \square$

Proof (of Theorem 4.3.6). By Theorem 4.1.2 there exists a nonempty Zariski open set $U' \subset \mathrm{GL}_n(K)$ such that $\mathrm{gin}_{<_{\mathrm{rev}}}(I) = \mathrm{in}_{<_{\mathrm{rev}}}(\varphi I)$ for all $\varphi \in U'$. Let $\sigma \in \mathrm{GL}_n(K)$ be the automorphism with $\sigma(x_i) = x_{n-i+1}$ for $i = 1, \ldots, n$. The set $U = \{\varphi^{-1}\sigma \colon \varphi \in U'\} \subset \mathrm{GL}_n(K)$ is a nonempty Zariski open subset of $\mathrm{GL}_n(K)$, see Lemma 4.3.8.

We claim that U satisfies the conditions of the theorem. To see this we first show that

$$\dim_K A_{i-1}(\gamma(\mathbf{x}); S/I)_j = \dim_K A_{i-1}(x_n, x_{n-1}, \ldots, x_1; S/\mathrm{gin}_{<_{\mathrm{rev}}}(I))_j$$

for all i and j and all $\gamma \in U$. In other words, we show that $A_{i-1}(\gamma(\mathbf{x}); S/I)$ and $A_{i-1}(\sigma(\mathbf{x}); S/\mathrm{gin}_{<_{\mathrm{rev}}}(I))$ have the same Hilbert function for all $\gamma \in U$. (See Chapter 6 for the definition and basic properties of Hilbert functions.) The equality of the Hilbert functions then also implies that $\gamma(\mathbf{x})$ is almost regular for all $\gamma \in U$, because by Theorem 4.2.10 and Proposition 4.3.3, $\sigma(\mathbf{x})$ is almost regular on $S/\mathrm{gin}_{<_{\mathrm{rev}}}(I)$ from which it follows that $A_{i-1}(\sigma(\mathbf{x}); S/\mathrm{gin}_{<_{\mathrm{rev}}}(I))$ and hence also $A_{i-1}(\gamma(\mathbf{x}); S/I)$ has finite length for all i.

In order to prove the equality of the Hilbert functions, let $\gamma = \varphi^{-1}\sigma \in U$ and set $y_i = \gamma(x_i)$ for $i = 1, \ldots, n$. Then $\varphi(y_i) = x_{n-i+1}$ for $i = 1, \ldots, n$, and hence

$$S/(I, y_1, \ldots, y_i) \cong S/(\varphi(I), x_n, x_{n-1}, \ldots, x_{n-i+1}) \quad \text{for all} \quad i.$$

In particular, $S/(I, y_1, \ldots, y_i)$ and $S/(\varphi(I), x_n, x_{n-1}, \ldots, x_{n-i+1})$ have the same Hilbert function for all i. By Corollary 6.1.5 the Hilbert functions of $S/(\varphi(I), x_n, x_{n-1}, \ldots, x_{n-i+1})$ and $S/\mathrm{in}_{<_{\mathrm{rev}}}(\varphi(I), x_n, x_{n-1}, \ldots, x_{n-i+1})$ coincide as well. Thus, since $\varphi \in U'$, we conclude from Lemma 4.3.7 that $S/(I, y_1, \ldots, y_i)$ and $S/(\mathrm{gin}_{<_{\mathrm{rev}}}(I), x_n, x_{n-1}, \ldots, x_{n-i+1})$ have the same Hilbert function for all i.

Since the Hilbert function is additive on short exact sequences, the exact sequence

$$0 \to A_{i-1}(\gamma(\mathbf{x}); S/I) \to S/(I, y_1, \ldots, y_{i-1})(-1)$$
$$\to S/(I, y_1, \ldots, y_{i-1}) \to S/(I, y_1, \ldots, y_i) \to 0,$$

implies that the Hilbert function $A_{i-1}(\gamma(\mathbf{x}); S/I)$ is determined by the Hilbert function of $S/(I, y_1, \ldots, y_{i-1})$ and that of $S/(I, y_1, \ldots, y_i)$. Correspondingly the Hilbert function of $A_{i-1}(\sigma(\mathbf{x}); S/\mathrm{gin}_{<_{\mathrm{rev}}}(I))$ is determined by those of

$$S/(\mathrm{gin}_{<_{\mathrm{rev}}}(I), x_n, x_{n-1}, \ldots, x_{n-j+1}) \quad \text{for} \quad j = i-1, i.$$

Hence our above considerations imply that

$$A_{i-1}(\gamma(\mathbf{x}); S/I) \quad \text{and} \quad A_{i-1}(\sigma(\mathbf{x}); S/\gin_{<_{\mathrm{rev}}}(I))$$

have the same Hilbert function, as desired.   □

Definition 4.3.9. A sequence $\mathbf{y} = \gamma(\mathbf{x})$ with $\gamma \in U$ and $U \subset \mathrm{GL}_n(K)$ as in Theorem 4.3.6 is called a **generic sequence** on S/I.

Since the annihilator number does not depend on the particular chosen generic sequence we set $\alpha_{ij}(S/I) = \alpha_{ij}(\mathbf{y}; S/I)$ for all i and j, where $\mathbf{y}$ is a generic sequence on S/I. The numbers $\alpha_{ij}(S/I)$ are called the **generic annihilator numbers** of S/I.

Remark 4.3.10. Let M be a finitely generated graded S-module. With the same arguments as used in the proof of Theorem 4.3.6 one can show that there exists a nonempty Zariski open subset $U \subset \mathrm{GL}(K)$ such that $\gamma(\mathbf{x})$ is almost regular on M for all $\gamma \in U$ and such that $\alpha_{ij}(\gamma(\mathbf{x}); M)$ is independent of $\gamma \in U$, and hence will be denoted by $\alpha_{ij}(M)$. Thus generic sequences and generic annihilator numbers can also be defined for modules. For this one just has to extend the concept of generic initial ideals to generic initial submodules. This can be done in an obvious way. We omit the details, since in this monograph we do not consider Gröbner basis theory for modules.

Of course we cannot expect that each K-basis of S_1 which is almost regular on a graded S-module M is also generic for this module. Indeed, if $I \subset S$ is a graded ideal such that S/I has Krull dimension 0, then any K-basis of S_1 is an almost regular sequence on S/I, since $\ell(S/I) < \infty$. On the other hand, not all K-bases of S_1 are generic sequences. For instance, if we choose the ideal $I = (x_1^4, x_1^3 x_2^2, x_2^3 + x_3^3, x_3^4)$ in $S = K[x_1, x_2, x_3]$. Then $\ell((I:_S x_1)/I) = \ell((I:_S x_2)/I) = 12$ and $\ell((I:_S x_3)/I) = 11$, while $\ell((\gin_{<_{\mathrm{rev}}}(I):_S x_3)/\gin_{<_{\mathrm{rev}}}(I)) = 10$, as can be easily checked with CoCoA. This shows that x_1, x_2, x_3 and none of its permutations are generic on S/I.

The following statement is almost tautological but of crucial importance.

Proposition 4.3.11. *Let $I \subset S$ be a graded ideal. Then*

$$\alpha_{ij}(S/I) = \alpha_{ij}(S/\gin_{<_{\mathrm{rev}}}(I)).$$

Proof. Let $i < n$. According to the definition of the generic annihilator numbers we have $\alpha_{ij}(S/I) = \dim_K A_i(x_n, x_{n-1}, \ldots, x_1; S/\gin_{<_{\mathrm{rev}}}(I))$, and

$$\alpha_{ij}(S/\gin_{<_{\mathrm{rev}}}(I)) = \dim_K A_i(x_n, x_{n-1}, \ldots, x_1; S/\gin_{<_{\mathrm{rev}}}(\gin_{<_{\mathrm{rev}}}(I)))$$
$$= \dim_K A_i(x_n, x_{n-1}, \ldots, x_1; S/\gin_{<_{\mathrm{rev}}}(I)).$$

The last equation follows from Corollary 4.2.7. Hence the conclusion.   □

4.3.2 Annihilator numbers and Betti numbers

Annihilator numbers of almost regular sequences and Betti numbers are intimately related to each other. The purpose of this subsection is to describe this relationship.

We shall use the convention that

$$
\binom{i}{-1} = \begin{cases} 0, & \text{if } i \neq -1, \\ 1, & \text{if } i = -1 \end{cases}
$$

The first basic fact is given in

Proposition 4.3.12. *Let M be a graded S-module and let $\mathbf{y} = y_1, \ldots, y_n$ be a K-basis of S_1 which is almost regular on M. Then*

$$
\beta_{i,i+j}(M) \leq \sum_{k=0}^{n-i} \binom{n-k-1}{i-1} \alpha_{kj}(\mathbf{y}; M) \quad \text{for all} \quad i \geq 0 \quad \text{and all} \quad j.
$$

Proof. After a suitable change of coordinates we may assume that $x_1, x_2, \ldots, x_n$ is almost regular on M.

To simplify notation we set $H_i(r)_j = H_i(x_1, \ldots, x_r; M)_j$ and $h_{ij}(r) = \dim_K H_i(r)_j$. Furthermore we set $A(r)_j = A_r(\mathbf{x}; M)_j$, $\alpha_{rj} = \dim_K A(r)_j$ and $\beta_{ij} = \beta_{ij}(M)$.

For $i = 0$ the assertion is trivially true. Now let $i > 0$. We will show by induction on i that

$$
h_{i,i+j}(r) \leq \sum_{k=0}^{r-i} \binom{r-k-1}{i-1} \alpha_{kj} \quad \text{for all} \quad r. \tag{4.1}
$$

Since $\beta_{i,i+j} = h_{i,i+j}(n)$, this will then prove the proposition.

For $i = 1$, the claim is that $h_{1,1+j}(r) \leq \sum_{k=0}^{r-1} \alpha_{kj}$. However, this follows easily by induction on r from the exact sequences

$$
H_1(r-1)_{1+j} \longrightarrow H_1(r)_{1+j} \longrightarrow A(r-1)_j \longrightarrow 0
$$

which yield the inequalities $h_{1,1+j}(r) \leq h_{1,1+j}(r-1) + \alpha_{r-1,j}$.

Now let $i > 1$. We proceed by induction on r. If $r < i$, then $h_{i,i+j}(r) = 0$, and the assertion is trivial. So now let $r \geq i$. Then the exact sequence

$$
H_i(r-1)_{i+j} \longrightarrow H_i(r)_{i+j} \longrightarrow H_{i-1}(r-1)_{i+j-1}
$$

and the induction hypothesis imply

$$
h_{i,i+j}(r) \leq h_{i,i+j}(r-1) + h_{i-1,i+j-1}(r-1)
$$

$$
\leq \sum_{k=0}^{r-i-1} \binom{r-2-k}{i-1} \alpha_{kj} + \sum_{k=0}^{r-i} \binom{r-2-k}{i-2} \alpha_{kj}
$$

$$= \sum_{k=0}^{r-i-1} [\binom{r-2-k}{i-1} + \binom{r-2-k}{i-2}]\alpha_{kj} + \alpha_{r-i,k}$$

$$= \sum_{k=0}^{r-i} \binom{r-k-1}{i-1}\alpha_{kj}.$$

$\square$

Chapter 7 discusses the conditions under which the inequalities in Proposition 4.3.12 become equalities.

Definition 4.3.13. Let M be a finitely generated graded S-module and let $\mathbf{y}$ be a K-basis of S_1 which is almost regular on M. Let α_{ij} be the annihilator numbers of M with respect to $\mathbf{y}$ and β_{ij} be the graded Betti numbers of M.

(a) An annihilator number $\alpha_{ij} \neq 0$ is called **extremal** if $\alpha_{k\ell} = 0$ for all pairs $(k,\ell) \neq (i,j)$ with $k \leq i$ and $\ell \geq j$.
(b) A Betti number $\beta_{i,i+j} \neq 0$ is called **extremal** if $\beta_{k,k+\ell} = 0$ for all pairs $(k,\ell) \neq (i,j)$ with $k \geq i$ and $\ell \geq j$.

Remark 4.3.14. We define a partial order on the set of pairs of integers by setting $(i,j) \leq (k,l)$ if $i \leq k$ and $j \leq l$. Let $\beta_{i_1,i_1+j_1}(M), \ldots, \beta_{i_r,i_r+j_r}(M)$ be the extremal Betti numbers of M. Then $\beta_{i,i+j}(M) = 0$ for all (i,j) such that $(i,j) \not\leq (i_k,j_k)$ for $k = 1, \ldots, r$. In particular, if $m = \max\{i_k + j_k : k = 1, \ldots, r\}$, then $\beta_{i,m}(M)$ is an extremal Betti number for all i such that $\beta_{i,m}(M) \neq 0$.

Figure 4.1 displays the α-diagram and Betti diagram of a finitely generated graded S-module. The entry with coordinates (i,j) in the α-diagram is the generic annihilator number α_{ij}. The outside corners of the dashed line give the positions of the extremal annihilator numbers.

Similarly, the entry with coordinates (i,j) in the Betti diagram is the graded Betti number $\beta_{i,i+j}$. Again the outside corners of the dashed line give the positions of the extremal Betti numbers.

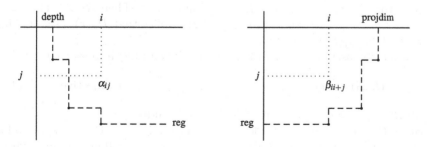

Fig. 4.1. An α-diagram and a Betti diagram

The following result is of central importance.

Theorem 4.3.15. *Let M be a graded S-module and let $\mathbf{y}$ be a K-basis of S_1 which is almost regular on M. Let α_{ij} be the annihilator numbers of M with respect to $\mathbf{y}$ and β_{ij} be the graded Betti numbers of M. Then $\beta_{i,i+j}$ is an extremal Betti number of M if and only if $\alpha_{n-i,j}$ is an extremal generic annihilator number of M. Moreover, if the equivalent conditions hold, then*

$$\beta_{i,i+j} = \alpha_{n-i,j}.$$

Proof. We adopt the simplified notation introduced in the proof of Proposition 4.3.12, and first treat the case $i \geq 1$.

Let $r \in [n]$. Just as for Betti numbers we say that $h_{i,i+j}(r)$ is extremal if $h_{i,i+j}(r) \neq 0$ and $h_{k,k+\ell}(r) = 0$ for all $(k, \ell) \neq (i, j)$ with $k \geq i$ and $\ell \geq j$.

The proof of the theorem for $i \geq 1$ is based on the following two observations:

(i) $h_{i,i+j}(r)$ is extremal for $r \geq 2$ if and only if $h_{i-1,i-1+j}(r-1)$ is extremal, and if the equivalent conditions hold, then $h_{i,i+j}(r) = h_{i-1,i-1+j}(r-1)$.

(ii) $h_{1,1+j}(r)$ is extremal if and only if $\alpha_{r-1,j}$ is extremal, and if the equivalent conditions hold, then $h_{1,1+j}(r) = \alpha_{r-1,j}$.

Assuming (i) and (ii) and recalling that $\beta_{i,i+j} = h_{i,i+j}(n)$ it follows for $i \geq 1$ that $\beta_{i,i+j}$ is extremal if and only if $h_{1,1+j}(n-i+1)$ is extremal which in turn is equivalent to the condition that $\alpha_{n-i,j}$ is extremal. Furthermore, (i) and (ii) imply that in this case $\beta_{i,i+j} = \alpha_{n-i,j}$, as desired.

Proof of (i): the long exact sequence of Koszul homology

$$\cdots \to H_k(r-1) \to H_k(r) \to H_{k-1}(r-1)(-1) \to H_{k-1}(r-1) \to$$

induces for each k and ℓ the exact sequence of vector spaces

$$\cdots \to H_k(r-1)_{k+\ell} \to H_k(r)_{k+\ell} \to \qquad\qquad (4.2)$$
$$H_{k-1}(r-1)_{k-1+\ell} \to H_{k-1}(r-1)_{(k-1)+\ell+1}$$

Suppose that $h_{i-1,i-1+j}(r-1)$ is extremal. Then $H_{i-1}(r-1)_{(i-1)+j+1} = 0$ and $H_{i-1}(r-1)_{(i-1)+j} \neq 0$. Therefore (4.2) implies that $H_i(r)_{i+j} \neq 0$. Next suppose that $(k, \ell) \neq (i, j)$ and that $k \geq i$ and $\ell \geq j$. Then $H_k(r-1)_{k+\ell} = 0$ and $H_{k-1}(r-1)_{k-1+\ell} = 0$, and hence (4.2) implies that $H_k(r)_{k+\ell} = 0$. This shows that $h_{i,i+j}(r)$ is extremal.

Conversely, assume $h_{i,i+j}(r)$ is extremal. Then (4.2) implies that

$$H_{k-1}(r-1)_{k-1+\ell} \to H_{k-1}(r-1)_{(k-1)+\ell+1} \quad \text{is injective} \qquad (4.3)$$

for all $(k-1, \ell) \neq (i-1, j)$ with $k-1 \geq i-1$ and $\ell \geq j$.

Since $H_{k-1}(r-1)$ has finite length, we have $H_{k-1}(r-1)_{(k-1)+\ell+1} = 0$ for $\ell \gg 0$. Then (4.3) implies that also $H_{k-1}(r-1)_{(k-1)+\ell} = 0$ in the range where the map is injective. Repeating this argument we conclude by induction that $H_{k-1}(r-1)_{k-1+\ell} = 0$ for all $(k-1, \ell) \neq (i-1, j)$ with $k-1 \geq i-1$ and $\ell \geq j$. In particular, $H_i(r-1)_{i+j} = 0$ and $H_{i-1}(r-1)_{(i-1)+j+1} = 0$, so that

$H_{i-1}(r-1)_{i-1+j} \simeq H_i(r)_{i+j} \neq 0$, by (4.2). This shows that $h_{i-1,(i-1)+j}(r-1)$ is extremal.

Proof of (ii): Let N be a graded S-module of finite length. We set

$$s(N) = \begin{cases} \max\{j: N_j \neq 0\}, & \text{if } N \neq 0, \\ -\infty, & \text{if } N = 0, \end{cases}$$

and set $s_i = s(A(i-1))$, $t_i = \max\{s(H_j(i)) - j: j \geq 1\}$ and $u_i = \max\{s(H_j(i)) - j: j \geq 2\}$. With these numbers so defined we have

(1) $\alpha_{r-1,j}$ is extremal if and only if $j = s_r$ and $s_i < s_r$ for $i < r$, and

(2) $h_{1,1+j}(r)$ is extremal if and only if $j = t_r$ and $u_r < t_r$.

We will show that

$$u_i = t_{i-1} \quad \text{for} \quad i > 1, \quad \text{and} \quad t_i = \max\{s_1, \ldots, s_i\} \quad \text{for} \quad i \geq 1. \quad (4.4)$$

Now if we suppose that $\alpha_{r-1,j}$ is extremal, then (1) and (4.4) imply that $j = s_r = t_r$, and that $u_r = t_{r-1} < t_r$. Hence $h_{1,1+j}(r)$ is extremal, by (2). Conversely, if we suppose that $h_{1,1+j}(r)$ is extremal, then (2) and (4.4) imply that $j = t_r = s_r$ and that $s_i < s_r$ for $i < r$. Hence $\alpha_{r-1,j}$ is extremal, by (1).

Finally, considering the equivalent conditions (1) and (2) and the exact sequence

$$\cdots \to H_1(r-1)_{1+t_r} \to H_1(r)_{1+t_r} \to A(r-1)_{s_r} \to 0,$$

we see that $H_1(r-1)_{1+t_r} = 0$ since $t_{r-1} < t_r$. It follow that $h_{1,1+j}(r) = \alpha_{r-1,j}$ for $j = t_r = s_r$, as desired.

It remains to prove (4.4): In order to show that $u_i = t_{i-1}$ for $i > 1$, we consider the exact sequence (4.2) with $r = i > 1$ and $k = j \geq 2$. Since for $\ell > t_{i-1}$ we have $H_j(i-1)_{j+\ell} = H_{j-1}(i-1)_{j-1+\ell} = 0$ it follows from (4.2) that $H_j(i)_{j+\ell} = 0$. This shows that $u_i \leq t_{i-1}$. For $\ell = t_{i-1}$ we have $H_{j-1}(i-1)_{j-1+t_{i-1}} \neq 0$ for some $j \geq 2$ and $H_{j-1}(i-1)_{j-1+t_{i-1}+1} = 0$. Hence (4.2) implies that $H_j(i)_{j+t_{i-1}} \neq 0$ for this j. This shows that $u_i \geq t_{i-1}$.

In order to complete the proof of (4.4) we show by induction on i that $t_i = \max\{t_{i-1}, s_i\}$, where we set $t_0 = -\infty$. For $i = 1$ we have $H_j(1) = 0$ for $j > 1$ and $H_1(1) \simeq A(0)(-1)$. Therefore, $t_1 = s_1$.

Now let $i > 1$. Our induction hypothesis implies that

$$\max\{s_1, \ldots, s_i\} = \max\{t_{i-1}, s_i\} = \max\{u_i, s_i\},$$

and by definition $t_i = \max\{u_i, c-1\}$ where $c = s(H_1(i))$.

It follows from the exact sequence

$$H_1(i-1)_{1+j} \to H_1(i)_{1+j} \to A(i-1)_j \to 0 \quad (4.5)$$

that $s_i \leq c - 1$. If $s_i = c - 1$, then $t_i = \max\{u_i, s_i\}$, as desired. On the other hand, if $s_i < c - 1$, then (4.5) yields the exact sequence

$$H_1(i-1)_c \to H_1(i)_c \to 0$$

with $H_1(i)_c \neq 0$. This implies that $H_1(i-1)_c \neq 0$. It follows that $s_i \leq c-1 \leq t_{i-1} = u_i$, and hence $\max\{u_i, s_i\} = u_i = \max\{u_i, c-1\} = t_i$, as desired.

At last we consider the case $i = 0$. If β_{0j} is an extremal Betti number of M, then $M = N \oplus S(-j)^{\beta_{0j}}$ with $\beta_{i,i+\ell}(N) = 0$ for all i and $\ell \geq j$. By what we have already proved in case $i > 0$, if follows that $\alpha_{k\ell}(N) = 0$ for $k < n$ and $\ell \geq j$. Since $\alpha_{k\ell} = \alpha_{k\ell}(M) = \alpha_{k\ell}(N)$ for $k < n$, we conclude that $\alpha_{k\ell} = 0$ for $(k, \ell) \neq (n, j)$ with $k \leq n$ and $\ell \geq j$. Moreover since $\alpha_{nj} = \beta_{0j} \neq 0$, we see that α_{nj} is an extremal generic annihilator number.

Conversely, if we assume that α_{nj} is an extremal generic annihilator number. Then $\alpha_{k\ell} = 0$ for $(k, \ell) \neq (n, j)$ with $k \leq n$ and $\ell \geq j$. Therefore, Proposition 4.3.12 implies that $\beta_{k,k+\ell} = 0$ for $(k, \ell) \neq (0, j)$ with $k \geq 0$ and $\ell \geq j$. Since $\beta_{0j} = \alpha_{nj} \neq 0$, it follows that β_{0j} is an extremal Betti number. $\qquad\square$

Theorem 4.3.15 has the following obvious consequence.

Corollary 4.3.16. *For any two almost regular sequences on M which form a K-basis of S_1, the extremal annihilator numbers coincide.*

We have seen in Chapter 3 that for a graded ideal $I \subset S$ and any monomial order $<$ on S one has $\beta_{ij}(I) \leq \beta_{ij}(\mathrm{in}_<(I))$ for all i and j, and indeed in most of the cases the inequalities are strict. However, we will see in a moment that the extremal Betti numbers remain unchanged when passing from I to $\mathrm{gin}_{<_{\mathrm{rev}}}(I)$. In fact, combining Theorem 4.3.15 and Proposition 4.3.11 we immediately obtain the following important result:

Theorem 4.3.17 (Bayer, Charalambous, Popescu). *Let $I \subset S$ be a graded ideal. Then for any two numbers $i, j \in \mathbb{N}$ one has:*

(a) *$\beta_{i,i+j}(I)$ is extremal if and only if $\beta_{i,i+j}(\mathrm{gin}_{<_{\mathrm{rev}}}(I))$ is extremal;*
(b) *if $\beta_{i,i+j}(I)$ is extremal, then $\beta_{i,i+j}(I) = \beta_{i,i+j}(\mathrm{gin}_{<_{\mathrm{rev}}}(I))$.*

This theorem yields at once the following fundamental results

Corollary 4.3.18 (Bayer, Stillman). *Let $I \subset S$ be a graded ideal. Then*

(a) *$\mathrm{proj\,dim}(I) = \mathrm{proj\,dim}(\mathrm{gin}_{<_{\mathrm{rev}}}(I))$;*
(b) *$\mathrm{depth}(S/I) = \mathrm{depth}(S/\mathrm{gin}_{<_{\mathrm{rev}}}(I))$;*
(c) *S/I is Cohen–Macaulay if and only if $S/\mathrm{gin}_{<_{\mathrm{rev}}}(I)$ is Cohen–Macaulay;*
(d) *$\mathrm{reg}(I) = \mathrm{reg}(\mathrm{gin}_{<_{\mathrm{rev}}}(I))$.*

Problems

4.1. Eliahou and Kervaire introduced stable ideals, a class of ideals which contains the class of strongly stable ideals but still shares most of the nice properties of this smaller class. The definition is this: let $u \in S$ be a monomial. We denote by $m(u)$ the maximal number j such that $x_j | u$. Then a monomial ideal $I \subset S$ is called a **stable ideal** if for all monomials $u \in I$ and all $i < m(u)$ one has $x_i(u/x_{m(u)}) \in I$.

Suppose that $u \in G(I)$ and all $i < m(u)$ one has $x_i(u/x_{m(u)}) \in I$. Show that I is stable.

4.2. A monomial ideal $I \subset S$ is called a **lexsegment ideal** if for all monomials $u \in I$ and all monomials v with $\deg v = \deg u$ and $v >_{\text{lex}} u$ one has $v \in I$. Show that lexsegment ideals are strongly stable, and give an example of an ideal which is strongly stable but not lexsegment and an example of an ideal which is stable but not strongly stable, as well as an example which is of Borel type but not stable.

4.3. Let I and J be monomial ideals of Borel type. Show that IJ is of Borel type.

4.4. Let $I = (x^2, y^2) \subset K[x, y]$. Compute $\text{gin}(I)$ for a base field of characteristic 2 and of characteristic $\neq 2$.

4.5. We say that a graded ideal $I \subset S$ generated in degree d has a linear resolution if $\beta_{ii+j} = 0$ for all $j \neq d$. Show that $\beta_{ii+j}(I) = \beta_{ii+j}(\text{gin}(I))$ if I has a linear resolution.

4.6. Let $I \subset S$ be a graded ideal such that S/I is Cohen–Macaulay. Then show that I has only one extremal Betti number.

4.7. Find an example of a graded ideal $I \subset S$ such that I has exactly two extremal Betti numbers.

4.8. Let $I \subset S$ be a strongly stable ideal. Compute the annihilator numbers of S/I with respect to the almost regular sequence $x_n, x_{n-1}, \ldots, x_1$.

Notes

The existence of generic initial ideals and their invariance under the action of the Borel subgroup of $\text{GL}_n(K)$ was first proved in characteristic 0 by Galligo [Gal74], and later by Bayer and Stillman [BS87b] for any order and any characteristic. In [BS87b] it is shown that Borel fixed ideals are strongly stable. The nature of Borel fixed ideals in positive characteristic was described by Pardue [Par94].

The remarkable result of Bayer and Stillman [BS87a], which says that a graded ideal and its generic initial ideal have the same regularity, has been nicely extended by Bayer, Charalambous and S. Popescu [BCP99] by showing that the position as well as the values of the extremal Betti numbers are preserved under the taking of generic initial ideals. For the proof of this result we followed the treatment given [AH00] by Aramova and Herzog, which uses generic annihilator numbers, since the same kind of arguments apply to exterior algebraic shifting which is discussed in Chapter 11. Comprehensive accounts of generic initial ideals are given in the book by Eisenbud [Eis95, Chapter 15] and in the article by Green [Gre98]. Additional interesting aspects of generic initial ideals can be found in the articles by Conca [Con04], Conca, Herzog and Hibi [CHH04] and Bigatti, Conca and Robbiano [BCR05]. The concept of almost regular sequences was first introduced by Schenzel, Trung and Tu Cuong in [STT78] under the name of "filter regular sequences".

5

The exterior algebra

In this chapter we consider graded modules over the exterior algebra. In particular, the exterior face ring of a simplicial complex is studied. Alexander duality is introduced as a special case of a more general duality for graded modules. Furthermore, the theory of Gröbner bases over the exterior algebra will be developed.

5.1 Graded modules over the exterior algebra

5.1.1 Basic concepts

Let K be a field and V a finite-dimensional K-vector space. We denote E the exterior algebra of V. The algebra E is a graded K-algebra with graded components $E_i = \bigwedge^i V$.

As a K-algebra E is standard graded with defining relations

$$v \wedge v = 0 \quad \text{for} \quad v \in V = E_1. \tag{5.1}$$

We fix a K-basis $e_1, e_2, \ldots, e_n$ of V. Then for all $i, j \in [n]$ one has

$$0 = (e_i + e_j) \wedge (e_i + e_j) = e_i \wedge e_i + e_i \wedge e_j + e_j \wedge e_i + e_j \wedge e_j$$
$$= e_i \wedge e_j + e_j \wedge e_i,$$

and so $e_i \wedge e_j = -e_j \wedge e_i$. From this fact one deduces that the elements $\mathbf{e}_F = e_{j_1} \wedge e_{j_2} \wedge \cdots \wedge e_{j_i}$ with $F = \{j_1 < j_2 < \cdots < j_i\} \subset [n]$ span the K-vector space E. By using that $e_1 \wedge \cdots \wedge e_n \neq 0$, we see that the elements $\mathbf{e}_F$ are linearly independent. Here we set $\mathbf{e}_F = 1$, if $F = \emptyset$. In particular we see that $E_i = 0$ for $i < 0$ and $i > n$, and that $\dim_K E_i = \binom{n}{i}$ for $i = 0, \ldots, n$.

The identity $e_i \wedge e_j = -e_j \wedge e_i$ can be easily generalized to obtain

$$\mathbf{e}_F \wedge \mathbf{e}_G = \begin{cases} (-1)^{\sigma(F,G)} \mathbf{e}_{F \cup G}, & \text{if } F \cap G = \emptyset, \\ 0, & \text{otherwise,} \end{cases} \tag{5.2}$$

J. Herzog, T. Hibi, *Monomial Ideals*, Graduate Texts in Mathematics 260, DOI 10.1007/978-0-85729-106-6_5, © Springer-Verlag London Limited 2011

where $\sigma(F, G) = |\{(i,j)\colon i \in F, \ j \in G, \ i > j\}|$.

The elements $\mathbf{e}_F$ with $F \subset [n]$ are called the **monomials** of E. There are only finitely many monomials in E, namely 2^n. An arbitrary element $f \in E$ is a unique K-linear combination $f = \sum_F a_F \mathbf{e}_F$ of monomials. We call $\mathrm{supp}(f) = \{\mathbf{e}_F\colon a_F \neq 0\}$ the **support** of f. The element f is called **homogeneous of degree** i, if $f \in E_i$. Since E is graded, f has a unique presentation $f = \sum_{i=0}^n f_i$ with $f_i \in E_i$. The summands f_i are called the **homogeneous components** of f.

Let f and g be homogeneous elements in E. By using (5.2) one obtains

$$f \wedge g = (-1)^{\deg f \deg g} g \wedge f. \tag{5.3}$$

Definition 5.1.1. A finite-dimensional K-vector space M is called a **graded** E-module, if

(1) $M = \bigoplus_i M_i$ is a direct sum of K-vector spaces M_i;
(2) M is a left and right E-module;
(3) for all integers i and j and all $f \in E_i$ and $x \in M_j$ one has $fx \in M_{i+j}$ and $fx = (-1)^{ij} xf$.

We note that for a graded E-module M we have $(fx)g = f(xg)$ for all $f, g \in E$ and all $x \in M$. In other words M is a bimodule.

Let M and N be graded E-modules. A map $\varphi\colon M \to N$ with $\varphi(fx) = f\varphi(x)$ for all $f \in E$ and all $x \in M$ such that $\varphi(M_j) \subset N_{j+i}$ for all j is called a **homogeneous** E-module homomorphism of degree i. For example, let $a \in E$ be homogeneous of degree i. Then the map $\varphi_a\colon E \to E$ with $\varphi_a(b) = b \wedge a$ is homogeneous of degree i.

We denote by $\mathcal{G}$ the category of graded E-modules whose morphisms are the homogeneous E-module homomorphisms of degree 0. The reader who is not so familiar with the language of categories and functors is referred to Appendix A.1, where the basic concepts are explained as much as is needed here and in the following subsections.

Let N be a graded E-module. A subset $M \subset N$ is called a **graded sub-module** of N, if M is a graded E-module and the inclusion map is a morphism in $\mathcal{G}$. If $M \subset N$ is a graded submodule, then the graded K-vector space N/M inherits a natural structure as a graded E-module.

A graded submodule of E is called a **graded ideal** of E. Let $J \subset E$ be a graded ideal. Since J is a two-sided ideal, the graded E-module E/J admits a natural structure as a graded K-algebra.

Homogeneous elements $f_1, \ldots, f_m \in J$ are called a **set of generators** of J, if for each (homogeneous) element $f \in J$, there exist (homogeneous) elements $g_i \in E$ such that $f = \sum_{i=1}^m g_i \wedge f_i$. (Because of the sign rule (3) in Definition 5.1.1 we need not to distinguish between right and left generators.)

5.1.2 The exterior face ring of a simplicial complex

Similarly as in the case of the polynomial ring we define monomial ideals in the exterior algebra, and introduce the exterior face ring of a simplicial complex, in complete analogy to the Stanley–Reisner ring, as a factor ring of a suitable monomial ideal.

A graded ideal $J \subset E$ is called a **monomial ideal** if J is generated by monomials. As for monomial ideals in the polynomial ring one shows that an ideal $J \subset E$ is a monomial ideal, if and only if the following condition is satisfied: $f \in J \Leftrightarrow \text{supp}(f) \subset J$.

Definition 5.1.2. Let Δ be a simplicial complex on the vertex set $[n]$, and let $J_\Delta \subset E$ be the monomial ideal generated by the monomials $\mathbf{e}_F$ with $F \notin \Delta$. The K-algebra $K\{\Delta\} = E/J_\Delta$ is called the **exterior face ring** of Δ.

Since J_Δ is a graded ideal, the exterior face ring $K\{\Delta\}$ is a graded K-algebra, and one has

$$\dim_K K\{\Delta\}_i = f_{i-1} \quad \text{for} \quad i = 0, \dots, d-1,$$

where $f_{-1} = 1$ and $(f_0, f_1, \dots, f_{d-1})$ is the f-vector of Δ. Indeed, the residue classes of the monomials $\mathbf{e}_F$ with $F \in \Delta$ form a K-basis of $K\{\Delta\}$.

5.1.3 Duality

We will show that E is an injective object in $\mathcal{G}$.

Let $M, N \in \mathcal{G}$, and let

$$^*\text{Hom}_E(M, N) = \bigoplus_i \text{Hom}_E(M, N)_i$$

where $\text{Hom}_E(M, N)_i$ is the set of homogeneous E-module homomorphisms $\varphi \colon M \to N$ of degree i. Then $^*\text{Hom}_E(M, N)$ is a graded E-module with left and right E-module structure defined as follows: for $f \in E$ and $\varphi \in {}^*\text{Hom}_E(M, N)$ we set $(f\varphi)(x) = \varphi(xf)$ and $(\varphi f)(x) = \varphi(x)f$ for all $x \in M$.

We check condition (3) in Definition 5.1.1: let $f \in E$ be homogeneous of degree i and $\varphi \in {}^*\text{Hom}_E(M, N)$ be homogeneous of degree j. Then for $x \in M_k$ we have

$$(f\varphi)(x) = \varphi(xf) = (-1)^{ik}\varphi(fx) = (-1)^{ik}f\varphi(x) = (-1)^{ik+i(j+k)}\varphi(x)f$$
$$= (-1)^{ij}(\varphi f)(x).$$

We set $M^\vee = {}^*\text{Hom}_E(M, E)$ and $M^* = {}^*\text{Hom}_K(M, K(-n))$. Then M^* is a graded E-module with graded components

$$(M^*)_j \cong \text{Hom}_K(M_{n-j}, K) \quad \text{for all } j.$$

The left E-module structure of M^* is defined similarly as for $^*\text{Hom}_E(M, N)$, while the right multiplication we define by the equation

$$\varphi f = (-1)^{ij}f\varphi \quad \text{for} \quad \varphi \in (M^*)_j \quad \text{and} \quad f \in E_i.$$

It is clear that $M \mapsto M^*$ is an exact functor.

Let $\varphi \in M^{\vee}$ and $x \in M$. Then $\varphi(x) = \sum_{F \subset [n]} \varphi_F(x) \mathbf{e}_F$ with $\varphi_F(x) \in K$ for all $F \subset [n]$. Thus for each $F \subset [n]$ we obtain a K-linear map $\varphi_F \colon M \to K\mathbf{e}_{[n]} = K(-n)$.

As the main result of this section we have

Theorem 5.1.3. *The map $M^{\vee} \to M^*$, $\varphi \mapsto \varphi_{[n]}$ is a functorial isomorphism of graded E-modules.*

Proof. For a subset $F \subset [n]$ we set $\bar{F} = [n] \setminus F$. We first consider the map $\alpha \colon E \to E^*$ of graded K-vector spaces given by

$$\alpha(\mathbf{e}_F)(\mathbf{e}_G) = \begin{cases} (-1)^{\sigma(\bar{F}, F)}, & \text{if } G = \bar{F}, \\ 0, & \text{otherwise.} \end{cases}$$

For each $G \subset [n]$ we define the element $\mathbf{e}_G^* \in E^*$ by

$$\mathbf{e}_G^*(\mathbf{e}_F) = \begin{cases} 1, & \text{if } G = F, \\ 0, & \text{otherwise.} \end{cases}$$

The elements $\mathbf{e}_G^*$ form a K-basis of E^* (namely the dual basis of the basis $\mathbf{e}_G$ with $G \subset [n]$), and we have $\alpha(\mathbf{e}_F) = (-1)^{\sigma(\bar{F}, F)} \mathbf{e}_{\bar{F}}^*$. This shows that α is an isomorphism of graded K-vector spaces.

We observe that

$$\mathbf{e}_H \mathbf{e}_G^* = \begin{cases} (-1)^{\sigma(G \setminus H, H)} \mathbf{e}_{G \setminus H}^*, & \text{if } H \subset G, \\ 0, & \text{otherwise.} \end{cases}$$

Next we notice that α is a morphism in the category $\mathcal{G}$ of graded E-modules. Indeed for all $F, G, H \subset [n]$ we have

$$\mathbf{e}_H \alpha(\mathbf{e}_F) = (-1)^{\sigma(\bar{F}, F)} \mathbf{e}_H \mathbf{e}_{\bar{F}}^* = (-1)^{\sigma(F, \bar{F}) + \sigma(\overline{(F \cup H)}, H)} \mathbf{e}_{\overline{(F \cup H)}}^*,$$

if $H \cap F = \emptyset$, and $\mathbf{e}_H \alpha(\mathbf{e}_F) = 0$ if $H \cap F \neq \emptyset$.

On the other hand, we have

$$\alpha(\mathbf{e}_H \wedge \mathbf{e}_F) = (-1)^{\sigma(H, F)} \alpha(\mathbf{e}_{H \cup F}) = (-1)^{\sigma(H, F) + \sigma(\overline{(H \cup F)}, H \cup F)} \mathbf{e}_{\overline{(H \cup F)}}^*,$$

if $H \cap F = \emptyset$, and $\alpha(\mathbf{e}_H \wedge \mathbf{e}_F) = 0$ if $H \cap F \neq \emptyset$.

Since $(\mathbf{e}_{\overline{(F \cup H)}} \wedge \mathbf{e}_H) \wedge \mathbf{e}_F = \mathbf{e}_{\overline{(F \cup H)}} \wedge (\mathbf{e}_H \wedge \mathbf{e}_F)$ we get

$$(-1)^{\sigma(\bar{F}, F) + \sigma(\overline{(F \cup H)}, H)} = (-1)^{\sigma(H, F) + \sigma(\overline{(H \cup F)}, H \cup F)}.$$

Thus the above calculations show that $\mathbf{e}_H \alpha(\mathbf{e}_F) = \alpha(\mathbf{e}_H \wedge \mathbf{e}_F)$, so that $\alpha \colon E \to E^*$ is an E-module homomorphism. Since α respects the grading and is bijective, it is indeed an isomorphism of graded E-modules.

Consider the functorial homomorphism

$$\psi : {}^*\mathrm{Hom}_E(M, {}^*\mathrm{Hom}_K(E, K)) \longrightarrow {}^*\mathrm{Hom}_K(M, K)$$

which is defined as $\psi(\rho)(x) = \rho(x)(1)$ for all $\rho \in {}^*\mathrm{Hom}_E(M, {}^*\mathrm{Hom}_K(E, K))$ and all $x \in M$. Note that ψ is an isomorphism of graded E-modules. Thus we obtain the desired isomorphism $M^\vee \to M^*$ with $\varphi \mapsto \varphi_{[n]}$ as the composition of the isomorphisms

$$ {}^*\mathrm{Hom}_E(M, E) \xrightarrow{\;{}^*\mathrm{Hom}(M,\alpha)\;} {}^*\mathrm{Hom}_E(M, E^*) \xrightarrow{\;\psi\;} {}^*\mathrm{Hom}_K(M, K). $$

$\square$

Corollary 5.1.4. (a) *The functor* $M \mapsto M^\vee$ *is contravariant and exact. In particular,* E *is an injective object in* $\mathcal{G}$.
(b) *For all* $M \in \mathcal{G}$ *one has* (i) $(M^\vee)^\vee \cong M$, *and* (ii) $\dim_K M = \dim_K M^\vee$.

Proof. All statements follow from Theorem 5.1.3 and the fact that the functor $M \mapsto M^*$ obviously has all the desired properties. $\square$

We apply the duality functor $M \mapsto M^\vee$ to face rings. Recall that for a simplicial complex Δ we denote by $\Delta^\vee$ the Alexander dual of Δ.

Proposition 5.1.5. *Let* Δ *be a simplicial complex on the vertex set* $[n]$. *Then one has*

(a) $0 :_E J_\Delta = J_{\Delta^\vee}$;
(b) $K\{\Delta\}^\vee = J_{\Delta^\vee}$ *and* $(J_\Delta)^\vee = K\{\Delta^\vee\}$.

Proof. (a) Since J_Δ is a monomial ideal, it follows that $0 :_E J_\Delta$ is again a monomial ideal. Then by using (5.2) we see that $\mathbf{e}_F \in 0 :_E J_\Delta$ if and only if $F \cap G \neq \emptyset$ for all $G \notin \Delta$. This is the case if and only if $G \not\subset [n] \setminus F$ for all $G \notin \Delta$. This is equivalent to saying that $[n] \setminus F \in \Delta$, which in turn implies that $F \notin \Delta^\vee$. Hence $\mathbf{e}_F \in 0 :_E J_\Delta$ if and only if $\mathbf{e}_F \in J_{\Delta^\vee}$. This yields the desired conclusion.

(b) We dualize the exact sequence

$$0 \longrightarrow J_\Delta \longrightarrow E \longrightarrow K\{\Delta\} \longrightarrow 0,$$

and obtain by Corollary 5.1.4 the exact sequence

$$0 \longrightarrow K\{\Delta\}^\vee \longrightarrow E^\vee \longrightarrow (J_\Delta)^\vee \longrightarrow 0.$$

Since the homogeneous elements of $E^\vee$ are given by right multiplication with homogeneous elements in E, $E^\vee$ can be identified with E and $K\{\Delta\}^\vee$ with $0 : J_\Delta$. Thus all assertions follow from (a) and the dualized exact sequence.

$\square$

5.1.4 Simplicial homology

Let $M \in \mathcal{G}$, where as before $\mathcal{G}$ is the category of graded E-modules, and let $v \in V = E_1$. Since $v \wedge v = 0$, left multiplication by v on M yields a finite complex of finitely generated K-vector spaces

$$(M, v): \quad \cdots \xrightarrow{v} M_{i-1} \xrightarrow{v} M_i \xrightarrow{v} M_{i+1} \xrightarrow{v} \cdots$$

We denote the ith cohomology of this complex by $H^i(M, v)$. Notice that $H(M, v) = \bigoplus_i H^i(M, v)$ is again an object in $\mathcal{G}$. Indeed,

$$H(M, v) = \frac{0 :_M v}{vM}.$$

It is clear that a short exact sequence

$$0 \longrightarrow U \longrightarrow M \longrightarrow N \longrightarrow 0,$$

of finitely generated graded E-modules induces the long exact cohomology sequence

$$\cdots \to H^i(U, v) \to H^i(M, v) \to H^i(N, v) \to H^{i+1}(U, v) \to \cdots$$

By taking the K-dual of the complex (M, v) we obtain a complex of K-vector spaces

$$(M, v)^* \quad \cdots \xrightarrow{v^*} \operatorname{Hom}_K(M_{i+1}, K) \xrightarrow{v^*} \operatorname{Hom}_K(M_i, K)$$
$$\xrightarrow{v^*} \operatorname{Hom}_K(M_{i-1}, K) \xrightarrow{v^*} \quad \cdots$$

whose ith homology we denote by $H_i(M, v)$.

Obviously there exist functorial isomorphisms

$$\operatorname{Hom}_K(H^i(M, v), K) \cong H_i(M, v), \quad \operatorname{Hom}_K(H_i(M, v), K) \cong H^i(M, v). \quad (5.4)$$

We have the following duality result.

Proposition 5.1.6. *Let M be a graded E-module. Then*

$$H^i(M^\vee, v) \cong H_{n-i}(M, v) \quad \text{for all} \quad i.$$

Proof. Consider the following diagram

$$
\begin{array}{ccc}
(M^\vee)_{i-1} & \xrightarrow{\alpha_{i-1}} & \operatorname{Hom}_K(M_{n-i+1}, K) \\
{\scriptstyle v} \downarrow & & \downarrow {\scriptstyle (-1)^{n-i} v^*} \\
(M^\vee)_i & \xrightarrow{\alpha_i} & \operatorname{Hom}_K(M_{n-i}, K),
\end{array}
$$

where the horizontal maps are the graded components of the isomorphism given in Theorem 5.1.3. This diagram is commutative (and this implies the assertion of the proposition). Indeed, let $\varphi \in (M^\vee)_{i-1}$ and $x \in M_{n-i+1}$. Then $\alpha_{i-1}(\varphi)(x) = c_x$, where $\varphi(x) = c_x \mathbf{e}_{[n]}$ with $c_x \in K$. It follows that

$$(-1)^{n-i} v^* (\alpha_{i-1}(\varphi)(x)) = (-1)^{n-i} \alpha_{i-1}(\varphi)(vx) = (-1)^{n-i} c_{vx},$$

for any $x \in M_{n-i}$. On the other hand, $\alpha_i(v\varphi)(x) = c_{xv}$, since $(v\varphi)(x) = \varphi(xv)$. Now since $\varphi(xv) = \varphi((-1)^{n-i} vx) = (-1)^{n-i} \varphi(vx)$, we see that $(-1)^{n-i} c_{vx} = c_{xv}$, and this yields the desired conclusion. $\qquad\square$

Definition 5.1.7. Let Δ be a simplicial complex on $[n]$, and let $e = e_1 + e_2 + \cdots + e_n \in V$. Then for all i we set $\tilde{H}_i(\Delta; K) = H_{i+1}(K\{\Delta\}, e)$ and $\tilde{H}^i(\Delta; K) = H^{i+1}(K\{\Delta\}, e)$. The vector spaces $\tilde{H}_i(\Delta; K)$ and $\tilde{H}^i(\Delta; K)$ are called the ith **reduced simplicial homology** and **cohomology** of Δ (with values in K).

We observe that there are functorial isomorphisms

$$\tilde{H}_i(\Delta; K) \cong \mathrm{Hom}_K(\tilde{H}^i(\Delta; K), K) \quad \text{and} \quad \tilde{H}^i(\Delta; K) \cong \mathrm{Hom}_K(\tilde{H}_i(\Delta; K), K).$$

In particular one has $\dim_K \tilde{H}_i(\Delta; K) = \dim_K \tilde{H}^i(\Delta; K)$ for all i.

For later applications we record

Proposition 5.1.8. *Let Δ_1 and Δ_2 be two simplicial complexes on $[n]$, and let $\Delta = \Delta_1 \cup \Delta_2$ and $\Gamma = \Delta_1 \cap \Delta_2$. Then*

(a) $J_\Delta = J_{\Delta_1} \cap J_{\Delta_2}$ *and* $J_\Gamma = I_{\Delta_1} + J_{\Delta_2}$.
(b) *There exists an exact sequence of the following form*

$$\cdots \longrightarrow \tilde{H}_k(\Gamma; K) \longrightarrow \tilde{H}_k(\Delta_1; K) \oplus \tilde{H}_k(\Delta_2; K) \longrightarrow \tilde{H}_k(\Delta; K)$$
$$\longrightarrow \tilde{H}_{k-1}(\Gamma; K) \longrightarrow \tilde{H}_{k-1}(\Delta_1; K) \oplus \tilde{H}_k(\Delta_2; K) \longrightarrow \tilde{H}_{k-1}(\Delta; K)$$
$$\longrightarrow \cdots.$$

This sequence is called the **reduced Mayer–Vietoris exact sequence.**

Proof. (a) One has $e_F \in J_\Delta$ if and only if $F \notin \Delta$, and this is the case if and only if $F \notin \Delta_1$ and $F \notin \Delta_2$. The last condition is equivalent to saying that $e_F \in J_{\Delta_1}$ and $e_F \in J_{\Delta_2}$, which in turn is equivalent to $e_F \in J_{\Delta_1} \cap J_{\Delta_2}$. The equality $J_\Gamma = J_{\Delta_1} + J_{\Delta_2}$ is proved similarly.
  (b) By using part (a) we obtain the short exact sequence

$$0 \to K\{\Delta\} \to K\{\Delta_1\} \oplus K\{\Delta_2\} \to K\{\Gamma\} \to 0,$$

and hence the short exact sequence of complexes

$$0 \to (K\{\Gamma\}, e)^* \to (K\{\Delta_1\}, e)^* \oplus (K\{\Delta_2\}, e)^* \to (K\{\Delta\}, e)^* \to 0.$$

The long exact homology sequence of this short exact sequence of complexes is the desired Mayer–Vietoris exact sequence. $\qquad\square$

It is customary to denote the complex $(K\{\Delta\}, e)^*$ by $\tilde{C}(\Delta; K)$ where

$$\tilde{C}_{j-1}(\Delta; K) = (K\{\Delta\}_j)^* \quad \text{for all} \quad j.$$

This complex is called the **augmented oriented chain complex**of Δ (with respect to K). A K-basis of $K\{\Delta\}_{j+1}$ is given by the elements $\mathbf{e}_F$ where $F \in \Delta$ and $|F| = j + 1$. The dual basis elements $(\mathbf{e}_F)^*$ establish then a K-basis of $\tilde{C}_j(\Delta; K)$. Usually one denotes for $F = \{i_0 < i_1 < \cdots < i_j\}$ the basis element $(\mathbf{e}_F)^*$ by $[i_0, i_1, \cdots, i_j]$. With this notation the chain map $\partial: \tilde{C}_j(\Delta; K) \to \tilde{C}_{j-1}(\Delta; K)$ is given by

$$\partial([i_0, i_1, \ldots, i_j]) = \sum_{k=0}^{j} (-1)^k [i_0, i_1, \ldots, i_{k-1}, i_{k+1} \ldots, i_j].$$

Example 5.1.9. (a) It is known from algebraic topology that $\tilde{H}^i(\Delta; K) = 0$ for all i, if the geometric realization of Δ is a contractible topological space. In particular, if Δ is a simplex, say, $\mathcal{F}(\Delta) = \{[n]\}$, then all reduced (co)homology of Δ vanishes. We can see this directly. Indeed, $\tilde{H}^i(\Delta; K) = H^i(E, e)$. After applying a linear automorphism, we may assume that $e = e_1$. Obviously the complex (E, e_1) is exact. Hence the conclusion.

Let Δ be a 1-dimensional simplicial complex on the vertex set $[n]$ forming a cycle of length n, that is, the facets of Δ are $\{i, i+1\}$, $i = 1, \ldots, n-1$ and $\{1, n\}$. Then $\dim_K \tilde{H}^1(\Delta; K) \neq 0$. Indeed, $\tilde{H}^1(\Delta; K)$ is the homology of the complex

$$(K\{\Delta\}, e): \quad 0 \longrightarrow K\{\Delta\}_0 \stackrel{e}{\longrightarrow} K\{\Delta\}_1 \stackrel{e}{\longrightarrow} K\{\Delta\}_2 \longrightarrow 0,$$

The map $e: K\{\Delta\}_0 \to K\{\Delta\}_1$ is injective. Since $K\{\Delta\}_0 = K$ and $K\{\Delta\}_1 = \bigoplus_{i=1}^{n} K e_i$ the cokernel of this injective map is a K-vector space of dimension $n - 1$. Thus the K-dimension of the image of $e: K\{\Delta\}_1 \to K\{\Delta\}_2$ is $\leq n - 1$. On the other hand, $\dim_K K\{\Delta\}_2$ is equal to the number of facets of Δ, which is n. Thus $H^2(K\{\Delta\}, e) \neq 0$, as desired.

Actually one has $H^2(K\{\Delta\}, e) \cong K$. To see this one just has to check that the above sequence is exact at $K\{\Delta\}_1$.

Now we can show

Proposition 5.1.10 (Alexander duality). *Let Δ be a simplicial complex on $[n]$. Then for each i one has a functorial isomorphism*

$$\tilde{H}^{i-2}(\Delta^\vee; K) \cong \tilde{H}_{n-i-1}(\Delta; K).$$

Proof. By using Proposition 5.1.5 and Proposition 5.1.6 we see that

$$\tilde{H}^{i-2}(\Delta^\vee; K) = H^{i-1}(K\{\Delta^\vee\}, e) \cong H^{i-1}((J_\Delta)^\vee, e) \cong H_{n-i+1}(J_\Delta, e).$$

Since $H^i(E, e) = 0$ for all i, the long exact cohomology sequence attached to the short exact sequence

$$0 \longrightarrow J_\Delta \longrightarrow E \longrightarrow K\{\Delta\} \longrightarrow 0$$

yields the isomorphisms $H^i(K\{\Delta\}, e) \cong H^{i+1}(J_\Delta, e)$. Taking the K-dual we obtain the isomorphisms $H_i(K\{\Delta\}, e) \cong H_{i+1}(J_\Delta, e)$. It follows that

$$H_{n-i+1}(J_\Delta, e) \cong H_{n-i}(K\{\Delta\}, e) = \tilde{H}_{n-i-1}(\Delta; K),$$

as desired.   □

We conclude this subsection by showing that simplicial (co)homology can be computed with any generic linear form. In what follows we will assume that K is an infinite field.

Definition 5.1.11. Let $M \in \mathcal{G}$. An element $v \in V$ is called **generic** on M if $\dim_K H^i(M, v) \leq \dim_K H^i(M, u)$ for all i and all $u \in V$.

We observe that the set of elements $v \in V$ which are generic on M is a nonempty Zariski open subset of V. We set $H^i(M) = H^i(M, v)$ if v is generic on M, and call $H^i(M)$ the ith **generalized simplicial cohomology** of M.

Lemma 5.1.12. *Let Δ be a simplicial complex. Then*

$$\tilde{H}^{i-1}(\Delta; K) \cong H^i(K\{\Delta\}) \quad \text{for all} \quad i.$$

Proof. The assertion follows once we can show that e is generic on $K\{\Delta\}$. The subset $U \subset V$ of elements $v = \sum_i^n a_i e_i \in E_1$ with $\prod_i^n a_i \neq 0$ is a nonempty Zariski open subset of V. We note that the complexes $(K\{\Delta\}, e)$ and $(K\{\Delta\}, v)$ are isomorphic for all $v \in U$. In fact, the isomorphism of complexes is induced by the algebra automorphism $\varphi \colon K\{\Delta\} \to K\{\Delta\}$ with $\varphi(e_i) = a_i e_i$ for $i = 1, \ldots, n$. It follows that

$$H^i(K\{\Delta\}, e) \cong H^i(K\{\Delta\}, v) \tag{5.5}$$

for all i and all $v \in U$. Let $W \subset V$ be the nonempty Zariski open subset of V of generic elements on $K\{\Delta\}$. Since $U \cap W \neq \emptyset$, we may choose $v \in U \cap W$. Thus (5.5) implies that $H^i(K\{\Delta\}, e)$ has minimal K-dimension for all i. In other words, e is generic on $K\{\Delta\}$.   □

5.2 Gröbner bases

Gröbner basis theory for the exterior algebra is very similar to that for the polynomial ring. We will sketch the main features of the theory and, if necessary, emphasize the differences. The fact that the exterior algebra has zero divisors is responsible for some modifications on the theory.

5.2.1 Monomial orders and initial ideals

Most of the concepts discussed in this subsection are completely analogous to those in the polynomial ring. Let, as before, K be a field, V a finite-dimensional vector space and E its exterior algebra. After having fixed a basis $e_1, \ldots, e_n$ of V, the elements $\mathbf{e}_F = e_{j_1} \wedge e_{j_2} \wedge \cdots \wedge e_{j_i}$ with $F = \{j_1 < j_2 < \cdots < j_i\}$ are the monomials of E. We define a **monomial order** on E as a total order $<$ on the set of monomials $\mathrm{Mon}(E)$ of E such that (i) $1 < u$ for all $1 \neq u \in \mathrm{Mon}(E)$; (ii) if $u, v \in \mathrm{Mon}(E)$ and $u < v$, then $u \wedge w < v \wedge w$ for all $w \in \mathrm{Mon}(E)$ such that $u \wedge w \neq 0 \neq v \wedge w$.

For each monomial $u = \mathbf{e}_F \in \mathrm{Mon}(E)$ we may consider the corresponding squarefee monomial $u^* = x_F \in S = K[x_1, \ldots, x_n]$ with $x_F = x_{j_1} x_{j_2} \cdots x_{j_i}$.

We define the lexicographic order on E induced by $e_1 > e_2 > \cdots > e_n$ as follows:

$$\mathbf{e}_F <_{\mathrm{lex}} \mathbf{e}_G \iff x_F <_{\mathrm{lex}} x_G.$$

The reverse lexicographic order is defined similarly. Indeed any monomial order on S induces, by restriction to the squarefree monomials, a monomial order on E. The converse is true as well: given a monomial order $<$ on E. Since there are only finitely many monomials in E, there exists a weight $\mathbf{w} = (w_1, \ldots, w_n) \in \mathbb{N}^n$ such that $\deg_{\mathbf{w}}(u) < \deg_{\mathbf{w}}(v)$ if and only $u^* < v^*$, where $\deg_{\mathbf{w}}(\mathbf{e}_F) = \sum_{i \in F} w_i$. This follows from Lemma 3.1.1. Now let $<'$ be any monomial order on S and consider the monomial order $<'_{\mathbf{w}}$ on S defined in Example 2.1.3. Then it is immediate that

$$u < v \quad \text{if and only if} \quad u^* <'_{\mathbf{w}} v^* \quad \text{for all} \quad u, v \in E.$$

The (reverse) lexicographic order on E induced by $e_1 > e_2 > \cdots > e_n$ can be described more directly as follows: for two subset $F, G \subset [n]$ one defines the symmetric difference as the set $F \triangle G = (F \setminus G) \cup (G \setminus F)$. Then one has

(i) $\mathbf{e}_F <_{\mathrm{lex}} \mathbf{e}_G$, if either $|F| < |G|$, or else $|F| = |G|$ and the *smallest* element in $F \triangle G$ belongs to F,

(ii) $\mathbf{e}_F <_{\mathrm{rev}} \mathbf{e}_G$, if either $|F| < |G|$, or else $|F| = |G|$ and the *largest* element in $F \triangle G$ belongs to G.

For the rest of this chapter we will assume that all monomial orders considered are induced by $e_1 > e_2 > \cdots > e_n$, unless otherwise stated.

Given a monomial order on E, and $0 \neq f \in E$. The **initial monomial** of f with respect to $<$, denoted $\mathrm{in}_<(f)$, is the biggest monomial (with respect to the given order) among the monomials belonging to $\mathrm{supp}(f)$.

Let $0 \neq g \in E$ be another element and $u \in \mathrm{Mon}(E)$. Then $\mathrm{in}_<(u \wedge f) = u \wedge \mathrm{in}_<(f)$, as long as $u \wedge \mathrm{in}_<(f) \neq 0$, and one has $\mathrm{in}_<(f \wedge g) = \mathrm{in}_<(f) \wedge \mathrm{in}_<(g)$, as long as $\mathrm{in}_<(f) \wedge \mathrm{in}_<(g) \neq 0$.

Let $J \subset E$ be a graded ideal. We define the **initial ideal** $\mathrm{in}_<(J)$ of J as the monomial ideal in E generated by all monomials $\mathrm{in}_<(f)$ with $0 \neq f \in J$.

Since J is graded one has

$$\text{in}_<(J) = (\{\text{in}_<(f)\colon 0 \neq f \in J,\ f \text{ homogeneous}\}).$$

In a similar way to Proposition 2.3.7 we have that $\text{in}_<(J)$ coincides with the K-vector space spanned by the monomials $\text{in}_<(f)$ with $0 \neq f \in J$, and f homogeneous. Let $f_1, \ldots, f_r \in J$ be homogeneous elements such that $\text{in}_<(f_1), \ldots, \text{in}_<(f_r)$ is a K-basis of $\text{in}_<(J)$. Then it follows that $f_1, \ldots, f_r$ is a homogeneous K-basis of J. In particular one has

$$\dim_K J_i = \dim_K \text{in}_<(J)_i \quad \text{for all} \quad i. \tag{5.6}$$

Definition 5.2.1. Let J be a nonzero ideal of E. A finite set $g_1, \ldots, g_s$ of elements of J is said to be a **Gröbner basis** of J with respect to $<$, if the initial ideal $\text{in}_<(J)$ of J is generated by $\text{in}_<(g_1), \text{in}_<(g_2), \ldots, \text{in}_<(g_s)$.

Example 5.2.2. Let E be the exterior algebra of the K-vector space V with basis e_1, e_2, e_3, e_4, and consider the ideal $J \subset E$ generated by the element $e_1 \wedge e_2 + e_3 \wedge e_4$. One has $J_i = E_i$ for $i = 3, 4$. Therefore if $<$ denotes the lexicographic order, then $\text{in}_<(J) = (e_1 \wedge e_2, e_1 \wedge e_3 \wedge e_4, e_2 \wedge e_3 \wedge e_4)$. Hence $e_1 \wedge e_2 + e_3 \wedge e_4, e_1 \wedge e_3 \wedge e_4, e_2 \wedge e_3 \wedge e_4$ is a Gröbner basis of J.

Thus, in contrast to the polynomial case, the initial ideal of a principal ideal in the exterior algebra need not be principal.

It is easy to see that a Gröbner basis of J is also a system of generators of J, compare Theorem 2.1.8 and its proof.

It is also clear that if we choose homogeneous elements $f_1, \ldots, f_r$ in J such that $\text{in}_<(f_1), \ldots, \text{in}_<(f_r)$ is a K-basis of $\text{in}_<(J)$, then $f_1, \ldots, f_r$ is Gröbner basis of J. Such a Gröbner basis is usually much too big. Therefore, as in the case of the polynomial ring, one defines: a Gröbner basis $\mathcal{G} = \{g_1, g_2, \ldots, g_s\}$ of J is **reduced**, if for all i the coefficient of $\text{in}_<(g_i)$ in g_i is 1, and if for all i and j with $i \neq j$, $\text{in}_<(g_i)$ divides none of the monomials of $\text{supp}(g_j)$.

In a similar way to ideals in the polynomial ring we have

Theorem 5.2.3. *For any graded ideal $J \subset E$ and any monomial order $<$ on E, a reduced Gröbner basis exists and is uniquely determined.*

Proof. Let $\{u_1, \ldots, u_s\}$ be the unique minimal set of monomial generators of $\text{in}_<(J)$, and choose homogeneous elements $f_i \in J$ with $\text{in}_<(f_i) = u_i$. Suppose that there exists $v \in \text{supp}(f_1)$, which is divisible by u_i for some $i \neq 1$, say $v = w \wedge u_i$. We may assume that v is the largest such element in $\text{supp}(f_1)$. Note that $v \neq u_1$, since $u_1, \ldots, u_s$ is a minimal system of generators of $\text{in}_<(J)$. There exists $a \in K$ such that v does not belong to the support of $f_1' = f_1 - aw \wedge f_i$. And of course we have $\text{in}_<(f_1') = \text{in}_<(f_1)$.

We replace f_1 by f_1'. If there is still an element $v' \in \text{supp}(f_1')$ which is divisible by u_i for some $i \neq 1$, then $v' < v$, and we may repeat the first step. Since there are only finitely many monomials in E, this procedure ends after a finite number of steps, and we obtain a homogeneous element $g_1 \in J$ with $\text{in}_<(g_1) = u_1$ and such that no u_i with $i \neq 1$ divides any $v \in \text{supp}(g_1), v \neq u_1$.

Now we modify f_2 in the same way. Thus induction on $|G(J)|$ guarantees the existence of a Gröbner basis $\{g_1, \ldots, g_s\}$ with the property that for all i and j with $i \neq j$, $\text{in}_<(g_i)$ divides none of the monomials of $\text{supp}(g_j)$. Dividing g_i by the coefficient of $\text{in}_<(g_i)$ for all i, we obtain a reduced Gröbner basis of J.

In order to prove uniqueness, we first notice that if $\mathcal{G} = \{g_1, \ldots, g_s\}$ is a reduced Gröbner basis of J, then the leading coefficients of the g_i are all 1 and the monomials $u_i = \text{in}_<(g_i)$ form the unique minimal system of monomial generators of $\text{in}_<(J)$. Thus if $\mathcal{G}' = \{g'_1, \ldots, g'_t\}$ is another reduced Gröbner basis of J, then we may assume that $u_i = \text{in}_<(g'_i)$ for all i. In particular, $s = t$. Suppose, $g_i \neq g'_i$. Since the leading coefficients of g_i and g'_i are both 1, it follows that the leading monomials cancel when we take the difference, so that either $g_i - g'_i = 0$, or $\text{in}_<(g_i - g'_i) \neq u_i$. Suppose the second case happens. Since $u_1, \ldots, u_s$ generate $\text{in}_<(J)$, it follows that there exists $j \neq i$ such that u_j divides $\text{in}_<(g_i - g'_i)$. This contradicts our hypothesis that $\mathcal{G}$ is a reduced Gröbner basis, because $\text{in}_<(g_i - g'_i) \in \text{supp}(g_i) \cup \text{supp}(g'_i)$.   □

5.2.2 Buchberger's criterion

As in the case of the polynomial ring, there is a division algorithm for elements in the exterior algebra. But before we formulate the exterior version of the division algorithm, let us pause for a moment and consider again Example 5.2.2, where $J = (g)$ with $g = e_1 \wedge e_2 + e_3 \wedge e_4$. Since each element in J is a multiple of g, each element in J has remainder 0 with respect to g in the sense of Chapter 2. Then by analogy with Theorem 2.3.2 one would expect that g is a Gröbner basis of J, which, as we have seen, is not the case. The crucial point is that $\text{in}_<(h \wedge g) = \text{in}_<(h) \wedge \text{in}_<(g)$ only if $\text{in}_<(h) \wedge \text{in}_<(g) \neq 0$. Hence we are forced to make the adequate modification in the formulation of the division algorithm.

In the previous section we have seen that any monomial order $<$ on E comes from a monomial order $<$ on S by restriction. With the notation introduced in Section 5.2.1 we have

$$\text{in}_<(f \wedge g)^* \leq \text{in}_<(f)^* \text{in}_<(g)^* \tag{5.7}$$

for all $f, g \in E$ with $f \wedge g \neq 0$. Equality holds in (5.7) if and only if $\text{in}_<(f) \wedge \text{in}_<(g) \neq 0$.

Theorem 5.2.4 (The division algorithm). *Fix a monomial order $<$ on E, and let $g_1, \ldots, g_s, f$ homogeneous nonzero elements of E. Then there exists a standard expression of f, i.e. homogeneous elements $h_1, \ldots, h_s, r \in E$ such that*

$$f = \sum_{i=1}^s h_i \wedge g_i + r,$$

with the property that no $v \in \text{supp}(r)$ belongs to $(\text{in}_<(g_1), \ldots, \text{in}_<(g_s))$, and whenever $h_i \wedge g_i \neq 0$, then $\text{in}_<(h_i)^* \text{in}_<(g_i)^* \leq \text{in}_<(f)^*$.

Proof. Let $J = (\operatorname{in}_<(g_1), \ldots, \operatorname{in}_<(g_s))$ and denote by $\operatorname{Mon}(J)$ the set of monomials belonging to J. We write $f = \sum_{u \in \operatorname{Mon}(E)} a_u u$ with $a_u \in K$ as

$$f = h + r_1 \quad \text{with} \quad h = \sum_{u \in \operatorname{Mon}(J)} a_u u \quad \text{and} \quad r_1 = \sum_{u \notin \operatorname{Mon}(J)} a_u u.$$

We may assume that $h \neq 0$, because otherwise $f = r_1$ is already the desired standard expression of f. Since $h \in J$, there exists i and a monomial w such that $\operatorname{in}_<(h) = w \wedge \operatorname{in}_<(g_i)$, and there exists $a \in K$, $a \neq 0$ such that h and $aw \wedge g_i$ have the same leading coefficient. We set $\tilde{f} = h - aw \wedge g_i$. Then

$$\operatorname{in}_<(\tilde{f}) < \operatorname{in}_<(h) \leq \operatorname{in}_<(f).$$

By using induction on $\operatorname{in}_<(f)$ we may assume that there exists a standard expression $\tilde{f} = \sum_{j=1}^{s} \tilde{h}_j \wedge g_j + r_2$ of $\tilde{f}$. Then $f = \sum_{j=1}^{s} h_j \wedge g_j + r$ with $h_i = \tilde{h}_i + aw$, $h_j = \tilde{h}_j$ for $j \neq i$ and $r = r_1 + r_2$.

We claim that this is a standard expression of f. By construction, no $v \in \operatorname{supp}(r)$ belongs to $(\operatorname{in}_<(g_1), \ldots, \operatorname{in}_<(g_s))$ and for $j \neq i$ with $h_j \wedge g_j \neq 0$, we have $\operatorname{in}_<(f)^* \geq \operatorname{in}_<(\tilde{f})^* \geq \operatorname{in}_<(h_j)^* \operatorname{in}_<(g_j)^*$. It remains to be shown that this last statement also holds for $j = i$. We have $w^* \operatorname{in}_<(g_i)^* = \operatorname{in}_<(h)^* \leq \operatorname{in}_<(f)^*$, and $\operatorname{in}_<(\tilde{h}_i)^* \operatorname{in}_<(g_i)^* \leq \operatorname{in}_<(\tilde{f})^* < \operatorname{in}_<(f)^*$. Hence, since $\operatorname{in}_<(h_i)^* \leq \max\{w^*, \operatorname{in}_<(\tilde{h}_i)^*\}$, it follows that $\operatorname{in}_<(h_i)^* \operatorname{in}_<(g_i)^* \leq \operatorname{in}_<(f)^*$. □

Let $J \subset E$ be a monomial ideal with $G(J) = \{\mathbf{e}_{F_1}, \ldots, \mathbf{e}_{F_s}\}$. Let $a_i = \deg \mathbf{e}_{F_i}$ for $i = 1, \ldots, s$, and consider the epimorphism of graded E-modules

$$\epsilon: \bigoplus_{i=1}^{s} E(-a_i) \longrightarrow J \quad \text{with} \quad b_i \mapsto \mathbf{e}_{F_i},$$

where $b_1, \ldots, b_s$ is the canonical homogeneous basis of $\bigoplus_{i=1}^{s} E(-a_i)$. The graded submodule $U = \operatorname{Ker} \epsilon$ of $\bigoplus_{i=1}^{s} E(-a_i)$ is called the **relation module** of J.

Lemma 5.2.5. *The relation module U of J with $G(J) = \{\mathbf{e}_{F_1}, \ldots, \mathbf{e}_{F_s}\}$ is generated by two types of relations:*

(i) *the S-relations $(-1)^{\sigma(F_j \setminus F_i, F_i)} \mathbf{e}_{F_j \setminus F_i} b_i - (-1)^{\sigma(F_i \setminus F_j, F_j)} \mathbf{e}_{F_i \setminus F_j} b_j$ for $i < j$, and*

(ii) *the T-relations $e_i b_j$ for all i and j with $i \in F_j$.*

Proof. Let $c \in U$; since $b_i \mapsto \mathbf{e}_{F_i}$, we can write $c = \sum_{F \subset [n]} c_F$ with each c_F of the form $c_F = \sum_i a_i \mathbf{e}_{G_i} b_i$ with $G_i \cup F_i = F$ for all i, where the sum is taken over those i with $F_i \subset F$ and where $a_i \in K$.

Since $\epsilon(c_F) = a \mathbf{e}_F$ for some $a \in K$, it follow that $\epsilon(c) = 0$ if and only if $\epsilon(c_F) = 0$ for all $F \subset [n]$. Thus we may as well assume that c is of the form $c = \sum_i a_i \mathbf{e}_{G_i} b_i$ with $G_i \cup F_i = F$. If for some i we have $G_i \cap F_i \neq \emptyset$, then $\mathbf{e}_{G_i} b_i$

is a multiple of a T-relation. Hence modulo T-relations we may assume that $G_i \cap F_i = \emptyset$, so that $G_i = F \setminus F_i$ for all i with $a_i \neq 0$. After a renumbering of the basis elements b_i we may assume that $a_i \neq 0$ for $i = 1, \ldots, r$ and that $a_i = 0$ for $i > r$. Then $c = \sum_{i=1}^{r} a_i \mathbf{e}_{F \setminus F_i} b_i$, and since hence $\epsilon(c) = 0$, we see that

$$0 = \sum_{i=1}^{r} a_i \mathbf{e}_{F \setminus F_i} \mathbf{e}_{F_i} = \sum_{i=1}^{r} (-1)^{\sigma(F \setminus F_i, F_i)} a_i \mathbf{e}_F.$$

Thus if we set $c_i = (-1)^{\sigma(F \setminus F_i, F_i)} a_i$, then $\sum_{i=1}^{r} c_i = 0$, so that

$$c = \sum_{i=1}^{r} c_i \left((-1)^{\sigma(F \setminus F_i, F_i)} \mathbf{e}_{F \setminus F_i} b_i \right)$$

$$= \sum_{i=2}^{r} c_i \left((-1)^{\sigma(F \setminus F_i, F_i)} \mathbf{e}_{F \setminus F_i} b_i - (-1)^{\sigma(F \setminus F_1, F_1)} \mathbf{e}_{F \setminus F_1} b_1 \right).$$

Now for any of the summands of c we have

$$(-1)^{\sigma(F \setminus F_i, F_i)} \mathbf{e}_{F \setminus F_i} b_i - (-1)^{\sigma(F \setminus F_1, F_1)} \mathbf{e}_{F \setminus F_1} b_1$$

$$= \mathbf{e}_{F \setminus (F_1 \cup F_i)} \left((-1)^{\sigma(F \setminus (F_1 \cup F_i), F_1 \setminus F_i) + \sigma(F \setminus F_i, F_i)} \mathbf{e}_{F_1 \setminus F_i} b_i \right.$$

$$\left. - (-1)^{\sigma(F \setminus (F_1 \cup F_i), F_i \setminus F_1) + \sigma(F \setminus F_1, F_1)} \mathbf{e}_{F_i \setminus F_1} b_1 \right)$$

$$= (-1)^{\sigma(F \setminus (F_1 \cup F_i), F_1 \cup F_i)} \mathbf{e}_{F \setminus (F_1 \cup F_i)} \left((-1)^{\sigma(F_1 \setminus F_i, F_i)} \mathbf{e}_{F_1 \setminus F_i} b_i \right. \quad (5.8)$$

$$\left. - (-1)^{\sigma(F_i \setminus F_1, F_1)} \mathbf{e}_{F_i \setminus F_1} b_1 \right).$$

Equation (5.8) follows from

$$(-1)^{\sigma(F \setminus (F_1 \cup F_i), F_1 \cup F_i) + \sigma(F \setminus F_i, F_i)} = (-1)^{\sigma(F_1 \setminus F_i, F_i) + \sigma(F \setminus (F_1 \cup F_i), F_1 \cup F_i)} \quad (5.9)$$

and the corresponding identity with the role of F_1 and F_i exchanged. The identity (5.9) itself is a consequence of the fact that

$$(\mathbf{e}_{F \setminus (F_1 \cup F_i)} \wedge \mathbf{e}_{F_1 \setminus F_i}) \wedge \mathbf{e}_{F_i} = \mathbf{e}_{F \setminus (F_1 \cup F_i)} \wedge (\mathbf{e}_{F_1 \setminus F_i} \wedge \mathbf{e}_{F_i}).$$

The calculations show that c is a linear combination of S-relations, as desired.
□

Given a sequence $g_1, g_2, \ldots, g_s$ of nonzero homogeneous elements in E with $\mathrm{in}_{<}(g_i) = \mathbf{e}_{G_i}$ and leading coefficient c_i. According to the two types of relations described in Lemma 5.2.5 we define the S- and T-**polynomials** of $g_1, \ldots, g_s$ as follows:

(i) for each $i < j$ let

$$S(g_i, g_j) = \frac{1}{c_i} (-1)^{\sigma(G_j \setminus G_i, G_i)} \mathbf{e}_{G_j \setminus G_i} \wedge g_i - \frac{1}{c_j} (-1)^{\sigma(G_i \setminus G_j, G_j)} \mathbf{e}_{G_i \setminus G_j} \wedge g_j,$$

and

(ii) for each i and j with $i \in G_j$ set $T(e_i, g_j) = \frac{1}{c_i} e_i \wedge g_j$.

Let $f \in E$ be some nonzero homogeneous element. We say that f **reduces to** 0 with respect to $g_1, g_2, \ldots, g_s$, if there exists a standard expression of the form $f = \sum_{i=1}^{s} h_i \wedge g_i$.

With this terminology introduced one has

Theorem 5.2.6 (Buchberger's criterion). *Let J be a nonzero graded ideal of E and $\mathcal{G} = \{g_1, \ldots, g_s\}$ a system of homogeneous generators of J. Then the following conditions are equivalent:*

(a) $\mathcal{G}$ *is a Gröbner basis of J.*
(b) *All S- and T-polynomials of $\mathcal{G}$ reduce to 0 with respect to $g_1, \ldots, g_s$.*

Proof. (a) $\Rightarrow$ (b): We may assume that the coefficients of $ini_<(g_i)$ in g_i are all 1. The S- and T-polynomials belong to J. Let $f \in J$ be any nonzero homogeneous element, and let $f = \sum_{i=1}^{s} h_i \wedge g_i + r$ be standard expression of f with respect to $g_1, g_2, \ldots, g_s$. Suppose that $r \neq 0$. Then $r \in J$, and hence $\mathrm{in}_<(r) \in \mathrm{in}_<(J)$. Since $\mathcal{G}$ is a Gröbner basis of J, there exists an integer i such that $\mathrm{in}_<(r)$ is a multiple of $\mathrm{in}_<(g_i)$. This a contradicts the property of a remainder.

(b) $\Rightarrow$ (a): Let $f \in J$ be a nonzero homogeneous element. Then $f = \sum_i h_i \wedge g_i$ and $\mathrm{in}_<(f)^* \leq \max_i \{\mathrm{in}_<(h_i \wedge g_i)^*\} \leq \max_i \{\mathrm{in}_<(h_i)^* \mathrm{in}_<(g_i)^*\}$, where the maximum is taken over those i for which $h_i \wedge g_i \neq 0$. If equality holds, then $\mathrm{in}_<(f)^* = \mathrm{in}_<(h_i)^* \mathrm{in}_<(g_i)^*$ for some i. This then implies that $\mathrm{in}_<(f) = \mathrm{in}_<(h_i) \wedge \mathrm{in}_<(g_i)$, and we are done.

Let $\delta = \max_i \{\mathrm{in}_<(h_i)^* \mathrm{in}_<(g_i)^*\}$, and suppose now that $\mathrm{in}(f)^* < \delta$. We will show that f can be rewritten as $f = \sum_i \tilde{h}_i \wedge g_i$ with $\mathrm{in}_<(\tilde{h}_i)^* \mathrm{in}_<(g_i)^* < \delta$ for all i with $\tilde{h}_i \wedge g_i \neq 0$. Proceeding by induction on δ completes then the proof.

We write f as

$$
\begin{aligned}
f &= \sum_{\mathrm{in}_<(h_i)^* \mathrm{in}_<(g_i)^* = \delta} h_i \wedge g_i + \sum_{\mathrm{in}_<(h_i)^* \mathrm{in}_<(g_i)^* < \delta} h_i \wedge g_i \\
&= \sum_{\mathrm{in}_<(h_i)^* \mathrm{in}_<(g_i)^* = \delta} a_i \mathrm{in}_<(h_i) \wedge g_i + \sum_{\mathrm{in}_<(h_i)^* \mathrm{in}_<(g_i)^* = \delta} (h_i - a_i \mathrm{in}_<(h_i)) \wedge g_i \\
&\quad + \sum_{\mathrm{in}_<(h_i)^* \mathrm{in}_<(g_i)^* < \delta} h_i \wedge g_i,
\end{aligned}
$$

where a_i is the leading coefficient of h_i. Since

$$
\mathrm{in}_<(h_i - a_i \mathrm{in}_<(h_i))^* \mathrm{in}_<(g_i)^* < \delta,
$$

it remains to consider the sum

$$
s = \sum_{\mathrm{in}_<(h_i)^* \mathrm{in}_<(g_i)^* = \delta} a_i \mathrm{in}_<(h_i) \wedge g_i.
$$

As $\text{in}_<(f)^* < \delta$, we also have $\text{in}_<(s)^* < \delta$. Hence it follows that

$$\sum_{\text{in}_<(h_i)^* \, \text{in}_<(g_i)^* = \delta} a_i \, \text{in}_<(h_i) \wedge \text{in}_<(g_i) = 0.$$

In other words,

$$u = \sum_{\text{in}_<(h_i)^* \, \text{in}_<(g_i)^* = \delta} a_i \, \text{in}_<(h_i) b_i \in U,$$

where U is the relation module of $(\text{in}_<(g_1), \ldots, \text{in}_<(g_s))$. By Lemma 5.2.5 the S- and T- relations generate U. Thus we can write $u = \sum_{j=1}^t c_j u_j$ where each u_j is either an S- or a T-relation. Since u is homogeneous of $\mathbb{Z}^n$-degree $\deg \delta$, we may assume that for each j we have the following equation of $\mathbb{Z}^n$-degrees

$$\deg(c_j u_j) = \deg(c_j) + \deg(u_j) = \deg \delta. \tag{5.10}$$

Let

$$\eta: \bigoplus_{i=1}^s E(-a_i) \longrightarrow J \quad \text{with} \quad b_i \mapsto g_i.$$

Then $\eta(u_k)$ is either as S- or T-polynomial. Therefore our assumption implies that $\eta(u_k) = \sum_{i=1}^s a_{ki} \wedge g_i$ with

$$\text{in}_<(\eta(u_k))^* \geq \text{in}_<(a_{ki})^* \, \text{in}_<(g_i)^* \tag{5.11}$$

whenever $a_{ki} \wedge g_i \neq 0$, and hence

$$s = \eta(u) = \sum_{k=1}^t c_k \wedge \eta(u_k)$$

$$= \sum_{i=1}^s (\sum_{k=1}^l c_k \wedge a_{ki}) \wedge g_i = \sum_{i=1}^s \tilde{h}_i \wedge g_i,$$

where $\tilde{h}_i = \sum_{k=1}^l c_k \wedge a_{ki}$. Since

$$\text{in}_<(\tilde{h}_i)^* \leq \max_k \{\text{in}_<(c_k \wedge a_{ki})^*\} = \text{in}_<(c_j \wedge a_{ji})^* \leq \text{in}_<(c_j)^* \, \text{in}_<(a_{ji})^*$$

for some j with $c_j \neq 0$, one obtains together with (5.11) that

$$\text{in}(\tilde{h}_i)^* \, \text{in}(g_i)^* \leq \text{in}_<(c_j)^* (\text{in}(a_{ji})^* \, \text{in}_<(g_i)^*)$$
$$\leq \text{in}_<(c_j)^* \, \text{in}_<(\eta(u_j))^* < \delta,$$

as desired. The last inequality follows from (5.10), since $\deg u_j > \deg \text{in}_<(\eta(u_j))^*$. $\qquad\square$

Remark 5.2.7. As we know from Lemma 5.2.5, the relation module U of $(\mathrm{in}_<(g_1), \ldots, \mathrm{in}_<(g_s))$ is generated by the S- and T-relations. This may of course not be a minimal set of generators for U. However, in the proof of the Buchberger criterion we have seen that, in order to show that $\mathcal{G}$ is a Gröbner basis of J, it is enough to check that the S- and T-polynomials corresponding to a *minimal* set of generators of U reduce to 0 with respect to $g_1, \ldots, g_s$.

Theorem 5.2.6 provides **Buchberger's algorithm** for computing a Gröbner basis:

(1) Start with a set of $\mathcal{G} = \{g_1, \ldots, g_s\}$ of homogeneous generators of J. If all S- and T-polynomials reduce to 0 with respect of $g_1, \ldots, g_s$, then $\mathcal{G}$ is a Gröbner basis of J, and the algorithm stops.

(2) If this is not the case, then one of the S- or T-polynomials has a remainder $r \neq 0$. Then replace $\mathcal{G}$ by $\mathcal{G}' = \mathcal{G} \cup \{r\}$ and proceed with (1) where $\mathcal{G}$ is replaced by $\mathcal{G}'$.

The algorithm terminates simply because there are only finitely many monomials in E.

5.2.3 Generic initial ideals and generic annihilator numbers in the exterior algebra

Throughout this section we will assume that K is an infinite field. We let as before V be an n-dimensional K-vector space with basis $e_1, \ldots, e_n$. We identify the elements in $\mathrm{GL}(n; K)$ with the automorphisms of V. Let $\alpha = (a_{ij}) \in \mathrm{GL}(n; K)$; then the corresponding automorphism is given by

$$\alpha(\sum_{i=1}^n x_i e_i) = \sum_{j=1}^n (\sum_{i=1}^n a_{ji} x_i) e_j.$$

This automorphism induces a K-algebra automorphism $E \to E$ which we denote again by α.

Let $<$ be a monomial order on E with $e_1 > e_2 > \cdots > e_n$ and J a graded ideal in E. Then one defines the generic initial ideal $\mathrm{gin}_<(J)$ as for ideals in the polynomial ring. The proof of its existence and the property of being Borel-fixed is verbatim the same as in the case of the polynomial ring.

The next two theorems summarize these facts.

Theorem 5.2.8. *Let $J \subset E$ be a graded ideal and $<$ a monomial order on E. Then there exists a nonempty open subset $U \subset \mathrm{GL}(n; K)$ such that $\mathrm{in}_<(\alpha J) = \mathrm{in}_<(\alpha' J)$ for all $\alpha, \alpha' \in U$.*

The ideal $\mathrm{in}_<(\alpha J)$ for $\alpha \in U$ is called the **generic initial ideal** of J with respect to $<$, and is denoted $\mathrm{gin}_<(J)$.

Theorem 5.2.9. *$\mathrm{gin}_<(J)$ is a Borel-fixed ideal, that is, $\mathrm{gin}_<(J)$ is stable under the action of the Borel subgroup $\mathcal{B}$ of $\mathrm{GL}(n; K)$, and $\mathrm{gin}_<(J) = J$ if J is Borel-fixed.*

A monomial ideal $J \subset E$ is called **strongly stable** if for each monomial $\mathbf{e}_F \in J$ and each $j \in F$ and $i < j$ one has that $e_i \wedge \mathbf{e}_{F \setminus \{j\}} \in J$.

In Proposition 4.2.6 we have seen that the generic initial ideal of a graded ideal in the polynomial ring is strongly stable (in the sense of monomial ideals in a polynomial ring), provided $\mathrm{char}(K) = 0$.

The corresponding result holds here. But we need no assumption on the characteristic of K.

Proposition 5.2.10. *The generic initial of a graded ideal $J \subset E$ is strongly stable.*

Proof. Suppose $\mathrm{gin}_<(J)$ is not strongly stable. Then there exists a monomial $\mathbf{e}_F \in \mathrm{gin}_<(J)$ and numbers $i < j$ with $j \in F$ such that $e_i \wedge \mathbf{e}_{F \setminus \{j\}} \notin \mathrm{gin}_<(J)$. Let $\alpha \in \mathcal{B}$ with $\alpha(e_j) = e_i + e_j$ and $\alpha(e_k) = e_k$ for $k \neq j$. Then $\alpha(\mathbf{e}_F) = \mathbf{e}_F \pm e_i \wedge \mathbf{e}_{F \setminus \{j\}}$, and hence does not belong to $\mathrm{gin}_<(J)$, contradicting Theorem 5.2.9. $\qquad\square$

Let $\mathbf{v} = v_1, \ldots, v_n$ be a K-basis of E_1 and M a graded E-module. Then for each i, the module $H(M/(v_1, \ldots, v_{i-1})M, v_i)$ is a graded E-module with jth graded components $H^j(M/(v_1, \ldots, v_{i-1})M, v_i)$. We set

$$\alpha_{ij}(\mathbf{v}; M) = \begin{cases} \dim_K H^j(M/(v_1, \ldots, v_{i-1})M, v_i), & \text{if } i < n, \\ \beta_{0j}(M), & \text{if } i = n, \end{cases}$$

and call these numbers the **annihilator numbers** of M with respect to $\mathbf{v}$.

Theorem 5.2.11. *Let $J \subset E$ be a graded ideal. With each $\gamma = (g_{ij}) \in \mathrm{GL}_n(K)$ we associate the sequence $\mathbf{v} = \gamma(\mathbf{e})$ with $v_j = \sum_{i=1}^n g_{ij} e_i$ for $j = 1, \ldots, n$. Then there exists a nonempty Zariski open subset $U \subset \mathrm{GL}_n(K)$ such that $\alpha_{ij}(\gamma(\mathbf{e}); E/J) = \alpha_{ij}(\mathbf{f}; E/\mathrm{gin}_{<_{\mathrm{rev}}}(J))$ for all i and j and all $\gamma \in U$, where $\mathbf{f} = e_n, e_{n-1}, \ldots, e_1$.*

In Subsection 5.1.4 we introduced generalized simplicial homology. As a consequence of Theorem 5.2.11 we have

Corollary 5.2.12. *Let $J \subset E$ be a graded ideal. Then*

$$H(E/J) = H(E/\mathrm{gin}_{<_{\mathrm{rev}}}(J), e_n) = H(E/\mathrm{gin}_{<_{\mathrm{rev}}}(J)).$$

Problems

5.1. Let V be a K-vector space with basis $e_1, \ldots, e_n$. By using the fact that $e_1 \wedge e_2 \wedge \cdots \wedge e_n \neq 0$, show that the elements $\mathbf{e}_F$, $F \subset [n]$ are linearly independent.

5.2. Let E be the exterior algebra of a finite-dimensional K-vector space, and M a graded E-module (cf. Definition 5.1.1). Show that $(fx)g = f(xg)$ for all $f, g \in E$ and all $x \in M$. (This then proves that M is an E-E bimodule.)

5.3. Let K be a field of characteristic $\neq 2$, V the K-vector space with basis e_1, e_2, e_3, e_4, E its exterior algebra and $f = e_1 + e_2 \wedge e_3 \in E$. Consider the left ideal $I = (\{g \wedge f : g \in E\})$ and the right ideal $J = (\{f \wedge g : g \in E\})$. Show that $I \neq J$.

5.4. Show that the map $E \to E^{\vee}$ which assigns to each homogeneous element $f \in E$ the element $g \mapsto g \wedge f$ in $E^{\vee}$ is an isomorphism of graded E-modules.

5.5. By referring to the definition of the oriented chain complex $\tilde{C}(\Delta; K)$ for a simplicial complex Δ as given in Subsection 5.1.4, show that the chain map $\partial \colon \tilde{C}_j(\Delta; K) \to \tilde{C}_{j-1}(\Delta; K)$ is defined by

$$\partial([i_0, i_1, \cdots, i_j]) = \sum_{k=0}^{j} (-1)^{k+1} [i_0, i_1, \cdots i_{k-1}, i_k \cdots, i_j],$$

where $F = \{i_0 < i_1 < \cdots < i_j\} \in \Delta$.

5.6. Let $J \subset E$ be a graded ideal and $<$ a monomial order on E. Show that $\mathrm{in}_<(J)$ coincides with the K-vector space spanned by the monomials $\mathrm{in}_<(f)$ with $0 \neq f \in J$, and f homogeneous.

5.7. Let $e = \sum_{i=1}^{n} e_i$ and $J = (e) \subset E$. Then for any monomial order, e is a Gröbner basis for J.

5.8. Let E be the exterior algebra of the K-vector space with basis e_1, e_2, e_3, and fix a monomial order $<$ with $e_1 > e_2 > e_3$. Compute $\mathrm{gin}_<(J)$ for $J = (e_2 \wedge e_3)$.

Notes

The exterior face ring was introduced by Gil Kalai [Kal84] in order to define exterior algebraic shifting, which will be treated in Chapter 11.

Shifting theory is defined by using generic initial ideals. For this purpose one has to develop Gröbner basis theory over the exterior algebra. In our presentation of this theory we followed the exposition in [AHH97].

Hilbert functions and resolutions

6

Hilbert functions and the theorems of Macaulay and Kruskal–Katona

The Hilbert function of a graded K-algebra R counts the vector space dimension of its graded components. It encodes important information on R such as its Krull dimension or its multiplicity. Hilbert's fundamental theorem tells us that the Hilbert function is a polynomial function for all large integers. The possible Hilbert functions are described by Macaulay's theorem.

The Hilbert function of the Stanley–Reisner ring of a simplicial complex Δ is determined by the f-vector of Δ, and vice versa. The possible f-vectors of a simplicial complex are characterized in the theorem of Kruskal–Katona. This theorem is the "squarefree" analogue of Macaulay's theorem.

6.1 Hilbert functions, Hilbert series and Hilbert polynomials

6.1.1 The Hilbert function of a graded R-module

Let K be a field and let $R = \bigoplus_{i \geq 0} R_i$ be graded K-algebra. R is called **standard graded** if R is a finitely generated K-algebra and all its generators are of degree 1. In other words, $R = K[R_1]$ and $\dim_K R_1 < \infty$. The archetype of a standard graded K-algebra is the polynomial ring $S = K[x_1, \ldots, x_n]$ with $\deg x_i = 1$ for $i = 1, \ldots, n$. Any other standard graded K-algebra is isomorphic to the polynomial ring modulo a graded ideal; that is, an ideal which is generated by homogeneous polynomials.

Let M be a finitely generated graded R-module. All the graded components M_i of M are finite-dimensional K-vector spaces. An element $x \in M$ is called **homogeneous** of degree i if $x \in M_i$. Any element in M can be uniquely written as a (finite) sum of homogeneous elements.

Definition 6.1.1. The numerical function

$$H(M, -) : \mathbb{Z} \longrightarrow \mathbb{Z}, \quad i \mapsto H(M, i) := \dim_K M_i$$

J. Herzog, T. Hibi, *Monomial Ideals*, Graduate Texts in Mathematics 260, DOI 10.1007/978-0-85729-106-6_6, © Springer-Verlag London Limited 2011

is called the **Hilbert function** of M. The formal Laurent series $H_M(t) = \sum_{i \in \mathbb{Z}} H(M, i) t^i$ is called the **Hilbert series** of M.

Example 6.1.2. Let $S = K[x_1, \ldots, x_n]$ be the polynomial ring in n variables. The monomials of degree i in S form a K-basis of S_i. It follows that

$$H(S, i) = \binom{n + i - 1}{i} = \binom{n + i - 1}{n - 1} \quad \text{and} \quad H_S(t) = \frac{1}{(1 - t)^n}.$$

Note that $H(S, -)$ is a polynomial function of degree $n - 1$ and that $H_S(t)$ is a rational function with exactly one pole at $t = 1$.

For an arbitrary graded R-module the Hilbert function and the Hilbert series are of the same nature as in the special case described in the example.

Theorem 6.1.3 (Hilbert). *Let K be a field, R a standard graded K-algebra and M a nonzero finitely generated graded R-module of dimension d. Then*

(a) *there exists a Laurent-polynomial $Q_M(t) \in \mathbb{Z}[t, t^{-1}]$ with $Q_M(1) > 0$ such that*

$$H_M(t) = \frac{Q_M(t)}{(1 - t)^d};$$

(b) *there exists a polynomial $P_M(x) \in \mathbb{Q}[x]$ of degree $d - 1$ (called the **Hilbert polynomial** of M) such that*

$$H(M, i) = P_M(i) \quad \text{for all} \quad i > \deg Q_M - d.$$

Proof. (a) After a base field extension we may assume that K is infinite. We proceed by induction on $\dim M$. If $\dim M = 0$, then $M_i = 0$ for $i \gg 0$, and the assertion is trivial.

Suppose now that $d = \dim M > 0$. We choose $y \in R_1$ such that $y \in \mathfrak{m} \setminus \bigcup_{P \in \mathrm{Ass}(M) \setminus \{\mathfrak{m}\}} P$, where $\mathfrak{m} = \bigoplus_{i > 0} R_i$. Then y is almost regular on M (cf. the proof of Lemma 4.3.1), and $\dim M/yM = d - 1$ since y does not belong to any minimal prime ideal of M.

The exact sequence

$$0 \longrightarrow N \longrightarrow M(-1) \overset{y}{\longrightarrow} M \longrightarrow M/yM \longrightarrow 0$$

with $N = (0 :_M y)(-1)$ yields the identity

$$H_{M/yM}(t) - H_M(t) + t H_M(t) - H_N(t) = 0.$$

In other words we have

$$H_M(t) = \frac{H_{M/yM}(t) - H_N(t)}{1 - t}.$$

By our induction hypothesis there exists a Laurent polynomial $Q_{M/yM}(t)$ with $Q_{M/yM}(1) > 0$ such that

$$H_{M/yM}(t) = \frac{Q_{M/yM}(t)}{(1-t)^{d-1}}.$$

Thus we see that

$$H_M(t) = \frac{Q_{M/yM}(t)/(1-t)^{d-1} - H_N(t)}{1-t} = \frac{Q_M(t)}{(1-t)^d}$$

with

$$Q_M(t) = Q_{M/yM}(t) - H_N(t)(1-t)^{d-1}. \tag{6.1}$$

Since $H_N(t)$ is a Laurent polynomial it follows from (6.1) that $Q_M(t)$ is a Laurent polynomial. Equation (6.1) also implies that $Q_M(1) = Q_{M/yM}(1) > 0$ if $d > 1$. Thus it remains to be shown that $Q_M(1) > 0$ if $d = 1$, or equivalently that $\ell(0 :_M y) < \ell(M/yM)$ if $\dim M = 1$.

Observe that M is a finitely generated $A = K[y]$-module, since M/yM has finite length. Since A is a principal ideal domain, and since M is a graded A-module of dimension 1, it follows that $M = A^r \oplus \bigoplus_{i=1}^s A/(y^{a_i})$ with $r > 0$. Thus we see that $\ell(M/yM) = r + s > s = \ell(0 :_M y)$.

(b) Let $Q_M(t) = \sum_{i=r}^s h_i t^i$. Then

$$H_M(t) = (\sum_{i=r}^s h_i t^i)/(1-t)^d = \sum_{i=r}^s h_i t^i \sum_{j \geq 0} \binom{d+j-1}{d-1} t^j.$$

By using the convention that $\binom{a}{i} = 0$ for $a < i$, we deduce from the preceding equation that

$$H(M, i) = \sum_{j=r}^s h_j \binom{d+(i-j)-1}{d-1}. \tag{6.2}$$

In particular, if we set $P_M(x) = \sum_{j=r}^s h_j \binom{x+d-j-1}{d-1}$, then $P_M(x)$ is a polynomial of degree $d-1$ with $H(M, i) = P_M(i)$ for $i > s - d$. $\quad\square$

Theorem 6.1.3 implies that the Krull dimension d of M is the pole order of the rational function $H_M(t)$ at $t = 1$. The **multiplicity** $e(M)$ of M is defined to be the positive number $Q_M(1)$. It follows from (6.2) that $e(M)/(d-1)!$ is the leading coefficient of the Hilbert polynomial $P_M(x)$ of M.

The a-**invariant** is the degree of the Hilbert series $H_M(t)$, that is, the number $\deg Q_M(t) - d$.

Let $Q_M(t) = \sum_{i=r}^s h_i t^i$. The coefficient vector $(h_r, h_{r+1}, \ldots, h_s)$ of $Q_M(t)$ is called the h-**vector** of M.

6.1.2 Hilbert functions and initial ideals

Let K be a field, $S = K[x_1, \ldots, x_n]$ the polynomial ring in n variables and $I \subset S$ an ideal. For a given monomial order $<$ there is a natural monomial K-basis of the residue class ring S/I. We denote by $\mathrm{Mon}(\mathrm{in}_<(I))$ the set of monomials in $\mathrm{in}_<(I)$.

Theorem 6.1.4 (Macaulay). *The set of monomials* $\mathrm{Mon}(S) \setminus \mathrm{Mon}(\mathrm{in}_<(I))$ *form a K-basis of S/I.*

Proof. Let $\mathcal{G} = \{g_1, \ldots, g_r\}$ be a Gröbner basis of I, and let $f \in S$. Then by Lemma 2.2.3, f has a unique remainder f' with respect to $\mathcal{G}$. The residue class of f modulo I is the same as that of f', and no monomial in the support of f' is divided by any of the monomials $\mathrm{in}_<(g_i)$. This shows that $\mathrm{Mon}(S) \setminus \mathrm{Mon}(\mathrm{in}_<(I))$ is a system of generators of the K-vector space S/I.

Assume there exists a set $\{u_1, \ldots, u_s\} \subset \mathrm{Mon}(S) \setminus \mathrm{Mon}(\mathrm{in}_<(I))$ and $a_i \in K \setminus \{0\}$ such that $h = \sum_{i=1}^s a_i u_i \in I$. We may assume that $u_1 = \mathrm{in}(h)$. Then $u_1 = \mathrm{in}_<(h) \in \mathrm{Mon}(\mathrm{in}_<(I))$, a contradiction. $\qquad\square$

As an immediate consequence we obtain the following important result

Corollary 6.1.5. *Let $I \subset S$ be a graded ideal and $<$ a monomial order on S. Then S/I and $S/\mathrm{in}_<(I)$ have the same Hilbert function, i.e. $H(S/I, i) = H(S/\mathrm{in}_<(I), i)$ for all i.*

We also obtain a Gröbner basis criterion.

Corollary 6.1.6. *Let $\mathcal{G} = \{g_1, \ldots, g_r\}$ be a homogeneous system of generators of I, and let $J = (\mathrm{in}_<(g_1), \ldots, \mathrm{in}_<(g_r))$. Then $\mathcal{G}$ is a Gröbner basis of I if and only if S/I and S/J have the same Hilbert function.*

Proof. We have $J \subset \mathrm{in}_<(I)$, so that $H(S/J, i) \geq H(S/\mathrm{in}_<(I), i) = H(S/I, i)$ for all i. Equality holds if and only if $J = \mathrm{in}_<(I)$. $\qquad\square$

6.1.3 Hilbert functions and resolutions

Let K be a field, $S = K[x_1, \ldots, x_n]$ the polynomial ring in n variables and M a finitely generated graded S-module. Let

$$\mathbb{F}: 0 \longrightarrow F_p \to F_{p-1} \longrightarrow \cdots \longrightarrow F_1 \longrightarrow F_0 \longrightarrow M \longrightarrow 0$$

be a graded minimal free S-resolution of M with

$$F_i = \bigoplus_j S(-j)^{\beta_{ij}} = \bigoplus_{j=1}^{\beta_i} S(-d_{ij}).$$

By using the fact that the Hilbert function is additive on short exact sequences and by using that $H_{S(-j)}(t) = t^j/(1-t)^n$, we deduce from the free S-resolution $\mathbb{F}$ of M the formula

$$H_M(t) = \frac{R_M(t)}{(1-t)^n}, \tag{6.3}$$

where $R_M(t) = \sum_{i=0}^p (-1)^i \sum_j \beta_{ij} t^j = \sum_{i=0}^p (-1)^i \sum_{j=1}^{\beta_i} t^{d_{ij}}$. A comparison with Theorem 6.1.3 shows that

$$Q_M(t)(1-t)^{n-d} = R_M(t)$$

From this equation we deduce that

$$R_M^{(n-d)}(1) = (-1)^{(n-d)}(n-d)! Q_M(1) = (-1)^{(n-d)}(n-d)! e(M),$$

and that $R_M^{(i)}(1) = 0$ for $i = 0, \ldots, n-d-1$. (Here $R_M^{(i)}$ denotes the ith formal derivative of R_M).

Thus

$$\sum_{i=0}^{p}(-1)^i \sum_{j=1}^{\beta_i} \prod_{r=0}^{k-1}(d_{ij} - r) = \begin{cases} 0, & \text{for } 0 \le k < n-d, \\ (-1)^{n-d}(n-d)! e(M), & \text{for } k = n-d. \end{cases}$$

This immediately implies the following useful formulas

Corollary 6.1.7. *With the notation introduced one has*

$$\sum_{i=0}^{p}(-1)^i \sum_{j=1}^{\beta_i} d_{ij}^k = \begin{cases} 0, & \text{for } 0 \le k < n-d, \\ (-1)^{n-d}(n-d)! e(M), & \text{for } k = n-d. \end{cases}$$

6.2 The h-vector of a simplicial complex

The h-vector of a module, as defined in Section 6.1 together with the Krull-dimension of the module encodes all the information provided by the Hilbert function. Let Δ be a $(d-1)$-dimensional simplicial complex on the vertex set $[n]$. In this section we want to relate the h-vector of the Stanley–Reisner ring $K[\Delta]$ to the f-vector $f(\Delta) = (f_0, f_1, \ldots, f_{d-1})$ of Δ. Here f_i denotes the number of faces of Δ of dimension i. Letting $f_{-1} = 1$, we defined in Chapter 1 the *h-vector* $h(\Delta) - (h_0, h_1, \ldots, h_d)$ of Δ by the formula

$$\sum_{i=0}^{d} f_{i-1}(t-1)^{d-i} = \sum_{i=0}^{d} h_i t^{d-i}.$$

Equivalently,

$$\sum_{i=0}^{d} f_{i-1} t^i (1-t)^{d-i} = \sum_{i=0}^{d} h_i t^i. \tag{6.4}$$

The following result justifies this definition with hindsight.

Proposition 6.2.1. *Let Δ be a simplicial complex of dimension $d-1$ with f-vector $(f_0, f_1, \ldots, f_{d-1})$. Then*

$$H_{K[\Delta]}(t) = \frac{\sum_{i=0}^{d} f_{i-1} t^i (1-t)^{d-i}}{(1-t)^d}.$$

Proof. Write $K[\Delta] = S/I_\Delta$ where $S = k[x_1, \ldots, x_n]$. By Corollary 1.1.4 the monomials not belonging to I_Δ form a K-basis of $K[\Delta]$. For a monomial $u = \mathbf{x}^{\mathbf{a}}$ we set $\operatorname{supp}(u) = \{i \in [n]: a_i \neq 0\}$. By Proposition 1.5.1, the monomials $u \in \operatorname{Mon}(S)$ with $\operatorname{supp}(u) \in \Delta$ form a K-basis of $K[\Delta]$.

Fix a face $F \in \Delta$. Then

$$\{u \in \operatorname{Mon}(S): \operatorname{supp}(u) = F\} = \{x_F v: \ v \in \operatorname{Mon}(K[\{x_i\}_{i \in F}])\}.$$

Since the disjoint union of the sets $\{u \in \operatorname{Mon}(S): \operatorname{supp}(u) = F\}$ with $F \in \Delta$ establishes a K-basis of $K[\Delta]$, we see that

$$H_{K[\Delta]}(t) = \sum_{F \in \Delta} \frac{t^{|F|}}{(1-t)^{|F|}},$$

and the desired formula for $H_{K[\Delta]}(t)$ follows. $\qquad\square$

Combining Proposition 6.2.1 with Theorem 6.1.3 we obtain

Corollary 6.2.2. *Let Δ be a simplicial complex of dimension $d - 1$ and K a field. Then $\dim K[\Delta] = d$.*

The number $\chi(\Delta) = \sum_{i=0}^{d-1} (-1)^i f_i$ is called the **Euler characteristic** of Δ. In terms of simplicial homology one has

$$-1 + \chi(\Delta) = \sum_{i=0}^{d} (-1)^{i-1} \dim_K K\{\Delta\}_i = \sum_{i=-1}^{d-1} (-1)^i \dim_K \tilde{H}^i(\Delta; K),$$

see Definition 5.1.7.

The Euler characteristic of Δ can also be expressed by the h-vector and the multiplicity of $K[\Delta]$ by the f-vector, as follows at once from (6.4).

Corollary 6.2.3. *With the notation introduced one has*

$$\chi(\Delta) = (-1)^{d-1} h_d + 1 \quad \text{and} \quad e(K[\Delta]) = f_{d-1}.$$

6.3 Lexsegment ideals and Macaulay's theorem

The purpose of the present section is to give the complete characterization of the possible Hilbert functions of a standard graded K-algebra R, where K as usual is a field.

We may write $R = S/I$ where $S = K[x_1, \ldots, x_n]$ is the polynomial ring with standard grading, and where $I \subset S$ is a graded ideal. By Theorem 6.1.5 we know that S/I and $S/\operatorname{in}_<(I)$ have the same Hilbert function for any monomial order on S. Therefore we may as well assume that $I \subset S$ is a monomial ideal. Since by Corollary 6.1.4 the monomials in S not belonging to I form a K-basis of S/I, and since this K-basis determines the Hilbert functions of S/I, it is then apparent that the Hilbert function of S/I does not depend on the base field K. Thus we assume that char $K = 0$. Now we pass from S/I to $S/\operatorname{gin}_<(I)$, again without changing the Hilbert function. By Proposition 4.2.6 we have that $\operatorname{gin}_<(I)$ is a strongly stable ideal.

Hence the problem of characterizing the Hilbert function of S/I is reduced to the case that I is a strongly stable ideal. A further reduction is needed to characterize the possible Hilbert functions.

Among the strongly stable monomial ideals (introduced in Subsection 4.2.2), there is a very distinguished class of monomial ideals, called lexsegment ideals.

We denote by $\operatorname{Mon}_d(S)$ the set of all monomials of S of degree d. A set $\mathcal{L} \subset \operatorname{Mon}_d(S)$ of monomials is called a **lexsegment** if there exists $u \subset \mathcal{L}$ such that $v \in \mathcal{L}$ for all $v \in \operatorname{Mon}_d(S)$ with $v \geq_{\text{lex}} u$.

One says that $\mathcal{L} \subset M_d(S)$ is **strongly stable**, if one has $x_i(u/x_j) \in \mathcal{L}$ for all $u \in \mathcal{L}$ and all $i < j$ such that x_j divides u.

For a monomial $u \in S$ we set $m(u) = \max\{i\colon x_i \text{ divides } u\}$, and call a set $\mathcal{L} \subset \operatorname{Mon}_d(S)$ **stable**, if $x_i(u/x_{m(u)}) \in \mathcal{L}$ for all $u \in \mathcal{L}$, and all $i < m(u)$.

As already defined before, a monomial ideal I is called a **lexsegment ideal**, or a **(strongly) stable monomial ideal**, if for each d the monomials of degree d in I form a lexsegment, or a (strongly) stable set of monomials, respectively.

Obviously one has the following implications:

$$\text{lexsegment} \Rightarrow \text{strongly stable} \Rightarrow \text{stable},$$

and all implications are strict.

A lexsegment in a polynomial ring may no longer be a lexsegment in a polynomial ring extension. For example, the set $\{x_1^2, x_1 x_2, x_2^2\}$ is a lexsegment in $K[x_1, x_2]$ but not in $K[x_1, x_2, x_3]$. A set which remains a lexsegment in all polynomial ring extensions is called a **universal lexsegment**.

The fundamental result of this section is the following

Theorem 6.3.1. *Let $I \subset S$ be a graded ideal. Then there exists a unique lexsegment ideal, denoted I^{lex}, such that S/I and S/I^{lex} have the same Hilbert function.*

The idea of the proof is simple: say, $I \subset S$ is a graded ideal. For each graded component I_j of I, and let I_j^{lex} be the K-vector space spanned by the (unique) lexsegment $\mathcal{L}_j$ with $|\mathcal{L}_j| = \dim_K I_j$. Then define $I^{\text{lex}} = \bigoplus_j I_j^{\text{lex}}$.

Obviously the I^{lex} so constructed is the only possible candidate meeting the requirements of the theorem. The only problem is, whether it is an ideal. If this is the case, then this is the unique lexsegment ideal with the same Hilbert function as I. It is clear that I^{lex} is indeed an ideal if and only if $\{x_1, \ldots, x_n\}\mathcal{L}_j \subset \mathcal{L}_{j+1}$.

Let $\mathcal{N} \subset \operatorname{Mon}(S)$ be any set of monomials. Then we call the set

$$\text{Shad}(\mathcal{N}) = \{x_1, \ldots, x_n\}\mathcal{N} = \{x_i u \colon u \in \mathcal{N}, i = 1, \ldots, n\}$$

the **shadow** of $\mathcal{N}$. Therefore I^{lex} is an ideal if and only if $\text{Shad}(\mathcal{L}_j) \subset \mathcal{L}_{j+1}$ for all j.

Note that if $\mathcal{N} \subset \text{Mon}_d(S)$ is stable, strongly stable or a lexsegment, then so is $\text{Shad}(\mathcal{N})$.

For stable ideals, the length of the shadow can be computed. Let $\mathcal{N} \subset \text{Mon}_d(S)$ be a set of monomials. We denote by $m_i(\mathcal{N})$ the number of elements $u \in \mathcal{N}$ with $m(u) = i$, and set $m_{\leq i}(\mathcal{N}) = \sum_{j=1}^{i} m_j(\mathcal{N})$. Then we have

Lemma 6.3.2. *Let* $\mathcal{N} \subset \text{Mon}_d(S)$ *be a stable set of monomials. Then* $\text{Shad}(\mathcal{N})$ *is again a stable set and*

(a) $m_i(\text{Shad}(\mathcal{N})) = m_{\leq i}(\mathcal{N})$;
(b) $|\text{Shad}(\mathcal{N})| = \sum_{i=1}^{n} m_{\leq i}(\mathcal{N})$.

Proof. (b) is of course a consequence of (a). For the proof of (a) we note that the map

$$\phi \colon \{u \in \mathcal{N} \colon m(u) \leq i\} \to \{u \in \text{Shad}(\mathcal{N}) \colon m(u) = i\}, \quad u \mapsto u x_i$$

is a bijection. In fact, ϕ is clearly injective. To see that ϕ is surjective, we let $v \in \text{Shad}(\mathcal{N})$ with $m(v) = i$. Since $v \in \text{Shad}(\mathcal{N})$, there exists $w \in \mathcal{N}$ with $v = x_j w$ for some $j \leq i$. It follows that $m(w) \leq i$. If $j = i$, then we are done. Otherwise, $j < i$ and $m(w) = i$. Then, since $\mathcal{N}$ is stable it follows that $u = x_j(w/x_i) \in \mathcal{N}$. The assertion follows, since $v = u x_i$.   □

Now Theorem 6.3.1 will be an easy consequence of

Theorem 6.3.3 (Bayer). *Let* $\mathcal{L} \subset \text{Mon}_d(S)$ *be a lexsegment, and* $\mathcal{N} \subset \text{Mon}_d(S)$ *be a strongly stable set of monomials with* $|\mathcal{L}| \leq |\mathcal{N}|$. *Then* $m_{\leq i}(\mathcal{L}) \leq m_{\leq i}(\mathcal{N})$ *for* $i = 1, \ldots, n$.

Proof. We first observe that $\mathcal{N} = \mathcal{N}_0 \cup \mathcal{N}_1 x_n \cup \cdots \cup \mathcal{N}_d x_n^d$ where each $\mathcal{N}_j$ is a strongly stable set of monomials of degree $d - j$ in the variables $x_1, \ldots, x_{n-1}$. Such a decomposition is unique. Similarly, one has the decomposition $\mathcal{L} = \mathcal{L}_0 \cup \mathcal{L}_1 x_n \cup \cdots \cup \mathcal{L}_d x_n^d$ where each $\mathcal{L}_j$ is a lexsegment.

We prove the theorem by induction on the number of variables. For $n = 1$, the assertion is trivial. Now let $n > 1$. The inequality

$$m_{\leq i}(\mathcal{L}) \leq m_{\leq i}(\mathcal{N}) \tag{6.5}$$

is trivial for $i = n$, since $|\mathcal{L}| = m_{\leq n}(\mathcal{L})$ and $|\mathcal{N}| = m_{\leq n}(\mathcal{N})$.

Note that $|\mathcal{L}_0| = m_{\leq n-1}(\mathcal{L})$ and that $|\mathcal{N}_0| = m_{\leq n-1}(\mathcal{N})$. Thus in order to prove (6.5) for $i = n - 1$ we have to show that

$$|\mathcal{L}_0| \leq |\mathcal{N}_0|. \tag{6.6}$$

Assume for a moment that (6.6) holds. Then by applying our induction hypothesis we obtain

$$m_{\leq i}(\mathcal{L}) = m_{\leq i}(\mathcal{L}_0) \leq m_{\leq i}(\mathcal{N}_0) = m_{\leq i}(\mathcal{N}) \quad \text{for} \quad i = 1, \ldots, n-1,$$

as desired. Thus it remains to prove the inequality (6.6).

For each j, let $\mathcal{N}_j^*$ be the lexsegment in $\text{Mon}_{d-j}(K[x_1, \ldots, x_{n-1}])$ with $|\mathcal{N}_j^*| = |\mathcal{N}_j|$ and set $\mathcal{N}^* = \mathcal{N}_0^* \cup \mathcal{N}_1^* x_n \cup \cdots \cup \mathcal{N}_d^* x_n^d$.

We claim that $\mathcal{N}^*$ is again a strongly stable set of monomials. Indeed, we need to show that $\{x_1, \ldots, x_{n-1}\}\mathcal{N}_j^* \subset \mathcal{N}_{j-1}^*$ for $j = 1, \ldots, d$. Since the sets $\{x_1, \ldots, x_{n-1}\}\mathcal{N}_j^*$ and $\mathcal{N}_{j-1}^*$ are lexsegments, it suffices to show that

$$|\{x_1, \ldots, x_{n-1}\}\mathcal{N}_j^*| \leq |\mathcal{N}_{j-1}^*|.$$

By using that $\mathcal{N}$ is a stable set of monomials we have that $\{x_1, \ldots, x_{n-1}\}\mathcal{N}_j \subset \mathcal{N}_{j-1}$ for $j = 1, \ldots, d$. Now we apply Lemma 6.3.2 and our induction hypothesis and obtain

$$|\{x_1, \ldots, x_{n-1}\}\mathcal{N}_j^*| = \sum_{i=1}^{n-1} m_{\leq i}(\mathcal{N}_j^*) \leq \sum_{i=1}^{n-1} m_{\leq i}(\mathcal{N}_j)$$
$$= |\{x_1, \ldots, x_{n-1}\}\mathcal{N}_j| \leq |\mathcal{N}_{j-1}| = |\mathcal{N}_{j-1}^*|.$$

This completes the proof of the fact that $\mathcal{N}^*$ is a strongly stable set of monomials.

Since $|\mathcal{N}^*| = |\mathcal{N}|$, we may replace $\mathcal{N}$ by $\mathcal{N}^*$ and thus may as well assume that $\mathcal{N}_0$ is a lexsegment.

For a set of monomials $\mathcal{S}$ we denote by $\min \mathcal{S}$ the lexicographically smallest element in $\mathcal{S}$. Since both $\mathcal{L}_0$ and $\mathcal{N}_0$ are lexsegments, inequality (6.6) will follow once we have shown that $\min \mathcal{L}_0 \geq \min \mathcal{N}_0$.

Given a monomial $m = \prod_{i=1}^{n} x_i^{a_i}$, we set $\bar{m} = (x_{n-1}/x_n)^{a_n} m$. This assignment is order preserving. In other words, if $m, n \in \text{Mon}_d(S)$ with $m \leq n$ (with respect to the lexicographic order), then $\bar{m} \leq \bar{n}$. We leave the verification of this simple fact to the reader.

Let $u = \min \mathcal{L}$ and $v = \min \mathcal{N}$. Since $\mathcal{N}$ is a strongly stable set of monomials it follows that $\bar{v} \in \mathcal{N}_0$. Hence $\min \mathcal{N}_0 \leq \bar{v}$. On the other hand, if $w = \min \mathcal{N}_0$, then $w \geq v$, and so $w = \bar{w} \geq \bar{v}$. In other words, we have $\min \mathcal{N}_0 = \bar{v}$. Similarly, we have $\min \mathcal{L}_0 = \bar{u}$.

Finally we observe that $u \geq v$ since $\mathcal{L}$ is a lexsegment and since $|\mathcal{L}| \leq |\mathcal{N}|$, by assumption. Hence we conclude that

$$\min \mathcal{L}_0 = \bar{u} \geq \bar{v} = \min \mathcal{N}_0,$$

as desired. □

Now we are ready to prove Theorem 6.3.1: based on our discussions following Theorem 6.3.1 it remains to be shown that if $I \subset S$ is a graded ideal and

$\mathcal{L}_j$ is the lexsegment with $|\mathcal{L}_j| = \dim_K I_j$, then $\operatorname{Shad}(\mathcal{L}_j) \subset \mathcal{L}_{j+1}$. As we have seen before, we may assume that I is strongly stable. Let $\mathcal{N}_j$ be the strongly stable set of monomials which spans the K-vector space I_j. Since $|\mathcal{L}_j| = |\mathcal{N}_j|$, Bayer's theorem together with Lemma 6.3.2 implies that

$$|\operatorname{Shad}(\mathcal{L}_j)| = \sum_{i=1}^{n} m_{\leq i}(\mathcal{L}_j) \leq \sum_{i=1}^{n} m_{\leq i}(\mathcal{N}_j) = |\operatorname{Shad}(\mathcal{N}_j)|.$$

On the other hand, since I is an ideal we clearly have that $\operatorname{Shad}(\mathcal{N}_j) \subset \mathcal{N}_{j+1}$. Hence

$$|\operatorname{Shad}(\mathcal{L}_j)| \leq |\operatorname{Shad}(\mathcal{N}_j)| \leq |\mathcal{N}_{j+1}| = |\mathcal{L}_{j+1}|.$$

Since, both $\operatorname{Shad}(\mathcal{L}_j)$ and $\mathcal{L}_{j+1}$ are lexsegments, this implies $\operatorname{Shad}(\mathcal{L}_j) \subset \mathcal{L}_{j+1}$, as desired.

We shall now use Theorem 6.3.1 to derive the conditions which characterize the Hilbert functions of standard graded K-algebras. To this end we introduce the so-called **binomial** or **Macaulay expansion** of a number. We first show

Lemma 6.3.4. *Let j be a positive integer. Then each positive integer a has a unique expansion*

$$a = \binom{a_j}{j} + \binom{a_{j-1}}{j-1} + \cdots + \binom{a_k}{k}$$

with $a_j > a_{j-1} > \cdots > a_k \geq k \geq 1$.

Proof. We choose a_j maximal such that $a \geq \binom{a_j}{j}$. If equality holds, then this is the desired expansion. Otherwise let $a' = a - \binom{a_j}{j}$. Then $a' > 0$ and by using induction on a, and since $a' < a$ we may assume that $a' = \binom{a_{j-1}}{j-1} + \cdots + \binom{a_k}{k}$ with $a_{j-1} > \cdots > a_k \geq k \geq 1$. Therefore, $a = \binom{a_j}{j} + \binom{a_{j-1}}{j-1} + \cdots + \binom{a_k}{k}$, and it remains to be shown that $a_j > a_{j-1}$. Since $\binom{a_j+1}{j} > a$ it follows that

$$\binom{a_j}{j-1} = \binom{a_j+1}{j} - \binom{a_j}{j} > a' \geq \binom{a_{j-1}}{j-1}.$$

Hence $a_j > a_{j-1}$. This proves the existence of a binomial expansion.

Next we show that if $a = \binom{a_j}{j} + \binom{a_{j-1}}{j-1} + \cdots + \binom{a_k}{k}$ with $a_j > a_{j-1} > \cdots > a_k \geq k \geq 1$, then a_j is the largest integer such that $a \geq \binom{a_j}{j}$. We prove this by induction on a. The assertion is trivial for $a = 1$. So now suppose that $a > 1$ and that $\binom{a_j+1}{j} \leq a$. Then

$$a' = \sum_{i=k}^{j-1} \binom{a_i}{i} \geq \binom{a_j+1}{j} - \binom{a_j}{j} = \binom{a_j}{j-1} \geq \binom{a_{j-1}+1}{j-1},$$

contradicting the induction hypothesis.

Now since the first summand in the expansion of a is uniquely determined, and since by induction on the length of the expansion we may assume that the expansion of a' is unique we conclude that also the expansion of a is unique.

□

Let $a = \binom{a_j}{j} + \binom{a_{j-1}}{j-1} + \cdots + \binom{a_k}{k}$ be the binomial expansion of a with respect to j. Then we define

$$a^{\langle j \rangle} = \binom{a_j + 1}{j + 1} + \binom{a_{j-1} + 1}{j} + \cdots + \binom{a_k + 1}{k + 1}.$$

For convenience we set $0^{\langle j \rangle}$ for all positive integers j. One has

Lemma 6.3.5. *Let $a \geq b$ and j be positive integers. Then $a^{\langle j \rangle} \geq b^{\langle j \rangle}$.*

Proof. We may assume that $a > b$. By the construction of the binomial expansions it follows that there exists an integer l such that

$$a_j = b_j, \quad a_{j-1} = b_{j-1}, \ldots, a_{l-1}b_{l-1}, a_l > b_l.$$

Since

$$\binom{a_l}{l} > \binom{b_l}{l} + \binom{b_{l-1}}{l-1} + \cdots + \binom{b_k}{k},$$

the assertion follows.   □

Binomial expansions naturally appear in the context of lexsegments. Indeed, let $u \in \mathrm{Mon}_j(S)$ and denote by $\mathcal{L}_u$ the lexsegment $\{v \in \mathrm{Mon}_j(S): v \geq u\}$. We also denote for any integer $1 \leq i \leq n$ by $\{x_i, \ldots, x_n\}^j$ the set of all monomials in degree j in the variables $x_i, \ldots, x_n$. Then we have

Lemma 6.3.6. *Let $u \in \mathrm{Mon}_j(S)$, $u = x_{k_1} x_{k_2} \cdots x_{k_j}$ with $k_1 \leq k_2 \leq \cdots \leq k_j$. Then*

$$\mathrm{Mon}_j(S) \setminus \mathcal{L}_u = \bigcup_{i=1}^{j} \{x_{k_i+1}, \ldots, x_n\}^{j-i+1} \prod_{r=1}^{i-1} x_{k_r}.$$

This union is disjoint and in particular we have

$$|\mathrm{Mon}_j(S) \setminus \mathcal{L}_u| = \sum_{i=1}^{j} \binom{a_i}{i} \quad \text{with} \quad a_i = n - k_{j-i+1} + i - 1.$$

Proof. We notice that

$$\mathrm{Mon}_j(S) \setminus \mathcal{L}_u = \{x_{k_1+1}, \ldots, x_n\}^j \cup (\mathrm{Mon}_{j-1}(S) \setminus \mathcal{L}_{ux_{k_1}^{-1}}) x_{k_1}.$$

By using induction on j, the assertion follows.   □

The definition of the binomial operator $a \mapsto a^{\langle j \rangle}$ is justified by the following result:

Proposition 6.3.7. *Let $\mathcal{L} \subset \mathrm{Mon}_j(S)$ be a lexsegment with $a = |\mathrm{Mon}_j(S) \setminus \mathcal{L}|$. Then*

$$|\mathrm{Mon}_{j+1}(S) \setminus \mathrm{Shad}(\mathcal{L})| = a^{\langle j \rangle}.$$

Proof. Let $u \in \mathrm{Mon}_j(S)$ be such that $\mathcal{L} = \mathcal{L}_u$. Then $\mathrm{Shad}(\mathcal{L}) = \mathcal{L}_{ux_n}$, and the desired equation follows immediately from Lemma 6.3.6.   □

As the final conclusion of all our considerations we now obtain

Theorem 6.3.8 (Macaulay). *Let $h \colon \mathbb{Z}_+ \to \mathbb{Z}_+$ be a numerical function. The following conditions are equivalent:*

(a) *h is the Hilbert function of a standard graded K-algebra;*
(b) *there exists an integer $n \geq 0$ and a lexsegment ideal $I \subset S = K[x_1, \ldots, x_n]$ such that $h(i) = H(S/I, i)$ for all $i \geq 0$;*
(c) *$h(0) = 1$, and $h(j+1) \leq h(j)^{\langle j \rangle}$ for all $j > 0$.*

Proof. By Theorem 6.3.1, h is the Hilbert function of a standard graded K-algebra if and only if it is the Hilbert function of an algebra S/I where each homogeneous component I_j is spanned by a lexsegment $\mathcal{L}_j$. This proves the equivalence of (a) and (b).

(b)$\Rightarrow$ (c): Let $h(j)$ be the Hilbert function of a lexsegment ideal of S/I, where I is a lexsegment ideal for which each homogeneous component I_j is spanned by a lexsegment $\mathcal{L}_j$. Since $\mathrm{Shad}(\mathcal{L}_j) \subset \mathcal{L}_{j+1}$ it follows from Proposition 6.3.7 that

$$H(S/I, j+1) = |\mathrm{Mon}_{j+1}(S) \setminus \mathcal{L}_{j+1}| \leq |\mathrm{Mon}_{j+1}(S) \setminus \mathrm{Shad}(\mathcal{L}_j)| = H(S/I, j)^{\langle j \rangle},$$

and of course we have $H(S/I, 0) = 1$. These are exactly the conditions given in (c).

(c)$\Rightarrow$ (b): Let $n = h(1)$, and set $S = K[x_1, \ldots, x_n]$. We first show by induction of j that $h(j) \leq \dim S_j = \binom{n+j-1}{j}$ for all j. The assertion is trivial for $j = 1$. Now assume that $h(j) \leq \binom{n+j-1}{j}$ for some $j \geq 1$. Then the statement in Lemma 6.3.5 implies that

$$h(j+1) \leq h(j)^{\langle j \rangle} \leq \binom{n+j-1}{j}^{\langle j \rangle} = \binom{n+j}{j+1},$$

as desired. It follows that $\dim_K S_j - h(j) \geq 0$ for all j. Now we let $\mathcal{L}_j \subset \mathrm{Mon}_j(S)$ be the unique lexsegment with $|\mathcal{L}_j| = \dim_K S_j - h(j)$ and let I_j be the K-vector space spanned by $\mathcal{L}_j$. We claim that $I = \bigoplus_{j \geq 0} I_j$ is an ideal. By construction, $H(S/I, j) = \dim S_j / I_j = h(j)$ for all j. Thus it remains to prove the claim, which amounts to show that $\mathrm{Shad}(\mathcal{L}_j) \subset \mathcal{L}_{j+1}$ for all $j > 0$, equivalently that $\mathrm{Mon}_{j+1}(S) \setminus \mathcal{L}_{j+1} \subset \mathrm{Mon}_{j+1}(S) \setminus \mathrm{Shad}(\mathcal{L}_j)$ for all $j > 0$. By Proposition 6.3.7 this is the case if and only if $h(j+1) = |\mathrm{Mon}_{j+1}(S) \setminus \mathcal{L}_{j+1}| \leq |\mathrm{Mon}_{j+1}(S) \setminus \mathrm{Shad}(\mathcal{L}_j)| = h(j)^{\langle j \rangle}$ for all $j > 0$. Thus the conclusion follows.   □

6.4 Squarefree lexsegment ideals and the Kruskal–Katona Theorem

When is a given sequence of integers $f = (f_0, f_1, \ldots, f_{d-1})$ the f-vector of a simplicial complex? The Kruskal–Katona theorem gives a complete answer to this question. The strategy for its proof is as follows: let Δ be a $(d-1)$-dimensional simplicial complex on the vertex set $[n]$, K a field and $K\{\Delta\}$ be the exterior face ring of Δ, as introduced in Chapter 5. Recall that $K\{\Delta\} = E/J_\Delta$, where E is the exterior algebra of the K-vector space $V = \bigoplus_{i=1}^n Ke_i$ and $J_\Delta \subset E$ is the graded ideal generated by all exterior monomials $e_F = e_{i_1} \wedge e_{i_2} \wedge \cdots \wedge e_{i_k}$ for which $F = \{i_1 < i_2 < \cdots < i_k\} \notin \Delta$. Then $K\{\Delta\}$ is a graded K-algebra and $H_{K\{\Delta\}}(t) = \sum_{i=0}^d f_{i-1}t^i$ where $f_{-1} = 1$ and $(f_0, f_1, \ldots, f_{d-1})$ is the f-vector of Δ. Thus we have to determine the possible Hilbert functions of graded algebras of the form E/J. The steps in solving this problem are completely analogous to those in the proof of Macaulay's theorem.

Let $J \subset E$ be a graded ideal. For the computation of the Hilbert function one may assume that the base field is infinite. Otherwise we choose a suitable extension of the base field. In Proposition 5.2.10 we have seen that $\mathrm{gin}(J)$ is a strongly stable ideal (strongly stable in the squarefree sense). Since $H_{E/J}(t) = H_{E/\mathrm{gin}(J)}$ we may as well assume that J itself is strongly stable.

Let $\mathrm{Mon}_j(E)$ denote the set of monomials of degree j in E. Lexsegments, stable and strongly stable subsets in $\mathrm{Mon}_j(E)$ as well as lexsegment ideals are defined in the obvious way. Naturally one defines the shadow of a subset $\mathcal{N} \subset \mathrm{Mon}_j(E)$ to be the set

$$\mathrm{Shad}(\mathcal{N}) = \{e_1, \ldots, e_n\}\mathcal{N} = \{e_i \wedge u : u \in \mathcal{N}, \ i = 1, \ldots, n\}.$$

Let $u = e_{i_1} \wedge e_{i_2} \wedge \cdots \wedge e_{i_j}$ be a monomial with $i_1 < i_2 < \cdots < i_j$. Then we set $m(u) = i_j$, and for a subset $\mathcal{N} \subset \mathrm{Mon}_j(E)$ and an integer $i \in [n]$ we let $m_i(\mathcal{N}) = |\{u \in \mathcal{N}: m(u) = i\}|$, and set $m_{\leq i}(\mathcal{N}) = \sum_{j=1}^i m_j(\mathcal{N})$.

The following series of statements then lead to the Kruskal–Katona theorem. At the end of this section we indicate where their proofs differ from the proofs of the corresponding statements in the previous section.

Lemma 6.4.1. *Let* $\mathcal{N} \subset \mathrm{Mon}_j(E)$ *be a stable set of monomials. Then* $\mathrm{Shad}(\mathcal{N})$ *is again a stable set and*

(a) $m_i(\mathrm{Shad}(\mathcal{N})) = m_{\leq i-1}(\mathcal{N})$;
(b) $|\mathrm{Shad}(\mathcal{N})| = \sum_{i=1}^{n-1} m_{\leq i}(\mathcal{N})$.

The exterior version of Bayer's theorem is the following

Theorem 6.4.2. *Let* $\mathcal{L} \subset \mathrm{Mon}_j(E)$ *be a lexsegment and* $\mathcal{N} \subset \mathrm{Mon}_j(E)$ *a strongly stable set of monomials with* $|\mathcal{L}| \leq |\mathcal{N}|$. *Then* $m_{\leq i}(\mathcal{L}) \leq m_{\leq i}(\mathcal{N})$ *for* $i = 1, \ldots, n$.

Lemma 6.4.1 and Theorem 6.4.2 then yield the result that, like for graded ideals in the polynomial ring, for each graded ideal J in the exterior algebra there exists a unique lexsegment ideal $J^{\text{lex}} \subset E$ such that $H_{E/J}(t) = H_{E/J^{\text{lex}}}(t)$. Thus it remains to understand the Hilbert series of a lexsegment ideal.

Let $a = \binom{a_j}{j} + \binom{a_{j-1}}{j-1} + \cdots + \binom{a_k}{k}$ be the binomial expansion of a with respect to j. Then we define the binomial operator $a \mapsto a^{(j)}$ by

$$a^{(j)} = \binom{a_j}{j+1} + \binom{a_{j-1}}{j} + \cdots + \binom{a_k}{k+1}.$$

Again for convenience we set $0^{(j)} = 0$ for positive integers j. The reader should compare this operator with the operator $a \mapsto a^{\langle j \rangle}$ defined in the previous section.

In analogy to Proposition 6.3.7 one has in the exterior case

Proposition 6.4.3. *Let $\mathcal{L} \subset \text{Mon}_j(E)$ be a lexsegment with $a = |\text{Mon}_j(E) \setminus \mathcal{L}|$. Then*

$$|\text{Mon}_{j+1}(E) \setminus \text{Shad}(\mathcal{L})| = a^{(j)}.$$

Combining all the results we finally get the algebraic version of the Kruskal–Katona theorem.

Theorem 6.4.4. *Let $(h_0, h_1, \ldots, h_n)$ be a sequence of integers. Then the following conditions are equivalent:*

(a) $\sum_{j=0}^{n} h_j t^j$ *is the Hilbert series of a graded K-algebra E/J;*
(b) *there exists a monomial ideal $J \subset E$ such that $\sum_{j=0}^{n} h_j t^j$ is the Hilbert series of E/J;*
(c) $h_0 = 1$ *and $0 \le h_{j+1} \le h_j^{(j)}$ for all j with $0 \le j < n$.*

Now if we apply Theorem 6.4.4 to the algebra $K\{\Delta\}$ and recall that $H_{K\{\Delta\}}(t) = \sum_{j=0}^{d} f_{j-1} t^j$ where $f = (f_0, f_1, \ldots, f_{d-1})$ is the f-vector of Δ, we obtain

Theorem 6.4.5 (Kruskal–Katona). *Let $f = (f_0, \ldots, f_{d-1})$ be a sequence of positive integers. Then the following conditions are equivalent:*

(a) *There exists a simplicial complex Δ with $f(\Delta) = f$;*
(b) $f_{j+1} \le f_j^{(j+1)}$ *for $0 \le j \le d - 2$.*

For the proof of Lemma 6.4.1 one observes that for a stable set of monomials $\mathcal{N} \subset \text{Mon}_j(E)$ the map

$$\phi \colon \{u \in \mathcal{N} \colon m(u) \le i - 1\} \to \{u \in \text{Shad}(\mathcal{N}) \colon m(u) = i\}, \quad u \mapsto u \wedge e_i$$

is bijective, cf. Lemma 6.3.2.

The proof of Proposition 6.4.3 is based on the fact that for $u \in \mathrm{Mon}_j(E)$ the complement of the lexsegment $\mathcal{L}_u = \{v \in \mathrm{Mon}_j(E): v \geq_{\mathrm{lex}} u\}$ can be decomposed as follows

$$\mathrm{Mon}_j(E) \setminus \mathcal{L}_u = \bigcup_{i=1}^{j} \{e_{k_i+1}, \dots, e_n\}^{j-i+1} e_{k_1} \wedge \dots \wedge e_{k_{i-1}},$$

where $u = e_{k_1} \wedge e_{k_2} \wedge \dots \wedge e_{k_j}$ with $k_1 < k_2 < \dots < k_j$, and where $\{e_{k_i+1}, \dots, e_n\}^{j-i+1}$ is the set of monomials of degree $j-i+1$ in the variables $e_{k_i+1}, \dots, e_n$, cf. Lemma 6.3.6.

Similarly to the proof of Theorem 6.3.3 one uses in the proof of Theorem 6.4.2 an order-preserving map $\alpha: \mathrm{Mon}_j(E) \to \mathrm{Mon}_j(E)$, this time defined as follows: let $u \in \mathrm{Mon}_j(E)$; if $m(u) < n$, then $\alpha(u) = u$, and if $m(u) = n$ and $u = u' \wedge e_n$, then $\alpha(u) = \pm e_k \wedge u'$, where $k < n$ is the largest integer such that $k \notin \mathrm{supp}(u')$. The sign of $\alpha(u)$ is chosen such that $\alpha(u)$, written in normal form, has coefficient $+1$. Then one shows that if $\mathcal{N} = \mathcal{N}' \cup \mathcal{N}'' \wedge e_n$ is a stable set of monomials in $\mathrm{Mon}_j(E)$, where $\mathcal{N}'$ and $\mathcal{N}''$ are monomials in $e_1, \dots, e_{n-1}$, then $\alpha(\min(\mathcal{N})) = \min(\mathcal{N}')$.

With this at hand, the proof of Theorem 6.4.2 reads as follows: we show by induction on n – the number of variables – that $m_{\leq i}(\mathcal{L}) \leq m_{\leq i}(\mathcal{N})$. For $i = n$ this is just our assumption. So now let $i < n$ and write $\mathcal{L} = \mathcal{L}' \cup \mathcal{L}'' \wedge e_n$ and $\mathcal{N} = \mathcal{N}' \cup \mathcal{N}'' \wedge e_n$ with $\mathcal{L}', \mathcal{L}'', \mathcal{N}'$ and $\mathcal{N}''$ sets of monomials in $e_1, e_2, \dots, e_{n-1}$. It is clear that $\mathcal{L}'$ is lexsegment, and that $\mathcal{N}'$ is stable. Hence if we show that $|\mathcal{L}'| \leq |\mathcal{N}'|$, we may apply our induction hypothesis, and the assertion follows immediately.

It may be assumed that $\mathcal{N}'$ and $\mathcal{N}''$ are lexsegments. In fact, let $\mathcal{N}^*, \mathcal{N}^{**}$ be the lexsegments in $e_1, e_2, \dots, e_{n-1}$ with $|\mathcal{N}^*| = |\mathcal{N}'|$ and $|\mathcal{N}^{**}| = |\mathcal{N}''|$ and set $\tilde{\mathcal{N}} = \mathcal{N}^* \cup \mathcal{N}^{**} \wedge e_n$. Then it is not hard to see that $\tilde{\mathcal{N}}$ is again stable.

Now we are in the following situation: $\mathcal{L} = \mathcal{L}' \cup \mathcal{L}'' \wedge e_n$ is a lexsegment, and $\mathcal{N} = \mathcal{N}' \cup \mathcal{N}'' \wedge e_n$ is stable as before, but in addition $\mathcal{N}'$ and $\mathcal{N}''$ are lexsegments. Assuming $|\mathcal{L}| \leq |\mathcal{N}|$, we want to show that $|\mathcal{L}'| \leq |\mathcal{N}'|$.

The required inequality follows, since

$$\min(\mathcal{N}') = \alpha(\min(\mathcal{N})) \leq_{\mathrm{lex}} \alpha(\min(\mathcal{L})) = \min(\mathcal{L}')$$

and since $\mathcal{L}'$ and $\mathcal{N}'$ are lexsegments.

Problems

6.1. Let R be a standard graded K-algebra, let M and N be graded S-modules and $\varphi: M \to N$ a homogeneous homomorphism, cf. Appendix A.2.
(a) Show that $\mathrm{Ker}(M \to N)$ is a graded R-module.
(b) Use (a) to show that R is isomorphic to S/I, where S is a polynomial ring over K and $I \subset S$ is a graded ideal.

6.2. Let $f_1, \ldots, f_k \in S = K[x_1, \ldots, x_n]$ be a regular sequence with $\deg f_i = a_i$, and let $I \subset S$ be the ideal generated by this regular sequence. Show:
(a) $H_{S/I}(t) = \prod_{i=1}^{k}(1 + t + \cdots + t^{a_i - 1})/(1 - t)^{n-k}$.
(b) $e(S/I) = \prod_{i=1}^{k} a_i$.

6.3. Let $R = S/I$ be a standard graded Cohen–Macaulay ring of codimension $s = \dim S - \dim R$. Then R is Gorenstein if and only if $\mathrm{Ext}_S^s(R, S) \cong R(a)$ for some integer a, see Corollary A.6.7. Use this characterization of a Gorenstein ring to show that if R is Gorenstein, then the h-vector of R is symmetric. In other words, if $h = (h_0, h_1, \ldots, h_c)$ is the h-vector of R, then $h_i = h_{c-i}$ for $i = 0, 1, \ldots, c$. How are the numbers a and c related to each other?

6.4. Let $S = K[x_1, \ldots, x_n]$ be the polynomial ring, and M a graded S-module. We say that M has a d-**linear resolution** if the graded minimal free resolution of M is of the form

$$0 \to S(-d-s)^{\beta_s} \to \cdots \to S(-d-1)^{\beta_1} \to S(-d)^{\beta_0} \to M \to 0.$$

Show that the ideal $I = (x_1, \ldots, x_n)^d$ has d-linear resolution. What is the multiplicity and the a-invariant of I? What are the Betti numbers of I?

6.5. Let Δ be a $(d-1)$-dimensional simplicial complex. Show that the h- and f-vectors of Δ satisfy the following identity $\sum_{i=0}^{d} h_i t^i (1+t)^{d-i} = \sum_{i=0}^{d} f_{i-1} t^i$.

6.6. Let Δ be a simplicial complex. Show that the a-invariant of $K[\Delta]$ is ≤ 0. Use this result to conclude that there exists no monomial order $<$ on $S = K[x_1, x_2, x_3, x_4]$ such that $\mathrm{in}_<(I)$ is a squarefree monomial ideal for the ideal $I = (x_1^2 - x_2 x_3, x_2^2 - x_3 x_4, x_3^2 - x_1 x_4)$.

6.7. Let Δ be Cohen–Macaulay simplicial complex with h-vector $(h_0, \ldots, h_s)$ and $h_s \neq 0$. Show that $h_i > 0$ for $i = 0, \ldots, s$. Find a simplicial complex with an h-vector such that $h_i < 0$ for some i.

6.8. Is the product of lexsegment ideals again a lexsegment ideal? Is the product of (strongly) stable ideals again (strongly) stable?

6.9. A squarefree monomial ideal $I \subset S = K[x_1, \ldots, x_n]$ is called **squarefree stable** if for all squarefree monomials $u \in I$ and for all $j < m(u)$ such that x_j does not divide u one has $x_j(u/x_{m(u)}) \in I$. The ideal I is called **squarefree strongly stable** if for all squarefree monomials $u \in I$ and for all $j < i$ such that x_i divides u and x_j does not divide u one has $x_j(u/x_i) \in I$. Finally, I is called a **squarefree lexsegment** ideal, if for all squarefree monomials $u \in I$ and all squarefree monomials v with $\deg u = \deg v$ and $u <_{\mathrm{lex}} v$ it follows that $v \in I$. Show that defining property for (strongly) squarefree stable and squarefree lexsegment ideals needs only be checked for the monomials in $G(I)$.

6.10. Find the Hilbert functions of all 0-dimensional graded ring $K[x_1, x_2, x_3]/I$ of multiplicity 6.

6.11. (a) Let $\mathcal{N} \subset \mathrm{Mon}_d(S)$ and let $\mathcal{L} \subset \mathrm{Mon}_d(S)$ the lexsegment with $|\mathcal{N}| = |\mathcal{L}|$. Show that $|\operatorname{Shad}(\mathcal{L})| \leq |\operatorname{Shad}(\mathcal{N})|$.
(b) Prove the corresponding results for monomial sets in the exterior algebra.

6.12. Let $n > 1$ be an odd number. Use Problem 6.11(b) and the Marriage Theorem (see Lemma 9.1.2) to prove the following statement: let $\mathcal{U}$ be the set of subsets of $[n]$ of cardinality $(n-1)/2$ and $\mathcal{V}$ the set of subsets of $[n]$ of cardinality $(n+1)/2$. Then there exists a bijection $\varphi\colon \mathcal{U} \to \mathcal{V}$ with the property that $A \subset \varphi(A)$ for all $A \subset \mathcal{U}$.

Notes

In 1927 Macaulay [Mac27] characterized the possible Hilbert functions of standard graded K-algebras. The essential part of our proof of Macaulay's theorem is based on Theorem 6.3.3 due to Bayer [Bay82]. On the other hand, in classical combinatorics on finite sets, Kruskal [Kru63] and Katona [Kat68] found the possible f-vectors of simplicial complexes. The h-vector of a simplicial complex, which is obtained by linear transformation from the f-vector, was introduced by McMullen [McM71]. However, an algebraic interpretation of the h-vector in terms of the Hilbert function of the Stanley–Reisner ring was given by Stanley in [Sta75]. Later Clement and Lindström [CL69] succeeded in generalizing Macaulay's theorem and the Kruskal–Katona theorem in a uniform way. It was also observed by Macaulay that the Hilbert function of a graded ideal and its initial ideal are the same. This provides an efficient method to compute Hilbert functions and related invariants, like dimension, multiplicity or the a-invariant of a standard graded K-algebra. This technique has been used in several papers to compute these invariants for determinantal rings; see for example [Stu90], [HTr92], [BH92] and [CH94]. The formula for the multiplicity in Corollary 6.1.7 is due to Peskine and Szpiro [PS74].

7

Resolutions of monomial ideals and the Eliahou–Kervaire formula

We introduce the Taylor complex, which for each monomial ideal provides a graded free resolution, but which in general is not minimal. Then we give a general upper bound for the graded Betti numbers of a graded module and discuss when this upper bound is reached. This happens to be the case for S/I when I is a stable monomial ideal. By means of Koszul homology the graded Betti numbers of stable monomial ideals are computed. The formulas which give these numbers are known as the Eliahou–Kervaire formulas. They are used to derive the Bigatti–Hullet theorem. We conclude this chapter with a squarefree version of the Eliahou–Kervaire formulas, and the comparison of the graded Betti numbers of a squarefree monomial ideal over the symmetric and exterior algebra.

7.1 The Taylor complex

Let I be a monomial ideal in the polynomial ring $S = K[x_1, \ldots, x_n]$ with $G(I) = \{u_1, \ldots, u_s\}$.

The **Taylor complex** $\mathbb{T}$ associated with the sequence $u_1, \ldots, u_s$ is a complex of free S-modules defined as follows: let T_1 be a free S-module with basis $e_1, \ldots, e_s$. Then

(1) $T_i = \bigwedge^i T_1$ for $i = 0 \ldots, s$. In particular, the elements $\mathbf{e}_F = e_{j_1} \wedge e_{j_2} \wedge \cdots \wedge e_{j_i}$ with $F = \{j_1 < j_2 < \cdots < j_i\} \subset [s]$ form a basis of T_i.
(2) the differential $\partial\colon T_i \to T_{i-1}$ is defined by

$$\partial(\mathbf{e}_F) = \sum_{i \in F}(-1)^{\sigma(F,i)}\frac{u_F}{u_{F\setminus\{i\}}}\mathbf{e}_{F\setminus\{i\}},$$

where for $G \subset [n]$, u_G denotes the least common multiple of the monomials u_i with $i \in G$, and where $\sigma(F,i) = |\{j \in F\colon j < i\}|$.

If we assign to each e_F the degree equal to $\deg u_F$, then the differential ∂ is a homogeneous map of graded free modules.

J. Herzog, T. Hibi, *Monomial Ideals*, Graduate Texts in Mathematics 260, DOI 10.1007/978-0-85729-106-6_7, © Springer-Verlag London Limited 2011

It is easily verified that $\partial \circ \partial = 0$, so that

$$\mathbb{T}: 0 \to T_s \to T_{s-1} \to \cdots \to T_2 \to T_1 \to T_0 \to 0$$

is a graded complex with rank $T_i = \binom{s}{i}$ for all i and $H_0(\mathbb{T}) = S/I$.

Theorem 7.1.1. *Let $I \subset S$ be monomial ideal with $G(I) = \{u_1, \ldots, u_s\}$. Then the Taylor complex $\mathbb{T}$ for the sequence $u_1, \ldots, u_s$ is acyclic, and hence a graded free S-resolution of S/I.*

Proof. We prove the theorem by induction on s. For $s = 1$ the assertion is trivial. So now let $s > 1$ and assume that the Taylor complex $\mathbb{T}'$ for the sequence $u_1, \ldots, u_{s-1}$ is acyclic. Note that $\mathbb{T}'$ can be identified with the subcomplex of $\mathbb{T}$ spanned by the basis elements e_F with $F \subset [s-1]$. Let $\mathbb{G} = \mathbb{T}/\mathbb{T}'$ be the quotient complex. Then $G_0 = 0$ and for each $i > 0$, the module G_i is free with basis $\mathbf{e}_{F \cup \{s\}}$ where $F \subset [s-1]$ and $|F| = i - 1$. The differential on $\mathbb{G}$ is given by

$$\partial(\mathbf{e}_{F \cup \{s\}}) = \sum_{i \in F} (-1)^{\sigma(F,i)} \frac{u_{F \cup \{s\}}}{u_{F \cup \{s\} \setminus \{i\}}} \mathbf{e}_{F \cup \{s\} \setminus \{i\}},$$

Hence $\mathbb{G}$ is isomorphic to the Taylor complex (homologically shifted by 1) for the sequence $v_1, \ldots, v_{s-1}$ with $v_i = \mathrm{lcm}(u_i, u_s)/u_s$ for $i = 1, \ldots, s-1$. In particular, we have $H_0(\mathbb{G}) = 0$, $H_1(\mathbb{G}) \cong S/(v_1, \ldots, v_{s-1})$, and our induction hypothesis implies that $H_i(\mathbb{G}) = 0$ for $i > 1$. Thus from the long exact homology sequence arising from the short exact sequence

$$0 \longrightarrow \mathbb{T}' \longrightarrow \mathbb{T} \longrightarrow \mathbb{G} \longrightarrow 0$$

we obtain the exact sequence

$$0 \longrightarrow H_1(\mathbb{T}) \longrightarrow H_1(\mathbb{G}) \to H_0(\mathbb{T}') \longrightarrow H_0(\mathbb{T}) \longrightarrow 0,$$

and $H_i(\mathbb{T}) = 0$ for $i > 1$.

We have $H_0(\mathbb{T}') = S/(u_1, \ldots, u_{s-1})$ and $H_0(\mathbb{T}) = S/(u_1, \ldots, u_s)$, and the homomorphism $H_0(\mathbb{T}') \to H_0(\mathbb{T})$ is just the canonical epimorphism $S/(u_1, \ldots, u_{s-1}) \to S/(u_1, \ldots, u_s)$, whose kernel is $(u_1, \ldots, u_s)/(u_1, \ldots, u_{s-1})$ which is isomorphic to $H_1(\mathbb{G}) = S/(v_1, \ldots, v_{s-1})$. This isomorphism is established by the connecting homomorphism $H_1(\mathbb{G}) \to H_0(\mathbb{T}')$, since under this homomorphism the homology class of e_s is mapped to the residue class of u_s modulo $(u_1, \ldots, u_{s-1})$. Therefore, $H_1(\mathbb{G}) \to H_0(\mathbb{T}')$ is injective and $H_1(\mathbb{T}) = 0$ as well. $\square$

Corollary 7.1.2. *Let $I \subset S$ be a monomial ideal minimally generated by s monomials. Then $\beta_i(S/I) \leq \binom{s}{i}$ for $i = 1, \ldots, s$.*

It should be noted that the Taylor resolution of a monomial ideal I is rarely a minimal resolution. For example, if $|G(I)| = s > n$, then the Taylor resolution can never be minimal because all minimal graded free resolutions have length at most n; see Appendix A.3.

7.2 Betti numbers of stable monomial ideals

7.2.1 Modules with maximal Betti numbers

Let $S = K[x_1, \ldots, x_n]$ denote the polynomial ring in n variables over an infinite field K and $\mathfrak{m} = (x_1, \ldots, x_n)$ its graded maximal ideal. Furthermore let M be a finitely generated graded S-module. In Proposition 4.3.12 we gave an upper bound for the graded Betti numbers of M in terms of the annihilator numbers of an almost regular sequence $\mathbf{y}$ on M which forms a K-basis of S_1. Thus in particular for generic annihilator numbers (cf. Remark 4.3.10) we have

$$\beta_{i,i+j}(M) \leq \sum_{k=0}^{n-i} \binom{n-k-1}{i-1} \alpha_{kj}(M) \quad \text{for all} \quad i \geq 0 \quad \text{and all} \quad j. \quad (7.1)$$

We say that M has **maximal** Betti numbers if equality holds in (7.1).

Theorem 7.2.1. *Let $\mathbf{y}$ be a generic sequence on M. Then the following conditions are equivalent:*

(a) *M has maximal Betti numbers.*
(b) *For all $j > 0$ and all i the multiplication maps*

$$y_i \colon H_j(y_1, \ldots, y_{i-1}; M)(-1) \longrightarrow H_j(y_1, \ldots, y_{i-1}; M)$$

are the zero maps.
(c) *For all $j > 0$ and all i one has $\mathfrak{m} H_j(y_1, \ldots, y_{i-1}; M) = 0$.*

If the equivalent conditions hold, then $\mathfrak{m} A_i(\mathbf{y}; M) = 0$ for all i.

Proof. (a) $\Leftrightarrow$ (b): To simplify notation we set $H_j(i)_k = H_j(y_1, \ldots, y_i; M)_k$ and $A(i)_k = A_i(\mathbf{y}; M)_k$ for all i, j and k. Inspecting the proof of Proposition 4.3.12 we see that equality holds in (7.1) if and only if the sequences

$$0 \to H_1(i-1)_k \longrightarrow H_1(i)_k \to A(i-1)_{k-1} \to 0, \quad (7.2)$$

and for $j > 0$ the sequences

$$0 \to H_j(i-1)_k \longrightarrow H_j(i)_k \to H_{j-1}(i-1)_{k-1} \to 0 \quad (7.3)$$

are all exact. However, this is the case if and only if the sequence $\mathbf{y}$ satisfies condition (b).

(a) $\Rightarrow$ (c): Let $U \subset \mathrm{GL}(n; K)$ be the nonempty Zariski open subset of $\mathrm{GL}(n; K)$ such that $\gamma(\mathbf{x})$ is generic for $\gamma \in U$. Then our given generic sequence $\mathbf{y} = y_1, \ldots, y_n$ is of the form $\mathbf{y} = \sigma(\mathbf{x})$ for some $\sigma \in U$. Let V be the set of $y \in S_1$ for which there exists $\gamma \in U$ with $\gamma(\mathbf{x}) = y_1, \ldots, y_{i-1}, y, y_{i+1}, \ldots, y_n$. Then $V \subset S_1$ is a Zariski open set of S_1 and it is nonempty, because $y_i \in V$.

For each $y \in V$ we have that $y_1, \ldots, y_{i-1}, y, y_{i+1}, \ldots, y_n$ is generic. Hence, since (a) $\Rightarrow$ (b), it follows that $y_1, \ldots, y_{i-1}, y, y_{i+1}, \ldots, y_n$ satisfies (b), which implies that the multiplication map $y \colon H_j(i-1)(-1) \to H_j(i-1)$ is the zero map for all $j > 0$, all i and all $y \in V$. Next we observe that the nonempty Zariski open set $V \subset S_1$ generates S_1 as a K-vector space. Indeed, if this were not the case V would be contained in a proper linear subspace $L \subset S_1$. Then this would imply that $V \cap S_1 \setminus L = \emptyset$ – a contradiction, since $S_1 \setminus L$ is a nonempty Zariski open subset of S_1.

The implication (c) $\Rightarrow$ (b) is trivial.

Since, as we have seen, the K-linear span of V is equal to S_1, it follows then that $\mathfrak{m} H_j(i-1) = 0$ for all $j > 0$ and all i, as desired.

Finally, (c) together with the exact sequences (7.2) yield that $\mathfrak{m} A_i(\mathbf{y}; M) = 0$ for all i. □

Suppose M has maximal Betti numbers, and set $\beta_{ij} = \beta_{ij}(M)$ and $\alpha_{ij} = \alpha_{ij}(M)$. Then for all i and j we have

$$\beta_{i,i+j} = \sum_{k=0}^{n-i} \binom{n-k-1}{i-1} \alpha_{kj}. \tag{7.4}$$

These equalities are equivalent to the following polynomial identities

$$\beta_{0j} s^j + \sum_{i=1}^{n} \beta_{i,i+j} t^i s^j = \alpha_{nj} s^j + \sum_{k=0}^{n-1} \alpha_{kj} t(1+t)^{n-k-1} s^j.$$

Substituting t by $u - 1$, we obtain the identities

$$\beta_{0j} s^j + \sum_{i=1}^{n} (-1)^i \beta_{i,i+j} (1-u)^i s^j = \alpha_{nj} s^j + \sum_{k=0}^{n-1} \alpha_{kj} (u-1) u^{n-k-1} s^j.$$

Expanding $(1-u)^i$ and comparing coefficients yields the following equations

$$\alpha_{n-r,j} - \alpha_{n-r-1,j} = \sum_{i=1}^{n} (-1)^{i+r} \binom{i}{r} \beta_{i,i+j} \quad \text{for} \quad r = 1, \ldots, n,$$

where we set $\alpha_{-1,j} = 0$ for all j. These equations finally imply that

$$\alpha_{n-r,j} = \sum_{k=0}^{r} \sum_{i=1}^{n} (-1)^{i+k} \binom{i}{r} \beta_{i,i+j} \quad \text{for } r = 0, \ldots, n \text{ and all } j. \tag{7.5}$$

Thus the generic annihilator numbers of a module with maximal Betti numbers are determined by its Betti numbers.

7.2.2 Stable monomial ideals

Strongly stable monomial ideals appear as generic initial ideals, as we have seen in Chapter 4. Here we want to compute the graded Betti numbers of such ideals. Indeed, they can even be computed for stable monomial ideals. Recall from Chapter 6 that a monomial ideal $I \subset S$ is called stable if for all monomials $u \in I$ and all $i < m(u)$ one has $x_i u' \in I$, where $u' = u/x_{m(u)}$ and $m(u)$ denotes the largest index j such that x_j divides u.

We use Koszul homology to compute the Betti numbers. Let $I \subset S$ be an ideal. For each i, $K_i(x_1, \ldots, x_n; S/I)$ is a free S/I-module with basis e_F with $F \subset [n]$ and $|F| = i$, where $e_F = e_{j_1} \wedge e_{j_2} \wedge \cdots \wedge e_{j_i}$ for $F = \{j_1 < j_2 < \cdots < j_{i-1} < j_i\}$.

By abuse of notation we denote the residue class modulo I of a monomial u in S again by u.

Theorem 7.2.2. *Let $I \subset S$ be a monomial ideal, and let $\mathfrak{m} = (x_1, \ldots, x_n)$ be the graded maximal ideal of S. Then the following conditions are equivalent:*

(a) *I is a stable monomial ideal.*
(b) *$\mathfrak{m} H_j(x_n, x_{n-1}, \ldots, x_i; S/I) = 0$ for $i, j = 1, \ldots, n$.*
(c) *$\mathfrak{m} H_1(x_n, x_{n-1}, \ldots, x_i; S/I) = 0$ for $i = 1, \ldots, n$.*

If the equivalent conditions hold, then $x_n, x_{n-1}, \ldots, x_1$ is a generic sequence for S/I and for each i, j, a basis of the K-vector space $H_j(x_n, x_{n-1}, \ldots, x_i; S/I)$ is given by the homology classes of the cycles

$$u' e_F \wedge e_{m(u)}, \quad u \in G(I), \quad |F| = j - 1, \quad i \leq \min F, \quad \max F < m(u).$$

Proof. The implication (b) $\Rightarrow$ (c) is trivial.

(c) $\Rightarrow$ (a): Since for each i, the annihilator module

$$((I, x_{i+1}, \ldots, x_n) :_S x_i)/(I, x_{i+1}, \ldots, x_n)$$

is a factor module of $H_1(x_n, \ldots, x_i; S/I)$, it follows that

$$\mathfrak{m}((I, x_{i+1}, \ldots, x_n) :_S x_i) \subset (I, x_{i+1}, \ldots, x_n) \quad \text{for} \quad i = 1, \ldots, n. \quad (7.6)$$

Now let $u \in I$ and suppose that $m(u) = i$. Then $u' = u/x_i$ and $u' \in (I, x_{i+1}, \ldots, x_n) :_S x_i$ and so $x_j u' \in (I, x_{i+1}, \ldots, x_n)$ for all j, by (7.6). Since x_k does not divide u' for $k \geq i + 1$ it follows that $x_j u' \in I$ for $j \leq i$. This shows that I is a stable monomial ideal.

(a) $\Rightarrow$ (b): We will show that $H_j(x_n, x_{n-1}, \ldots, x_i; S/I)$ has the specified basis. Assuming this, let $c = [u' e_F \wedge e_{m(u)}]$ be a homology class in $H_j(x_n, x_{n-1}, \ldots, x_i; S/I)$. Then of course $x_j c = 0$ for $j = i, \ldots, n$. But we also have $x_j c = 0$ for $j < i$, since $x_j c = [x_j u' e_F \wedge e_{m(u)}]$ and since $x_j u' \in I$, because I is a stable monomial ideal. Thus it follows that $\mathfrak{m} H_j(x_n, x_{n-1}, \ldots, x_i; S/I) = 0$, as desired.

For our further discussions we set $H_j(i) = H_j(x_n, x_{n-1}, \ldots, x_i; S/I)$ and $A(i) = ((I, x_{i+1}, \ldots, x_n) :_S x_i)/(I, x_{i+1}, \ldots, x_n)$. In order to prove the statement concerning the basis of $H_j(i)$ we proceed by induction on $n - i$. If $i = n$, we only have to consider $H_1(n)$ which is obviously minimally generated by the homology classes of the elements $u'e_n$ with $u \in G(I)$ such that $m(u) = n$.

Now assume that $i < n$ and that the assertion is proved for $i + 1$. Then for $H_j(i+1)$ we have a basis as described in the theorem and hence $\mathfrak{m}H_j(i+1) = 0$ for all $j \geq 1$, as we have seen before. We also have $\mathfrak{m}A(i) = 0$, since I is a stable monomial ideal. Thus the standard long exact sequence of Koszul homology (see Theorem A.3.3) splits into the short exact sequence

$$0 \to H_1(i+1) \to H_1(i) \to A(i) \to 0,$$

and for $j > 0$ into the short exact sequences

$$0 \to H_{j+1}(i+1) \to H_{j+1}(i) \to H_j(i+1) \to 0.$$

For each j, the map $H_{j+1}(i+1) \to H_{j+1}(i)$ is just the inclusion map, while $H_{j+1}(i) \to H_j(i+1)$ is the homomorphism induced by the map $K_{j+1}(x_n, \ldots, x_i; S/I) \to K_j(x_n, \ldots, x_{i+1}; S/I)$ which assigns to each element $a_0 + a_1 \wedge e_i \in K_{j+1}(x_n, \ldots, x_i; S/I)$ with $a_0 \in K_{j+1}(x_n, \ldots, x_{i+1}; S/I)$ and $a_1 \in K_j(x_n, \ldots, x_{i+1}; S/I)$ the element a_1, see Appendix A3. The end terms in these exact sequences are K-vector spaces with the specified bases, according to our induction hypothesis.

A K-basis of $A(i)$ is given be the residue classes of the elements u' with $u \in G(I)$ and $m(u) = i$. A preimage of u' under the map $H_1(i) \to A(i)$ is the homology class $[u'e_i]$. These homology classes together with the basis elements of $H_1(i+1)$ establish the desired basis for $H_1(i)$.

Similarly, let $[u'e_F \wedge e_{m(u)}]$ be a basis element in $H_j(i+1)$ with $u \in G(I)$ and the conditions on F as described in the theorem. Then $u'e_i \wedge e_F \wedge e_{m(u)}$ is a cycle in $H_{j+1}(i)$ whose homology class is mapped to $[u'e_F \wedge e_{m(u)}]$ under the homomorphism $H_{j+1}(i) \to H_j(i+1)$. Thus the homology classes of these cycles together with the basis elements of $H_{j+1}(i+1)$ form the basis of $H_{j+1}(i)$, as asserted.

Finally, assuming that I is strongly stable, Proposition 4.2.4(c) implies that I is Borel-fixed which according to Proposition 4.2.6(b) implies that $\mathrm{gin}_{<_{\mathrm{rev}}}(I) = I$. Hence it arises directly from the definition of a generic sequence that $x_n, x_{n-1}, \ldots, x_1$ is generic on S/I. □

By the preceding theorem we can compute $\dim_K H_i(x_1, \ldots, x_n; S/I)_{i+j}$ for a stable ideal just by counting the basis elements given there, and observing that a homology class $[u'e_F \wedge e_{m(u)}]$ in $H_i(x_1, \ldots, x_n; S/I)$ is of degree $i+j-1$ if and only if u is of degree j. Thus if we denote by $G(I)_j$ the set of elements of $G(I)$ which are of degree j, we obtain the following important result:

Corollary 7.2.3 (Eliahou–Kervaire). *Let $I \subset S$ be a stable ideal. Then*

(a) $\beta_{i,i+j}(I) = \sum_{u \in G(I)_j} \binom{m(u)-1}{i}$;

(b) $\operatorname{proj\,dim} S/I = \max\{m(u) \colon u \in G(I)\}$;

(c) $\operatorname{reg}(I) = \max\{\deg(u) \colon u \in G(I)\}$.

For a stable monomial ideal $I \subset S$, let m_{kj} be the number of monomials in $u \in G(I)_j$ with $m(u) = k$. Then for $i > 0$ the Eliahou–Kervaire formula for the Betti numbers implies

$$\beta_{i,i+j}(S/I) = \beta_{i-1,i-1+j}(I) \tag{7.7}$$

$$= \sum_{u \in G(I)_j} \binom{m(u)-1}{i-1} = \sum_{k=1}^{n} \binom{k-1}{i-1} m_{kj}.$$

Since S/I has maximal Betti numbers, as follows from Theorem 7.2.1 and Theorem 7.2.2, we may compare (7.7) with formula (7.4) and obtain

Corollary 7.2.4. *Let $I \subset S$ be a stable monomial ideal with generic annihilator numbers α_{ij}. Then $\alpha_{ij} = m_{n-i,j}$ for $i = 0, \ldots, n-1$ and all j.*

7.3 The Bigatti–Hulett theorem

In Chapter 6 we have seen that for any graded ideal $I \subset S$ there is a unique lexsegment ideal I^{lex} such that S/I and S/I^{lex} have the same Hilbert function. Now we will present the following important property of I^{lex}.

Theorem 7.3.1 (Bigatti–Hulett). *Let K be a field of characteristic 0, $S = K[x_1, \ldots, x_n]$ the polynomial ring in n variables and $I \subset S$ a graded ideal. Then*

$$\beta_{i,i+j}(I) \leq \beta_{i,i+j}(I^{\mathrm{lex}}) \quad \text{for all } i \text{ and } j.$$

In other words, among all ideals with the same Hilbert function, the lexsegment ideal has the largest Betti numbers.

Proof. By Corollary 6.1.5, S/I and $S/\operatorname{gin}_<(I)$ have the same Hilbert function, and by Corollary 3.3.3 we have $\beta_{i,i+j}(I) \leq \beta_{i,i+j}(\operatorname{gin}_<(I))$ for all i and j. Thus we may replace I by $\operatorname{gin}_<(I)$. Moreover, $\operatorname{gin}_<(I)$ is a strongly stable monomial ideal, as we have seen in Proposition 4.2.6. Hence we may as well assume that I is a stable monomial ideal. By Eliahou–Kervaire (Corollary 7.2.3) we then have

$$\beta_{i,i+j}(I) = \sum_{u \in G(I)_j} \binom{m(u)-1}{i}. \tag{7.8}$$

We denote by $I_{\langle j \rangle}$ the ideal generated by all elements of degree j in I, and set $m_k(I_{\langle j \rangle}) = m_k(G(I_{\langle j \rangle}))$ and $m_{\leq k}(I_{\langle j \rangle}) = m_{\leq k}(G(I_{\langle j \rangle}))$, cf. Subsection 6.3. Then $G(I)_j = G(I_{\langle j \rangle}) \setminus G(\mathfrak{m} I_{\langle j-1 \rangle})$. Accordingly, we write the right-hand side of (7.8) as a difference $A - B$, where A is the sum of the binomials $\binom{m(u)-1}{i}$ taken over all $u \in G(I_{\langle j \rangle})$ and B is the sum of the same binomials taken over all $u \in G(\mathfrak{m}I_{\langle j-1 \rangle})$. Then

$$\beta_{i,i+j}(I) = A - B$$

with

$$A = \sum_{u \in G(I_{\langle j \rangle})} \binom{m(u)-1}{i} = \sum_{k=1}^{n} m_k I_{\langle j \rangle} \binom{k-1}{i}$$

$$= \sum_{k=1}^{n} (m_{\le k}(I_{\langle j \rangle}) - m_{\le k-1}(I_{\langle j \rangle})) \binom{k-1}{i}$$

$$= \sum_{k=1}^{n} m_{\le k}(I_{\langle j \rangle}) \binom{k-1}{i} - \sum_{k=0}^{n-1} m_{\le k}(I_{\langle j \rangle}) \binom{k}{i}$$

$$= m_{\le n}(I_{\langle j \rangle}) \binom{n-1}{i} + \sum_{k=1}^{n-1} m_{\le k}(I_{\langle j \rangle}) \left(\binom{k-1}{i} - \binom{k}{i} \right)$$

$$= m_{\le n}(I_{\langle j \rangle}) \binom{n-1}{i} - \sum_{k=1}^{n-1} m_{\le k}(I_{\langle j \rangle}) \binom{k-1}{i-1},$$

and

$$B = \sum_{u \in G(\mathfrak{m}I_{\langle j-1 \rangle})} \binom{m(u)-1}{i} = \sum_{k=1}^{n} m_k(\mathfrak{m}I_{\langle j-1 \rangle}) \binom{k-1}{i}$$

$$= \sum_{k=1}^{n} m_{\le k}(I_{\langle j-1 \rangle}) \binom{k-1}{i}.$$

The last equation results from Lemma 6.3.2.

Now the theorem follows at once from the above presentation of $\beta_{i,i+j}(I)$ as the difference of the terms A and B, if we observe that $m_{\le n}(I_{\langle j \rangle}) = \dim_K I_j = \dim_K (I^{\text{lex}})_j = m_{\le n}((I^{\text{lex}})_{\langle j \rangle})$, and that, according to Theorem 6.3.3, one has $m_{\le k}((I^{\text{lex}})_{\langle \ell \rangle}) \le m_{\le k}(I_{\langle \ell \rangle})$ for all k and ℓ.   □

7.4 Betti numbers of squarefree stable ideals

In this section we study squarefree stable ideals. The ultimate goal of this section is to derive Eliahou–Kervaire type formulas for the graded Betti numbers of squarefree stable ideals. This will enable us in Chapter 11 to prove a theorem for squarefree monomial ideals analogous to the Bigatti–Hulett theorem.

A squarefree monomial ideal $I \subset S = K[x_1, \ldots, x_n]$ is called **squarefree stable** if for all squarefree monomials $u \in I$ and for all $j < m(u)$ such that x_j does not divide u one has $x_j(u/x_{m(u)}) \in I$.

We denote by **x** the sequence $x_1, \ldots, x_n$, and also denote for simplicity the residue class modulo I of a monomial u in S again by u. Then we have

Theorem 7.4.1. *Let $I \subset S$ be a squarefree stable ideal. Then for each $i > 0$, a basis of the homology classes of $H_j(\mathbf{x}; S/I)$ is given by the homology classes of the cycles $u' e_F \wedge e_{m(u)}$ with*

$$u \in G(I), \quad |F| = j - 1, \quad \max(F) < m(u) \quad \text{and} \quad F \cap \mathrm{supp}(u) = \emptyset.$$

Proof. A minimal free S-resolution of S/I is $\mathbb{Z}^n$-graded; in other words, the differentials are homogeneous homomorphisms and, for each i, we have $F_i = \bigoplus_j S(-\mathbf{a}_{ij})$ with $\mathbf{a}_{ij} \in \mathbb{Z}^n$. Moreover, by virtue of Theorem 8.1.1, all shifts $\mathbf{a}_{ij}$ are squarefree, i.e. $\mathbf{a}_{ij} \in \mathbb{Z}^n$ is of the form $\sum_{t \in F} \epsilon_t$, where $F \subset [n]$, and where $\epsilon_1, \epsilon_2, \ldots, \epsilon_n$ is the canonical basis of $\mathbb{Z}^n$. Thus, due to Corollary A.3.5 it follows that $H_j(\mathbf{x}; S/I)$ is a multigraded K-vector space with $H_j(\mathbf{x}; S/I)_\mathbf{a} = 0$, if $\mathbf{a} \in \mathbb{Z}^n$ is not squarefree. Hence if we want to compute the homology module $H_j(\mathbf{x}; S/I)$ it suffices to consider its squarefree multigraded components.

In order to simplify notation we set $H_j(i) = H_j(x_n, x_{n-1}, \ldots, x_i; S/I)$. Then for each $0 < j < n$, there is an exact sequence whose graded part for each $\mathbf{a} \in \mathbb{Z}^n$ yields the long exact sequence of vector spaces

$$\cdots \xrightarrow{x_i} H_j(i+1)_\mathbf{a} \longrightarrow H_j(i)_\mathbf{a} \longrightarrow H_{j-1}(i+1)_{\mathbf{a}-\epsilon_i} \xrightarrow{x_i} H_{j-1}(i+1)_\mathbf{a}$$
$$\longrightarrow H_{j-1}(i)_\mathbf{a} \longrightarrow \cdots .$$

We now show the following more precise result: for all $j > 0$, all $0 < i \leq n$ and all squarefree $\mathbf{a} \in \mathbb{Z}^n$, $H_j(i)_\mathbf{a}$ is generated by the homology classes of the cycles

$$u' \mathbf{e}_F \wedge e_{m(u)}, \quad u \in G(I), \quad |F| = j - 1$$

with

$$i \leq \min(F), \quad \max(F) < m(u), \quad F \cap \mathrm{supp}(u) = \emptyset \quad \text{and}$$
$$F \cup \mathrm{supp}(u) = \{i \colon a_i \neq 0\}.$$

The proof is achieved by induction on $n - i$. The assertion is obvious for $i = n$. We now suppose that $i < n$. For such i, but $j = 1$, the assertion is again obvious. Hence we assume in addition that $j > 1$. We first claim that

$$H_{j-1}(i+1)_{\mathbf{a}-\epsilon_i} \xrightarrow{x_i} H_{j-1}(i+1)_\mathbf{a}$$

is the zero map. Since $\mathbf{a} \in \mathbb{Z}^n$ is squarefree, the components of $\mathbf{a}$ are either 0 or 1. If the ith component of $\mathbf{a}$ is 0, then $\mathbf{a} - \epsilon_i$ has a negative component; hence $H_{j-1}(i+1)_{\mathbf{a}-\epsilon_i} = 0$. Thus we may assume the ith component of $\mathbf{a}$ is 1. Then $\mathbf{a} - \epsilon_i$ is squarefree and, by induction hypothesis, $H_{j-1}(i+1)_{\mathbf{a}-\epsilon_i}$ is generated by the homology classes of cycles of the form $u' \mathbf{e}_F \wedge e_{m(u)}$ with

$i \notin \operatorname{supp}(u)$. Such an element is mapped to the homology class of $u'x_i\mathbf{e}_F \wedge e_{m(u)}$ in $H_{j-1}(i+1)_\mathbf{a}$. However, since I is squarefree stable, we have $u'x_i \in I$, so that $u'x_i\mathbf{e}_F \wedge e_{m(u)} = 0$.

From these observations we deduce for all $j > 1$ we have the short exact sequences

$$0 \longrightarrow H_j(i+1)_\mathbf{a} \longrightarrow H_j(i)_\mathbf{a} \longrightarrow H_{j-1}(i+1)_{\mathbf{a}-\epsilon_i} \longrightarrow 0.$$

The first map $H_j(i+1)_\mathbf{a} \to H_j(i)_\mathbf{a}$ is simply induced by the natural inclusion map of the corresponding Koszul complexes, while the second map $H_j(i)_\mathbf{a} \to H_{j-1}(i+1)_{\mathbf{a}-\epsilon_i}$ is a connecting homomorphism. Given the homology class of a cycle $z = u'\mathbf{e}_F \wedge e_{m(u)}$ in $H_{j-1}(i+1)_{\mathbf{a}-\epsilon_i}$, it is easy to see that, up to a sign, the homology class of the cycle $u'e_i \wedge \mathbf{e}_F \wedge e_{m(u)}$ in $H_j(i)_\mathbf{a}$ is mapped to $[z]$. This implies all of our assertions. $\square$

As an immediate consequence we obtain

Corollary 7.4.2. *Let $I \subset S$ be a squarefree stable ideal. Then*

(a) $\beta_{i,i+j}(I) = \sum_{u\in G(I)_j} \binom{m(u)-j}{i}$;
(b) $\operatorname{proj\,dim} S/I = \max\{m(u) - \deg(u) + 1 : u \in G(I)\}$;
(c) $\operatorname{reg}(I) = \max\{\deg(u) : u \in G(I)\}$.

Let I be a squarefree stable monomial ideal. We denote by $I_{[j]}$ the ideal generated by all squarefree monomials of degree j in I, and set $m_k(I_{[j]}) = m_k(G(I_{[j]}))$ and $m_{\leq k}(I_{[j]}) = m_{\leq k}(G(I_{[j]}))$, cf. Subsection 6.3. Then $G(I)_j = G(I_{[j]}) \setminus G(I_{[j]} \cap \mathfrak{m}I_{[j-1]})$.

A monomial I is called **squarefree strongly stable** if for all squarefree monomials $u \in I$ and for all $j < i$ such that x_i divides u and x_j does not divide u one has $x_j(u/x_i) \in I$, and that I is called **squarefree lexsegment** if for all squarefree monomials $u \in I$ and all squarefree monomials v with $\deg u = \deg v$ and $u <_{\mathrm{lex}} v$ it follows that $v \in I$. By Lemma 6.4.1 together with Theorem 6.4.2 it follows that for each squarefree monomial ideal I there exists a unique squarefree lexsegment ideal, denoted I^{sqlex}, with the property that S/I and S/I^{sqlex} have the same Hilbert function.

In Corollary 11.3.15 we prove the squarefree version of the Bigatti–Hullet theorem. For its proof we will need the following.

Theorem 7.4.3. *Let $I \subset S$ be a squarefree strongly stable ideal. Then $\beta_{i,i+j}(I) \leq \beta_{i,i+j}(I^{\mathrm{sqlex}})$ for all i and j.*

Proof. By Corollary 7.4.2 we have $\beta_{ii+j}(I) = C - D$, where

$$C = \sum_{u\in G(I_{[j]})} \binom{m(u) - j}{i},$$

and

$$D = \sum_{u \in G(I_{[j]} \cap \mathfrak{m} I_{[j-1]})} \binom{m(u) - j}{i}.$$

Furthermore,

$$C = \sum_{k=1}^{n} (m_{\le k}(I_{[j]}) - m_{\le k-1}(I_{[j]})) \binom{k-j}{i}$$

$$= \sum_{k=1}^{n} m_{\le k}(I_{[j]}) \binom{k-j}{i} - \sum_{k=1}^{n} m_{\le k-1}(I_{[j]}) \binom{k-j}{i}$$

$$= m_{\le n}(I_{[j]}) \binom{n-j}{i} + \sum_{k=1}^{n-1} m_{\le k}(I_{[j]}) \binom{k-j}{i} - \sum_{k=1}^{n-1} m_{\le k}(I_{[j]}) \binom{k-j+1}{i}$$

$$= m_{\le n}(I_{[j]}) \binom{n-j}{i} - \sum_{k=1}^{n-1} m_{\le k}(I_{[j]}) \left(\binom{k-j+1}{i} - \binom{k-j}{i} \right)$$

$$= m_{\le n}(I_{[j]}) \binom{n-j}{i} - \sum_{k=j}^{n-1} m_{\le k}(I_{[j]}) \binom{k-j}{i-1}.$$

On the other hand, Lemma 6.4.1 implies that

$$D = \sum_{k=j}^{n} \sum_{\substack{u \in G(I_{[j]} \cap \mathfrak{m} I_{[j-1]}) \\ m(u)=k}} \binom{k-j}{i} = \sum_{k=j}^{n} m_{\le k-1}(I_{[j-1]}) \binom{k-j}{i}.$$

Thus for all i and j we obtain

$$\beta_{i,i+j}(I) = m_{<n}(I_{[j]}) \binom{n-j}{i} - \sum_{k=j}^{n-1} m_{\le k}(I_{[j]}) \binom{k-j}{i-1} \tag{7.9}$$

$$- \sum_{k=j}^{n} m_{\le k-1}(I_{[j-1]}) \binom{k-j}{i}.$$

It follows from Theorem 6.4.2 that $m_{\le k}(I_{[\ell]}^{\mathrm{sqlex}}) \le m_{\le k}(I_{[\ell]})$ for all k and ℓ. Using this fact, the assertion follows from formula (7.9). $\qquad\square$

7.5 Comparison of Betti numbers over the symmetric and exterior algebra

Let $I \subset S$ be a squarefree monomial ideal, and E the exterior algebra of the K-vector space V with basis $e_1, \ldots, e_n$. We denote by $J \subset E$ the corresponding squarefree monomial ideal in the exterior algebra E, that is, the ideal $J \subset E$ with $e_{i_1} \wedge \cdots \wedge e_{i_k} \in G(J)$ if and only if $x_{i_1} \cdots x_{i_k} \in G(I)$.

Since S/I is a $\mathbb{Z}^n$-graded module, it admits a minimal multigraded free S-resolution

$$\mathbb{F}: \cdots \longrightarrow F_2 \longrightarrow F_1 \longrightarrow F_0 \longrightarrow S/I \longrightarrow 0,$$

where $F_i = \bigoplus_{a \in \mathbb{Z}^n} S(-a)^{\beta_{i,a}(S/I)}$; see the corresponding statements and proofs for graded modules in Appendix A.3. The numbers $\beta_{i,\mathbf{a}}(S/I) = \dim_K \operatorname{Tor}_i^S(K, S/I)_{\mathbf{a}}$ are called the **multigraded Betti numbers** of S/I. Similarly, E/J has a multigraded free E-resolution. The purpose of this subsection is to compare the multigraded Betti numbers of E/J with those of S/I.

For $\mathbf{a} = (a_1, \ldots, a_n) \in \mathbb{Z}^n$ we set $|\mathbf{a}| = \sum_{i=1}^n a_i$ and $\operatorname{supp}(\mathbf{a}) = \{i \colon a_i \neq 0\}$. The following theorem yields an interpretation of the $\mathbb{Z}^n$-graded components of $\operatorname{Tor}_i^E(K, J_\Delta)$ in terms of reduced simplicial homology.

Theorem 7.5.1. *Let Δ be a simplicial complex on $[n]$, $a \in \mathbb{N}^n$ and $W = \operatorname{supp}(\mathbf{a})$. Then, for all $i \geq 0$, we have*

$$\operatorname{Tor}_i^E(K, J_\Delta)_{\mathbf{a}} \cong \tilde{H}^{|\mathbf{a}|-i-2}(\Delta_W; K).$$

Proof. By Theorem A.8.2, $\operatorname{Tor}_{i+1}^E(K, K\{\Delta\})_{\mathbf{a}}$ may be identified with the component of multidegree $\mathbf{a}$ of the Cartan homology $H_{i+1}(e_1, \ldots, e_n; K\{\Delta\})$. A basis of $C_{i+1}(e_1, \ldots, e_n; K\{\Delta\})_{\mathbf{a}}$ is given by

$$e_F \mathbf{x}^{(\mathbf{a}_F)}, \quad F \in \Delta_W, \quad |\mathbf{a}_F| = i + 1,$$

where $\mathbf{a}_F = (a_1', \ldots, a_n')$ with $a_j' = a_j$ for $j \notin F$ and $a_j' = a_j - 1$ for $j \in F$.

Recall from Section 5.1.4 that $(K\{\Delta_W\}, e)$ is the complex of K-vector spaces

$$\cdots \xrightarrow{e} K\{\Delta_W\}_{i-1} \xrightarrow{e} K\{\Delta_W\}_i \xrightarrow{e} K\{\Delta_W\}_{i+1} \xrightarrow{e} \cdots$$

with $e = e_1 + e_2 + \cdots + e_n$. We define a K-linear map

$$\varphi_{i+1} \colon C_{i+1}(e_1, \ldots, e_n; K\{\Delta\})_{\mathbf{a}} \to (K\{\Delta_W\}, e)_{|\mathbf{a}|-i-1}$$

by setting $\varphi(e_F \mathbf{x}^{(\mathbf{a}_F)}) = e_F$. We observe that φ_{i+1} is an isomorphism of K-vector spaces, and that the family $\varphi = (\varphi_i)$ of maps is compatible with the differential of both complexes. Therefore, in view of Definition 5.1.7 we obtain the isomorphisms

$$\operatorname{Tor}_i^E(K, J_\Delta)_{\mathbf{a}} \cong \operatorname{Tor}_{i+1}^E(K, K\{\Delta\})_{\mathbf{a}}$$
$$\cong H^{|\mathbf{a}|-i-1}(K\{\Delta_W\}, e) \cong \tilde{H}^{|\mathbf{a}|-i-2}(\Delta_W; K),$$

as desired. □

Let M be a $\mathbb{Z}^n$-graded A-module where $A = S$ or $A = E$. The **multigraded Poincaré series** of M over A is defined by

$$P_M^A(t, \mathbf{s}) = \sum_{i \geq 0} \sum_{\mathbf{a} \in \mathbb{Z}^n} \beta_{i,\mathbf{a}}^A(M) t^i s^{\mathbf{a}}.$$

A comparison of Theorem 7.5.1 and Theorem 8.1.1 now immediately yields

Corollary 7.5.2. *Let $I \subset S$ be a squarefree monomial ideal and $J \subset E$ the corresponding monomial ideal in the exterior algebra. Then*

$$P_{E/J}^E(t, \mathbf{s}) = \sum_{i \geq 0} \sum_{\mathbf{a} \in \mathbb{Z}^n} \beta_{i,\mathbf{a}}^S(S/I) \frac{t^i \mathbf{s^a}}{\prod_{j \in \text{supp}(\mathbf{a})} (1 - ts_j)}.$$

From this identity of formal power series we deduce

Corollary 7.5.3. *The ideal I has a d-linear resolution over S if and only if J has a d-linear resolution over E.*

Problems

7.1. Show that the ideal $I = (x_1^3, x_1 x_2 x_3, x_2^2 x_3^2)$ in $K[x_1, x_2, x_3]$ has a minimal Taylor resolution.

7.2. Show that a monomial ideal generated by m elements has projective dimension at most $m - 1$. Give an example of a graded ideal of projective dimension ≥ 3 which is generated by 3 elements.

7.3. Let $u \in S$ be a monomial. The **principal stable ideal** generated by u is the smallest stable monomial ideal containing u. Let I be the principal stable ideal generated by $u = x_1^{a_1} x_2^{a_2} \cdots x_n^{a_n}$. Describe the elements of $G(I)$ and compute the graded Betti numbers of I.

7.4. Give an example of a monomial ideal I with property that $\beta_{i,i+j}(I) = \beta_{i,i+j}(I^{\text{lex}})$ for all i, j which is not a lexsegment ideal, even after permutation of the variables.

7.5. Let I be a stable or squarefree stable ideal, and suppose that $\beta_{i,i+j}(I) \neq 0$. Show that $\beta_{k,k+j}(I) \neq 0$ for $k = 0, \ldots, i$.

7.6. Compute I^{lex} for $I = (x_1^2, x_2^2, x_3^2)$ and compare their graded Betti numbers.

7.7. Is the polarization of a (strongly) stable ideal again (strongly) stable?

7.8. Compute the extremal Betti numbers of a stable monomial ideal in terms of the numbers $m(u)$ and $\deg(u)$ with $u \in G(I)$.

7.9. Let M be a graded E-module (see Chapter 5). Show that $\text{proj dim } M < \infty$ if and only if M is free.

Notes

Taylor introduced in her thesis [Tay66] in 1966 a complex resolving monomial ideals, which nowadays is called the Taylor complex. This complex is in general not minimal. Eliahou and Kervaire [EK90] succeeded in describing the minimal free resolution of the important class of stable monomial ideals. Bigatti [Big93] and Hulett [Hul93] used this Eliahou–Kervaire resolution and independently proved that, for the case of characteristic 0, among the graded ideals with a fixed Hilbert function, the lexsegment ideal possesses the maximal graded Betti numbers. For positive characteristics a similar result was obtained by Pardue [Par94] by using the technique of polarizations. The squarefree versions of these theorems were studied in [AHH98] and [AHH00a]. The result referring to the comparison of Betti numbers over the exterior algebra and the polynomial is taken from [AAH00].

8

Alexander duality and resolutions

The Alexander dual of a simplicial complex plays an essential role in combinatorics and commutative algebra. One of the fundamental results is the Eagon–Reiner theorem, which says that the Stanley–Reisner ideal of a simplicial complex has a linear resolution if and only if its Alexander dual is Cohen–Macaulay. After discussing this theorem in detail, we introduce the notion of componentwise linear ideals and sequentially Cohen–Macaulay simplicial complexes, and explain the relationship of these concepts with shellability.

8.1 The Eagon–Reiner theorem

8.1.1 Hochster's formula

A very useful result to compute the graded Betti numbers of the Stanley–Reisner ideal of simplicial complex is the so-called Hochster formula. To state the formula we introduce some notation and terminologies.

Let Δ be a simplicial complex on $[n]$. For a face F of Δ, the **link** of F in Δ is the subcomplex

$$\mathrm{link}_\Delta F = \{G \in \Delta : F \cup G \in \Delta,\ F \cap G = \emptyset\}.$$

Thus in particular $\mathrm{link}_\Delta \emptyset = \Delta$. For a subset W of $[n]$ the **restriction** of Δ on W is the subcomplex

$$\Delta_W = \{F \in \Delta : F \subset W\}.$$

Finally, the notation $\tilde{H}_q(\Delta; K)$ stands for the qth **reduced homology group** of Δ with coefficient K, where K is a field; see Chapter 5.

Let $S = K[x_1, \ldots, x_n]$ denote the polynomial ring in n variables over a field K with each $\deg x_i = 1$. Let Δ be a simplicial complex on $[n]$ and

I_Δ its Stanley–Reisner ideal. We observe that I_Δ is $\mathbb{Z}^n$-graded, so I_Δ admits a minimal $\mathbb{Z}^n$-graded free S-resolution, simply because the kernel of a $\mathbb{Z}^n$-graded homomorphism is again $\mathbb{Z}^n$-graded. (cf. Appendix A.2 where it is shown that a graded module has a minimal graded free resolution). It follows that $\mathrm{Tor}_i(K, I_\Delta)$ is a $\mathbb{Z}^n$-graded K-vector space.

Since the Koszul complex $K(\mathbf{x}; I_\Delta)$ is a complex of $\mathbb{Z}^n$-graded modules it follows that the Koszul homology modules $H_i(\mathbf{x}; I_\Delta)$ are $\mathbb{Z}^n$-graded K-vector spaces and for all $\mathbf{a} \in \mathbb{Z}^n$ one has

$$\mathrm{Tor}_i(K, I_\Delta)_\mathbf{a} \cong H_i(\mathbf{x}; I_\Delta)_\mathbf{a} \quad \text{for all} \quad \mathbf{a} \in \mathbb{Z}^n. \tag{8.1}$$

The corresponding isomorphism for graded modules is given in Corollary A.3.5.

The numbers

$$\beta_{i,\mathbf{a}}(I_\Delta) = \dim_K \mathrm{Tor}_i(K, I_\Delta)_\mathbf{a}$$

are called the **multigraded** or $\mathbb{Z}^n$-**graded Betti numbers** of I_Δ.

An element $\mathbf{a} \in \mathbb{Z}^n$ is called **squarefree** if $\mathbf{a}$ has only the integers 0 and 1 as possible entries. We set $\mathrm{supp}(\mathbf{a}) = \{i: a_i \neq 0\}$.

The following fundamental theorem of Hochster gives a very useful description of the $\mathbb{Z}^n$-graded Betti numbers of a Stanley–Reisner ideal.

Theorem 8.1.1 (Hochster). *Let Δ be a simplicial complex and $a \in \mathbb{Z}^n$. Then we have:*

(a) $\mathrm{Tor}_i^S(K, I_\Delta)_\mathbf{a} = 0$ *if $\mathbf{a}$ is not squarefree;*
(b) *if $\mathbf{a}$ is squarefree and $W = \mathrm{supp}(\mathbf{a})$, then*

$$\mathrm{Tor}_i^S(K, I_\Delta)_\mathbf{a} \cong \tilde{H}^{|W|-i-2}(\Delta_W; K) \quad \text{for all} \quad i$$

Proof. We compute $\mathrm{Tor}_i^S(K, I_\Delta)_\mathbf{a}$ by means of formula (8.1). For $F \subset [n]$, $F = \{j_0 < j_1 < \cdots < j_i\}$, we set $\mathbf{e}_F = e_{j_0} \wedge e_{j1} \wedge \cdots \wedge e_{j_i}$. The elements $\mathbf{e}_F$ with $F \subset [n]$ and $|F| = i$ form a basis of the free S-module $K_i(\mathbf{x}; S)$. The $\mathbb{Z}^n$-degree of $\mathbf{e}_F$ is $\epsilon(F) \in \mathbb{Z}^n$, where $\epsilon(F)$ is the $(0, 1)$-vector with $\mathrm{supp}(\epsilon(F)) = F$.

A K-basis of $K_i(\mathbf{x}; I_\Delta)_\mathbf{a}$ is given by

$$\mathbf{x}^\mathbf{b} \mathbf{e}_F, \quad \mathbf{b} + \epsilon(F) = \mathbf{a}, \quad \mathrm{supp}(\mathbf{b}) \notin \Delta.$$

We define the simplicial complex

$$\Delta_\mathbf{a} = \{F \subset [n]: F \subset \mathrm{supp}(\mathbf{a}), \, \mathrm{supp}(\mathbf{a} \setminus \epsilon(F)) \notin \Delta\}.$$

Let $\tilde{C}(\Delta_\mathbf{a}; K)[-1]$ be the oriented augmented chain complex of $\Delta_\mathbf{a}$ shifted by -1 in homological degree, cf. Subsection 5.1.4. Then we obtain an isomorphism of complexes

$$\alpha: \tilde{C}(\Delta_\mathbf{a}; K)[-1] \longrightarrow K_i(\mathbf{x}; I_\Delta)_\mathbf{a},$$

where

$$\alpha_i: \tilde{C}_{i-1}(\Delta_\mathbf{a}; K) \longrightarrow K_i(\mathbf{x}; I_\Delta)_\mathbf{a}, \quad F = [j_0, j_1, \cdots, j_{i-2}] \mapsto \mathbf{x}^{\mathbf{a}-\epsilon(F)} \mathbf{e}_F.$$

It follows that

$$H_i(\mathbf{x}; I_\Delta)_\mathbf{a} \cong \tilde{H}_{i-1}(\Delta_\mathbf{a}; K). \tag{8.2}$$

We begin with the proof of (a): Suppose $\mathbf{a}$ is not squarefree. Then there exists j such that $a_j > 1$. We define $\mathbf{a}(r) = (a_1, \ldots, a_j + r, \ldots, a_n)$ for $r \geq 0$. It follows from the definition of $\Delta_\mathbf{a}$ that $\Delta_\mathbf{a} = \Delta_{\mathbf{a}(r)}$ for all $r \geq 0$. Since $H_i(\mathbf{x}; I_\Delta)$ has only finitely many nonzero graded components there exists $r \gg 0$ such that $H_i(\mathbf{x}; I_\Delta)_{\mathbf{a}(r)} = 0$. Thus by (8.2) we have

$$H_i(\mathbf{x}; I_\Delta)_\mathbf{a} \cong \tilde{H}_{i-1}(\Delta_\mathbf{a}; K) = \tilde{H}_{i-1}(\Delta_{\mathbf{a}(r)}; K) \cong H_i(\mathbf{x}; I_\Delta)_{\mathbf{a}(r)} = 0.$$

Proof of (b): Let $\mathbf{a} \in \mathbb{Z}^n$ squarefree with $W = \mathrm{supp}(\mathbf{a})$. Then $F \in \Delta_\mathbf{a}$ if and only if $F \subset W$ and $W \setminus F \notin \Delta_W$. This is equivalent to saying that $F \in (\Delta_W)^\vee$.

Thus (8.1), (8.2) together with Proposition 5.1.10 yield

$$\mathrm{Tor}_i^S(K, I_\Delta)_\mathbf{a} \cong \tilde{H}_{i-1}((\Delta_W)^\vee; K) \cong \tilde{H}^{|W|-i-2}(\Delta_W; K), \tag{8.3}$$

as desired. $\qquad\qquad\qquad\qquad\qquad\qquad\qquad\qquad\qquad\qquad\qquad\qquad\square$

Example 8.1.2. Let Δ be a simplicial complex on the vertex set $\{1, 2, 3, 4, 5\}$ with the facets $\{1, 2, 3, 4\}$, $\{2, 5\}$ and $\{4, 5\}$. Let $\mathbf{a} = (1, 0, 1, 0, 1)$; then $\mathrm{supp}(\mathbf{a}) = W = \{1, 3, 5\}$. One has $\tilde{H}^{|W|-i-2}(\Delta_W; K) = 0$ unless $i = 1$ and $\tilde{H}^{|W|-3}(\Delta_W; K) = K$. Hence $\beta_{i,\mathbf{a}}(I_\Delta) = 0$ unless $i = 1$ and $\beta_{1,\mathbf{a}}(I_\Delta) = 1$.

The $\mathbb{Z}^n$-graded components of $\mathrm{Tor}_i^S(K, I_\Delta)$ can also be expressed in terms of certain links. For this we need

Lemma 8.1.3. *Let Δ be a simplicial complex on $[n]$ and $W \subset [n]$ with $W \notin \Delta$. Let $F = [n] \setminus W \in \Delta^\vee$. Then*

$$\mathrm{link}_{\Delta^\vee} F = (\Delta_W)^\vee.$$

Proof. Each of $\mathrm{link}_{\Delta^\vee} F$ and $(\Delta_W)^\vee$ is a simplicial complex on W. Let $G \subset W$. Then $G \in (\Delta_W)^\vee$ if and only if $W \setminus G \notin \Delta$. On the other hand, $G \in \mathrm{link}_{\Delta^\vee} F$ if and only if $F \cup G \in \Delta^\vee$. In other words, $G \in \mathrm{link}_{\Delta^\vee} F$ if and only if $[n] \setminus (F \cup G) = W \setminus G$ does not belong to Δ. $\qquad\qquad\square$

Corollary 8.1.4. *Let Δ be a simplicial complex, $\mathbf{a} \in \mathbb{Z}^n$ be squarefree and $F = [n] \setminus \mathrm{supp}(\mathbf{a})$. Then*

$$\mathrm{Tor}_i^S(K, I_\Delta)_\mathbf{a} \cong \tilde{H}_{i-1}(\mathrm{link}_{\Delta^\vee} F; K) \quad \text{for all} \quad i$$

In particular it follows that the graded Betti number $\beta_{ij}(I_\Delta)$ of I_Δ can be computed by the formula

$$\beta_{ij}(I_\Delta) = \sum_{F \in \Delta^\vee, |F| = n - j} \dim_K \tilde{H}_{i-1}(\mathrm{link}_{\Delta^\vee} F; K).$$

Proof. Lemma 8.1.3 and (8.3) yield the desired isomorphism. The formula for $\beta_{ij}(I_\Delta)$ follows from the first part since $\mathrm{link}_{\Delta^\vee} F = \emptyset$, if $F \notin \Delta^\vee$. $\qquad\square$

8.1.2 Reisner's criterion and the Eagon–Reiner theorem

The K-algebra $K[\Delta] = S/I_\Delta$ is called the **Stanley–Reisner ring** of Δ. We say that Δ is **Cohen–Macaulay over** K if $K[\Delta]$ is Cohen–Macaulay.

Lemma 8.1.5. *Every Cohen–Macaulay simplicial complex is pure.*

Proof. Let Δ be Cohen–Macaulay over K. According to Lemma 1.5.4 the minimal prime ideals of I_Δ correspond to the facets of Δ. Hence Δ is pure if and only if all minimal prime ideals of I_Δ have the same height. However this is guaranteed by the assumption that $K[\Delta]$ is Cohen–Macaulay; see Appendix A.5. □

The following result is known as the Reisner criterion for the Cohen–Macaulay property of the Stanley–Reisner ring.

Theorem 8.1.6 (Reisner). *A simplicial complex Δ is Cohen–Macaulay over K if and only if, for all faces F of Δ including the empty face $\emptyset$ and for all $i < \dim \mathrm{link}_\Delta F$, one has $\tilde{H}_i(\mathrm{link}_\Delta F; K) = 0$.*

Proof. We use local cohomology to prove the theorem. Let $\mathbf{a} \in \mathbb{Z}^n$. By Theorem A.7.3 we have

$$H_\mathfrak{m}^i(K[\Delta])_\mathbf{a} = 0 \quad \text{if} \quad a_i > 0 \quad \text{for some} \quad i, \tag{8.4}$$

and

$$H_\mathfrak{m}^i(K[\Delta])_\mathbf{a} = \tilde{H}_{i-|F|-1}(\mathrm{link}_\Delta F; K) \tag{8.5}$$

with $F = \mathrm{supp}(\mathbf{a})$ if $a_i \leq 0$ for all i.

Let $\dim \Delta = d - 1$. By virtue of Corollary A.7.2 and Theorem A.7.1 in Appendix A.7 it then follows that Δ is Cohen–Macaulay over K if and only if

$$\tilde{H}_{i-|F|-1}(\mathrm{link}_\Delta F; K) = 0 \quad \text{for all} \quad F \in \Delta \quad \text{and all} \quad i < d. \tag{8.6}$$

Assume now that Δ is Cohen–Macaulay over K. Then Δ is pure and hence $\dim \mathrm{link}_\Delta F = d - |F| - 1$. Therefore, (8.6) implies that $\tilde{H}_i(\mathrm{link}_\Delta F; K) = 0$ for $i < \dim \mathrm{link}_\Delta F$.

Conversely, assume that for all $F \in \Delta$ including the empty face $\emptyset$ and for all $i < \dim \mathrm{link}_\Delta F$, one has $\tilde{H}_i(\mathrm{link}_\Delta F; K) = 0$. Let $F \in \Delta$, set $\Gamma = \mathrm{link}_\Delta F$ and let $G \in \Gamma$. Then $\mathrm{link}_\Gamma G = \mathrm{link}_\Delta(F \cup G)$, and so $\tilde{H}_i(\mathrm{link}_\Gamma G; K) = 0$ for all $i < \dim \mathrm{link}_\Gamma G$. Thus proceeding by induction on the dimension of Δ, we may assume that all proper links of Δ are Cohen–Macaulay over K. In particular, the link of each vertex of Δ is pure. Thus all facets containing a given vertex have the same dimension.

We may assume that $\dim \Delta > 0$, since Δ is Cohen–Macaulay over K if $\dim \Delta = 0$. Indeed, in this case $K[\Delta]$ is a 1-dimensional reduced standard graded K-algebra, and hence Cohen–Macaulay. We then observe that Δ is connected. This follows from the fact that $\tilde{H}_0(\Delta; K) = \tilde{H}_0(\text{link}_\Delta \emptyset; K) = 0$, because $1 + \dim_K \tilde{H}_0(\Delta; K)$ coincides with the number of connected components of Δ. Next we conclude that Δ is pure. Indeed, let F and G be two facets of Δ. Since Δ is connected, there exist facets $F_1, \ldots, F_m$ with $F = F_1$ and $G = F_m$ and such that $F_i \cap F_{i+1} \neq \emptyset$ for $i = 1, \ldots, m-1$. Since for each i, F_i and F_{i+1} have a vertex in common, it follows that $\dim F_i = \dim F_{i+1}$ for all i, as we have above. In particular $\dim F = \dim G$, as asserted.

Now, as we know that Δ is pure, it follows that $\dim \text{link}_\Delta F + 1 + |F| = d$ for all $F \in \Delta$. This implies that $i - |F| - 1 = (i - d) + \dim \text{link}_\Delta F < \dim \text{link}_\Delta F$ for $i < d$. Thus our hypothesis implies (8.6), and shows that Δ is Cohen–Macaulay over K.   □

In the course of the proof of Reisner's theorem we showed

Corollary 8.1.7. *Every Cohen–Macaulay simplicial complex is connected.*

Corollary 8.1.8. *Let Δ be a Cohen–Macaulay complex and F is a face of Δ. Then $\text{link}_\Delta F$ is Cohen–Macaulay.*

Proof. Let G be a face of $\text{link}_\Delta F$. Then

$$\text{link}_{\text{link}_\Delta F} G = \text{link}_\Delta(F \cup G).$$

Hence Reisner's criterion says that $\text{link}_\Delta F$ is Cohen–Macaulay.   □

We are now in the position to prove

Theorem 8.1.9 (Eagon–Reiner). *Let Δ be a simplicial complex on $[n]$ and let K be a field. Then the Stanley–Reisner ideal $I_\Delta \subset K[x_1, \ldots, x_n]$ has a linear resolution if and only if $K[\Delta^\vee]$ is Cohen–Macaulay.*

More precisely, I_Δ has a q-linear resolution if and only if $K[\Delta^\vee]$ is Cohen–Macaulay of dimension $n - q$.

Proof. Let $K[\Delta^\vee]$ be Cohen–Macaulay with $\dim \Delta^\vee = d - 1$. Let F be a face of $\Delta^\vee$ with $|F| = n - j$. Reisner's theorem says that $\tilde{H}_{i-1}(\text{link}_{\Delta^\vee} F; K) = 0$ unless $i - 1 = \dim \text{link}_{\Delta^\vee} F$. Since $\Delta^\vee$ is pure, one has $\dim \text{link}_{\Delta^\vee} F = d - (n - j) - 1$. Thus by using Corollary 8.1.4 it follows that $\beta_{ij}(I_\Delta) = 0$ unless $j - i = n - d$. Hence I_Δ has a $(n - d)$-linear resolution.

Conversely, suppose that I_Δ has a q-linear resolution. Then every minimal nonface of Δ is a q-element subset of $[n]$. Hence $\Delta^\vee$ is pure of dimension $n - q - 1$. Let F be a face of $\Delta^\vee$ with $|F| = n - j$. Again by using Corollary 8.1.4 it follows that $\tilde{H}_{i-1}(\text{link}_{\Delta^\vee} F; K) = 0$ unless $j = i + q$. In other words, $\tilde{H}_i(\text{link}_{\Delta^\vee} F; K) = 0$ unless $i = j - q - 1$. Since $\dim \text{link}_{\Delta^\vee} F = (n - q) - (n - j) - 1 = j - q - 1$, the homology group $\tilde{H}_i(\text{link}_{\Delta^\vee} F; K) = 0$ vanishes unless $i = \dim \text{link}_{\Delta^\vee} F$. Thus Reisner's theorem guarantees that $\Delta^\vee$ is Cohen–Macaulay, as desired.   □

We conclude this subsection with the following complement to the Eagon–Reiner theorem.

Proposition 8.1.10. *Let Δ be a simplicial complex. Then*

$$\operatorname{proj\,dim} I_\Delta = \operatorname{reg} K[\Delta^\vee].$$

Proof. The regularity of a finitely generated graded S-module in terms of local cohomology is given by

$$\operatorname{reg}(M) = \max\{j \colon H_{\mathfrak{m}}^i(M)_{j-i} \neq 0 \text{ for some } i\},$$

see Appendix A.7. Then (8.4) and (8.5) applied to $\Delta^\vee$ imply that

$$H_{\mathfrak{m}}^i(K[\Delta^\vee])_{j-i} = \sum_{F \in \Delta, |F| = i-j} \dim_K \tilde{H}_{j-1}(\operatorname{link}_{\Delta^\vee} F; K),$$

so that $\operatorname{reg}(K[\Delta^\vee]) = \max\{j \colon \tilde{H}_{j-1}(\operatorname{link}_{\Delta^\vee} F; K) \neq 0 \text{ for some } F \in \Delta^\vee\}$.

By comparing this with the formula for $\beta_{ij}(I_\Delta)$ in Corollary 8.1.4, the assertion follows.   □

8.2 Componentwise linear ideals

We first begin with the study of a special class of ideals with linear resolution.

8.2.1 Ideals with linear quotients

In general it is not so easy to find ideals with linear resolution. However, a big class of such ideals are those with linear quotients. Let $I \subset S$ be a graded ideal. We say that I has **linear quotients**, if there exists a system of homogeneous generators $f_1, f_2, \ldots, f_m$ of I such that the colon ideal $(f_1, \ldots, f_{i-1}) : f_i$ is generated by linear forms for all i.

Proposition 8.2.1. *Suppose $I \subset S$ is a graded ideal generated in degree d and that I has linear quotients. Then I has a d-linear resolution.*

Proof. Let $f_1, \ldots, f_m$ be a system of generators of I where each f_j is of degree d, and assume that for all k, $L_k = (f_1, \ldots, f_{k-1}) : f_k$ is generated by linear forms. We show by induction on k that $I_k = (f_1, \ldots, f_k)$ has a d-linear resolution. The assertion is obvious for $k = 1$. Suppose now that $k > 1$ and let $\ell_1, \ldots, \ell_r$ be linear forms generating L_k minimally. Observe that $\ell_1, \ldots, \ell_r$ is a regular sequence. Indeed, if we complete $\ell_1, \ldots, \ell_r$ to a K-basis $\ell_1, \ldots, \ell_n$ of S_1. Then $\varphi \colon S \to S$ with $\varphi(x_i) = \ell_i$ for $i = 1, \ldots, n$ is a K-automorphism. Since $x_1, \ldots, x_r$ is a regular sequence it follows that $\ell_1 = \varphi(x_1), \ldots, \ell_r = \varphi(x_r)$ is a regular sequence as well.

Now since $\ell_1, \ldots, \ell_r$ is a regular sequence, the Koszul complex $K(\ell_1, \ldots, \ell_r; S)$ provides a minimal graded free resolution of S/L_k; cf. Theorem A.3.4. This implies that

$$\mathrm{Tor}_i^S((S/L_k)(-d), K)_{i+j} \cong \mathrm{Tor}_i^S(S/L_k, K)_{i+(j-d)} = 0 \quad \text{for} \quad j \neq d.$$

We want to show that $\mathrm{Tor}_i(I_k, K)_{i+j} = 0$ for all i and all $j \neq d$. Observe that $I_k/I_{k-1} \cong (S/L_k)(-d)$, so that we have the following short exact sequence

$$0 \longrightarrow I_{k-1} \longrightarrow I_k \longrightarrow (S/L_k)(-d) \longrightarrow 0.$$

This sequence yields the long exact sequence

$$\mathrm{Tor}_i^S(I_{k-1}, K)_{i+j} \longrightarrow \mathrm{Tor}_i^S(I_k, K)_{i+j} \longrightarrow \mathrm{Tor}_i^S((S/L_k)(-d), K)_{i+j} \quad (8.7)$$

By applying our induction hypothesis we see that both ends in this exact sequence vanish for $j \neq d$. Thus this also holds for the middle term, as desired. $\qquad \square$

Analyzing the proof of the previous proposition we see that the Betti numbers of I can be computed once we know for each k the number of generators of $L_k = (f_1, \ldots, f_{k-1}) : f_k$. Let this number be r_k.

Corollary 8.2.2. *Let $I \subset S$ be a graded ideal with linear quotients generated in one degree. Then with the notation introduced one has*

$$\beta_i(I) = \sum_{k=1}^{n} \binom{r_k}{i}.$$

In particular it follows that $\mathrm{proj}\dim(I) = \max\{r_1, r_2, \ldots, r_n\}$.

Proof. In the long exact sequence (8.7) for $j = d$

$$\to \mathrm{Tor}_{i+1}^S((S/L_k)(-d), K)_{(i+1)+(d-1)} \to \mathrm{Tor}_i^S(I_{k-1}, K)_{i+d} \to \mathrm{Tor}_i^S(I_k, K)_{i+d}$$

$$\to \mathrm{Tor}_i^S((S/L_k)(-d), K)_{i+d} \to \mathrm{Tor}_{i-1}^S(I_{k-1}, K)_{(i-1)+(d+1)} \to$$

the end terms vanish, so that we obtain the short exact sequence

$$0 \to \mathrm{Tor}_i^S(I_{k-1}, K)_{i+d} \to \mathrm{Tor}_i^S(I_k, K)_{i+d} \to \mathrm{Tor}_i^S((S/L_k)(-d), K)_{i+d} \to 0,$$

from which we deduce that $\beta_i(I_k) = \beta_i(I_{k-1}) + \binom{r_k}{i}$. Induction on k completes the proof. $\qquad \square$

8.2.2 Monomial ideals with linear quotients and shellable simplicial complexes

What does it mean that a monomial ideal I does have linear quotients with respect to a monomial system of generators?

Suppose $G(I) = \{u_1, \ldots, u_m\}$. Then we have

Lemma 8.2.3. *The monomial ideal I has linear quotients with respect to the monomial generators $u_1, u_2, \ldots, u_m$ of I if and only if for all $j < i$ there exists an integer $k < i$ and an integer ℓ such that*

$$\frac{u_k}{\gcd(u_k, u_i)} = x_\ell \quad \text{and} \quad x_\ell \text{ divides } \frac{u_j}{\gcd(u_j, u_i)}.$$

Proof. The assertion follows immediately from the fact that $(u_1, \ldots, u_{i-1}) : u_i$ is generated by the monomials $u_j / \gcd(u_j, u_i)$, $j = 1, \ldots, i-1$, see Proposition 1.2.2. □

As an immediate consequence we obtain

Corollary 8.2.4. *Let I be a squarefree monomial ideal with $G(I) = \{u_1, u_2, \ldots, u_m\}$, and let $F_i = \operatorname{supp}(u_i)$ for $i = 1, \ldots, m$. Then I has linear quotients with respect to $u_1, u_2, \ldots, u_m$ if and only if for all i and all $j < i$ there exists an integer $\ell \in F_j \setminus F_i$ and an integer $k < i$ such that $F_k \setminus F_i = \{\ell\}$.*

We will now relate linear quotients of squarefree monomial ideals to shellability of simplicial complexes.

Let Δ be a simplicial complex on $[n]$. We say that Δ is **(nonpure) shellable** if its facets can be ordered $F_1, F_2, \ldots, F_m$ such that, for all $2 \le m$, the subcomplex

$$\langle F_1, \ldots, F_{j-1} \rangle \cap \langle F_j \rangle$$

is pure of dimension $\dim F_j - 1$. An order of the facets satisfying this conditions is called a **shelling order**.

To say that $F_1, F_2, \ldots, F_m$ is a shelling order of Δ is equivalent to saying that for all i and all $j < i$, there exists $\ell \in F_i \setminus F_j$ and $k < i$ such that $F_i \setminus F_k = \{\ell\}$. Thus we obtain

Proposition 8.2.5. *Let Δ be a simplicial complex. The following conditions are equivalent:*

(a) *I_Δ has linear quotients with respect to a monomial system of generators;*
(b) *the Alexander dual $\Delta^\vee$ of Δ is shellable.*

More precisely, if $G(I_\Delta) = \{u_1, u_2, \cdots, u_m\}$ and $F_i = \operatorname{supp}(u_i)$ for $i = 1, \ldots, m$, then I has linear quotients with respect to $u_1, \ldots, u_m$ if and only if $\bar{F}_1, \bar{F}_2, \cdots, \bar{F}_m$ is a shelling order of $\Delta^\vee$, where $\bar{F}$ is the complement of F is $[n]$.

Proof. It follows from Lemma 1.5.3 that $\bar{F}_1, \bar{F}_2, \cdots, \bar{F}_m$ are the facets of $\Delta^\vee$. Since $\bar{F}_r \setminus \bar{F}_s = F_s \setminus F_r$ for all r and s, all assertions follow from Corollary 8.2.4. □

The following result gives a useful combinatorial condition for the Cohen–Macaulay property of simplicial complexes.

Theorem 8.2.6. *A pure shellable simplicial complex is Cohen–Macaulay over an arbitrary field.*

Proof. By Proposition 8.2.5 the simplicial complex Δ is shellable if $I_{\Delta^\vee}$ is generated in one degree and has linear quotients with respect to a monomial system of generators. This property is independent of the characteristic of the base field. Thus, if Δ is shellable, then $I_{\Delta^\vee}$ has a linear resolution over an arbitrary base field, and hence the desired result follows by Eagon–Reiner (Theorem 8.1.9). □

For later applications we give a different characterization of shellability in terms of partitions. Let Δ be a simplicial complex and $G \subset F$ faces of Δ. The set

$$[G, F] = \{H \in \Delta \colon G \subset H \subset F\}$$

is called an **interval**. A disjoint union

$$\Delta = \bigcup_{i=1}^{m} [G_i, F_i]$$

of intervals is called a **partition** of Δ.

Let $F_1, \ldots, F_m$ be a shelling of Δ. This shelling gives rise to the following partition of Δ: we let $\Delta_j = \langle F_1, \ldots, F_j \rangle$, and define the **restriction** of the facet F_k by

$$\mathcal{R}(F_k) = \{i \in F_k \colon F_k \setminus \{i\} \in \Delta_{k-1}\}.$$

Proposition 8.2.7. *Let $F_1, \ldots, F_m$ be a shelling of Δ. Then*

$$\Delta = \bigcup_{k=1}^{m} [\mathcal{R}(F_k), F_k]$$

is a partition of Δ.

Proof. Let $F \in \Delta$, and let k be the smallest integer such that $F \subset F_k$. We claim that $\mathcal{R}(F_k) \subset F$. Indeed, let $i \in \mathcal{R}(F_k)$ and suppose that $i \notin F$. Since $F_k \setminus \{i\} \in \Delta_{k-1}$, it follows that $F \in \Delta_{k-1}$, a contradiction. This implies that Δ is the union of the intervals $[\mathcal{R}(F_k), F_k]$.

Suppose this union is not disjoint. Then there exist integers $j < k$ such that $[\mathcal{R}(F_j), F_j] \cap [\mathcal{R}(F_k), F_k] \neq \emptyset$. This implies that $\mathcal{R}(F_k) \subset F_j$. In other words, the elements $i \in F_k$ such that $F_k \setminus \{i\} \in \Delta_{k-1}$ belong to F_j. Hence $F_k \in \Delta_{k-1}$, a contradiction. □

Next we characterize the partitions which arise from shellings:

Proposition 8.2.8. *Given an ordering of $F_1, \ldots, F_m$ of the facets of Δ and a map $\mathcal{R} \colon \{F_1, \ldots, F_m\} \to \Delta$, the following conditions are equivalent:*

(i) *$F_1, \ldots, F_m$ is a shelling and $\mathcal{R}$ is its restriction map;*

(ii) (α) $\Delta = \bigcup_{k=1}^{m} [\mathcal{R}(F_k), F_k]$ *is a partition, and*
     (β) $\mathcal{R}(F_i) \subset F_j$ *implies* $i \leq j$ *for all* i, j.

Proof. (i) $\Rightarrow$ (ii) follows from the definition of the restriction map attached to a shelling. In order to prove the implication (ii) $\Rightarrow$ (i), we show that

$$\Delta_{k-1} \cap \langle F_k \rangle = \langle F_k \setminus \{i\} \colon\ i \in \mathcal{R}(F_k) \rangle,$$

which then yields (i).

The conditions (α) and (β) imply that $\langle F_k \rangle \setminus \Delta_{k-1} = [\mathcal{R}(F_k), F_k]$, so that $\Delta_{k-1} \cap \langle F_k \rangle = \langle F_k \rangle \setminus [\mathcal{R}(F_k), F_k]$. Since

$$\langle F_k \rangle \setminus [\mathcal{R}(F_k), F_k] = \langle F_k \setminus \{i\} \colon\ i \in \mathcal{R}(F_k) \rangle,$$

the assertion follows. $\square$

In the definition of shellability no statement is made about the dimension of the facets in the shelling order. However, as we shall see, the facets in a shelling can always be arranged such that they appear in order of decreasing dimension.

Proposition 8.2.9. *Let* $F_1, \ldots, F_m$ *be a shelling of the* $(d-1)$-*dimension simplicial complex* Δ *with restriction map* $\mathcal{R}$. *Let* $F_{i_1}, F_{i_2}, \cdots, F_{i_m}$ *be the rearrangement obtained by taking first all facets of dimension* $d-1$ *in the induced order, then all facets of dimension* $d-2$ *in the induced order, and continuing this way in order of decreasing dimension. Then this rearrangement is also a shelling, and its restriction map* $\mathcal{R}'$ *is the same, that is,* $\mathcal{R}'(F) = \mathcal{R}(F)$ *for all facets* F.

Proof. By using Proposition 8.2.8 it suffices to show that

$$\mathcal{R}(F_{i_j}) \subset F_{i_k} \quad \text{implies} \quad j \leq k.$$

Suppose this condition is not satisfied. Then there exist integers r and s such that

$$r < s, \quad |F_r| \leq |F_s|, \quad \text{and} \quad \mathcal{R}(F_r) \subset F_s. \qquad (8.8)$$

We choose r and s in (8.8) with s minimal. Observe that $\mathcal{R}(F_r) \neq F_s$, because otherwise we would have that $F_r \subset F_s$, a contradiction. Then there exists $i \in F_s$ such that $\mathcal{R}(F_r) \subset F_s \setminus \{i\}$. The shelling property of $F_1, \ldots, F_m$ implies that there exists $t < s$ such that $F_s \setminus \{i\} \subset F_t$. It follows that $\mathcal{R}(F_r) \subset F_t$, so that $|F_t| \geq |F_s|$. Moreover, Proposition 8.2.8 implies that $r < t$. This contradicts the choice of s. $\square$

8.2.3 Componentwise linear ideals

Let I be a graded ideal of $S = K[x_1, \ldots, x_n]$ and $\mathfrak{m} = (x_1, \ldots, x_n)$ the graded maximal ideal of S. If I is a graded ideal of S, then we write $I_{\langle j \rangle}$ for the ideal generated by all homogeneous polynomials of degree j belonging to I. Moreover, we write $I_{\leq k}$ for the ideal generated by all homogeneous polynomials of I whose degree is less than or equal to k.

We say that a graded ideal $I \subset S$ is **componentwise linear** if $I_{\langle j \rangle}$ has a linear resolution for all j. Typical examples of componentwise linear ideals are stable monomial ideals.

Ideals with linear resolution are componentwise linear, as follows from

Lemma 8.2.10. *If $I \subset S$ is a graded ideal with linear resolution, then $\mathfrak{m}I$ has again a linear resolution.*

Proof. Say that I is generated in degree d. Then the least shift in the ith position of the graded minimal free resolution of $\mathfrak{m}I$ is at least $i + d + 1$. This implies that $\mathrm{Tor}_i^S(K, I)_{i+j} = 0$ for all $i \geq 0$ and $j < d + 1$. Consider the long exact Tor sequence arising from the short exact sequence $0 \to \mathfrak{m}I \to I \to I/\mathfrak{m}I \to 0$. Since $I/\mathfrak{m}I \cong K(-)^b$ for some b, we obtain the exact sequence

$$\mathrm{Tor}_{i+1}^S(K, K(-d)^b)_{(i+1)+(j-1)} \to \mathrm{Tor}_i^S(K, \mathfrak{m}I)_{i+j} \to \mathrm{Tor}_i^S(K, I)_{i+j}.$$

For $j > d+1$, we have $\mathrm{Tor}_i^S(K, I)_{i+j} = 0$ and $\mathrm{Tor}_{i+1}^S(K, K(-d)^b)_{(i+1)+(j-1)} = 0$, since I and $K(-d)$ have d-linear resolutions. It follows that $\mathrm{Tor}_i^S(K, \mathfrak{m}I)_{i+j} = 0$ for all $i \geq 0$ and $j > d + 1$. Thus $\mathfrak{m}I$ has $(d + 1)$-linear resolution. $\square$

Another interesting class of componentwise linear ideals are the ideals with linear quotients, as we shall see later in this section.

In the analysis of componentwise linear ideals, we begin with a simple fact which says that the part $I_{\leq k}$ of I determines already a certain range of its graded Betti numbers. We first observe

Lemma 8.2.11. *Let $I \subset$ be a componentwise linear ideal. Then $I_{\leq j}$ is componentwise linear for all j.*

Proof. Let $k \leq j$, then $(I_{\leq j})_{\langle k \rangle} = I_{\langle k \rangle}$. Therefore $(I_{\leq j})_{\langle k \rangle}$ has a linear resolution for $k \leq j$. Let $k > j$, then $(I_{\leq j})_{\langle k \rangle} = \mathfrak{m}(I_{\leq j})_{\langle k-1 \rangle}$. Thus, by using Lemma 8.2.10 and induction on $k - j$ it follows that $(I_{\leq j})_{\langle k \rangle}$ has a linear resolution for $k > j$, as well. Hence $I_{\leq j}$ is componentwise linear. $\square$

Lemma 8.2.12. *Let $I \subset S$ be a graded ideal. Then, for all k and for all $j \leq k$, one has*

$$\beta_{i,i+j}(I) = \beta_{i,i+j}(I_{\leq k}).$$

Proof. Let $H_i(\mathbf{x}; I)$ denote the Koszul homology of I with respect to the sequence $\mathbf{x} = x_1, x_2, \ldots, x_n$ of the variables. By using an isomorphism of graded K-vector space $\operatorname{Tor}_i^S(K, I) \cong H_i(\mathbf{x}; I)$ (see Corollary A.3.5), it follows that

$$\beta_{i,i+j}(I) = \dim_K H_i(\mathbf{x}; I)_{i+j}.$$

A homogeneous cycle c of degree $i + j$ representing a homology class in $H_i(\mathbf{x}; I)_{i+j}$ is a linear combination $\sum_F a_F \mathbf{e}_F$ of the canonical basis elements $\mathbf{e}_F = e_{k_1} \wedge \cdots \wedge e_{k_i}$ with coefficients $a_F \in I_j$. Thus c also represents a cycle in $H_i(\mathbf{x}; I_{\leq k})_{i+j}$ provided $j \leq k$. Similarly, the i-boundaries of the Koszul complex for I and for $I_{\leq k}$ coincides whenever $j \leq k$. Hence

$$H_i(\mathbf{x}; I)_{i+j} \cong H_i(\mathbf{x}; I_{\leq k})_{i+j}$$

for $j \leq k$. This proves the assertion. $\square$

By using Lemma 8.2.12 we show that the graded Betti numbers of a componentwise linear ideal can be determined by the graded Betti numbers of its components.

Proposition 8.2.13. *Suppose that the graded ideal $I \subset S$ is componentwise linear. Then*

$$\beta_{i,i+j}(I) = \beta_i(I_{\langle j \rangle}) - \beta_i(\mathfrak{m}I_{\langle j-1 \rangle})$$

for all j.

Proof. Let t denote the highest degree of generators of a minimal set of generators of I. Our proof will be done by induction on t. Let $t = 1$. Then I is generated by linear form, and hence has a linear resolution. Since $I_{\langle j \rangle} = \mathfrak{m}I_{\langle j-1 \rangle}$ if $j > 1$ and since $\beta_{i,i+j}(I) = 0$ if $j > 1$, the assertion is true for $j > 1$. On the other hand, since $I = I_{\langle 1 \rangle}$ and $\mathfrak{m}I_{\langle 0 \rangle} = 0$, the assertion is obvious for $j = 1$.

Now, suppose that $t > 1$ and consider the exact sequence

$$0 \longrightarrow I_{\leq t-1} \longrightarrow I \longrightarrow I_{\langle t \rangle}/\mathfrak{m}I_{\langle t-1 \rangle} \longrightarrow 0$$

which for each j yields the long exact sequence

$$\operatorname{Tor}_i(K, I_{\leq t-1})_{i+j} \longrightarrow \operatorname{Tor}_i(K, I)_{i+j} \longrightarrow \operatorname{Tor}_i(K, I_{\langle t \rangle}/\mathfrak{m}I_{\langle t-1 \rangle})_{i+j}. \quad (8.9)$$

Since $I_{\leq t-1}$ is generated in degree $\leq t-1$, one has $(I_{\leq t-1})_{\langle j \rangle} = \mathfrak{m}(I_{\leq t-1})_{\langle j-1 \rangle}$ for $j \geq t$. Since $I_{\leq t-1}$ is componentwise linear (Lemma 8.2.11), our induction hypothesis guarantees that $\beta_{i,i+j}(I_{\leq t-1}) = 0$ for $j \geq t$. Hence by the long exact sequence (8.9) one has

$$\operatorname{Tor}_i(K, I)_{i+j} = \operatorname{Tor}_i(K, I_{\langle t \rangle}/\mathfrak{m}I_{\langle t-1 \rangle})_{i+j} \quad (8.10)$$

for $j \geq t$.

Now, we show our formula for $\beta_{i,i+j}(I)$. By Lemma 8.2.12 one has $\beta_{i,i+j}(I) = \beta_{i,i+j}(I_{\leq t-1})$ for $j \leq t - 1$. Thus by induction hypothesis our formula is true for $j \leq t - 1$. Let $j \geq t$ and consider the exact sequence

$$0 \longrightarrow \mathfrak{m}I_{\langle t-1 \rangle} \longrightarrow I_{\langle t \rangle} \longrightarrow I_{\langle t \rangle}/\mathfrak{m}I_{\langle t-1 \rangle} \longrightarrow 0$$

which for each j yields the long exact sequence

$$\operatorname{Tor}_{i+1}(K, I_{\langle t \rangle}/\mathfrak{m}I_{\langle t-1 \rangle})_{i+j} \longrightarrow \operatorname{Tor}_i(K, \mathfrak{m}I_{\langle t-1 \rangle})_{i+j} \longrightarrow \operatorname{Tor}_i(K, I_{\langle t \rangle})_{i+j}$$
$$\longrightarrow \operatorname{Tor}_i(K, I_{\langle t \rangle}/\mathfrak{m}I_{\langle t-1 \rangle})_{i+j} \longrightarrow \operatorname{Tor}_{i-1}(K, \mathfrak{m}I_{\langle t-1 \rangle})_{i+j}.$$

Since $I_{\langle t-1 \rangle}$ has a $(t-1)$-linear resolution, it follows that $\mathfrak{m}I_{\langle t-1 \rangle}$ has a t-linear resolution (Lemma 8.2.10). Hence $\operatorname{Tor}_{i-1}(K, \mathfrak{m}I_{\langle t-1 \rangle})_{i+j} = 0$ for $j \geq t$. On the other hand, since the graded module $I_{\langle t \rangle}/\mathfrak{m}I_{\langle t-1 \rangle}$ is generated in degree t, it follows that $\operatorname{Tor}_{i+1}(K, I_{\langle t \rangle}/\mathfrak{m}I_{\langle t-1 \rangle})_{i+j} = 0$ for $j = t$. Thus by using (8.10) our formula is true for $j = t$. Finally, let $j > t$. Then $\operatorname{Tor}_i(K, I_{\langle t \rangle})_{i+j} = 0$ and $\operatorname{Tor}_{i-1}(K, \mathfrak{m}I_{\langle t-1 \rangle})_{i+j} = 0$. Thus $\operatorname{Tor}_i(K, I_{\langle t \rangle}/\mathfrak{m}I_{\langle t-1 \rangle})_{i+j} = 0$. In view of (8.10) one has $\beta_{i,i+j}(I) = 0$. Since $\mathfrak{m}I_{\langle j-1 \rangle} = I_{\langle j \rangle}$, our formula is true. $\quad\square$

As an immediate consequence of the preceding result we obtain

Corollary 8.2.14. *Let $I \subset S$ be a componentwise linear ideal. Then the regularity of I is equal to the highest degree of a generator in a minimal set of generators of I.*

8.2.4 Ideals with linear quotients and componentwise linear ideals

As an extension of Proposition 8.2.1 we have

Theorem 8.2.15. *Let $I \subset S$ be a graded ideal which has linear quotients with respect to a minimal homogeneous system of generators of I. Then I is componentwise linear.*

Proof. Let $f_1, \ldots, f_m$ be a minimal homogeneous system of generators of I such that $(f_1, \ldots, f_{i-1}): f_i$ is generated by linear forms for $i = 1, \ldots, m$. Proceeding by induction on m, we may assume that $J = (f_1, \ldots, f_{m-1})$ is componentwise linear.

Now we show the following: let $J = (f_1, \ldots, f_{m-1})$ be any graded ideal which is componentwise linear, and let $f \in S$ be a homogeneous element of degree d such that $J: f$ is generated by linear forms. Assume further that $f_1, \ldots, f_{m-1}, f$ is a minimal system of generators of $I = (f_1, \ldots, f_{m-1}, f)$. Then I is componentwise linear.

In order to prove this statement we proceed by induction on

$$s(J, f) = \max\{0, p - d\},$$

where p is the maximal degree of the f_i. Suppose $s(J, f) = 0$. Then $d \geq p$. Observe that $I_{\langle j \rangle} = J_{\langle j \rangle}$ for $j < d$, and so $I_{\langle j \rangle}$ has a linear resolution for $j < d$. Next observe that $J_{\langle d \rangle}: f = J : f$. Obviously, $J_{\langle d \rangle}: f \subset J : f$. Conversely, let $J: f = L$, where $L = (\ell_1, \ldots, \ell_r)$. Then for each i we have $\ell_i f \in J$. Since I is minimally generated by $f_1, \ldots, f_{m-1}, f$, it follows that $\ell_i f$ is a linear combination of $f_1, \ldots f_{m-1}$, whose nonzero coefficients are of positive degree. Since $\deg \ell_i f = d + 1$, only those f_i with degree $\leq d$ can occur in this linear combination. Thus we see that $\ell_i f \in J_{\leq d}$. Since $(J_{\langle d \rangle})_{d+1} = (J_{\leq d})_{d+1}$, it follows that $\ell_i f \in J_{\langle d \rangle}$, and hence $\ell_i \in J_{\langle d \rangle}$.

The above considerations show that $J_{\langle d \rangle} : f$ is generated by linear forms. Since $J_{\langle d \rangle} + (f) = I_{\langle d \rangle}$, the arguments in the proof of Proposition 8.2.1 show that $I_{\langle d \rangle}$ has a linear resolution. If follows from Lemma 8.2.10 that $I_{\langle d+j \rangle}$ has a linear resolution for all $j \geq 0$, and hence I is componentwise linear.

Now we assume that $s(J, f) > 0$. We complete the system of generators $\ell_1, \ldots, \ell_r$ of L by the linear forms $\ell_{r+1}, \ldots, \ell_n$ to obtain a minimal set of generators of the graded maximal ideal $\mathfrak{m} = (x_1, \ldots, x_n)$ of S, and set $g_i = \ell_i f$ for $i = r + 1, \ldots, n$. Since $(J + (g_{r+1}, \ldots, g_n))_{\langle d+j \rangle} = I_{\langle d+j \rangle}$ for all ≥ 1, and since $I_{\langle j \rangle}$ is componentwise linear for $j \leq d$ (independent of s), as we have seen before, it suffices to show that $(J + (g_{r+1}, \ldots, g_n))_j$ is componentwise linear. In order to prove this we show: For all $i = 1, \ldots, n - r$,

(1) the elements $f_1, \ldots, f_{m-1}, g_{r+1}, \ldots, g_{r+j}$ form a minimal set of generators of $I_j = (f_1, \ldots, f_{m-1}, g_{r+1}, \ldots, g_{r+j})$;
(2) $J + (g_{r+1}, \ldots, g_{r+j-1}) : g_{r+j}$ is generated by linear forms.

Suppose (1) and (2) are correct. Since $\deg g_1 = d + 1$, it follows that $s(J, g_{r+1}) < s(J, f) - 1$. Hence our induction hypothesis implies that I_1 is componentwise linear. Since $s(I_1, g_{r+2}) = 0$, our induction hypothesis implies again that I_2 is componentwise linear. Proceeding in this way, we see that $J + (g_{r+1}, \ldots, g_n)$ is componentwise linear.

Proof of (1): Suppose we can omit some f_i. Then f_i can be expressed by the remaining f_j and the g_i which are all multiples of f. This implies that $f_1, \ldots, \hat{f_i}, \ldots, f_{m-1}, f$ is a minimal set of generators of I: a contradiction.

On the other hand, if we can omit g_{r+i}, then $g_{r+i} = g + \sum_{\substack{j=1 \\ j \neq i}}^{n-r} h_j g_{r+j}$ with $h_j \in S$ and $g \in J$. It follows that $(\ell_{r+i} - \sum_{\substack{j=1 \\ j \neq i}}^{n-r} h_j \ell_{r+j}) f \in J$, and hence $\ell_{r+i} - \sum_{\substack{j=1 \\ j \neq i}}^{n-r} h_j \ell_{r+j} \in L$: a contradiction.

Proof of (2): Let $h \in J + (g_{r+1}, \ldots, g_{r+j-1}) : g_{r+j}$. Then $\ell_{r+j} f h \in J + (g_{r+1}, \ldots, g_{r+j-1})$. Therefore there exists $g \in J$ and $h_i \in S$ such that $\ell_{r+j} f h = g + \sum_{i=1}^{j-1} h_i \ell_{r+i} f$. This implies that $\ell_{r+j} h - \sum_{i=1}^{j-1} h_i \ell_{r+i} \in J : f = (\ell_1, \ldots, \ell_r)$, and hence $\ell_{r+j} h \in (\ell_1, \ldots, \ell_{r+j-1})$. Since the sequence $\ell_1, \ldots, \ell_{r+j}$ is a regular sequence, we conclude that $h \in (\ell_1, \ldots, \ell_{r+j-1})$, as desired. $\square$

Example 8.2.16. In Theorem 8.2.15 the condition that I has linear quotients with respect to a *minimal* system of homogeneous generators cannot be omitted. Indeed, let $I = (x^2, y^2)$. Then I is not componentwise linear, but I has linear quotients with respect the nonminimal system of generators x^2, xy^2, y^2 of I.

8.2.5 Squarefree componentwise linear ideals

Let $I \subset S$ be a squarefree monomial ideal. Then, for each degree j, we write $I_{[j]}$ for the ideal generated by the squarefree monomials of degree j belonging to I. We say that I is **squarefree componentwise linear** if $I_{[j]}$ has a linear resolution for all j.

Proposition 8.2.17. *A squarefree monomial ideal $I \subset S$ is componentwise linear if and only if I is squarefree componentwise linear.*

Proof. Suppose that I is componentwise linear. Fix $j > 0$. Then $I_{\langle j \rangle}$ has a linear resolution. The exact sequence

$$0 \longrightarrow I_{[j]} \longrightarrow I_{\langle j \rangle} \longrightarrow I_{\langle j \rangle}/I_{[j]} \longrightarrow 0$$

gives rise to the long exact sequence

$$\rightarrow \mathrm{Tor}_{i+1}(K, I_{\langle j \rangle}/I_{[j]}) \xrightarrow{\ \alpha_i\ } \mathrm{Tor}_i(K, I_{[j]}) \longrightarrow \mathrm{Tor}_i(K, I_{\langle j \rangle}) \rightarrow .$$

Since the ideals under consideration are monomial ideals, it follows that all the Tor-groups in the long exact sequence are multigraded K-vector spaces. Now, Hochster's formula (Theorem 8.1.1) says that $\mathrm{Tor}_i(K, I_{[j]})$ has only squarefree components; in other words, $\mathrm{Tor}_i(K, I_{[j]})_a = 0$ if one entry of the vector a is > 1. On the other hand, since all generators of $I_{\langle j \rangle}/I_{[j]}$ have non-squarefree degrees, it follows that $\mathrm{Tor}_i(K, I_{\langle j \rangle}/I_{[j]})$ has only non-squarefree components. Since α_i is multihomogeneous, α_i must be the zero map. Thus for each i the map $\mathrm{Tor}_i(K, I_{[j]}) \rightarrow \mathrm{Tor}_i(K, I_{\langle j \rangle})$ is injective. Since $I_{\langle j \rangle}$ has a linear resolution, the graded K-vector space $\mathrm{Tor}_i(K, I_{\langle j \rangle})$ is concentrated in degree $i + j$. Thus $\mathrm{Tor}_i(K, I_{[j]})$ is concentrated in degree $i + j$. Hence $I_{[j]}$ has a linear resolution, as required.

Conversely, suppose that $I_{[j]}$ has a linear resolution for all j. We will show by using induction on j that $I_{\langle j \rangle}$ has a linear resolution for all j.

Let t denote the lowest degree for which $I_t \neq 0$. Since $I_{\langle t \rangle} = I_{[t]}$, $I_{\langle t \rangle}$ has a linear resolution. Suppose that $I_{\langle j \rangle}$ has a linear resolution for some $j \geq t$. Then $\mathfrak{m} I_{\langle j \rangle}$ also has a linear resolution. The first part of the proof shows that the squarefree part L of $\mathfrak{m} I_{\langle j \rangle}$ has a linear resolution. Since L is contained in $I_{[j+1]}$, we get the exact sequence

$$0 \longrightarrow L \longrightarrow \mathfrak{m} I_{\langle j \rangle} \oplus I_{[j+1]} \longrightarrow I_{\langle j+1 \rangle} \longrightarrow 0, \qquad (8.11)$$

where $u \in L$ is mapped to $(u, -u) \in \mathfrak{m} I_{\langle j \rangle} \oplus I_{[j+1]}$. We have already noted that both L and $\mathfrak{m} I_{\langle j \rangle}$ have a linear resolution. Furthermore, $I_{[j+1]}$ has a linear resolution.

From the long exact Tor sequence which is derived from (8.11) we deduce that $I_{\langle j+1 \rangle}$ has a linear resolution once it is shown that

$$\mathrm{Tor}_i(K, L) \longrightarrow \mathrm{Tor}_i(K, \mathfrak{m} I_{\langle j \rangle}) \oplus \mathrm{Tor}_i(K, I_{[j+1]})$$

is injective for all i. But this is clear since already the first component of this map is injective as we have seen in the first part of the proof, because L is the squarefree part of $\mathfrak{m} I_{\langle j \rangle}$. $\qquad\square$

8.2.6 Sequentially Cohen–Macaulay complexes

We now turn to the discussion of the combinatorics on squarefree componentwise linear ideals.

Let Δ be a simplicial complex on $[n]$ of dimension $d - 1$. Recall that, for each $0 \leq i \leq d - 1$, the ith **skeleton** of Δ is the simplicial complex $\Delta^{(i)}$ on $[n]$ whose faces are those faces F of Δ with $|F| \leq i + 1$. In addition, for each $0 \leq i \leq d - 1$, we define the **pure ith skeleton** of Δ to be the pure subcomplex $\Delta(i)$ of Δ whose facets are those faces F of Δ with $|F| = i + 1$.

We say that a simplicial complex Δ is **sequentially Cohen–Macaulay** if $\Delta(i)$ is Cohen–Macaulay for all i.

Theorem 8.2.18 (Björner–Wachs). *Let Δ be a shellable simplicial complex. Then all skeletons and pure skeletons of Δ are shellable.*

Proof. Let $\dim \Delta = d-1$, and let $0 \leq s \leq d-1$ be an integer. We want to show that $\Delta^{(s)}$ and $\Delta(s)$ are shellable. Applying Proposition 8.2.9 the shellability of $\Delta^{(s)}$ guarantees the shellability of $\Delta(s)$. Thus it suffices to show that $\Delta^{(s)}$ is shellable. By Proposition 8.2.9 we may assume that the shelling $F_1, \ldots, F_m$ has the property that $|F_i| \leq |F_j|$ for all $i \geq j$. Let k be the largest integer for which $|F_k| \geq s + 1$. The subcomplex of Δ generated by the facets $F_1, \ldots, F_k$ is again shellable and has the same sth skeleton. Thus we may assume as well that $|F_i| \geq s + 1$ for all i. Since $\Delta^{(i)} = (\Delta^{(i+1)})^{(i)}$, a simple induction argument shows that we may assume that $s = d - 2$.

Let

$$\Delta = \bigcup_{j=1}^{m} [R_j, F_j] \tag{8.12}$$

be the partition of Δ induced by the shelling, see Proposition 8.2.7.

Let $F_1, \ldots, F_\ell$ be the facets of dimension $d - 1$. For each j, $1 \leq j \leq \ell$, choose an ordering $i_1, \ldots, i_{q_j}$ of the elements of $F_j \setminus R_j$, and let

$$R_{j,k} = R_j \cup \{i_1, \ldots, i_{k-1}\}, \quad \text{and} \quad F_{j,k} = F_j \setminus \{i_k\} \quad \text{for} \quad k = 1, \ldots, q_j.$$

Then we obtain for each j the partition

$$[R_j, F_j] = [F_j, F_j] \cup \bigcup_{k=1}^{q_j} [R_{j,k}, F_{j,k}]$$

of the interval $[R_j, F_j]$, and hence the partition

$$\Delta^{(d-2)} = \bigcup_{j=1}^{\ell} \bigcup_{k=1}^{q_j} [R_{j,k}, F_{j,k}] \cup \bigcup_{j=\ell+1} [R_j, F_j]. \tag{8.13}$$

Since the partition (8.12) of Δ is induced by a shelling, it satisfies condition (β) of Proposition 8.2.8. From this it is easy to see that the partition (8.13) of $\Delta^{(d-2)}$ satisfies condition (β) of Proposition 8.2.8 as well, and hence is induced by a shelling of $\Delta^{(d-2)}$. □

Corollary 8.2.19. *Any shellable simplicial complex is sequentially Cohen–Macaulay.*

We now come to one of the main results of this section, which is a generalization of the Eagon–Reiner theorem.

Theorem 8.2.20. *Let Δ be a simplicial complex on $[n]$. Then $I_\Delta \subset S$ is componentwise linear if and only if $\Delta^\vee$ is sequentially Cohen–Macaulay.*

Proof. Let $I = I_\Delta$. Then by Proposition 8.2.17 I is componentwise linear if and only if I is squarefree componentwise linear. Let Δ_j denote the simplicial complex on $[n]$ with $I_{\Delta_j} = I_{[j]}$. Let $F \subset [n]$. Then $F \notin \Delta_j$ if and only if there is a subset $G \subset [n]$ such that $G \subset F$, $|G| = j$ and $G \notin \Delta$. In other words, $[n] \backslash F \in (\Delta_j)^\vee$ if and only if there is a subset $G \subset [n]$ such that $[n] \backslash F \subset [n] \backslash G$, $|[n] \backslash G| = n - j$ and $[n] \backslash G \in \Delta^\vee$. Hence $(\Delta_j)^\vee = \Delta^\vee(n - j - 1)$. By virtue of Theorem 8.1.9 it follows that $I_{[j]}$ has a linear resolution if and only if $\Delta^\vee(n - j - 1)$ is Cohen–Macaulay. Hence I_Δ is componentwise linear if and only if $\Delta^\vee(i)$ is Cohen–Macaulay for all i. □

As a consequence of the above results one obtains an alternative and simple proof of Theorem 8.2.15 in the case of squarefree monomial ideals.

Corollary 8.2.21. *Let I be a squarefree monomial ideal with linear quotients. Then I is componentwise linear.*

Proof. Let Δ be the simplicial complex with $I = I_\Delta$. Proposition 8.2.5 says $\Delta^\vee$ is nonpure shellable, and hence by Corollary 8.2.19 $\Delta^\vee$ is sequentially Cohen–Macaulay. Thus the assertion follows from Theorem 8.2.20. □

8.2.7 Ideals with stable Betti numbers

We now prove a fundamental result on componentwise linear ideals. Let $<_{\mathrm{rev}}$ denote the reverse lexicographic monomial order on S induced by the ordering $x_1 > \cdots > x_n$ of the variables.

Theorem 8.2.22. *Suppose that the base field K is of characteristic 0. Then a graded ideal $I \subset S$ is componentwise linear if and only if $\beta_{i,i+j}(I) = \beta_{i,i+j}(\mathrm{gin}_{<_{\mathrm{rev}}}(I))$ for all i and j.*

In order to prove Theorem 8.2.22, the following result concerning the homological data of generic initial ideals will be required.

Theorem 8.2.23. *Let I be a graded ideal generated in degree d. Then*

(a) *If $\beta_{i,i+j}(\mathrm{gin}(I)) \neq 0$, then $\beta_{i',i'+j}(\mathrm{gin}(I)) \neq 0$ for all $i' < i$;*
(b) *If $\beta_{0,j}(\mathrm{gin}(I)) \neq 0$, then $\beta_{0,j'}(\mathrm{gin}(I)) \neq 0$ for all $d \leq j' < j$.*

Proof. Statement (a) follows from the Eliahou–Kervaire formula in Corollary 4.2.6, since by Proposition 7.2.3 the generic initial ideal is strongly stable.

Let $g_1, \ldots, g_m \in I_d$ be the generators of I. Suppose that $\beta_{0,j-1}(\mathrm{gin}(I)) = 0$. Then consider the ideal $I_{\geq j-2}$. Since $\mathrm{gin}(I_{\geq j-2}) = \mathrm{gin}(I)_{\geq j-2}$, we may assume that $\beta_{0,d+1}(\mathrm{gin}(I)) = 0$ and have to show that $\mathrm{gin}(I)$ is generated in degree d.

Since $(\mathrm{in}(g_1), \ldots, \mathrm{in}(g_m))$ is a strongly stable ideal with all generators of degree d, it follows from Theorem 7.2.2 that $(\mathrm{in}(g_1), \ldots, \mathrm{in}(g_m))$ has a linear resolution. In particular, the first syzygy module of this ideal is generated in degree $d + 1$. Now since $\beta_{0,d+1}(\mathrm{gin}(I)) = 0$ it follows that all S-polynomials of degree $d + 1$ reduce to 0 with respect to $\{g_1, \ldots, g_m\}$. Thus Proposition 2.3.5 implies that $\{g_1, \ldots, g_m\}$ is a Gröbner basis of I, equivalently, $\mathrm{gin}(I)$ is generated in degree d. $\square$

Before starting our proof of Theorem 8.2.22, we state the following:

Lemma 8.2.24. *Let I and J be graded ideals of S generated in degree d with the same graded Betti numbers. Then $I_{\geq d+1}$ and $J_{\geq d+1}$ have the same graded Betti numbers.*

Proof. The exact sequence

$$0 \longrightarrow I_{\geq d+1} \longrightarrow I \longrightarrow K(-d)^{\beta_{0,d}} \longrightarrow 0$$

induces the long exact sequence

$$\rightarrow \mathrm{Tor}_{i+1}(I_{\geq d+1})_{(i+1)+(j-1)} \rightarrow \mathrm{Tor}_{i+1}(I)_{(i+1)+(j-1)} \rightarrow \mathrm{Tor}_{i+1}(K)_{(i+1)+j-(d+1)}^{\beta_{0,d}}$$

$$\rightarrow \mathrm{Tor}_i(I_{\geq d+1})_{i+j} \rightarrow \mathrm{Tor}_i(I)_{i+j} \rightarrow \mathrm{Tor}_i(K)_{i+j-d}^{\beta_{0,d}} \rightarrow \cdots$$

It then follows that $\beta_{i,i+j}(I_{\geq d+1}) = \beta_{i,i+j}(I)$ for all i and for all $j \neq d, d+1$. Also, $\beta_{i,i+j}(I_{\geq d+1}) = 0$ if $j \leq d$. Now, if $j = d + 1$, then the above long exact sequence becomes

$$\rightarrow \mathrm{Tor}_{i+1}(I)_{i+1+d} \rightarrow \mathrm{Tor}_{i+1}(K)_{i+1}^{\beta_{0,d}} \rightarrow \mathrm{Tor}_i(I_{\geq d+1})_{i+d+1} \rightarrow \mathrm{Tor}_i(I)_{i+d+1} \rightarrow 0.$$

Hence, $\beta_{i,i+d+1}(I_{\geq d+1}) = \beta_{i,i+d+1}(I) + \binom{n}{i+1}\beta_{0,d}(I) - \beta_{i+1,i+1+d}(I)$.

The same formulas are valid for $\beta_{i,i+j}(J)$. This completes the proof. $\square$

We are now in the position to give a proof of Theorem 8.2.22.

Proof (Proof of 8.2.22). First, suppose that I is componentwise linear. By Proposition 8.2.13 we have

$$\beta_{i,i+j}(I) = \beta_i(I_{\langle j \rangle}) - \beta_i(\mathfrak{m}I_{\langle j-1 \rangle})$$

where $\mathfrak{m}$ is the irrelevant maximal ideal $(x_1, \ldots, x_n)$ of S. Since a strongly stable ideal is componentwise linear and since $\mathrm{gin}(I)$ is strongly stable, the same formula is valid for $\mathrm{gin}(I)$. Therefore, it suffices to prove that $\beta_i(I_{\langle j\rangle}) = \beta_i(\mathrm{gin}(I)_{\langle j\rangle})$ and $\beta_i(\mathfrak{m}I_{\langle j-1\rangle}) = \beta_i(\mathfrak{m}\,\mathrm{gin}(I)_{\langle j-1\rangle})$.

Since $I_{\langle j\rangle}$ has a linear resolution, it follows from Corollary 4.3.18 that $\mathrm{gin}(I_{\langle j\rangle})$ has a linear resolution, so that $\mathrm{gin}(I_{\langle j\rangle}) = \mathrm{gin}(I)_{\langle j\rangle}$. Since $I_{\langle j\rangle}$ and $\mathrm{gin}(I_{\langle j\rangle})$ have the same Hilbert function, and since the Betti numbers of a module with linear resolution are determined by its Hilbert function, the first equality follows. To prove the second one, we note that $\mathfrak{m}I_{\langle j-1\rangle}$ has again a linear resolution and that, by the same reason as before, $\mathfrak{m}\,\mathrm{gin}(I)_{\langle j-1\rangle} = \mathrm{gin}(\mathfrak{m}I_{\langle j-1\rangle})$.

Second, suppose that I and $\mathrm{gin}(I)$ have the same graded Betti numbers. Let $\max(I)$ (resp. $\min(I)$) denote the maximal (resp. minimal) degree of a homogeneous generator of I. To show that I is componentwise linear, we work with induction on $r - \max(I) - \min(I)$. Set $d = \min(I)$.

Let $r = 0$. Since I and $\mathrm{gin}(I)$ have the same graded Betti numbers, it follows that $\mathrm{gin}(I)$ is generated in degree d. Since $\mathrm{gin}(I)$ is a strongly stable ideal, we have that $\mathrm{gin}(I)$ has a linear resolution, and hence by Theorem 3.3.4 I has a linear resolution.

Now, suppose that $r > 0$. Since $\mathrm{gin}(I_{\geq d+1}) = \mathrm{gin}(I)_{\geq d+1}$, our induction hypothesis and Lemma 8.2.24 imply that $I_{\geq d+1}$ is componentwise linear. Thus, it suffices to prove that $I_{\langle d\rangle}$ has a linear resolution. Suppose this is not the case. Then, by Corollary 4.3.18, $\mathrm{gin}(I_{\langle d\rangle})$ has regularity $> d$. Moreover, since $\mathrm{gin}(I_{\langle d\rangle})$ is strongly stable, its regularity equals $\max(\mathrm{gin}(I_{\langle d\rangle}))$. It follows from Theorem 8.2.23 that $\mathrm{gin}(I_{\langle d\rangle})$ has a generator of degree $d + 1$. Now,

$$\beta_{0,d+1}(I) = \dim I_{d+1} - \dim(\mathfrak{m}I_{\langle d\rangle})_{d+1}$$
$$= \dim I_{d+1} - \dim(I_{\langle d\rangle})_{d+1},$$

and

$$\beta_{0,d+1}(\mathrm{gin}(I)) = \dim \mathrm{gin}(I)_{d+1} - \dim(\mathfrak{m}\,\mathrm{gin}(I)_{\langle d\rangle})_{d+1}$$
$$= \dim \mathrm{gin}(I)_{d+1} - \dim(\mathfrak{m}\,\mathrm{gin}(I_{\langle d\rangle}))_{d+1}$$
$$> \dim \mathrm{gin}(I)_{d+1} - \dim \mathrm{gin}(I_{\langle d\rangle})_{d+1},$$

because $(\mathfrak{m}\,\mathrm{gin}(I_{\langle d\rangle}))_{d+1}$ is properly contained in $\mathrm{gin}(I_{\langle d\rangle})_{d+1}$. Hence

$$\beta_{0,d+1}(\mathrm{gin}(I)) > \beta_{0,d+1}(I);$$

a contradiction. This completes our proof. $\square$

Problems

8.1. By using Hochster's formula 8.1.1 compute the graded Betti numbers of the Stanley–Reisner ideal of the simplicial complex of Figure 1.1.

8.2. Let Δ be a simplicial complex and K a field. Show that Δ is connected if and only if $\tilde{H}_0(\Delta; K) = 0$.

8.3. Show that Proposition 8.1.10 implies Theorem 8.1.9.

8.4. Let I be a stable or squarefree stable ideal Then I is componentwise linear.

8.5. Find a componentwise linear monomial ideal I which is not stable.

8.6. Let I be a componentwise linear ideal. Then show that $\operatorname{reg}(I) = \max\{i\colon \beta_{0,i}(I) \neq 0\}$.

8.7. Let $I_{n,d}$ be the ideal in $S = K[x_1, \ldots, x_n]$ generated by all squarefree monomials of degree d. This ideal is called **squarefree Veronese of type** (n, d).
(a) Show that $I_{n,d}$ has linear quotients.
(b) Let Δ be the simplicial complex on the vertex set $[n]$ with $I_\Delta = I_{n,d}$. Show that Δ is the $(d-2)$th skeleton of the $(n-1)$-simplex.
(c) Use Alexander duality to show that all skeletons of the $(n-1)$-simplex are shellable.

8.8. Let $I \subset S$ be a graded ideal which has linear quotients with respect to a homogeneous system of generators $f_1, \ldots, f_m$ of I with $\deg f_1 \leq \deg f_2 \leq \cdots \leq \deg f_m$.
(a) Prove the following generalization of Corollary 8.2.2:

$$\beta_{i,i+j}(I) = \sum_{\substack{k=1 \\ \deg f_k = j}}^{m} \binom{r_k}{i},$$

where r_k is the number of generators $(f_1, \ldots, f_{k-1}) : f_k$
(b) Let I be a (squarefree) stable ideal with $G(I) = \{u_1, \ldots, u_m\}$ such that for $i < j$ either $\deg u_i < \deg u_j$, or $\deg u_i = \deg u_j$ and $u_j <_{\text{lex}} u_i$. Show that I has linear quotients with respect to $u_1, \ldots, u_m$.
(c) Use (a) and (b) to give a new proof of Corollary 7.2.3(a) and Corollary 7.4.2.

8.9. For a monomial u let $\min(u)$ be the smallest number i such that x_i divides u, and $m(u) = \max(u)$ the maximal such number.
Let $I \subset S$ be a stable monomial ideal.
(a) Prove that $\operatorname{height}(I) = \max\{\min(u)\colon u \in G(I)\}$ and $\operatorname{proj\,dim}(S/I) = \max\{\max(u)\colon u \in G(I)\}$.
(b) Show that the following conditions are equivalent: (i) S/I is Cohen–Macaulay; (ii) I has no embedded prime ideal; (iii) $|\operatorname{Ass}(S/I)| = 1$.

8.10. Let $I \subset S$ be a squarefree stable monomial ideal.
(a) Let Δ be the simplicial complex with $I_\Delta = I$. Show that $I_{\Delta^\vee}$ is again squarefree stable.
(b) Prove: S/I is Cohen–Macaulay if and only if all minimal prime ideals of I have the same height.

8.11. Let Δ be the simplex on $[n]$. We know from Theorem 8.2.18 that the ith skeleton is shellable. Give an explicit shelling of the ith skeleton.

Notes

The technique of Alexander duality was first applied in [TH96] to show that the first graded Betti numbers of a Stanley–Reisner ideal are independent of the characteristic of the base field. A far-reaching application of Alexander duality is given by the Eagon–Reiner theorem [ER98], which provides a powerful tool in the study of Cohen–Macaulay simplicial complexes. Hochster's theorem [Hoc77] to compute the graded Betti numbers as well as Reisner's Cohen–Macaulay criterion [Rei76] play essential roles in the proof of this theorem. A generalization of the Eagon–Reiner theorem, relating the projective dimension of a Stanley–Reisner ideal to the regularity of the Stanley–Reisner ring of the Alexander dual, is due to Terai. Further generalizations by Bayer, Charalambous and S. Popescu [BCP99] concern the extremal (multigraded) Betti numbers of a simplicial complex and its dual. Easier to prove, but also useful, is the fact that a simplicial complex is shellable if and only if the Stanley–Reisner ideal of its Alexander dual is generated in one degree and has linear quotients. Ideals with linear quotients were first considered in [HT02]. Alexander duality for arbitrary monomial ideals and duality functors introduced by Miller [Mil00a], [Mil98], [Mil00b], as well as Alexander duality for squarefree modules introduced by Römer [Roe01] and Yanagawa [Yan00], has many more interesting applications.

Our presentation of shellability and some of its fundamental properties follows the article by Björner and Wachs [BW97]. The fact that pure shellability implies Cohen–Macaulayness was discovered by [Gar80] and [KK79]. Componentwise linear ideals were introduced in [HH99] to generalize the Eagon–Reiner theorem in a different direction. It turned out that the property of being componentwise linear corresponds to being sequentially Cohen–Macaulay via Alexander duality; see [IIRW99]. The concept of sequentially Cohen–Macaulay simplicial complexes was introduced by Stanley [Sta95]. Componentwise linear ideals are distinguished by the remarkable property, observed in [AHH00b], that in characteristic 0, the graded Betti numbers of such an ideal and that of its generic initial ideal coincide. The proof Theorem 8.2.15 is taken from [SV08].

Part III

Combinatorics

9

Alexander duality and finite graphs

Alexander duality in combinatorics is studied. We demonstrate how Alexander duality is effective to develop the algebraic combinatorics related with finite partially ordered sets and finite graphs. Topics discussed include classification of Cohen–Macaulay bipartite graphs and Cohen–Macaulay chordal graphs together with algebraic aspects of Dirac's classical theorem on chordal graphs.

9.1 Edge ideals of finite graphs

We introduce edge ideals of finite graphs and study the algebraic properties of edge ideals of bipartite graphs. Certain monomial ideals arising from finite partially ordered sets and Alexander duality will play an important role for the classification of Cohen–Macaulay bipartite graphs.

9.1.1 Basic definitions

Let G be a **finite simple graph** on the vertex set $V(G)$ with edge set $E(G)$. In other words, $|V(G)| < \infty$ and $E(G) \subset V(G) \times V(G) \setminus \{\{v, v\} : v \in V(G)\}$.

All graphs considered in this book are finite simple graphs, which henceforth will simply be called graphs.

Without loss of generality we may assume that $V(G) = [n]$, where $[n] = \{1, 2, \ldots, n\}$.

Given a subset W of $[n]$ we define the **induced subgraph** of G on W to be the subgraph G_W on W consisting of those edges $\{i, j\} \in E(G)$ with $\{i, j\} \subset W$. A **complete graph** on $[n]$ is the finite graph G on $[n]$ for which $\{i, j\} \in E(G)$ for all $i \in [n]$ and $j \in [n]$ with $i \neq j$. The **complementary graph** of a finite graph G on $[n]$ is the finite graph $\overline{G}$ on $[n]$ whose edge set $E(\overline{G})$ consists of those 2-element subsets $\{i, j\}$ of $[n]$ for which $\{i, j\} \notin E(G)$.

A finite graph can be viewed as 1-dimensional simplicial complex.

J. Herzog, T. Hibi, *Monomial Ideals*, Graduate Texts in Mathematics 260,   153
DOI 10.1007/978-0-85729-106-6_9, © Springer-Verlag London Limited 2011

A **walk** of G of length q between i and j, where i and j are vertices of G, is a sequence of edges of the form $\{\{i_0, i_1\}, \{i_1, i_2\}, \ldots, \{i_{q-1}, i_q\}\}$, where $i_0, i_1, \ldots, i_q$ are vertices of G with $i_0 = i$ and $i_q = j$.

A **cycle** of G of length q is a subgraph C of G such that

$$E(C) = \{\{i_1, i_2\}, \{i_2, i_3\}, \ldots, \{i_{q-1}, i_q\}, \{i_q, i_1\}\},$$

where $i_1, i_2, \ldots, i_q$ are vertices of G and where $i_j \neq i_k$ if $j \neq k$.

A graph G on $[n]$ is **connected** if, for any two vertices i and j of G, there is a walk between i and j.

A **forest** is a finite graph with no cycle. A **tree** is a forest which is connected.

A graph G on $[n]$ is called **bipartite** if there is a decomposition $[n] = V_1 \cup V_2$ such that every edge of G is of the form $\{i, j\}$ with $i \in V_1$ and $j \in V_2$.

Bipartite graphs can be characterized as follows:

Lemma 9.1.1. *A finite graph G is bipartite if and only if every cycle of G is of even length. In particular every forest is bipartite.*

Proof. First, suppose that G is a bipartite graph with the decomposition $U \cup V$ of its vertices. Let $C = \{\{v_1, v_2\}, \{v_2, v_3\}, \ldots, \{v_{q-1}, v_q\}, \{v_q, v_1\}\}$ be a cycle of length q of G with $v_1 \in U$. Then $v_2 \in V$ and $v_3 \in U$. In general, one has $v_i \in U$ if i is odd and $v_i \in V$ if i is even. Since $v_q \in V$, it follows that q is even.

In order to prove the converse, we may assume that G is connected. Suppose that every cycle of G is of even length. Let u and v be vertices of G. Let W be a walk of G of length q between u and v and W' a walk of G of length q' between u and v. Since every cycle of G is of even length, it follows easily that $q + q'$ is even. In other words, either (i) both q and q' are even or (ii) both q and q' are odd.

Now, fix a vertex v_0 of G. Let U (resp. V) denote the vertices w of G such that there is a walk of even (resp. odd) length between v_0 and w. Then $U \cap V = \emptyset$ with $v_0 \in U$. Let $w, w' \in U$ with $w \neq w'$ and with $\{w, w'\} \in E(G)$. Since $w \in U$, there is a walk of even length between v_0 and w. It follows that there is a walk of odd length between v_0 and w', a contradiction. Thus $\{w, w'\} \notin E(G)$ for $w, w' \in U$ with $w \neq w'$. Similarly, $\{w, w'\} \notin E(G)$ for $w, w' \in V$ with $w \neq w'$. Hence every edge of G is of the form $\{u, v\}$ with $u \in U$ and $v \in V$. Thus G is bipartite, as desired.   □

The following well-known theorem on classical graph theory is indispensable for the classification of Cohen–Macaulay bipartite graphs.

Lemma 9.1.2 (The Marriage Theorem). *Let G be a bipartite graph on the vertex set $W \cup W'$ with $|W| = |W'|$. For each $U \subset W$ we write $N(U)$ for the set of those $j \in W'$ such that $\{i, j\} \in E(G)$ for some $i \in U$. Suppose that $|N(U)| \geq |U|$ for all subset $U \subset W$. Then there is a bijection $f : W \to W'$ such that $\{i, f(i)\}$ is an edge of G for all $i \in W$.*

Proof. First, suppose that $|N(U)| \geq |U| + 1$ for all nonempty proper subsets $U \subset W$. Fix an arbitrary edge $\{a, b\}$ of G with $a \in W$ and $b \in W$. Since $|N(U) \setminus \{b\}| \geq |U|$ for all subsets $U \subset W \setminus \{a\}$, the induction hypothesis guarantees the existence of a bijection $f_0 : W \setminus \{a\} \to W' \setminus \{b\}$ such that $\{i, f_0(i)\}$ is an edge of G for all $i \in W \setminus \{a\}$. Now, let $f_0(a) = b$. Then $f_0 : W \to W'$ is a bijection such that $\{i, f_0(i)\}$ is an edge of G for all $i \in W$.

Second, suppose that there is a nonempty proper subset $U_0 \subset W$ with $|U_0| = |N(U_0)|$. If $V \subset U_0$, then $N(V) \subset N(U_0)$ and $|N(V)| \geq |V|$ for all subset $V \subset U_0$. Thus there is a bijection $f_0 : U_0 \to N(U_0)$ such that $\{i, f_0(i)\}$ is an edge of G for all $i \in U_0$.

Let V' be a subset of $W \setminus U_0$. We claim $|N(V') \setminus N(U_0)| \geq |V'|$. In fact, since $|N(V' \cup U_0)| \geq |V'| + |U_0|$ and since $N(V' \cup U_0) = (N(V') \setminus N(U_0)) \cup N(U_0)$, it follows that

$$|N(V') \setminus N(U_0)| \geq |V'| + |U_0| - |N(U_0)| = |V'|.$$

Hence there is a bijection $f_1 : W \setminus U_0 \to W' \setminus N(U_0)$ such that $\{i, f_1(i)\}$ is an edge of G for all $i \in W \setminus U_0$.

Now, gluing f_0 and f_1 yields a desired bijection $f : W \to W'$. $\square$

A **chord** of a cycle C is an edge $\{i, j\}$ of G such that i and j are vertices of C with $\{i, j\} \notin E(C)$. A **chordal graph** is a finite graph each of whose cycles of length > 3 has a chord. Every induced subgraph of a chordal graph is again chordal.

A subset C of $[n]$ is called a **clique** of G if for all i and j belonging to C with $i \neq j$ one has $\{i, j\} \in E(G)$. The **clique complex** of a finite graph G on $[n]$ is the simplicial complex $\Delta(G)$ on $[n]$ whose faces are the cliques of G.

A simplicial complex Δ on $[n]$ is called **flag** if every minimal nonface of Δ is a 2-elements subset of $[n]$.

Lemma 9.1.3. *A simplicial complex Δ is flag if and only if Δ is the clique complex of a finite graph.*

Proof. Let G be a finite graph on $[n]$ and $\Delta(G)$ its clique complex. A subset $F \subset [n]$ is a nonface of $\Delta(G)$ if and only if F is not a clique of G. Thus if F is a nonface of $\Delta(G)$, then there are i and j belonging to F with $\{i, j\} \notin E(G)$. Since $\{i, j\}$ is a nonface of $\Delta(G)$ which is contained in F, it follows that every minimal nonface of Δ is a 2-element subset of $[n]$. Thus $\Delta(G)$ is a flag complex.

Conversely, suppose that Δ is a flag complex and that G is the 1-skeleton of Δ. Let $\Delta(G)$ be the clique complex of G. In general, one has $\Delta \subset \Delta(G)$. Let F be a clique of G. Since every 2-element subset of F is an edge of G, it follows that every 2-element subset of F belongs to Δ. Since Δ is a flag complex, one has $F \in \Delta$. Thus $\Delta(G) \subset \Delta$. Hence $\Delta = \Delta(G)$, as desired. $\square$

Let, as usual, $S = K[x_1, \ldots, x_n]$ be the polynomial ring in n variables over a field K. We associate each edge $e = \{i, j\}$ of G with the monomial $u_e = x_i x_j$

of S. The **edge ideal** of G is the monomial ideal $I(G)$ of S which is generated by all quadratic monomials u_e with $e \in E(G)$.

It follows that the edge ideal of G coincides with the Stanley–Reisner ideal of the clique complex of the complementary graph of G, i.e. $I(G) = I_{\Delta(\overline{G})}$.

Let K be a field. We say a graph G is **(sequentially) Cohen–Macaulay** or **Gorenstein** over K if $S/I(G)$ has this property, and we say that G is **shellable** or **vertex decomposable** if $\Delta(\overline{G})$ has this property. In general these properties do depend on K. If this is not the case, then we call G simply (sequentially) Cohen–Macaulay or Gorenstein, without referring to K. Finally we say that G is **of type** k if $S/I(G)$ is Cohen–Macaulay of type k, cf. Appendix A.6.

A **vertex cover** of a graph G on $[n]$ is a subset $C \subset [n]$ such that $\{i, j\} \cap C \neq \emptyset$ for all $\{i, j\} \in E(G)$. A vertex cover C is called **minimal** if C is a vertex of G, and no proper subset of C is a vertex cover of G. We denote by $\mathcal{M}(G)$ the set of minimal vertex covers of G.

An **independent set** of G is a set $S \subset [n]$ such that $\{i, j\} \notin E(G)$ for all $i, j \in S$. Obviously, S is an independent set of G if and only if $[n] \setminus S$ is a vertex cover of G. Thus the maximal independent sets of G correspond to the minimal vertex covers of G.

Lemma 9.1.4. *Let G be a graph on $[n]$. A subset $C = \{i_1, \dots, i_r\} \subset [n]$ is a vertex cover of G if and only if the prime ideal $P_C = (x_{i_1}, \dots, x_{i_r})$ contains $I(G)$. In particular, C is a minimal vertex cover of G if and only if P_C is a minimal prime ideal of $I(G)$.*

Proof. A generator $x_i x_j$ of $I(G)$ belongs to P_C, if and only if x_{i_k} divides $x_i x_j$ for some $i_k \in C$. This is the case if and only if $C \cap \{i, j\} \neq \emptyset$. Thus $I(G) \subset P_C$ if and only if C is a vertex cover of G.

The second statement follows from the fact that all minimal prime ideals of a monomial ideal are monomial prime ideals (Corollary 1.3.9). $\square$

Let G be a graph. We write I_G for the Alexander dual $I(G)^\vee$ of $I(G)$, and call it the **vertex cover ideal** of G. This naming is justified by the following

Corollary 9.1.5. *The ideal I_G is minimally generated by those monomials x_C for which $C \in \mathcal{M}(G)$.*

Proof. The proof is an immediate consequence of Lemma 9.1.4 together with Corollary 1.5.5. $\square$

9.1.2 Finite partially ordered sets

A partially ordered set will be called a **poset**. A subset $C \subset P$ is called a **chain** of P if C is a totally ordered subset with respect to the induced order. The length of a chain C is $|C| - 1$. The **rank** of P is the maximal length of a chain in P.

A subset P' of a poset P is called a **subposet** if, for $a, b \in P'$, one has $a < b$ in P' if and only $a < b$ in P.

Let P and Q be posets. A map $\varphi : P \to Q$ is an **order-preserving** map of posets if $a, b \in P$ with $a \leq b$ in P, then $\varphi(a) \leq \varphi(b)$ in Q. We say that P is isomorphic to Q if there exists a bijection $\varphi : P \to Q$ such that both φ and its inverse φ^{-1} are order-preserving.

A **lattice** is a partially ordered set L such that, for any two elements a and b belonging to L, there is a unique greatest lower bound $a \wedge b$, called the **meet** of a and b, and there is a unique least upper bound $a \vee b$, called the **join** of a and b. Thus in particular a finite lattice possesses both a unique minimal element $\hat{0}$ and a unique maximal element $\hat{1}$. A subposet L' of a lattice L is called a **sublattice** of L if L' is a lattice and, for $a, b \in L'$, the meet of a and b in L' coincides with that in L and the join of a and b in L' coincides with that in L.

Example 9.1.6. (a) Let $\mathcal{L}_n$ denote the set of all subsets of $[n]$, ordered by inclusion. Then $\mathcal{L}_n$ is a lattice, called the **boolean lattice** of rank n.

(b) Let $n > 0$ be an integer and $\mathcal{D}_n$ the set of all divisors of n, ordered by divisibility. Then $\mathcal{D}_n$ is a lattice, called the **divisor lattice** of n. Thus in particular a boolean lattice is a divisor lattice.

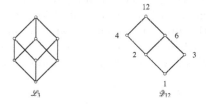

Fig. 9.1. A boolean lattice and a divisor lattice.

Let $P = \{p_1, \ldots, p_n\}$ be a finite poset with a partial order $\leq$. A **poset ideal** of P is a subset α of P with the property that if $a \in \alpha$ and $b \in P$ with $b \leq a$, then $b \in \alpha$. In particular the empty set as well as P itself is a poset ideal. Let $\mathcal{J}(P)$ denote the set of poset ideals of P. If α and β are poset ideals of P, then each of the $\alpha \cap \beta$ and $\alpha \cup \beta$ is again a poset ideal. Hence $\mathcal{J}(P)$ is a lattice ordered by inclusion.

A lattice L is called **distributive** if, for all a, b, c belonging to L, one has

$$(a \vee b) \wedge c = (a \wedge c) \vee (b \wedge c),$$
$$(a \wedge b) \vee c = (a \vee c) \wedge (b \vee c).$$

For example, every divisor lattice is a distributive lattice and, in particular, every boolean lattice is a distributive lattice.

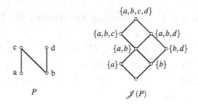

Fig. 9.2. A poset and its lattice of poset ideals.

On the other hand, for an arbitrary finite poset P, the lattice $\mathcal{J}(P)$ is a distributive lattice. Birkhoff's fundamental structure theorem for finite distributive lattices guarantees the converse.

Theorem 9.1.7 (Birkhoff). *Given a finite distributive lattice L there is a unique finite poset P such that L is isomorphic to $\mathcal{J}(P)$.*

Proof. Let L be a finite distributive lattice. An element $a \in L$ with $a \neq \hat{0}$ is called **join-irreducible** if whenever $a = b \vee c$ with $b, c \in L$, then one has either $a = b$ or $a = c$. Let P denote the subposet of L consisting of all join-irreducible elements of L.

We claim L is isomorphic to $\mathcal{J}(P)$. We define the map $\varphi : \mathcal{J}(P) \to L$ by setting $\varphi(\alpha) = \bigvee_{a \in \alpha} a$, where $\alpha \in \mathcal{J}(P)$. Thus in particular $\varphi(\emptyset) = \hat{0}$. Clearly, φ is order-preserving. Since each element $a \in L$ is the join of the join-irreducible elements b with $b \leq a$ in L, it follows that φ is surjective.

The highlight of the proof is to show that φ is injective. Let α and β be poset ideals of P with $\alpha \neq \beta$, say, $\beta \not\subset \alpha$. Let b^* be a maximal element of β with $b^* \not\in \alpha$. We show $\varphi(\alpha) \neq \varphi(\beta)$. Suppose, on the contrary, that $\varphi(\alpha) = \varphi(\beta)$. Thus

$$\bigvee_{a \in \alpha} a = \bigvee_{b \in \beta} b. \tag{9.1}$$

Since L is distributive, it follows that

$$(\bigvee_{a \in \alpha} a) \wedge b^* = \bigvee_{a \in \alpha} (a \wedge b^*).$$

Since $a \vee b^* < b$ and since b^* is join-irreducible, it follows that $(\bigvee_{a \in \alpha} a) \wedge b^* < b^*$. However, since $b^* \in \beta$, one has

$$(\bigvee_{b \in \beta} b) \wedge b^* = \bigvee_{b \in \beta} (b \wedge b^*) = b^*.$$

This contradicts (9.1). Hence φ is injective.

Now, the inverse map φ^{-1} is defined as follows: For each element $c \in L$, $\varphi^{-1}(c)$ is the set of join-irreducible elements $a \in L$ with $a \leq c$. Clearly, $\varphi^{-1}(c) \in \mathcal{J}(P)$ and φ^{-1} is order-preserving.

Consequently, L is isomorphic to $\mathcal{J}(P)$ with the bijective order-preserving map φ.

Finally, since P is isomorphic to the subposet consisting of all join-irreducible elements of the distributive lattice $\mathcal{J}(P)$, it follows that, for two finite posets P and Q, if $\mathcal{J}(P)$ is isomorphic to $\mathcal{J}(Q)$, then P is isomorphic to Q. In other words, the existence of a finite poset P such that L is isomorphic to $\mathcal{J}(P)$ is unique. $\qquad\square$

Let $K[\mathbf{x}, \mathbf{y}] = K[x_1, \ldots, x_n, y_1, \ldots, y_n]$ denote the polynomial ring in $2n$ variables over a field K. For each poset ideal α of P we associate the squarefree monomial

$$u_\alpha = (\prod_{p_i \in \alpha} x_i)(\prod_{p_j \in P \setminus \alpha} y_j)$$

of $K[\mathbf{x}, \mathbf{y}]$. Let H_P denote the squarefree monomial ideal of $K[\mathbf{x}, \mathbf{y}]$ which is generated by all u_α with $\alpha \in \mathcal{J}(P)$:

$$H_P = (\{u_\alpha\}_{\alpha \in \mathcal{J}(P)}).$$

We now come to the crucial result of this section.

Theorem 9.1.8. *The squarefree monomial ideal H_P has linear quotients. Thus in particular H_P has a linear resolution.*

Proof. Fix a total order $<$ on $G(H_P)$ with the property that if $\gamma \subset \alpha$, then $u_\gamma < u_\alpha$. Let $\gamma \subset \alpha$ with $\gamma \neq \alpha$. Then there is $p \in \alpha \setminus \gamma$ such that $\delta = \alpha \setminus \{p\}$ is a poset ideal of P. Since $u_\delta = y_p u_\alpha / x_p$, one has $y_p = u_\delta / \gcd(u_\delta, u_\alpha)$. Since $p \notin \gamma$, it follows that y_p divides u_γ. Thus Lemma 8.2.3 says that H_P has linear quotients, as desired. $\qquad\square$

Lemma 9.1.9. *$H_P^\vee$ is minimally generated by those squarefree quadratic monomials $x_i y_j$ for which $p_i \leq p_j$.*

Proof. Let Δ_P denote the simplicial complex on the vertex set

$$V_n = \{x_1, \ldots, x_n, y_1, \ldots, y_n\}$$

whose Stanley–Reisner ideal I_{Δ_P} coincides with H_P.

Let $w = x_1 \cdots x_n y_1 \cdots y_n$. If u is a squarefree monomial of $K[\mathbf{x}, \mathbf{y}]$, then we write F_u for the set of those variables x_i and y_j which divide u. Since $\{F_{u_\alpha} : \alpha \in \mathcal{J}(P)\}$ is the set of minimal nonfaces of Δ_P, it follows that $\{F_{w/u_\alpha} : \alpha \in \mathcal{J}(P)\}$ is the set of facets of $\Delta_P^\vee$. Thus a subset $F \subset V_n$ is a face of $\Delta_P^\vee$ if and only if there is a poset ideal α of P such that $F \subset F_{w/u_\alpha}$.

Given a subset $F \subset V_n$ we set $F_x = F \cap \{x_1, \ldots, x_n\}$ and $F_y = \{x_j : y_j \in F\}$. Note that, for a poset ideal α of P, one has $(F_{w/u_\alpha})_x = \{x_1, \ldots, x_n\} \setminus \{x_i : p_i \in \alpha\}$ and $(F_{w/u_\alpha})_y = \{x_j : p_j \in \alpha\}$. Thus a subset $F \subset V_n$ is a face of $\Delta_P^\vee$ if and only if there exists a poset ideal α of P such that $F_x \cap \{x_i : p_i \in \alpha\} = \emptyset$ and with $F_y \subset \{x_j : p_j \in \alpha\}$. In particular $\{x_i, y_j\}$ is a minimal nonface of

$\Delta_P^\vee$ if $p_i \leq p_j$. We claim that every minimal nonface of $\Delta_P^\vee$ is of the form $\{x_i, y_j\}$ with $p_i \leq p_j$.

Since $w/u_\emptyset = x_1 \cdots x_n$ and $w/u_P = y_1 \cdots y_n$, both $\{x_1, \ldots, x_n\}$ and $\{y_1, \ldots, y_n\}$ are facets of $\Delta_P^\vee$. Let $F \subset V_n$ be a nonface of $\Delta_P^\vee$. Then $F_x \neq \emptyset$ and $F_y \neq \emptyset$. Since F is a nonface, there exists no poset ideal α of P with $F_x \cap \{x_i : p_i \in \alpha\} = \emptyset$ and with $F_y \subset \{x_j : p_j \in \alpha\}$. Let γ denote the poset ideal generated by $\{p_j : x_j \in F_y\}$ (i.e. γ is the smallest poset ideal of P which contains $\{p_j : x_j \in F_y\}$). Since $F_y \subset \{x_j : p_j \in \gamma\}$, it follows that $\{x_i : p_i \in \gamma\}$ must intersect F_x. We choose $x_i \in F_x$ such that $p_i \in \gamma$. Let p_j be a maximal element in γ with $p_i \leq p_j$. Then $x_j \in F_y$ by the choice of γ. Thus $\{x_i, y_j\} \subset F$ with $p_i \leq p_j$, as desired. $\square$

9.1.3 Cohen–Macaulay bipartite graphs

A vertex $i \in [n]$ of G is called an **isolated vertex**, if G has no edge of the form $\{i, j\}$. Suppose i is an isolated vertex of G, and G' the induced subgraph of G on $[n] \setminus \{i\}$. Then obviously G is Cohen–Macaulay over K if and only if G' is Cohen–Macaulay over K. Thus, throughout this subsection, we will always assume that G has no isolated vertices. Our final goal in this section is to classify all Cohen–Macaulay bipartite graphs.

A finite graph G is called **unmixed** if all minimal vertex covers of G have the same cardinality.

Lemma 9.1.10. *Every Cohen–Macaulay graph is unmixed.*

Proof. Recall that, for a subset $C \subset [n]$, the notation P_C stands for the monomial prime ideal of $S = K[x_1, \ldots, x_n]$ generated by those variables x_i with $i \in C$. Let $C_1, \ldots, C_s$ be the minimal vertex covers of G. By using Lemma 9.1.4 it follows that $P_{C_1}, \ldots, P_{C_s}$ are the minimal prime ideals of $I(G)$. Since $S/I(G)$ is Cohen–Macaulay, all minimal prime ideals of $I(G)$ have the same height. In other words, all C_i have the same cardinality. $\square$

Let $P = \{p_1, \ldots, p_n\}$ be a finite poset with a partial order $\leq$ and $V_n = \{x_1, \ldots, x_n, y_1, \ldots, y_n\}$. We write $G(P)$ for the bipartite graph on V_n whose edges are those 2-element subset $\{x_i, y_j\}$ such that $p_i \leq p_j$.

In general, we say that a bipartite graph G on $W \cup W'$ **comes from a poset** if $|W| = |W'|$ and if there is a finite poset on $[n]$, where $n = |W|$, such that after relabelling of the vertices of G one has $G(P) = G$.

Lemma 9.1.11. *A bipartite graph coming from a poset is Cohen–Macaulay.*

Proof. Let $P = \{p_1, \ldots, p_n\}$ be a finite poset and $V_n = \{x_1, \ldots, x_n, y_1, \ldots, y_n\}$. The edge ideal $I(G(P))$ is generated by those 2-element subsets $\{x_i, y_j\}$ with $p_i \leq p_j$. By Lemma 9.1.9 we have that $I(G(P)) = H_P^\vee$. Since Theorem 9.1.8 says that H_P has a linear resolution, Theorem 8.1.9 guarantees that $I(G(P))$ is Cohen–Macaulay, as desired. $\square$

A pure simplicial complex of dimension $d-1$ is called **connected in codimension one** if, for any two facets F and G of Δ, there exists a sequence of facets $F = F_0, F_1, ..., F_{q-1}, F_q = G$ such that $|F_i \cap F_{i+1}| = d-1$.

For the proof of the main theorem of this section we need the following result.

Lemma 9.1.12. *Every Cohen–Macaulay complex is connected in codimension one.*

Proof. Let Δ be a Cohen–Macaulay complex of dimension $d-1$. If $d-1 = 0$, the assertions are trivial. Therefore we now assume that $d-1 > 0$. Let F and G be two facets of Δ. Since Δ is connected (Lemma 8.1.7), there exists a sequence of facets $F = F_0, F_1, \ldots, F_{q-1}, F_q = G$ such that $F_i \cap F_{i+1} \neq \emptyset$. Let y_i be a vertex belonging to $F_i \cap F_{i+1}$. Since Δ is Cohen–Macaulay, $\mathrm{link}_\Delta(\{y_i\})$ is Cohen–Macaulay (Lemma 8.1.8). By working with induction on the dimension of Δ, we may assume that $\mathrm{link}_\Delta(\{y_i\})$ is connected in codimension one. Moreover there exists a sequence of facets $F_i = H_0, H_1, \cdots, H_{r-1}, H_r = F_{i+1}$ of Δ, where all H_j contain y_i with $|H_i| = d$, such that $|H_j \cap H_{j+1}| = d-1$. Composing all these sequences of facets which we have between each F_i and F_{i+1} yields the desired sequence between F and G. $\square$

We now come to the main result of the present section.

Theorem 9.1.13. *A bipartite graph G is Cohen–Macaulay if and only if G comes from a finite poset.*

Proof. The "if" part is already proved by Lemma 9.1.11. To see why the "only if" part is true, suppose that G is Cohen–Macaulay.

Step 1: Let $W \cup W'$ be the partition of the vertex set of G. Since each of W and W' is a minimal vertex cover of G, it follows from Lemma 9.1.10 that $|W| = |W'|$. Let $W = \{x_1, \ldots, x_n\}$ and $W' = \{y_1, \ldots, y_n\}$.

Step 2: For each $U \subset W$ we write $N(U)$ for the set of those $y_j \in W'$ such that $\{x_i, y_j\} \in E(G)$ for some $x_i \in U$. We claim $|U| \leq |N(U)|$. Since $(W \setminus U) \cup N(U)$ is a vertex cover of G and since G is unmixed, it follows that $|(W \setminus U) \cup N(U)| \geq |W|$. Thus $|U| \leq |N(U)|$. Lemma 9.1.2 stated below enables us to assume that $\{x_i, y_i\} \in E(G)$ for $i = 1, \ldots, n$.

Step 3: Let Δ be the simplicial complex on $W \cup W'$ whose Stanley–Reisner ideal coincides with the edge ideal $I(G)$. Since each of W and W' is a facet of Δ and since Δ is connected in codimension one, it follows that there is a sequence of facets $F_0, F_1, \ldots, F_s$ of Δ with $F_0 = W'$ and $F_s = W$ such that $|F_{k-1} \cap F_k| = n-1$ for $k = 1, \ldots, s$. Let $F_1 = (W' \setminus \{y_i\}) \cup \{x_j\}$. Since $\{x_\ell, y_\ell\} \in E(G)$ for $\ell = 1, \ldots, n$, one has $i = j$. Let, say,

$$F_1 = (W' \setminus \{y_n\}) \cup \{x_n\} = \{y_1, \ldots, y_{n-1}, x_n\}.$$

Since Δ is Cohen–Macaulay, it follows that $\mathrm{link}_\Delta(\{x_n\})$ is Cohen–Macaulay. Now, since each of $\{x_1, \ldots, x_{n-1}\}$ and $\{y_1, \ldots, y_{n-1}\}$ is a facet of $\mathrm{link}_\Delta(\{x_n\})$ and since $\mathrm{link}_\Delta(\{x_n\})$ is connected in codimension 1, one may assume that

$$\{y_1, \ldots, y_{n-2}, x_{n-1}\}$$

is a facet of $\mathrm{link}_\Delta(\{x_n\})$. Thus

$$\{y_1, \ldots, y_{n-2}, x_{n-1}, x_n\}$$

is a facet of Δ.

Now, repeated applications of this argument enables us to assume that

$$\{y_1, \ldots, y_i, x_{i+1}, \ldots, x_n\}$$

is a facet of Δ for each $i = 1, \ldots n$. In particular $\{x_i, y_j\}$ cannot be an edge of G if $j < i$. In other words, if $\{x_i, y_j\}$ is an edge of G, then $i \leq j$.

Step 4: Let $i < j < k$. We now claim that if each of $\{x_i, y_j\}$ and $\{x_j, y_k\}$ is an edge of G, then $\{x_i, y_k\}$ is an edge of G. Suppose, on the contrary, that $\{x_i, y_k\}$ is not an edge of G. In other words, $\{x_i, y_k\}$ is a face of Δ. Since Δ is pure of dimension $n - 1$, there is an n-element subset $F \subset W \cup W'$ with $\{x_i, y_k\} \subset F$ such that no 2-element subset of F is an edge of G. Since each of $\{x_i, y_j\}$ and $\{x_j, y_k\}$ is an edge of G, one has $y_j \notin F$ and $x_j \notin F$.

On the other hand, since $\{x_\ell, y_\ell\}$ is an edge of G for each $1 \leq \ell \leq n$, it follows that, for each $1 \leq \ell \leq n$, one has either $x_\ell \notin F$ or $y_\ell \notin F$. Hence if $F = \{x_{i_1}, \ldots, x_{i_r}, y_{j_1}, \ldots, y_{j_s}\}$, then

$$\{i_1, \ldots, i_r, j_1, \ldots, j_s\} = [n] \quad \text{and} \quad \{i_1, \ldots, i_r\} \cap \{j_1, \ldots, j_s\} = \emptyset.$$

This implies that $x_j \in F$ or $y_j \in F$; a contradiction.

Step 5: Let $P = \{p_1, \ldots, p_n\}$ be a finite set and define the binary operation $\leq_P$ on P by setting $p_i \leq_P p_j$ if $\{x_i, y_j\} \in E(G)$. It then follows from second, third and fourth steps that $\leq_P$ defines a partial order on P. Clearly, one has $G = G(P)$. Thus G comes from a poset, as desired. □

As an immediate consequence of Theorem 9.1.13 we have

Corollary 9.1.14. *Let G be a bipartite graph with vertex partition $V \cup V'$. Then the following conditions are equivalent:*

(i) *G is a Cohen–Macaulay graph;*
(ii) *$|V| = |V'|$ and the vertices $V = \{x_1, \ldots, x_n\}$ and $V' = \{y_1, \ldots, y_n\}$ can be labelled such that:*
    (α) $\{x_i, y_i\}$ are edges for $i = 1, \ldots, n$;
    (β) if $\{x_i, y_j\}$ is an edge, then $i \leq j$;
    (γ) if $\{x_i, y_j\}$ and $\{x_j, y_k\}$ are edges, then $\{x_i, y_k\}$ is an edge.

Proof. By Theorem 9.1.13, the graph G is Cohen–Macaulay if and only if $G = G(P)$ for some poset $P = \{p_1, \ldots, p_n\}$.

(i) $\Rightarrow$ (ii): We may assume that $p_i \leq p_j$ implies that $i \leq j$. With this labelling (ii) follows from (i).

(ii) $\Rightarrow$ (i): Let $P = \{p_1, \ldots, p_n\}$ be the poset with $p_i \leq p_j$ if and only if $\{x_i, y_j\} \in E(G)$. The conditions in (ii) imply that P is indeed a poset. Moreover, $G = G(P)$. □

Figure 9.3 shows of a bipartite graph which satisfies the conditions of Corollary 9.1.14, and hence is Cohen–Macaulay.

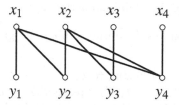

$$x_1 \qquad x_2 \qquad x_3 \qquad x_4$$

$$y_1 \qquad y_2 \qquad y_3 \qquad y_4$$

Fig. 9.3. A Cohen–Macaulay bipartite graph.

Let $P = \{p_1, \ldots, p_n\}$ be a poset. For each poset ideal α of P, we set $\alpha_x = \{x_i \colon p_i \in \alpha\}$ and $\alpha_y = \{y_j \colon p_j \notin \alpha\}$.

Corollary 9.1.15. *Let P be a finite poset. Then the minimal vertex covers of $G(P)$ are precisely the sets $\alpha_x \cup \alpha_y$ with $\alpha \in \mathcal{J}(P)$.*

Proof. According to Lemma 9.1.4 the monomial vertex covers of $G(P)$ correspond to the generators of $I(G(P))^{\vee}$. By Lemma 9.1.9 we have $H_P^{\vee} = I(G(P))$. Therefore, $I(G(P))^{\vee} = (H_P^{\vee})^{\vee} = H_P$. Thus the assertion follows from the definition of H_P. $\qquad \square$

Corollary 9.1.16. *An unmixed bipartite graph G is Cohen–Macaulay if and only if it is shellable.*

Proof. If G is Cohen–Macaulay, then we may assume by Theorem 9.1.13 that $G = G(P)$ for some finite poset P. Since $I(G(P)) = H_P^{\vee}$ and since by Theorem 9.1.8, H_P has linear quotients, the assertion follows from Proposition 8.2.5. $\qquad \square$

9.1.4 Unmixed bipartite graphs

Let G be a bipartite graph without isolated vertices and let

$$V(G) = \{x_1, \ldots, x_m\} \cup \{y_1, \ldots, y_n\}$$

denote the set of vertices of G. Suppose that G is unmixed. In Step 1 and Step 2 in the proof of Theorem 9.1.13, it is shown that $m = n$ and that one may assume that $\{x_i, y_i\} \in I(G)$ for $i = 1, \ldots, n$. For these two steps in the proof we use only the fact that G is unmixed.

It follows that each minimal vertex cover of G is of the form

$$\{x_{i_1}, \ldots, x_{i_s}, y_{i_{s+1}}, \ldots, y_{i_n}\}$$

where $\{i_1, \ldots, i_n\} = [n]$.

For a minimal vertex cover $C = \{x_{i_1}, \ldots, x_{i_s}, y_{i_{s+1}}, \ldots, y_{i_n}\}$ of G, we set $\overline{C} = \{x_{i_1}, \ldots, x_{i_s}\}$. Let $\mathcal{L}_n$ denote the Boolean lattice on the set $\{x_1, \ldots, x_n\}$ and let

$$\mathcal{L}_G = \{\overline{C} \mid C \text{ is a minimal vertex cover of } G\} \subset \mathcal{L}_n.$$

Remark 9.1.17. Let G and G' be unmixed bipartite graphs on $\{x_1, \ldots, x_m\} \cup \{y_1, \ldots, y_n\}$.

(i) Both $\emptyset$ and $\{x_1, \ldots, x_n\}$ belong to $\mathcal{L}_G$.

(ii) If $G \neq G'$, then $I(G) \neq I(G')$. Hence $\mathcal{L}_G \neq \mathcal{L}_{G'}$, since for different squarefree monomial ideals the set of minimal prime ideals differs.

The first result of this subsection is

Theorem 9.1.18. *Let $\mathcal{L}$ be a subset of $\mathcal{L}_n$. Then there exists a (unique) unmixed bipartite graph G on $\{x_1, \ldots, x_n\} \cup \{y_1, \ldots, y_n\}$ such that $\mathcal{L} = \mathcal{L}_G$ if and only if $\emptyset$ and $\{x_1, \ldots, x_n\}$ belong to $\mathcal{L}$, and $\mathcal{L}$ is a sublattice of $\mathcal{L}_n$.*

Proof. Let G be an unmixed bipartite graph with $\mathcal{L} = \mathcal{L}_G$. Let $C = \{x_{i_1}, \ldots, x_{i_s}, y_{i_{s+1}}, \ldots, y_{i_n}\}$ and $C' = \{x_{j_1}, \ldots, x_{j_t}, y_{j_{t+1}}, \ldots, y_{j_n}\}$ be minimal vertex covers of G. Then

$$\{y_k \mid x_k \notin \overline{C} \cap \overline{C'}\} = \{y_{i_{s+1}}, \ldots, y_{i_n}\} \cup \{y_{j_{t+1}}, \ldots, y_{j_n}\},$$

$$\{y_k \mid x_k \notin \overline{C} \cup \overline{C'}\} = \{y_{i_{s+1}}, \ldots, y_{i_n}\} \cap \{y_{j_{t+1}}, \ldots, y_{j_n}\}.$$

First, we show that $\overline{C} \cap \overline{C'} \in \mathcal{L}_G$, that is, $C_1 = (\overline{C} \cap \overline{C'}) \cup \{y_k \mid x_k \notin \overline{C} \cap \overline{C'}\}$ is a minimal vertex cover of G. Suppose that an edge $\{x_i, y_j\}$ of G satisfies $y_j \notin \{y_k \mid x_k \notin \overline{C} \cap \overline{C'}\} = \{y_{i_{s+1}}, \ldots, y_{i_n}\} \cup \{y_{j_{t+1}}, \ldots, y_{j_n}\}$. Since C (resp. C') is a vertex cover of G, one has $x_i \in \overline{C}$ (resp. $x_i \in \overline{C'}$). Hence $x_i \in \overline{C} \cap \overline{C'}$. Thus C_1 is a minimal vertex cover of G.

Second, we show that $\overline{C} \cup \overline{C'} \in \mathcal{L}_G$, that is, $C_2 = (\overline{C} \cup \overline{C'}) \cup \{y_k \mid x_k \notin \overline{C} \cup \overline{C'}\}$ is a minimal vertex cover of G. Suppose that an edge $\{x_i, y_j\}$ of G satisfies $x_i \notin \overline{C} \cup \overline{C'}$. Since C (resp. C') is a vertex cover of G, one has $y_j \in \{y_{i_{s+1}}, \ldots, y_{i_n}\}$ (resp. $y_j \in \{y_{j_{t+1}}, \ldots, y_{j_n}\}$). Thus $y_j \in \{y_{i_{s+1}}, \ldots, y_{i_n}\} \cap \{y_{j_{t+1}}, \ldots, y_{j_n}\} = \{y_k \mid x_k \notin \overline{C} \cup \overline{C'}\}$ and hence C_2 is a minimal vertex cover of G.

Conversely, suppose that $\emptyset$ and $\{x_1, \ldots, x_n\}$ belong to $\mathcal{L}$, and $\mathcal{L}$ is a sublattice of $\mathcal{L}_n$. For each element $\alpha \in \mathcal{L}$, let α^* denote the set $\{y_j \mid x_j \notin S\}$. Let I be the ideal $\bigcap_{\alpha \in \mathcal{L}}(\alpha \cup \alpha^*)$. We will show that there exists an unmixed bipartite graph G such that $I = (x_i y_j \mid \{x_i, y_j\} \in E(G))$.

Since $\emptyset \in \mathcal{L}$ and $\{x_1, \ldots, x_n\} \in \mathcal{L}$, it follows that $I \subset (x_i y_j \mid 1 \leq i, j \leq n)$. Suppose that a monomial v of degree ≥ 3 belongs to the minimal set of generators of I.

If $v = x_i x_j u$ where $i \neq j$ and u is a (squarefree) monomial, then there exist $\alpha, \beta \in \mathcal{L}$ such that $x_i \in \alpha \setminus \beta$, $x_j \in \beta \setminus \alpha$, $u \notin \langle \alpha \cup \alpha^* \rangle$ and $u \notin \langle \beta \cup \beta^* \rangle$. Since $\mathcal{L}$ is a sublattice of $\mathcal{L}_n$, one has $\alpha \cap \beta \in \mathcal{L}$. Note that $(\alpha \cap \beta)^* = \alpha^* \cup \beta^*$.

Hence $I \subset I_1$, where I_1 is the ideal generated $(\alpha \cap \beta) \cup (\alpha^* \cup \beta^*)$. However, none of the variables in v appears in the set $(\alpha \cap \beta) \cup (\alpha^* \cup \beta^*)$.

If $v = y_i y_j u$ where $i \neq j$ and u is a (squarefree) monomial, then there exist $\alpha, \beta \in \mathcal{L}$ such that $y_i \in \alpha^* \setminus \beta^*$, $y_j \in \beta^* \setminus \alpha^*$, $u \notin \langle \alpha \cup \alpha^* \rangle$ and $u \notin \langle \beta \cup \beta^* \rangle$. Since $\mathcal{L}$ is a sublattice of $\mathcal{L}_n$, $\alpha \cup \beta \in \mathcal{L}$. Note that $(\alpha \cup \beta)^* = \alpha^* \cap \beta^*$. Hence $I \subset I_2$, where I_2 is generated by $(\alpha \cup \beta) \cup (\alpha^* \cap \beta^*)$. However, none of the variables in v appears in the set $(\alpha \cup \beta) \cup (\alpha^* \cap \beta^*)$.

Thus the minimal set of generators of I is a subset of $\{x_i y_j \mid 1 \leq i, j \leq n\}$. Hence there exists a bipartite graph G such that $I = I(G)$. Since the primary decomposition of the edge ideal $I(G)$ of G is $I = \bigcap_{C \in \mathcal{M}(G)}(C)$, where $\mathcal{M}(G)$ is the set of minimal vertex covers of G, one has $\mathcal{M}(G) = \{\alpha \cup \alpha^* \mid \alpha \in \mathcal{L}\}$. Thus $\mathcal{L} = \mathcal{L}_G$. Since the cardinality of each $\alpha \cup \alpha^*$ with $\alpha \in \mathcal{L}$ is n, it follows that G is unmixed, as desired. $\qquad\square$

The next theorem tells which sublattices of $\mathcal{L}_n$ correspond to Cohen–Macaulay bipartite graphs. A sublattice $\mathcal{L}$ of $\mathcal{L}_n$ is called **full** if $\mathrm{rank}\,\mathcal{L} = n$.

Theorem 9.1.19. *A subset $\mathcal{L}$ of $\mathcal{L}_n$ is a full sublattice of $\mathcal{L}_n$ if and only if there exists a Cohen–Macaulay bipartite graph G on $\{x_1, \ldots, x_n\} \cup \{y_1, \ldots, y_n\}$ with $\mathcal{L} = \mathcal{L}_G$.*

Proof. Let G be a Cohen–Macaulay bipartite graph on the set $\{x_1, \ldots, x_n\} \cup \{y_1, \ldots, y_n\}$. Theorem 9.1.13 guarantees the existence of a finite poset P with $G = G(P)$, where $|P| = n$. Corollary 9.1.15 says that $\mathcal{L}_G$ coincides with $\mathcal{J}(P)$. Thus $\mathcal{L}_G$ is a full sublattice of $\mathcal{L}_n$.

Conversely, suppose that $\mathcal{L}$ is a full sublattice of $\mathcal{L}_n$. One has $\mathcal{L} = \mathcal{J}(P)$ for a unique poset P with $|P| = n$ (Theorem 9.1.7). Let $G = G(P)$. Then G is a Cohen–Macaulay bipartite graph. Corollary 9.1.15 says that $\mathcal{L}_G$ coincides with $\mathcal{J}(P)$. Thus $\mathcal{L}_G = \mathcal{L}$, as required. $\qquad\square$

9.1.5 Sequentially Cohen–Macaulay bipartite graphs

The main purpose of this subsection is to show the following extension of Corollary 9.1.16.

Theorem 9.1.20. *A bipartite graph is sequentially Cohen–Macaulay if and only if it is shellable.*

For a vertex v of a graph G, let $N_G(v)$ denote the set of vertices w of G such that $\{v, w\} \in E(G)$. The number $\deg v = |N_G(v)|$ is called the **degree** of v.

For the proof of the theorem we need the following two lemmata.

Lemma 9.1.21. *Let G be a bipartite graph with bipartition $\{x_1, \ldots, x_m\}$ and $\{y_1, \ldots, y_n\}$. If G is sequentially Cohen–Macaulay, then there exists a vertex $v \in V(G)$ with $\deg v = 1$.*

Proof. We may assume that $m \leq n$, and that G has no isolated vertices. Let $I_G = I(G)^\vee$ be the vertex cover ideal of G, and let $L = (I_G)_{[n]}$ be the squarefree part of the nth component of I_G. Since G is sequentially Cohen–Macaulay, it follows from Theorem 8.2.20 that I_G is componentwise linear. Hence Theorem 8.2.17 says that L has a linear resolution.

Let $g_1, \ldots, g_q$ be a minimal monomial set of generators of L. We may assume that $g_1 = y_1 \cdots y_n$ and $g_2 = x_1 \cdots x_m y_1 \cdots y_{n-m}$. Then $x_1 \cdots x_m g_1 - y_1 \cdots y_m g_2 = 0$ is a relation of g_1 and g_2. Since L has linear relations, the previous relation is a linear combination of linear relations. Therefore there exists x_i and g_k such that $x_i g_1 - v g_k = 0$, where v is a variable. It follows that $v = y_j$ for some j and $g_k = x_i y_1 \cdots y_{j-1} y_{j+1} \cdots y_n$. Since $\{x_i, y_1, \ldots, y_{j-1}, y_{j+1}, \ldots, y_n\}$ is vertex cover of G, it follows that $\{x_k, y_j\}$ cannot be an edge of G for $k \neq i$. Thus $\deg y_j = 1$, as desired. $\qquad\square$

For the next lemma and its proof we introduce the following notation: let G be a graph and $U \subset V(G)$ a subset of the vertex set of G. Then we write $G \setminus U$ for the induced subgraph of G on $V(G) \setminus U$.

Lemma 9.1.22. *Let x be a vertex of G, and let $G' = G \setminus (\{x\} \cup N_G(x))$. If G is sequentially Cohen–Macaulay, then G' is sequentially Cohen–Macaulay.*

Proof. Let Δ be the simplicial complex of independent sets of G and Δ' the simplicial complex of independent sets of G'. We first show that

$$\Delta' = \text{link}_\Delta\{x\}. \tag{9.2}$$

Let $F \in \text{link}_\Delta\{x\}$. Then $x \notin F$ and $F \cup \{x\} \in \Delta$. Hence $(F \cup \{x\}) \cap N_G(x) = \emptyset$. Thus $F \subset V(G')$. Hence $F \in \Delta'$, since F is an independent set of G'.

Conversely, if $F \in \Delta'$, then F is an independent set of G'. Since $F \cap N_G(x) = \emptyset$, it follows that $F \cup \{x\}$ is an independent set of G.

Now let $F_1, \ldots, F_s$ be the facets of Δ. We may assume that the $F_1, \ldots, F_r$ are precisely those facets which contain x. It then follows from (9.2) that $F_1' = F_1 \setminus \{x\}, \ldots, F_r' = F_r \setminus \{x\}$ are the facets of Δ'.

Next observe that

$$\Delta'(i) = \text{link}_{\Delta(i+1)}\{x\} \tag{9.3}$$

for all $i \leq d$, where $d = \dim \Delta'$.

In fact, if F_i' is a face of Δ' with $\dim F_i' = d$, then $F_i' \cup \{x\} \in \Delta(d+1)$. Therefore $\{x\} \in \Delta(i+1)$ for all $i \leq d+1$. Suppose now F is a facet of $\Delta'(i)$. Then $\dim F = i$ and $F \subset F_j \cup \{x\}$. That is, $F \cup \{x\} \subset F_j$. It follows that $F \cup \{x\}$ is a facet of $\Delta(i+1)$, and hence $F \in \text{link}_{\Delta(i+1)}\{x\}$. Conversely, let $F \in \text{link}_{\Delta(i+1)}\{x\}$ be a facet. Then $\dim F = i$ and $F \cup \{x\} \in \Delta$. Therefore, $F \cup \{x\} \subset F_j$ for some j. In other words, $F \subset F_j \setminus \{x\}$. This implies that F is a facet of Δ'.

The definition of sequentially Cohen–Macaulay says that $\Delta(i)$ is Cohen–Macaulay for all i. By (9.3) and Corollary 8.1.8, it follows that $\Delta'(i)$ is Cohen–Macaulay for all i. Hence Δ' is sequentially Cohen–Macaulay for all i. $\qquad\square$

Proof (Proof of Theorem 9.1.20). If G is shellable, then G is sequentially Cohen–Macaulay (Corollary 8.2.19). Conversely, we prove by induction on the number of vertices of G that G is shellable, if G is sequentially Cohen–Macaulay. By Lemma 9.1.21 there exists a vertex x of G with $\deg x = 1$. Let y be the vertex adjacent to x. Applying Lemma 9.1.22 we see that $G_1 = G \backslash (\{x\} \cup N_G(x))$ and $G_2 = G \backslash (\{y\} \cup N_G(y))$ are sequentially Cohen–Macaulay. By our induction hypothesis we assume that G_1 and G_2 are shellable.

Let $F_1', \ldots, F_r'$ be a shelling of $\Delta_1 = \Delta(\bar{G}_1)$ and $H_1', \ldots, H_s'$ be a shelling of $\Delta_2 = \Delta(\bar{G}_2)$. By Lemma 9.1.21, $F_1' \cup \{x\}, \ldots, F_r' \cup \{x\}$ are the facets of $\Delta = \Delta(\bar{G})$ which contain x, and $H_1' \cup \{x\}, \ldots, H_r' \cup \{x\}$ are the facets of Δ which contain y. Since an independent set of G cannot contain both x and y, these two sets of facets are disjoint. On the other hand, each maximal independent set contains either x or y, since $\deg x = 1$. Thus

$$F_1' \cup \{x\}, \ldots, F_r' \cup \{x\}, H_1' \cup \{x\}, \ldots, H_r' \cup \{x\}$$

is precisely the set of facets of Δ.

We now show that this is a shelling of Δ. In fact, let $F = F_i' \cup \{x\}$ and $H = H_j \cup \{y\}$. Since $H_j' \cup \{x\}$ is an independent set of G, there exists a facet of Δ containing it. This facet must be of the type $F' = F_\ell' \cup \{x\}$ for some ℓ. It follows that $F \setminus F' = \{y\}$ and that F' comes before H in the shelling order.

The remaining cases to be considered follow directly from the above shellings of Δ_1 and Δ_2. $\qquad\square$

9.2 Dirac's theorem on chordal graphs

We discuss the algebraic aspects of Dirac's theorem by using Alexander duality.

9.2.1 Edge ideals with linear resolution

We say that a finite graph G on $[n]$ is **decomposable** if there exist proper subsets P and Q of $[n]$ with $P \cup Q = [n]$ such that $P \cap Q$ is a clique of G and that $\{i, j\} \notin E(G)$ for all $i \in P \setminus Q$ and for all $j \in Q \setminus P$.

Lemma 9.2.1. *Every chordal graph which is not complete is decomposable.*

Proof. Let G be a chordal graph on $[n]$ which is not complete. Let a and b be vertices of G with $\{a, b\} \notin E(G)$.

A subset $A \subset [n]$ is called a separator of G if there exist subsets P and Q of $[n]$ with $P \cup Q = [n]$ and $P \cap Q = A$ and that $\{i, j\} \notin E(G)$ for all $i \in P \setminus Q$ and for all $j \in Q \setminus P$. Clearly $[n] \setminus \{a, b\}$ is a separator of G. Let $B \subset [n] \setminus \{a, b\}$ be a minimal separator (with respect to inclusion) of G.

What we must prove is that B is a clique of G. Let $|B| > 1$ and $x, y \in B$ with $x \neq y$. Let P and Q be subset of $[n]$ with $P \cup Q = [n]$ and $P \cap Q = B$

and that $\{i, j\} \notin E(G)$ for all $i \in P \setminus Q$ and for all $j \in Q \setminus P$. Let $C_1, \ldots, C_s$ be the connected components of the induced subgraph of G on $P \setminus B$ and $D_1, \ldots, D_t$ the connected components of the induced subgraph of G on $Q \setminus B$.

We claim that there is a vertex x_0 of C_1 such that $\{x_0, x\} \in E(G)$. To see why this is true, suppose that $\{z, x\} \notin E(G)$ for all vertices z of C_1. Let V denote the set of vertices of C_1 and $B' = B \setminus \{x\}$. Let $P_0 = V \cup B'$ and $Q_0 = [n] \setminus V$. Then $P_0 \cup Q_0 = [n]$ and $P_0 \cap Q_0 = B'$. Since $\{z, x\} \notin E(G)$ for all $z \in V$, it follows that $\{i, j\} \notin E(G)$ for all $i \in P_0 \setminus Q_0$ and for all $j \in Q_0 \setminus P_0$. Hence B' is a separator of G, which contradicts the minimality of B. Consequently, there is a vertex x_0 of C_1 such that $\{x_0, x\} \in E(G)$. Similarly, there is a vertex y_0 of C_1 such that $\{y_0, y\} \in E(G)$. In addition, there is a vertex x_1 of D_1 such that $\{x_1, x\} \in E(G)$, and there is a vertex y_1 of D_1 such that $\{y_1, y\} \in E(G)$.

Now, let W_1 be a walk of minimal length between x and y whose vertices belong to $V \cup \{x, y\}$ and W_2 a walk of minimal length between x and y whose vertices belong to $V' \cup \{x, y\}$, where V' is the set of vertices of D_1. Then combining W_1 and W_2 yields a cycle C of G of length > 3. The minimality of W_1 and W_2 together with the fact that $\{i, j\} \notin E(G)$ for all $i \in V$ and $j \in V'$ guarantees that, except for $\{x, y\}$, the cycle C has no chord. Since G is chordal, the edge $\{x, y\}$ must belong to G.

Hence $\{x, y\} \in E(G)$ for all x and y of B with $x \neq y$. Thus B is a clique of G, as desired. $\square$

Corollary 9.2.2. *Let G be a chordal graph on $[n]$ and $\Delta(G)$ its clique complex. Then $\tilde{H}_i(\Delta(G); K) = 0$ for all $i \neq 0$.*

Proof. We work with induction on the number of vertices on G. If G is a complete graph, then $\Delta(G)$ is the simplex on $[n]$. Thus $\tilde{H}_i(\Delta(G); K) = 0$ for all i, see Example 5.1.9. Suppose that G is not complete. Lemma 9.2.1 says that G is decomposable. Let P and Q be proper subsets of $[n]$ with $P \cup Q = [n]$ such that $P \cap Q$ is a clique of G and that $\{i, j\} \notin E(G)$ for all $i \in P \setminus Q$ and for all $j \in Q \setminus P$. Let $\Delta = \Delta(G)$, $\Delta_1 = \Delta(G_P)$, $\Delta_2 = \Delta(G_Q)$ and $\Gamma = \Delta(G_{P \cap Q})$. Then $\Delta = \Delta_1 \cup \Delta_2$ and $\Gamma = \Delta_1 \cap \Delta_2$. Since each of G_P and G_Q is chordal and since each of P and Q is a proper subset of $[n]$, it follows from the induction hypothesis that $\tilde{H}_i(\Delta_1; K) = 0$ for all $i \neq 0$ and $\tilde{H}_i(\Delta_2; K) = 0$ for all $i \neq 0$. In addition, since $P \cap Q$ is a clique of G, Γ is a simplex. Thus $\tilde{H}_i(\Gamma; K) = 0$ for all $i \neq 0$.

Now, the reduced Mayer–Vietoris exact sequence of Δ_1 and Δ_2 (Proposition 5.1.8) is given by

$$\cdots \longrightarrow \tilde{H}_k(\Gamma; K) \longrightarrow \tilde{H}_k(\Delta_1; K) \oplus \tilde{H}_k(\Delta_2; K) \longrightarrow \tilde{H}_k(\Delta; K)$$
$$\longrightarrow \tilde{H}_{k-1}(\Gamma; K) \longrightarrow \tilde{H}_{k-1}(\Delta_1; K) \oplus \tilde{H}_k(\Delta_2; K) \longrightarrow \tilde{H}_{k-1}(\Delta; K)$$
$$\longrightarrow \cdots .$$

Hence $\tilde{H}_i(\Delta; K) = 0$ for all $i \neq 0$, as required. $\square$

Theorem 9.2.3 (Fröberg). *The edge ideal $I(G)$ of a finite graph G has a linear resolution if and only if the complementary graph $\overline{G}$ of G is chordal.*

Proof. Since $I(G) = I_{\Delta(\overline{G})}$, what we must prove is that the Stanley–Reisner ideal of the clique complex $\Delta(\overline{G})$ of $\overline{G}$ has a linear resolution if and only if $\overline{G}$ is chordal.

Hochster's formula (Theorem 8.1.1) says that $I_{\Delta(\overline{G})}$ has a 2-linear resolution if and only if, for any subset $W \subset [n]$, one has $\tilde{H}_i(\Delta(\overline{G})_W; K) = 0$ unless $i = 0$.

Suppose that $\overline{G}$ is not chordal. Then there is a cycle of $\overline{G}$ of length $q > 3$ with no chord. Let W be the set of vertices of C. Then $\Delta(\overline{G})_W$ coincides with C and $\dim_K \tilde{H}_1(C; K) \neq 0$, see Example 5.1.9. Thus $I_{\Delta(\overline{G})}$ cannot have a 2-linear resolution.

On the other hand, suppose that $\overline{G}$ is chordal. Let W be a subset of $[n]$. Then $\Delta(\overline{G})_W$ is the clique complex of the induced subgraph $\overline{G}_W$ of $\overline{G}$ on W. It is clear that every induced subgraph of a chordal graph is again chordal. In particular $\overline{G}_W$ is chordal. Thus by Corollary 9.2.2, one has $\tilde{H}_i(\Delta(\overline{G})_W; K) = 0$ for all $i \neq 0$. Thus $I_{\Delta(\overline{G})}$ has a 2-linear resolution. $\qquad\square$

9.2.2 The Hilbert–Burch theorem for monomial ideals

Let $I \subset S$ be a monomial ideal with $G(I) = \{u_1, \ldots, u_s\}$, where $s \geq 2$. We introduce the $\binom{s}{2} \times s$ matrix

$$A_I = (a_k^{(i,j)})_{1 \leq i < j \leq s, 1 \leq k \leq s}$$

whose entries $a_k^{(i,j)} \in S$ are

$$a_i^{(i,j)} = u_j / \gcd(u_i, u_j), \quad a_j^{(i,j)} = -u_i / \gcd(u_i, u_j), \quad \text{and} \quad a_k^{(i,j)} = 0 \text{ if } k \notin \{i, j\}$$

for all $1 \leq i < j \leq s$ and for all $1 \leq k \leq s$.

If the monomials $u_1, \ldots, u_s$ have the greatest common divisor $w \neq 1$, then one has $A_I = A_{I'}$, where I' is the monomial ideal with $G(I') = \{u_1/w, \ldots, u_s/w\}$.

For an arbitrary matrix C we denote by $C(j)$ the matrix which is obtained from C by omitting the jth column. With this notation introduced one has

Lemma 9.2.4 (Hilbert–Burch). *Let $I \subset S$ be a monomial ideal with $G(I) = \{u_1, \ldots, u_s\}$, where $s \geq 2$, and suppose that the monomials $u_1, \ldots, u_s$ have no common factor. Then the following conditions are equivalent:*

(a) $\operatorname{proj dim} I = 1$

(b) *The matrix A_I contains an $(s-1) \times s$ submatrix $A_I^\sharp$ with the property that, after relabelling the rows of $A_I^\sharp$ if necessary, one has*

$$(-1)^j \det(A_I^\sharp(j)) = u_j \quad \text{for all} \quad j.$$

Moreover, when one has such an $(s-1) \times s$ submatrix $A_I^\sharp$, a minimal graded free resolution of I is

$$0 \longrightarrow \bigoplus_{j=1}^{s-1} S(-b_j) \xrightarrow{A_I^\sharp} \bigoplus_{j=1}^{s} S(-\deg u_j) \longrightarrow I \longrightarrow 0,$$

where $b_j = \deg(u_\mu u_\nu / \gcd(u_\mu, u_\nu))$ if the jth row of $A_I^\sharp$ is the (μ, ν)th row of A_I.

Proof. (a) $\Rightarrow$ (b): We first observe that the matrix A_I describes the map $T_2 \to T_1$ of the Taylor complex for the sequence $u_1, \ldots, u_s$ (Theorem 7.1.1). Therefore the first syzygy module $G \subset T_1 = \bigoplus_{j=1}^{s} Se_j$ of S/I in the Taylor complex is generated by the elements $a_i^{(i,j)} e_i + a_j^{(i,j)} e_j$. By using that $\operatorname{proj dim} I = 1$ and counting ranks, we see that G is free of rank $s-1$. In particular, G is minimally generated by $s-1$ homogeneous elements. The graded version of Nakayama's lemma implies that we can choose a minimal set of generators of G among the Taylor relations. In other words, we can choose $s-1$ of the relations $a_i^{(i,j)} e_i + a_j^{(i,j)} e_j$ which form a basis of the free module G. The submatrix of A_I whose rows correspond to these elements will be the desired matrix $A_I^\sharp$. Set $F = T_1$ and let $\alpha : G \to F$ be the linear map defined by $A_I^\sharp$. Then we obtain the exact sequence

$$0 \longrightarrow G \xrightarrow{\ \alpha\ } F \xrightarrow{\ \varphi\ } S \longrightarrow S/I \longrightarrow 0, \qquad (9.4)$$

where $\varphi(e_i) = u_i$ for $i = 1, \ldots, s$.

By the Hilbert–Burch theorem as presented in [BH98, Theorem 1.4.17] it follows that

$$0 \longrightarrow G \xrightarrow{\ \alpha\ } F \xrightarrow{\ \psi\ } S \longrightarrow S/I \longrightarrow 0 \qquad (9.5)$$

is exact, where $\psi(e_i) = (-1)^i \det(A_I^\sharp(i))$ for $i = 1, \ldots, s$.

The condition that the monomials $u_1, \ldots, u_s$ have no common factor implies that height $I \geq 2$, so that $\dim S/I \leq n - 2$. Since $\operatorname{proj dim} S/I = 2$, the Auslander–Buchsbaum formula (see Corollary A.4.3) shows that depth $S/I = n - 2$. Since one always has that depth $S/I \leq \dim S/I$, it follows that depth $S/I = \dim S/I$. In other words, S/I is Cohen–Macaulay of dimension $n - 2$.

Since S is Cohen–Macaulay, one has grade $I =$ height $I = 2$ ([BH98, Corollary 2.1.4]). Therefore, $\operatorname{Ext}_S^i(S/I, S) = 0$ for $i < 2$. This implies that the S-duals of the exact sequences (9.4) and (9.5) are acyclic. That is, one has exact sequences

$$0 \longrightarrow S \xrightarrow{\ \varphi^*\ } F^* \xrightarrow{\ \alpha^*\ } G^*$$

and

$$0 \longrightarrow S \xrightarrow{\ \psi^*\ } F^* \xrightarrow{\ \alpha^*\ } G^*$$

We conclude that $\operatorname{Im}\varphi^* = \operatorname{Im}\psi^*$. Since both these modules are cyclic their generators differ only by a unit of S, that is, by an element $c \in K \setminus \{0\}$. Hence

$$cu_i = (-1)^i \det(A_I^\sharp(i)) \quad \text{for} \quad i = 1, \dots, s.$$

Since the entries of $A_I^\sharp$ are all of the form $\pm u$ with $u \in \operatorname{Mon}(S)$, we conclude that c is an integer. Suppose that $c \neq \pm 1$. Then there exists a prime number p which divides c. The ideal I is defined over any field and our considerations so far did not depend on the base field K. Thus we may as well assume that $K = \mathbb{Z}/(p)$. This leads to a contradiction, since then the ideal is generated by the maximal minors of $A_I^\sharp$ is the zero ideal and is not I, as it should be. Thus $c = \pm 1$. If $c = -1$, then we exchange two rows of $A_I^\sharp$.

(b) $\Rightarrow$ (a): Since the generators of I have no common factor, condition (b) implies that the ideal of maximal minors of $A_I^\sharp$ has grade ≥ 2. Then a direct application of [BH98, Theorem 1.4.7] yields the desired conclusion. □

Any submatrix of A_I as in Lemma 9.2.4(b) is called a **Hilbert–Burch matrix** of I. After fixing the order of the generators of I, we consider two Hilbert–Burch matrices to be equal if they coincide up to permutation of rows.

In general, there is not a unique Hilbert–Burch matrix.

Example 9.2.5. Let $n = 6$ and $I = (x_4x_5x_6, x_1x_5x_6, x_1x_2x_6, x_1x_2x_5)$. Thus the 6×4 matrix A_I is

$$\begin{bmatrix} x_1 & -x_4 & 0 & 0 \\ x_1x_2 & 0 & -x_4x_5 & 0 \\ 0 & x_2 & -x_5 & 0 \\ x_1x_2 & 0 & 0 & -x_4x_6 \\ 0 & x_2 & 0 & -x_6 \\ 0 & 0 & x_5 & -x_6 \end{bmatrix}.$$

By using Lemma 9.2.4 one has $\operatorname{proj\,dim} I = 1$. In fact, I has three Hilbert–Burch matrices

$$\begin{bmatrix} x_1 & -x_4 & 0 & 0 \\ 0 & x_2 & -x_5 & 0 \\ 0 & x_2 & 0 & -x_6 \end{bmatrix}, \quad \begin{bmatrix} x_1 & -x_4 & 0 & 0 \\ 0 & x_2 & -x_5 & 0 \\ 0 & 0 & x_5 & -x_6 \end{bmatrix} \text{ and } \begin{bmatrix} x_1 & -x_4 & 0 & 0 \\ 0 & x_2 & 0 & -x_6 \\ 0 & 0 & x_5 & -x_6 \end{bmatrix}.$$

Thus, for example,

$$0 \longrightarrow S(-4)^3 \xrightarrow{\begin{bmatrix} x_1 & -x_4 & 0 & 0 \\ 0 & x_2 & -x_5 & 0 \\ 0 & 0 & x_5 & -x_6 \end{bmatrix}} S(-3)^4 \longrightarrow I \longrightarrow 0$$

is a minimal graded free resolution of I.

9.2.3 Chordal graphs and quasi-forests

Let G be a finite graph on $[n]$. A **perfect elimination ordering** of G is an ordering $i_n, \ldots, i_2, i_1$ of the vertices $1, 2, \ldots, n$ of G such that, for each $1 < j \leq n$, $C_{i_j} = \{i_k \in [n] : 1 \leq k < j, \{i_k, i_j\} \in E(G)\}$ is a clique of G. In 1961 G. A. Dirac proved that a finite graph G is chordal if and only if G has a perfect elimination ordering.

Let Δ be a simplicial complex. A facet $F \in \mathcal{F}(\Delta)$ is said to be a **leaf** of Δ if (either F is the only facet of Δ, or) there exists a facet $G \in \mathcal{F}(\Delta)$ with $G \neq F$, called a **branch** F, such that $H \cap F \subset G \cap F$ for all $H \in \mathcal{F}(\Delta)$ with $H \neq F$. Observe for a leaf F and a branch G of F, the subcomplex Δ' with $\mathcal{F}(\Delta') = \mathcal{F}(\Delta) \setminus \{F\}$ coincides with the restriction $\Delta_{[n]\setminus(F\setminus(G\cap F))}$. A vertex i of Δ is called a **free vertex** of Δ if i belongs to exactly one face. Note that every leaf has at least one free vertex.

The following example displayed in Figure 9.4 shows that a facet with a free vertex need not be a leaf.

Fig. 9.4. A nonleaf with a free vertex.

A **quasi-forest** is a simplicial complex such that there exists a labelling $F_1, \ldots, F_q$ of the facets of Δ, called a **leaf order**, such that for each $1 < i \leq m$ the facet F_i is a leaf of the subcomplex $\langle F_1, \ldots F_i \rangle$. A **quasi-tree** is a quasi-forest which is connected.

Lemma 9.2.6. *A finite graph G has a perfect elimination ordering if and only if the clique complex $\Delta(G)$ of G is a quasi-forest.*

Proof. "Only if": Let G be a graph on $[n]$ and suppose (for simplicity) that the ordering $n, n-1, \ldots, 1$ is a perfect elimination ordering. Thus, for each $1 < j \leq n$, $F_j = \{k \in [n] : 1 \leq k < j, \{k, j\} \in E(G)\} \cup \{j\}$ belongs to $\Delta(G)$. Let $F \in \Delta(G)$ and j the largest integer for which $j \in F$. Since F is a clique of G, one has $\{k, j\} \in E(G)$ for all $k \in F$ with $k \neq j$. Hence $F \subset F_j$. In particular, $\Delta(G) = \langle F_1, \ldots, F_n \rangle$. Let $1 < i \leq n$ and j the largest integer $< i$ for which $j \in F_i$. Then $F_k \cap F_i \subset F_j \cap F_i$ for all $k < i$. In fact, if $a \in F_k \cap F_i$ with $a \neq j$, then $a \leq k < i$ with $a \in F_i$. Thus $a < j$ and $\{a, j\} \in E(G)$. Hence $a \in F_j$.

We will show, in general, that if a simplicial complex Δ can be obtained from faces $F_1, \ldots, F_m$ such that, for each i, there is $j < i$ with $F_k \cap F_i \subset F_j \cap F_i$ for all $k < i$, then Δ is a quasi-forest. Our proof will be done by working with induction on m. Hence we may assume that the simplicial complex Γ obtained from $F_1, \ldots, F_{m-1}$ is a quasi-forest with $\Gamma \neq \Delta$. Thus there is a leaf order $G_1, \ldots, G_r$ of the quasi-forest Γ, where $\mathcal{F}(\Gamma) = \{G_1, \ldots, G_r\} \subset$

$\{F_1, \ldots, F_{m-1}\}$. Now, there is $j_0 < m$ such that $F_k \cap F_m \subset F_{j_0} \cap F_m$ for all $k < m$. It then follows that each vertex belonging to $F_m \setminus F_{j_0}$ is a free vertex. Let $F_{j_0} \subset G_s$. If it happens that $G_s \subset F_m$, then $G_1, \ldots, G_{s-1}, F_m, G_{s+1}, \ldots, G_r$ is a leaf order of Δ. Because, since each vertex $F_m \setminus G_s$ is a free vertex, one has $G_s \cap G_i = F_m \cap G_i$ for all i. On the other hand, if $G_s \not\subset F_m$, then $G_k \cap F_m \subset F_{j_0} \cap F_m = G_s \cap F_m$ for all $k \leq r$. Thus F_m is a leaf of Δ and $G_1, \ldots, G_r, F_m$ is a leaf order of Δ.

"If": Suppose that $\Delta(G)$ is a quasi-forest with a leaf order $F_1, \ldots, F_q$. We may suppose that n is a free vertex of F_q. Let $F_q' = F_q \setminus \{n\}$. If F_q has more than one free vertex, then $F_1, \ldots, F_{q-1}, F_q'$ is a leaf order of $\Delta(G')$ where G' is the induced subgraph of G on $[n-1]$. Thus $\Delta(G')$ is again a quasi-forest. By using induction, we may assume that, say, $n-1, n-2, \ldots, 1$ is a perfect elimination ordering of $\Delta(G')$. Then $n, n-1, \ldots, 1$ is a perfect elimination ordering of $\Delta(G)$. On the other hand, if n is the only free vertex of F_q, then $F_1, F_2, \ldots, F_{q-1}$ is a leaf order of $\Delta(G')$, and as before we see that $\Delta(G)$ has a perfect elimination ordering. $\qquad\square$

Lemma 9.2.7. *A quasi-forest is a flag complex.*

Proof. Let Δ be a quasi-forest on $[n]$ and $F_1, \ldots, F_q$ its leaf order. We work with induction on q. Let $q > 1$. Since $\Delta' = \langle F_1, \ldots, F_{q-1} \rangle$ is a quasi-forest, it follows that Δ' is a flag complex. Let F_i with $i < q$ be a branch of F_q. Thus $\Delta' = \Delta_{[n] \setminus (F_q \setminus (F_q \cap F_i))}$.

Suppose there exists a minimal nonface H of Δ having at least three elements of $[n]$. Since Δ' is flag, $H \notin \Delta'$, and therefore there exists $b \in F_q$ with $b \in H$, and since H is a nonface, there exists $a \in H$ with $a \notin F_q$.

Since $|H| > 2$, one has $\{a, b\} \in \Delta$. Thus there is F_j with $j \neq q$ such that $\{a, b\} \in F_j$. Hence $b \in F_j \cap F_q$. Thus $b \in F_i$. Hence $H \cap (F_q \setminus (F_q \cap F_i)) = \emptyset$. This shows that H is a minimal nonface of Δ', a contradiction. $\qquad\square$

Lemma 9.2.8. *Let G be a finite graph on $[n]$ and Γ a simplicial complex on $[n]$ such that G is the 1-skeleton of Γ. Then $\Gamma = \Delta(G)$ if and only if Γ is a flag complex.*

Proof. Let $\binom{[n]}{2}$ denote the set of 2-element subsets of $[n]$ and $\mathcal{N}(\Gamma)$ the set of minimal nonfaces of Γ. If F is a face of Γ, then F is a clique of G. Thus $\Gamma \subset \Delta(G)$. Moreover, since $\mathcal{N}(\Gamma) \cap \binom{[n]}{2} = E(\bar{G})$ and since $\mathcal{N}(\Delta(G)) = E(\bar{G})$, it follows that $\Gamma = \Delta(G)$ if and only if $\mathcal{N}(\Gamma) \subset \binom{[n]}{2}$, i.e. Γ is a flag complex. $\qquad\square$

Corollary 9.2.9. *A finite graph G has a perfect elimination ordering if and only if G is the 1-skeleton of a quasi-forest.*

Proof. Since G is the 1-skeleton of $\Delta(G)$, it follows that G is the 1-skeleton of a quasi-forest if $\Delta(G)$ is a quasi-forest. Conversely, if G is the 1-skeleton of a quasi-forest Γ, then by Lemma 9.2.7 and Lemma 9.2.8 one has $\Gamma = \Delta(G)$.

Hence $\Delta(G)$ is a quasi-forest. Consequently, G is the 1-skeleton of a quasi-forest if and only if $\Delta(G)$ is a quasi-forest. Thus Lemma 9.2.6 guarantees that G has a perfect elimination ordering if and only if G is the 1-skeleton of a quasi-forest, as required.   □

Given a simplicial complex Δ on $[n]$ with $\mathcal{F}(\Delta) = \{F_1, \ldots, F_q\}$, we introduce the $\binom{q}{2} \times q$ matrix

$$M_\Delta = (a_k^{(i,j)})_{1 \le i < j \le q, 1 \le k \le q}$$

whose entries $a_k^{(i,j)} \in S = K[x_1, \ldots, x_n]$ are

$$a_i^{(i,j)} = x_{F_i \setminus F_j}, \quad a_j^{(i,j)} = -x_{F_j \setminus F_i}, \quad \text{and } a_k^{(i,j)} = 0 \text{ if } k \notin \{i,j\}$$

for all $1 \le i < j \le q$ and for all $1 \le k \le q$.

Given a simplicial complex Δ on $[n]$, we introduced in Chapter 1 the simplicial complex $\bar{\Delta} = \langle [n] \setminus F : F \in \mathcal{F}(\Delta) \rangle$ on $[n]$. Let $I(\bar{\Delta})$ denote the facet ideal of $\bar{\Delta}$. Let $A_{I(\bar{\Delta})}$ denote the $\binom{q}{2} \times q$ matrix associated with the monomial ideal $I(\bar{\Delta})$; see Subsection 9.2.2. One has $M_\Delta = A_{I(\bar{\Delta})}$, because

$$F_i \setminus F_j = ([n] \setminus F_j) \setminus ([n] \setminus (F_i \cup F_j)) = ([n] \setminus F_j) \setminus (([n] \setminus F_i) \cap ([n] \setminus F_j)).$$

The quasi-forest can be characterized in terms of the matrix M_Δ. In fact,

Lemma 9.2.10. *A simplicial complex $\Delta = \langle F_1, \ldots, F_q \rangle$ on $[n]$ is a quasi-forest if and only if the matrix M_Δ contains a Hilbert–Burch matrix for the ideal $(x_{[n]}/x_{F_1}, \ldots, x_{[n]}/x_{F_q})$.*

Proof. "Only if": Let Δ be a quasi-forest on $[n]$ and fix a leaf order $F_1, \ldots, F_q$ of the facets of Δ. Let $q > 1$. Let F_k with $k \ne q$ be a branch of F_q and Δ' the simplicial complex on $[n] \setminus (F_q \setminus F_k)$ with $\mathcal{F}(\Delta') = \mathcal{F}(\Delta) \setminus \{F_q\}$. Since Δ' is a quasi-forest, it follows that M_Δ contains a $(q-2) \times q$ submatrix M', where none of the (i,q)th rows, $1 \le i < q$, of M_Δ belongs to M', with the property that, for each $1 \le j < q$, if $M'(j,q)$ is the $(q-2) \times (q-2)$ submatrix of M' obtained by removing the jth and qth columns from M', then $|\det(M'(j,q))| = x_{[n] \setminus (F_q \setminus F_k)}/x_{F_j}$. Let $M_\Delta^\sharp$ denote the $(q-1) \times q$ submatrix of M_Δ obtained by adding the (k,q)th row of M_Δ to M'. Since $a_q^{(k,q)} = -x_{F_q \setminus F_k}$, one has $|\det(M_\Delta^\sharp(j))| = x_{[n]}/x_{F_j}$ for each $1 \le j < q$. Moreover, since $|\det(M_\Delta^\sharp(q))| = x_{F_k \setminus F_q} \det(M'(k,q)$ and since $|\det(M'(k,q))| = x_{[n] \setminus (F_q \setminus F_k)}/x_{F_k}$, one has $|\det(M_\Delta^\sharp(q))| = x_{[n]}/x_{F_q}$.

"If": Suppose that the matrix M_Δ contains a $(q-1) \times q$ submatrix $M_\Delta^\sharp$ with the property that, for each $1 \le j \le q$, if $M_\Delta^\sharp(j)$ is the $(q-1) \times (q-1)$ submatrix of $M_\Delta^\sharp$ obtained by removing the jth column from $M_\Delta^\sharp$, then $|\det(M_\Delta^\sharp(j))| = x_{[n]}/x_{F_j}$. Let Ω denote the finite graph on $[n]$ whose edges are those $\{i,j\}$ with $1 \le i < j \le n$ such that the (i,j)th row of M_Δ belongs to $M_\Delta^\sharp$. We claim that Ω contains no cycle. To see why this is true, if C is a cycle of Ω with $\{i_0, j_0\} \in E(G)$, then in the matrix $M_\Delta^\sharp(j_0)$ the (i,j)th rows with $\{i, j\} \in E(C)$ are linearly dependent. Thus $\det(M_\Delta^\sharp(j_0)) = 0$, which is impossible. Hence Ω contains no cycle. Since the number of edges of Ω is $q-1$, it follows that Ω is a tree, i.e. a connected graph without cycle. Since Ω has an end vertex, i.e. a vertex which joins with exactly one vertex, it follows that there is a column of $M_\Delta^\sharp$ which contains exactly one nonzero entry. Suppose, say, that the qth column contains exactly one nonzero entry and (k, q)th row of M_Δ appears in $M_\Delta^\sharp$. Then, for each $1 \le j < q$, the monomial $x_{F_q \setminus F_k}$ divides $\det(M_\Delta^\sharp(j))$. Hence $F_q \setminus F_k \subset [n] \setminus F_j$. Thus $(F_q \setminus F_k) \cap F_j = \emptyset$ for all $1 \le j < q$. In other words, $F_j \cap F_q \subset F_k \cap F_q$ for all $1 \le j < q$. Hence F_q is a leaf of Δ and F_k is a branch of F_q. Let Δ' be the simplicial complex on $[n] \setminus (F_q \setminus F_k)$ with $\mathcal{F}(\Delta') = \mathcal{F}(\Delta) \setminus \{F_q\}$. and $M_{\Delta'}^\sharp$ the $(q-2) \times (q-1)$ submatrix of $M_{\Delta'}$ which is obtained by removing the (k, q)th row and qth column from $M_\Delta^\sharp$. Since Δ' is a simplicial complex on $[n] \setminus (F_q \setminus F_k)$ and since $x_{F_q \setminus F_k} x_{[n] \setminus (F_q \setminus F_k)} / x_{F_j} = x_{[n]} / x_{F_j}$ for each $1 \le j < q$, by working with induction on q, it follows that Δ' is a quasi-forest. Hence Δ is a quasi-forest, as desired. $\qquad\qquad\square$

The tree Ω which appears in the proof of "If" part of Lemma 9.2.10 is called a **relation tree** of a quasi-forest Δ. A relation tree of a quasi-forest is, in general, not unique. The inductive technique done in the "Only if" part suggests the way how to find all relation trees of a quasi-forest.

In Example 9.2.5 one has $I = I(\bar\Delta)$, where Δ is the quasi-forest with the facets $F_1 = \{1, 2, 3\}$, $F_2 = \{2, 3, 4\}$, $F_3 = \{3, 4, 5\}$ and $F_4 = \{3, 4, 6\}$. Each of the three 3×4 submatrices of A_I is a relation tree of Δ.

Since $M_\Delta = A_{I(\bar\Delta)}$, by using Lemma 9.2.10 together with Lemma 9.2.4, we now establish our crucial

Corollary 9.2.11. *Let Δ be a simplicial complex on $[n]$ and $I(\bar\Delta)$ the facet ideal of $\bar\Delta$. Then Δ is a quasi-forest if and only if $\operatorname{proj\,dim} I(\bar\Delta) = 1$.*

9.2.4 Dirac's theorem on chordal graphs

It turns out that, by using Alexander duality, the algebraic mechanism behind Dirac's theorem is quite rich.

Theorem 9.2.12. *Given a finite graph G on $[n]$ with $E(G) \ne \binom{[n]}{2}$, the following conditions are equivalent:*

(i) *G is chordal;*
(ii) *$I_{\Delta(G)}$ has a linear resolution;*
(iii) *$\operatorname{reg} I_{\Delta(G)} = 2$;*
(iv) *$\operatorname{proj\,dim} I_{\Delta(G)^\vee} = 1$;*
(v) *$\Delta(G)$ is a quasi-forest;*

(vi) *G is the 1-skeleton of a quasi-forest;*
(vii) *G has a perfect elimination ordering.*

Proof. First, Theorem 9.2.3 says that G is chordal if and only if $I_{\Delta(G)}$ has a linear resolution. Second, the ideal $I_{\Delta(G)} = I(\bar{G})$ is generated by quadratic monomials since $\Delta(G)$ is flag. Thus $I_{\Delta(G)}$ has a linear resolution if and only if $\operatorname{reg} I_{\Delta(G)} = 2$. Moreover, since $\operatorname{reg} I_{\Delta(G)} = \operatorname{proj dim} I_{\Delta(G)^\vee} + 1$ (Proposition 8.1.10), one has $\operatorname{reg} I_{\Delta(G)} = 2$ if and only if $\operatorname{proj dim} I_{\Delta(G)^\vee} = 1$. In addition, since $I_{\Delta(G)^\vee} = I(\Delta(\bar{G}))$ (Lemma 1.5.3), Corollary 9.2.11 guarantees that $\operatorname{proj dim} I_{\Delta(G)^\vee} = 1$ if and only if $\Delta(G)$ is a quasi-forest. On the other hand, as was shown in the proof of Corollary 9.2.9, the clique complex $\Delta(G)$ is a quasi-forest if and only if G is the 1-skeleton of a quasi-forest. Finally, Corollary 9.2.9 says that G is the 1-skeleton of a quasi-forest if and only if G has a perfect elimination ordering. $\square$

9.3 Edge ideals of chordal graphs

In this section we classify all Cohen–Macaulay chordal graphs and in addition show that all chordal graphs are shellable.

9.3.1 Cohen–Macaulay chordal graphs

Let Δ be a simplicial complex.

Theorem 9.3.1. *Let K be a field, and let G be a chordal graph on the vertex set $[n]$. Let $F_1, \ldots, F_m$ be the facets of $\Delta(G)$ which admit a free vertex. Then the following conditions are equivalent:*

 (i) *G is Cohen–Macaulay;*
 (ii) *G is Cohen–Macaulay over K;*
(iii) *G is unmixed;*
(iv) *$[n]$ is the disjoint union of $F_1, \ldots, F_m$.*

For the proof of our main theorem we need the following algebraic fact:

Lemma 9.3.2. *Let R be a Noetherian ring, $S = R[x_1, \ldots, x_n]$ the polynomial ring over R, k an integer with $0 \leq k < n$, and J the ideal*

$$(I_1 x_1, \ldots, I_k x_k, \{x_i x_j\}_{1 \leq i < j \leq n}) \subset S,$$

where $I_1, \ldots, I_k$ are ideals in R. Then the element $x = \sum_{i=1}^n x_i$ is a nonzero divisor on S/J.

Proof. For a subset $T \subset [n]$ we let L_T be the ideal generated by all monomials $x_i x_j$ with $i, j \in T$ and $i < j$, and we set $I_T = \sum_{j \in T} I_j$ and $X_T = (\{x_j\}_{j \in T})$.

One has

$$L_T = \bigcap_{\ell \in T} X_{T \setminus \{\ell\}}.$$

Hence we get

$$J = (I_1 x_1, \ldots, I_k x_k, L_{[n]}) = \bigcap_{T \subset [k]} (I_T, X_{[k] \setminus T}, L_{[n]})$$

$$= \bigcap_{T \subset [k]} (I_T, X_{[k] \setminus T}, L_{[n] \setminus ([k] \setminus T)})$$

$$= \bigcap_{\substack{T \subset [k] \\ \ell \in [n] \setminus ([k] \setminus T)}} (I_T, X_{[k] \setminus T}, X_{([n] \setminus ([k] \setminus T)) \setminus \{\ell\}})$$

$$= \bigcap_{\substack{T \subset [k] \\ \ell \in [n] \setminus ([k] \setminus T)}} (I_T, X_{[n] \setminus \{\ell\}}).$$

Thus in order to prove that x is a nonzero divisor modulo J it suffices to show that x is a nonzero divisor modulo each of the ideals $(I_T, X_{[n] \setminus \{\ell\}})$. To see this, we first pass to the residue class ring modulo I_T, and hence if we replace R by R/I_T it remains to be shown that x is a nonzero divisor on $R[x_1, \ldots, x_n]/(x_1, \ldots, x_{\ell-1}, x_{\ell+1}, \ldots x_n)$. But this is obviously the case. $\qquad\square$

Proof (of Theorem 9.3.1). (i)$\Rightarrow$ (ii) is trivial.

(ii)$\Rightarrow$ (iii) follows from Lemma 9.1.10.

(iii)$\Rightarrow$ (iv): Let G be a unmixed chordal graph on $[n]$. Let $F_1, \ldots, F_m$ be those facets of $\Delta(G)$ which have a free vertex. Fix a free vertex v_i of F_i and set $W = \{v_1, \ldots, v_m\}$. Suppose that $B = [n] \setminus (\bigcup_{i=1}^m F_i) \neq \emptyset$, and write $G|_B$ for the induced subgraph of G on B. If $X \subset B$ is a minimal vertex cover of $G|_B$, then $X \cup ((\bigcup_{i=1}^m F_i) \setminus W)$ is a minimal vertex cover of G, because $\{v_i, b\} \notin E(G)$ for all $1 \leq i \leq m$ and for all $b \in B$. In particular $G|_B$ is unmixed. Since the induced subgraph $G|_B$ is again chordal, by working with induction on the number of vertices, it follows that if $H_1, \ldots, H_s$ are the facets of $\Delta(G|_B)$ with free vertices, then B is the disjoint union $B = \bigcup_{j=1}^s H_j$. Let v'_j be a free vertex of H_j and set $W' = \{v'_1, \ldots, v'_s\}$. Since $((\bigcup_{i=1}^m F_i) \setminus W) \cup (B \setminus W')$ is a minimal vertex cover of G and since G is unmixed, every minimal vertex cover of G consists of $n - (m + s)$ vertices.

We claim that $F_i \cap F_j = \emptyset$ for $i \neq j$. In fact, if, say, $F_1 \cap F_2 \neq \emptyset$ and if $w \in [n]$ satisfies $w \in F_i$ for all $1 \leq i \leq \ell$, where $\ell \geq 2$, and $w \notin F_i$ for all $\ell < i \leq m$, then $Z = (\bigcup_{i=1}^\ell F_i) \setminus \{w, v_{\ell+1}, \ldots, v_m\}$ is a minimal vertex cover of the induced subgraph $G' = G|_{[n] \setminus B}$ on $[n] \setminus B$. Let Y be a minimal vertex cover of G with $Z \subset Y$. Since $Y \cap B$ is a vertex cover of $G|_B$, one has $|Y \cap B| \geq |B| - s$. Moreover, $|Y \cap ([n] \setminus B)| \geq n - |B| - (m - \ell + 1) > n - |B| - m$. Hence $|Y| > n - (m + s)$, a contradiction.

Consequently, a subset Y of $[n]$ is a minimal vertex cover of G if and only if $|Y \cap F_i| = |F_i| - 1$ for all $1 \leq i \leq m$ and $|Y \cap H_j| = |H_j| - 1$ for all $1 \leq j \leq s$.

Now, since $\Delta(G|_B)$ is a quasi-forest (Theorem 9.2.12), one of the facets $H_1, \ldots, H_s$ must be a leaf of $\Delta(G|_B)$. Let, say, H_1 be a leaf of $\Delta(G|_B)$. Let δ and δ', where $\delta \neq \delta'$, be free vertices of H_1 with $\{\delta, a\} \in E(G)$ and $\{\delta', a'\} \in E(G)$, where a and a' belong to $[n] \setminus B$. If $a \neq a'$ and if $\{\delta, a'\} \in E(G)$, then one has either $\{\delta, a'\} \in E(G)$ or $\{\delta', a\} \in E(G)$, because G is chordal and $\{\delta, \delta'\} \in E(G)$. Hence there exists a subset $A \subset [n] \setminus B$ such that

(1) $\{a, b\} \notin E(G)$ for all $a, b \in A$ with $a \neq b$,
(2) for each free vertex δ of H_1, one has $\{\delta, a\} \in E(G)$ for some $a \in A$, and
(3) for each $a \in A$, one has $\{\delta, a\} \in E(G)$ for some free vertex δ of H_1.

In fact, it is obvious that a subset $A \subset [n] \setminus B$ satisfying (2) and (3) exists. If $\{a, a'\} \in E(G)$, $\{\delta, a\} \in E(G)$ and $\{\delta, a'\} \notin E(G)$ for some $a, a' \in A$ with $a \neq a'$ and for a free vertex δ of H_1, then every free vertex δ' of H_1 with $\{\delta', a'\} \in E(G)$ must satisfy $\{\delta', a\} \in E(G)$. Hence $A \setminus \{a'\}$ satisfies (2) and (3). Repeating this technique yields a subset $A \subset [n] \setminus B$ satisfying (1), (2) and (3), as required.

If $s > 1$, then H_1 has a branch. Let $w_0 \notin H_1$ be a vertex belonging to a branch of the leaf H_1 of $\Delta(G|_B)$. Thus $\{\xi, w_0\} \in E(G)$ for all nonfree vertices ξ of H_1. We claim that either $\{a, w_0\} \notin E(G)$ for all $a \in A$, or one has $a \in A$ with $\{a, \xi\} \in E(G)$ for every nonfree vertices ξ of H_1. To see why this is true, if $\{a, w_0\} \in E(G)$ and $\{\delta, a\} \in E(G)$ for some $a \in A$ and for some free vertex δ of H_1, then one has a cycle (a, δ, ξ, w_0) of length four for every nonfree vertex ξ of H_1. Since $\{\delta, w_0\} \notin E(G)$, one has $\{a, \xi\} \in E(G)$.

Let X be a minimal vertex cover of G such that $X \subset [n] \setminus (A \cup \{w_0\})$ (resp. $X \subset [n] \setminus A$) if $\{a, w_0\} \notin E(G)$ for all $a \in A$ (resp. if one has $a \in A$ with $\{a, \xi\} \in E(G)$ for every nonfree vertices ξ of H_1.) Then, for each vertex γ of H_1, there is $w \notin X$ with $\{\gamma, w\} \in E(G)$. Hence $H_1 \subset X$, in contrast to our considerations before. This contradiction guarantees that $B = \emptyset$. Hence $[n]$ is the disjoint union $[n] = \bigcup_{i=1}^m F_i$, as required.

Finally suppose that $s = 1$. Then H_1 is the only facet of $\Delta(G|_B)$. Then $X = \bigcup_{i=1}^m (F_i \setminus v_i)$ is a minimal free vertex cover G with $H_1 \subset X$, a contradiction.

(iv) $\Rightarrow$ (iii): Let $F_1, \ldots, F_m$ denote the facets of $\Delta(G)$ with free vertices and, for each $1 \leq i \leq m$, write F_i for the set of vertices of F_i. Given a minimal vertex cover $X \subset [n]$ of G, one has $|X \cap F_i| \geq |F_i| - 1$ for all i since F_i is a clique of G. If, however, for some i, one has $|X \cap F_i| = |F_i|$, i.e. $F_i \subset X$, then $X \setminus \{v_i\}$ is a vertex cover of G for any free vertex v_i of F_i. This contradicts the fact that X is a minimal vertex cover of G. Thus $|X \cap F_i| = |F_i| - 1$ for all i. Since $[n]$ is the disjoint union $[n] = \bigcup_{i=1}^m F_i$, it follows that $|X| = n - m$ and G is unmixed, as desired.

(iii) and (iv) $\Rightarrow$ (i): We know that G is unmixed. Moreover, if $v_i \in F_i$ is a free vertex, then $[n] \setminus \{v_1, \ldots, v_m\}$ is a minimal vertex cover of G. In particular it follows that $\dim S/I(G) = m$.

For $i = 1, \ldots, m$, we set $y_i = \sum_{j \in F_i} x_j$. We will show that $y_1, \ldots, y_m$ is a regular sequence on $S/I(G)$. This then yields that G is Cohen–Macaulay.

Let $F_i = \{i_1, \ldots, i_k\}$, and assume that $i_{\ell+1}, \ldots, i_k$ are the free vertices of F_i. Let $G' \subset G$ be the induced subgraph of G on the vertex set $[n] \setminus \{i_1, \ldots, i_k\}$. Then $I(G) = (I(G'), J_1 x_{i_1}, J_2 x_{i_2}, \ldots, J_\ell x_{i_\ell}, J)$, where $J_j = (\{x_r : \{r, i_j\} \in E(G)\})$ for $j = 1, \ldots, \ell$, and where $J = (\{x_{i_r} x_{i_s} : 1 \leq r < s \leq k\})$.

Since $[n]$ is the disjoint union of $F_1, \ldots, F_m$ it follows that all generators of the ideal $(I(G'), y_1, \ldots, y_{i-1})$ belong to $K[\{x_i\}_{i \in [n] \setminus F_i}]$. Thus if we set

$$R = K[\{x_i\}_{i \in [n] \setminus F_i}] / (I(G'), y_1, \ldots, y_{i-1}),$$

then $(S/I(G))/(y_1, \ldots, y_{i-1})(S/I(G))$ is isomorphic to

$$R[x_{i_1}, \ldots, x_{i_k}] / (I_1 x_{i_1}, \ldots, I_\ell x_{i_\ell}, \{x_{i_r} x_{i_s} : 1 \leq r < s \leq k\}),$$

where for each j, the ideal I_j is the image of J_j under the residue class map onto R. Therefore it follows from Lemma 9.3.2 that the element y_i is regular on $(S/I(G))/(y_1, \ldots, y_{i-1})(S/I(G))$. $\square$

Corollary 9.3.3. *Let G be a chordal graph, and let $F_1, \ldots, F_m$ be the facets of $\Delta(G)$ which have a free vertex. Let i_j be a free vertex of F_j for $j = 1, \ldots, m$, and let G' be the induced subgraph of G on the vertex set $[n] \setminus \{i_1, \ldots, i_m\}$. Then*

(a) *the type of G is the number of maximal independent subsets of G';*
(b) *G is Gorenstein if and only if G is a disjoint union of edges.*

Proof. (a) Let $F \subset [n]$ and $S = K[x_1, \ldots, x_n]$. If J is the ideal generated by the set of monomials $\{x_i x_j : i, j \in F \text{ and } i < j\}$ and if $x = \sum_{i \in F} x_i$, then for any $i \in F$ one has

$$(S/J)/x(S/J) \cong S_i/(\{x_j : j \in F, j \neq i\})^2,$$

where $S_i = K[x_1, \ldots, x_{i-1}, x_{i+1}, \ldots, x_n]$.

Thus if we factor by a maximal regular sequence as in the proof of Theorem 9.3.1 we obtain a 0-dimensional ring of the form

$$A = T/(P_1^2, \ldots, P_m^2, I(G'')).$$

Here $P_j = (\{x_k : k \in F_j, \ k \neq i_j\})$, G'' is the subgraph of G consisting of all edges which do not belong to any F_j, and T is the polynomial ring over K in the set of variables $X = \{x_k : k \in [n], k \neq i_j \text{ for all } j = 1, \ldots, m\}$. It is obvious that A is obtained from the polynomial ring T by factoring out the squares of all variables of T and all $x_i x_j$ with $\{i, j\} \in E(G')$. Therefore A has a K-basis of squarefree monomials corresponding to the independent subsets of G', and the socle of A is generated as a K-vector space by the monomials corresponding to the maximal independent subsets of G'.

(b) If G is a disjoint union of edges, then $I(G)$ is a complete intersection, and hence Gorenstein.

Conversely, suppose that G is Gorenstein. Then A is Gorenstein. Since A is a 0-dimensional ring with monomial relations, A is Gorenstein if and only if A is a complete intersection, see A.6.5. This is the case only if $E(G') = \emptyset$, in which case G is a disjoint union of edges. $\qquad\square$

9.3.2 Chordal graphs are shellable

In this section we present a remarkable extension of the fact that a chordal graph is Cohen–Macaulay if and only if it is unmixed, as was shown in Theorem 9.3.1.

Theorem 9.3.4. *Every chordal graph is shellable.*

Proof. Let G be a chordal graph on $[n]$. We prove the theorem by induction on n. First we observe that if each connected component of G is shellable, then G is shellable. If fact, if Δ_1 and Δ_2 are shellable simplicial complexes on disjoint sets of vertices, and if $F_1, \ldots, F_r$ is a shelling of Δ_1 and $H_1, \ldots, H_s$ is a shelling of Δ_2, then

$$F_1 \cup H_1, \ldots, F_1 \cup H_s, F_2 \cup H_1, \ldots, F_2 \cup H_s, \cdots, F_r \cup H_1, \ldots, F_r \cup H_s$$

is a shelling of $\Delta_1 \cup \Delta_2$. Thus we may assume that G is connected. Then $\Delta(G)$ is a quasi-tree (Theorem 9.2.12), say, with leaf order $H_1, \ldots, H_m$ where $m > 1$. Let x be a free vertex of H_m. Then $G_{\{x_1\} \cup N_G(x_1)} = H_m$. We may assume that $V(H_m) = \{x_1, \ldots, x_r\}$. By our induction hypothesis, the chordal subgraphs $G_i = G \setminus (\{x_i\} \cup N_G(x_i))$ $(i = 1, \ldots, r)$ are shellable. Let $F_{i1}, \ldots, F_{is_i}$ be a shelling of G_i for $i = 1, \ldots, r$.

We have to show that $\Delta(\bar{G})$ is shellable. Since G_{H_m} is a complete subgraph of G, and since H_m has a free vertex, it follows that each facet of $\Delta(\bar{G})$ intersects H_m in exactly one vertex. By using (9.2) it follows that

$$F_{11} \cup \{x_1\}, \ldots, F_{1s_1} \cup \{x_1\}, F_{21} \cup \{x_2\}, \ldots, F_{2s_2} \cup \{x_2\}, \ldots$$
$$\ldots, F_{r1} \cup \{x_r\}, \ldots, F_{rs_r} \cup \{x_r\}$$

is the complete list of the facets of $\Delta(\bar{G})$. We claim that this is a shelling order of $\Delta(\bar{G})$. Let $F' < F$ be two facets of $\Delta(\bar{G})$. Suppose $F' = F_{ik} \cup \{x_i\}$ and $F = F_{j\ell} \cup \{x_j\}$ where $i < j$. Notice that $F_{j\ell} \cup \{x_1\}$ is an independent set of G because $F_{j\ell} \cap H_m = \emptyset$. Thus $F_{j\ell} \cup \{x_1\}$ is a face of $\Delta(\bar{G})$, hence $F_{j\ell} \cup \{x_1\} \subset F_{1t} \cup \{x_1\}$ for some $1 \leq t \leq s_1$. Set $F'' = F_{1t} \cup \{x_1\}$. Then $x_j \in F \setminus F'$, $F \setminus F'' = \{x_j\}$ and $F'' < F$.

If $F' = F_{ik} \cup \{x_i\}$ and $F = F_{i\ell} \cup \{x_i\}$, then the shelling property for these facets follows from the shellability of the G_i. $\qquad\square$

Problems

9.1. Give an example of a finite graph G such that G is isomorphic to $\bar{G}$, i.e. G coincides with $\bar{G}$ after relabelling the vertices of G.

9.2. (a) Find a simplicial complex which cannot be flag.
(b) What is the clique complex of the complete graph on $[n]$?
(c) What is the clique complex of the cycle on $[n]$ of length n?

9.3. (a) Draw the finite graph G on $[2n]$ with the edge ideal

$$(x_1x_2, x_3x_4, \ldots, x_{n-1}x_n).$$

(b) Draw the finite graph G on $[5]$ with the edge ideal

$$(x_1x_2, x_1x_3, x_1x_5, x_1x_6, x_2x_3, x_2x_4, x_2x_6, x_3x_4, x_3x_5).$$

9.4. Give an example of a Gorenstein graph on $[n]$.

9.5. Find the vertex cover ideal of each of the finite graphs of (a) and (b) of Problem 9.3.

9.6. Let $P = \{p_1, p_2, p_3, p_4\}$ be the finite poset with $p_1 < p_3, p_1 < p_4$ and $p_2 < p_4$. Find all poset ideals of P and compute H_P, Δ_P together with $\Delta_P^\vee$.

9.7. Find an unmixed graph which is not Cohen–Macaulay.

9.8. By using Theorem 9.1.13 classify all Cohen–Macaulay trees.

9.9. Let G be the bipartite graph obtained from Figure 9.3 by adding the edge $\{x_3, y_2\}$. Is G Cohen–Macaulay?

9.10. Let G be the chordal graph with the edges

$$\{1,2\}, \{1,3\}, \{1,5\}, \{2,3\}, \{3,4\}, \{3,5\}, \{4,5\}, \{5,6\}.$$

(a) Find a perfect elimination ordering of G.
(b) Find a leaf order of the quasi-tree $\Delta(G)$.
(c) Find a relation tree of $\Delta(G)$.
(d) Compute $\Delta(G)^\vee$ and $I_{\Delta(G)^\vee}$.
(e) Compute the minimal graded free resolution of $I_{\Delta(G)^\vee}$.

9.11. (a) Is the chordal graph of Problem 9.10 Cohen–Macaulay?
(b) Find the vertex cover ideal of the chordal graph of Problem 9.10.

9.12. Draw all Cohen–Macaulay chordal graphs with at most 5 vertices.

Notes

An important characterization of chordal graphs which has many applications was given by Dirac [Dir61] in 1961. The algebraic aspects of Dirac's theorem which are summarized in Theorem 9.2.12 appeared in [HHZ04b]. It turned out that the Hilbert–Burch Theorem [Bur68] plays an important role in this context. In commutative algebra chordal graphs first appeared in Fröberg's theorem [Fro90] in which he characterized squarefree monomial ideals with 2-linear resolution.

Villarreal [Vil90] was the first to study edge ideals of a finite graph systematically. One of the central problems in this theory is to classify all Cohen–Macaulay finite graphs. Such a classification is given for chordal graphs [HHZ06] and for bipartite graphs [HH05]. The results presented here, which assert that a bipartite graph is sequentially Cohen–Macaulay if and only it is shellable, and that each chordal graph is shellable, are due to [VV08]. The fact that a bipartite graph is Cohen–Macaulay if and only it is pure shellable has been shown already in [EV97], and that a chordal graph is sequentially Cohen–Macaulay was first shown in [FT07]. Our characterization of unmixed bipartite graphs is taken from [HHO09]. Another characterization of unmixed bipartite graphs is given in [Vil07].

10

Powers of monomial ideals

We collect several topics on powers of monomial ideals, including powers of
monomial ideals with linear resolution as well as depth and normality of powers of monomial ideals. One of the main results presented in this chapter says
that if a monomial ideal generated in degree 2 has a linear resolution, then
all powers of the ideal have a linear resolution. In order to prove this and
other results some techniques on toric ideals will be required. It is shown that
the depth of the powers of a graded ideal is constant for all high powers of
the ideal. For special classes of monomial ideals we compute their limit depth
explicitly. The limit depth of normally torsionfree squarefree monomial ideals
can be expressed in terms of their analytic spread. It is shown that the facet
ideal of a simplicial complex is normally torsionfree if and only if it is a Mengerian simplicial complex. In particular, one obtains the precise limit depth
for bipartite graphs as well as for simplicial forests.

10.1 Toric ideals and Rees algebras

10.1.1 Toric ideals

Since the Gröbner basis techniques on toric ideals will be required to develop
the theory of powers of monomial ideals, we quickly discuss fundamental materials on toric ideals together with typical examples.

A **monomial configuration** of $S = K[x_1, \ldots, x_n]$ is a finite set $\mathcal{A} = \{u_1, \ldots, u_m\}$ of monomials of S.

The **toric ring** of $\mathcal{A}$ is the subring $K[\mathcal{A}] = K[u_1, \ldots, u_m]$ of S. Let $R = [t_1, \ldots, t_m]$ denote the polynomial ring in m variables over K and define the surjective homomorphism

$$\pi : R \to K[\mathcal{A}]$$

by setting $\pi(t_i) = u_i$ for $i = 1, \ldots, m$. The **toric ideal** of $\mathcal{A}$ is the kernel of
π. In other words, the toric ideal of $\mathcal{A}$ is the defining ideal of the toric ring
$K[\mathcal{A}]$. We write $I_\mathcal{A}$ for the toric ideal of $\mathcal{A}$. Every toric ideal is a prime ideal.

J. Herzog, T. Hibi, *Monomial Ideals*, Graduate Texts in Mathematics 260,   183
DOI 10.1007/978-0-85729-106-6_10, © Springer-Verlag London Limited 2011

A **binomial** of R is a polynomial of the form $u - v$, where u and v are monomials of R. A **binomial ideal** is an ideal which is generated by binomials.

Proposition 10.1.1. *The toric ideal I_A of A is spanned by those binomials $u - v$ of R with $\pi(u) = \pi(v)$. In particular I_A is a binomial ideal.*

Proof. Let $f = c_1 u_1 + \cdots + c_r u_r$ be a polynomial belonging to I_A, where u_i is a monomial of R with $c_i \in K$. Suppose that $\pi(u_1) = \cdots = \pi(u_k)$ and $\pi(u_1) \neq \pi(u_\ell)$ for $k < \ell \leq r$. The monomials belonging to the toric ring $K[A]$ is a K-basis of $K[A]$. Thus, since $\pi(f) = 0$, it follows that $c_1 + \cdots + c_k = 0$. Hence $c_1 = -(c_2 + \cdots + c_k)$. Thus

$$c_1 u_1 + \cdots + c_r u_r = c_2(u_2 - u_1) + \cdots + c_k(u_k - u_1),$$

where $\pi(u_i) = \pi(u_1)$ for $i = 2, \ldots, k$. Since $f - (c_1 u_1 + \cdots + c_r u_r) \in I_A$, working with induction on r yields the desired result. □

A binomial $f = u - v$ belonging to I_A is called **primitive** if there exists no binomial $g = u' - v' \in I_A$, $g \neq f$, such that u' divides u and v' divides v. Every primitive binomial is irreducible.

Proposition 10.1.2. *A reduced Gröbner basis of I_A consists of primitive binomials.*

Proof. If f and g are binomials, then their S-polynomial $S(f, g)$ is a binomial. If $f_1, \ldots, f_s$ and g are binomials, then every remainder of g with respect to $f_1, \ldots, f_s$ is a binomial. Since I_A is generated by binomials, Buchberger algorithm yields a Gröbner basis of I_A consisting of binomials. It then follows from the discussion appearing in the first half of the proof of Theorem 2.2.7 that a reduced Gröbner basis of I_A consists of binomials.

Let $\mathcal{G}$ denote the reduced Gröbner basis of I_A with respect to a monomial order $<$. Let $f = u - v$ be a binomial belonging to $\mathcal{G}$ with u its initial monomial and suppose that $g = u' - v' \in I_A$ with $f \neq g$ is a binomial for which u' divides u and v' divides v. Since u belongs to the minimal system of the monomial generators of $\text{in}_<(I_A)$, in the case that u' is the initial monomial of g one has $u = u'$. Thus $f = g$; a contradiction. Thus v' is the initial monomial of g. Since v' divides v, it follows that v belongs to $\text{in}_<(I_A)$. Hence there is a binomial $h = u'' - v'' \in \mathcal{G}$ with u'' its initial monomial such that u'' divides v. This is impossible, because $\mathcal{G}$ is a reduced Gröbner basis. □

Two typical examples of toric ideals arising from combinatorics are now studied.

Let $P = \{p_1, \ldots, p_n\}$ be a finite poset and write $K[\mathbf{x}, \mathbf{y}]$ for the polynomial ring $K[x_1, \ldots, x_n, y_1, \ldots, y_n]$ in $2n$ variables over a field K. Recall that we associate each poset ideal α of P with the squarefree monomial $u_\alpha = (\prod_{p_i \in \alpha} x_i)(\prod_{p_j \in P \setminus \alpha} y_j)$ of $K[\mathbf{x}, \mathbf{y}]$. Let $A_P = \{u_\alpha : \alpha \in \mathcal{J}(P)\}$, where

$\mathcal{J}(P)$ is the set of poset ideals of P, and $K[\mathcal{A}_P]$ the toric ring of $\mathcal{A}_P$. Let $K[\mathbf{t}] = K[\{t_\alpha : \alpha \in \mathcal{J}(P)\}]$ denote the polynomial ring in $|\mathcal{J}(P)|$ variables over K and $I_{\mathcal{A}_P}$ the toric ideal of $\mathcal{A}_P$. Thus $I_{\mathcal{A}_P}$ is the kernel of the surjective homomorphism $\pi : K[\mathbf{t}] \rightarrow K[\mathcal{A}_P]$ defined by setting $\pi(t_\alpha) = u_\alpha$ for each $\alpha \in \mathcal{J}(P)$. Fix a total order $<$ of the variables of $K[\mathbf{t}]$ with the property that $t_\alpha \leq t_\beta$ if $\alpha \subset \beta$ and write $<_{\mathrm{rev}}$ for the reverse lexicographic order on $K[\mathbf{t}]$ induced by the ordering $<$.

Theorem 10.1.3. *The reduced Gröbner basis of the toric ideal $I_{\mathcal{A}_P}$ with respect to $<_{\mathrm{rev}}$ consists of those binomials*

$$t_\alpha t_\beta - t_{\alpha \cap \beta} t_{\alpha \cup \beta}$$

with neither $\alpha \subset \beta$ nor $\beta \subset \alpha$.

Proof. If α and β are poset ideals of P, then each of $\alpha \cap \beta$ and $\alpha \cup \beta$ is a poset ideal of P. It is clear that the binomial $t_\alpha t_\beta - t_{\alpha \cap \beta} t_{\alpha \cup \beta}$ belongs to $I_{\mathcal{A}_P}$ and, in the case of neither $\alpha \subset \beta$ nor $\beta \subset \alpha$, its initial monomial is $t_\alpha t_\beta$.

Once we show that the set of those binomials $t_\alpha t_\beta - t_{\alpha \cap \beta} t_{\alpha \cup \beta}$ with neither $\alpha \subset \beta$ nor $\beta \subset \alpha$ is a Gröbner basis of $I_{\mathcal{A}_P}$ with respect to $<_{\mathrm{rev}}$, it follows immediately that such a set of binomials is the reduced Gröbner basis of $I_{\mathcal{A}_P}$ with respect to $<_{\mathrm{rev}}$.

Let $\mathcal{G}$ denote the reduced Gröbner basis of $I_{\mathcal{A}_P}$ with respect to $<_{\mathrm{rev}}$. Let $\prod_{j=1}^q t_{\alpha_j} - \prod_{j=1}^q t_{\beta_j}$ be a binomial belonging to $\mathcal{G}$ with $\prod_{j=1}^q t_{\alpha_j}$ its initial monomial. What we must prove is that there are k and ℓ such that one has neither $\alpha_k \subset \alpha_\ell$ nor $\alpha_\ell \subset \alpha_k$. Suppose on the contrary that $\alpha_1 \subset \alpha_2 \subset \cdots \subset \alpha_q$. Then $(\prod_{p_i \in \alpha_1} x_i)^q$ divides $\pi(\prod_{i=1}^q t_{\alpha_i})$. Hence $(\prod_{p_i \in \alpha_1} x_i)^q$ must divide $\pi(\prod_{i=1}^q t_{\beta_i})$. Hence $\alpha_1 \subset \beta_j$ for all $1 \leq j \leq q$. Since $\alpha_1 \neq \beta_j$ for all $1 \leq j \leq q$, it follows that $t_{\alpha_1} <_{\mathrm{rev}} t_{\beta_j}$ for all $1 \leq j \leq q$. Hence $\prod_{j=1}^q t_{\alpha_j} <_{\mathrm{rev}} \prod_{j=1}^q t_{\beta_j}$. This contradicts the fact that $\prod_{j=1}^q t_{\alpha_j}$ is the initial monomial of $\prod_{j=1}^q t_{\alpha_j} - \prod_{j=1}^q t_{\beta_j}$. $\square$

Let G be a finite graph on $[n]$ with no loop and no multiple edge. Let $E(G) = \{e_1, \ldots, e_m\}$ denote the set of edges G. For each edge $e = \{i, j\}$ of G we associate the quadratic monomial $x_e = x_i x_j$ of S. Let $\mathcal{A}_G = \{x_{e_1}, \ldots, x_{e_m}\}$ and $K[\mathcal{A}_G]$ its toric ring. Let $K[\mathbf{t}] = K[t_1, \ldots, t_m]$ denote the polynomial ring in m variables over K and $I_{\mathcal{A}_G} \subset K[\mathbf{t}]$ the toric ideal of $\mathcal{A}_G$. Thus $I_{\mathcal{A}_G}$ is the kernel of the surjective homomorphism $\pi : K[\mathbf{t}] \rightarrow K[\mathcal{A}_G]$ defined by setting $\pi(t_i) = x_{e_i}$ for $i = 1, \ldots, m$.

Recall that a walk of G of length q is a subgraph W of G such that $E(W) = \{\{v_0, v_1\}, \{v_1, v_2\}, \ldots, \{v_{q-1}, v_q\}\}$, where $v_0, v_1, \ldots, v_q$ are vertices of G. An **even** walk is a walk of even length. A walk W with $E(W) = \{\{v_0, v_1\}, \ldots, \{v_{q-1}, v_q\}\}$ is called **closed** if $v_0 = v_q$.

Given an even closed walk W of G with

$$E(W) = \{\{v_0, v_1\}, \{v_1, v_2\}, \ldots, \{v_{2q-2}, v_{2q-1}\}, \{v_{2q-1}, v_0\}\} \qquad (10.1)$$

of length $2q$, we introduce the binomial

$$f_W = \prod_{j=1}^{q} t_{e_{i_{2j-1}}} - \prod_{j=1}^{q} t_{e_{i_{2j}}}$$

of $K[\mathbf{t}]$, where $e_{i_j} = \{v_{j-1}, v_j\}$ for $1 \leq j < 2q$ and where $e_{i_{2q}} = \{v_{2q-1}, v_0\}$. It is clear that the binomial f_W belongs to I_A.

An even closed walk W' of G with

$$E(W') = \{\{u_0, u_1\}, \{u_1, v_2\}, \ldots, \{u_{2p-2}, u_{2p-1}\}, \{u_{2p-1}, u_0\}\}$$

is called a **subwalk** of W, where W is an even closed walk (10.1) of G, if, for each $1 \leq k \leq p$, there exist $1 \leq \ell \leq q$ and $1 \leq \ell' \leq q$ with

$$\{u_{2k-2}, u_{2k-1}\} = \{v_{2\ell-2}, v_{2\ell-1}\}, \quad \{u_{2k-1}, u_{2k}\} = \{v_{2\ell'-1}, v_{2\ell'}\}.$$

An even closed walk W of G is called **primitive** if no even closed walk W' of G with $W \neq W'$ is a subwalk of W.

Every even cycle is a primitive even closed walk. Every primitive even closed walk of a bipartite graph G is an even cycle.

Lemma 10.1.4. *If $f \in I_{A_G}$ is a primitive binomial, then there is a primitive even closed walk W of G with $f = f_W$.*

Proof. If the binomial f_W arising from an even closed walk W of G is primitive, then clearly W is a primitive even closed walk of G. Thus what we must prove is that, for every primitive binomial f of I_{A_G}, there is an even closed walk W of G with $f = f_W$.

Let $f = \prod_{k=1}^{q} t_{i_k} - \prod_{k=1}^{q} t_{j_k}$ be a primitive binomial of I_{A_G}. Let, say, $\pi(t_{i_1}) = x_1 x_2$. Since $\pi(\prod_{k=1}^{q} t_{i_k}) = \pi(\prod_{k=1}^{q} t_{j_k})$, one has $\pi(t_{j_m}) = x_2 x_r$ for some $1 \leq m \leq q$ with $r \neq 1$. Say $m = 1$ and $r = 3$, i.e. $\pi(t_{j_1}) = t_2 t_3$. Then $\pi(t_{i_\ell}) = t_3 t_s$ for some $2 \leq \ell \leq q$ with $s \neq 2$. Repeated application of such procedure enables us to find an even closed walk W of G such that f_W is of the form $f_W = f_W^{(+)} - f_W^{(-)}$, where $f_W^{(+)}$ is a monomial which divides $\prod_{k=1}^{q} t_{i_k}$ and where $f_W^{(-)}$ is a monomial which divides $\prod_{k=1}^{q} t_{j_k}$. Since f is primitive, it follows that $f = f_W$, as desired. □

Corollary 10.1.5. *A reduced Gröbner basis of I_{A_G} consists of binomials of the form f_W, where W is a primitive even closed walk of G.*

10.1.2 Rees algebras and the x-condition

Let I be a graded ideal of $S = K[x_1, \ldots, x_n]$ generated by homogeneous polynomials $f_1, \ldots, f_m$ with $\deg f_1 = \cdots = \deg f_m$. Let t be a variable over S. The graded subalgebra

$$\mathcal{R}(I) = \bigoplus_{j=0}^{\infty} I^j t^j = S[f_1 t, \ldots, f_m t]$$

of $S[t]$ is called the **Rees algebra** of I. We regard $\mathcal{R}(I)$ to be a bigraded algebra with $\deg(x_i) = (1,0)$ for $i = 1, \ldots, n$ and $\deg(f_i t) = (0,1)$ for $i = 1, \ldots, m$.

Let $R = S[y_1, \ldots, y_m]$ be the polynomial ring over S in the variables $y_1, \ldots, y_m$ and regard R to be a bigraded algebra with $\deg(x_i) = (1,0)$ for $i = 1, \ldots, n$ and $\deg(y_j) = (0,1)$ for $j = 1, \ldots, m$. Then a natural surjective homomorphism of bigraded K-algebras

$$\varphi : R \to \mathcal{R}(I)$$

arises by setting $\varphi(x_i) = x_i$ for $i = 1, \ldots, n$ and $\varphi(y_j) = f_j t$ for $j = 1, \ldots, m$.

If the bigraded minimal free R-resolution of $\mathcal{R}(I)$ is given by

$$\mathbb{F}: 0 \longrightarrow F_p \longrightarrow \cdots \longrightarrow F_1 \longrightarrow F_0 \longrightarrow \mathcal{R}(I) \longrightarrow 0,$$

where $F_i = \bigoplus_j R(-a_{ij}, -b_{ij})$ for $i = 0, \ldots, p$, then the x-**regularity** of $\mathcal{R}(I)$ is defined to be the nonnegative integer

$$\mathrm{reg}_x(\mathcal{R}(I)) = \max_{i,j}\{a_{ij} - i\}.$$

Proposition 10.1.6. *Suppose that $I \subset S$ is a graded ideal generated in degree d. Then*

$$\mathrm{reg}(I^k) \le kd + \mathrm{reg}_x(\mathcal{R}(I)).$$

In particular if $\mathrm{reg}_x(\mathcal{R}(I)) = 0$, then each power of I has a linear resolution.

Proof. The bigraded minimal free R-resolution $\mathbb{F}$ of $\mathcal{R}(I)$ gives the exact sequence

$$0 \to (F_p)_{(*,k)} \longrightarrow \cdots \longrightarrow (F_1)_{(*,k)} \longrightarrow (F_0)_{(*,k)} \longrightarrow \mathcal{R}(I)_{(*,k)} \to 0 \quad (10.2)$$

of graded S-modules for all k. Since $\mathcal{R}(I)_{(*,k)} = I^k(dk)$ and $R(-a, -b)_{(*,k)} \cong \bigoplus_{|u|=k-b} S(-a)y^u$, the exact sequence (10.2) is a (possibly nonminimal) graded free S-resolution of $I^k(dk)$. Thus $\mathrm{reg}(I^k(dk)) \le \mathrm{reg}_x(\mathcal{R}(I))$ so $\mathrm{reg}(I^k) \le kd + \mathrm{reg}_x(\mathcal{R}(I))$, as desired. $\square$

We say that I satisfies the x-**condition** if $\mathrm{reg}_x(R(I)) = 0$.

Corollary 10.1.7. *Let $I \subset S$ be a graded ideal generated in one degree and $\mathcal{R}(I) = R/P$. Suppose that there exists a monomial order $<$ on R such that the defining ideal P of $\mathcal{R}(I)$ has a Gröbner basis $\mathcal{G}$ whose elements are at most linear in the variables $x_1, \ldots, x_n$, i.e. $\deg_x(f) \le 1$ for all $f \in \mathcal{G}$. Then each power of I has a linear resolution.*

Proof. The initial ideal $\mathrm{in}_<(P)$ is generated by monomials $u_1, \ldots, u_m$ with each $\deg_x(u_j) \leq 1$. Let $\mathbb{T}$ be the Taylor resolution of $\mathrm{in}_<(P)$; see Section 7.1. Recall that the module T_i has the basis e_F with $F = \{j_1 < j_2 < \cdots < j_i\} \subset [m]$ and that each basis element e_F has the multidegree (a_F, b_F), where $x^{a_F} y^{b_F} = \mathrm{lcm}\{u_{j_1}, \ldots, u_{j_i}\}$. Thus $\deg_x(e_F) \leq i$ for all $e_F \in T_i$. Since the shifts of $\mathbb{T}$ bound the shifts of a minimal multigraded resolution of $\mathrm{in}_<(P)$, it follows that $\mathrm{reg}_x(R/\mathrm{in}_<(P)) = 0$. Since $\mathrm{reg}_x(R/P) \leq \mathrm{reg}_x(R/\mathrm{in}_<(P))$ (which is the bigraded version of Theorem 3.3.4(c)), one has $\mathrm{reg}_x(R/P) = 0$. $\qquad\square$

Together with Corollary 10.1.7 the following result is also useful to show that a given ideal I has the property that all powers of I have linear resolution.

Corollary 10.1.8. *Let $I \subset S$ be a graded ideal generated in one degree and $\mathcal{R}(I) = R/P$. Suppose that there exists a monomial order $<$ on R such that the defining ideal P of $\mathcal{R}(I)$ has a Gröbner basis $\mathcal{G}$ consisting of polynomials of degree 2. Then each power of I has a linear resolution.*

Proof. Let $\mathcal{G} = \{g_1, \ldots, g_s\}$ be a Gröbner basis with $\deg g_i = 2$ for all i. Since the defining ideal of $\mathcal{R}(I)$ is bihomogeneous, we may assume that each g_i is bihomogeneous. Suppose $u \in \mathrm{supp}(g_i)$ for some $u \in S$. Then $\deg g_i = (2, 0)$, and it follows that $g_i \in S$, which is impossible. Therefore for each $u \in \mathrm{supp}(g_i)$ we have $\deg_x(u) \leq 1$, as desired. $\qquad\square$

In case that $I \subset S$ be a monomial ideal generated in one degree, there is a refinement of the x-condition which guarantees that all powers of I have linear quotients. As usual, let $G(I)$ be the minimal system of monomial generators of I. Then the Rees algebra of I is of the form

$$\mathcal{R}(I) = K[x_1, \ldots, x_n, \{ut\}_{u \in G(I)}] \subset S[t].$$

Let $S = K[x_1, \ldots, x_n, \{y_u\}_{u \in G(I)}]$ denote the polynomial ring in $n + |G(I)|$ variables over K with each $\deg x_i = \deg y_u = 1$. The toric ideal of $\mathcal{R}(I)$ is the kernel $J_{\mathcal{R}(I)}$ of the surjective homomorphism $\pi : R \to \mathcal{R}(I)$ defined by setting $\pi(x_i) = x_i$ for all $1 \leq i \leq n$ and $\pi(y_u) = ut$ for all $u \in G(I)$.

The important point in the following discussions is that we will choose a special monomial order to make sure that we are able to control the linear quotients of the powers of I.

Let $<_{\mathrm{lex}}$ denote the lexicographic order on S induced by $x_1 > x_2 > \cdots > x_n$. Fix an arbitrary monomial order $<^\#$ on $K[\{y_u\}_{u \in G(I)}]$. We then introduce the new monomial order $<^\#_{\mathrm{lex}}$ on R defined as follows: For monomials $(\prod_{i=1}^n x_i^{a_i})(\prod_{u \in G(I)} y_u^{a_u})$ and $(\prod_{i=1}^n x_i^{b_i})(\prod_{u \in G(I)} y_u^{b_u})$ belonging to R, one has

$$(\prod_{i=1}^n x_i^{a_i})(\prod_{u \in G(I)} y_u^{a_u}) <^\#_{\mathrm{lex}} (\prod_{i=1}^n x_i^{b_i})(\prod_{u \in G(I)} y_u^{b_u})$$

if either

(i) $\prod_{u \in G(I)} y_u^{a_u} <^{\#} \prod_{u \in G(I)} y_u^{b_u}$ or

(ii) $\prod_{u \in G(I)} y_u^{a_u} = \prod_{u \in G(I)} y_u^{b_u}$ and $\prod_{i=1}^{n} x_i^{a_i} <_{\text{lex}} \prod_{i=1}^{n} x_i^{b_i}$.

Let $\mathcal{G}(J_{\mathcal{R}(I)})$ denote the reduced Gröbner basis of $J_{\mathcal{R}(I)}$ with respect to $<^{\#}_{\text{lex}}$.

Theorem 10.1.9. *Suppose that I satisfies the x-condition with respect to $<^{\#}_{\text{lex}}$. Then each power of I has linear quotients.*

Proof. Fix $k \geq 1$. Each $w \in G(I^k)$ has a unique expression, called the *standard expression*, of the form $w = u_1 \cdots u_k$ with each $u_i \in G(I)$ such that $y_{u_1} \cdots y_{u_k}$ is a standard monomial of R with respect to $<^{\#}$, that is, a monomial which does not belong to the initial ideal of $J_{\mathcal{R}(I)}$. Let w^* denote the standard monomial $y_{u_1} \cdots y_{u_k}$. Let $G(I^k) = \{w_1, \ldots, w_s\}$ with $w_1^* <^{\#} \cdots <^{\#} w_s^*$.

We claim that I^k has linear quotients with the ordering $w_1, \ldots, w_s$ of its generators. Let f be a monomial belonging to the colon ideal $(w_1, \ldots, w_{j-1}) : w_j$. Thus $fw_j = gw_i$ for some $i < j$ and for some monomial g. Let $w_j = u_1 \cdots u_k$ and $w_i = v_1 \cdots v_k$ be the standard expressions of w_j and w_i. The binomial $f y_{u_1} \cdots y_{u_s} - g y_{v_1} \cdots y_{v_s}$ belongs to $J_{\mathcal{R}(I)}$. Since $y_{v_1} \cdots y_{v_s} <^{\#} y_{u_1} \cdots y_{u_s}$, it follows that the initial monomial of $f y_{u_1} \cdots y_{u_s} - g y_{v_1} \cdots y_{v_s}$ is $f y_{u_1} \cdots y_{u_s}$. Hence there is a binomial $h^{(+)} - h^{(-)}$ belonging to $\mathcal{G}(J_{\mathcal{R}(I)})$ whose initial monomial $h^{(+)}$ divides $f y_{u_1} \cdots y_{u_s}$. Since $y_{u_1} \cdots y_{u_s}$ is a standard monomial with respect to $<^{\#}$, it follows from the definition of the monomial order $<^{\#}_{\text{lex}}$ that it remains to be a standard monomial with respect to $<^{\#}_{\text{lex}}$. Hence the initial monomial of none of the binomials belonging to $\mathcal{G}(J_{\mathcal{R}(I)})$ can divide $y_{u_1} \cdots y_{u_s}$. As a consequence, the initial monomial $h^{(+)}$ must be divided by some variable, say, x_a. Since $h^{(+)}$ is at most linear in the variables $x_1, \ldots, x_n$, one has $h^{(+)} = x_a y_{u_{p_1}} \cdots y_{u_{p_t}}$; then x_a divides f and where $y_{u_{p_1}} \cdots y_{u_{p_t}}$ divides $y_{u_1} \cdots y_{u_s}$. Let $h^{(-)} = x_b y_{v_{q_1}} \cdots y_{v_{q_t}}$, where $y_{v_{q_1}} \cdots y_{v_{q_t}} <^{\#} y_{u_{p_1}} \cdots y_{u_{p_t}}$. One has $x_a u_{p_1} \cdots u_{p_t} = x_b v_{q_1} \cdots v_{q_t}$.

To complete our proof, we show that $x_a \in (w_1, \ldots, w_{j-1}) : w_j$. Since $y_{u_{p_1}} \cdots y_{u_{p_t}}$ divides $y_{u_1} \cdots y_{u_s}$, one has $y_{u_1} \cdots y_{u_s} = y_{u_{p_1}} \cdots y_{u_{p_t}} y_{u_{p_{t+1}}} \cdots y_{u_{p_k}}$. Since $y_{v_{q_1}} \cdots y_{v_{q_t}} <^{\#} y_{u_{p_1}} \cdots y_{u_{p_t}}$, it follows that

$$y_{v_{q_1}} \cdots y_{v_{q_t}} y_{u_{p_{t+1}}} \cdots y_{u_{p_k}} <^{\#} y_{u_1} \cdots y_{u_k} = w_j^*.$$

Let $w_{i_0} = v_{q_1} \cdots v_{q_t} u_{p_{t+1}} \cdots u_{p_k} \in G(I^k)$. Then $x_a w_j = x_b w_{i_0}$. Since $w_{i_0}^* \leq^{\#} y_{v_{q_1}} \cdots y_{v_{q_t}} y_{u_{p_{t+1}}} \cdots y_{u_{p_k}}$, one has $w_{i_0}^* <^{\#} w_j^*$. Hence $i_0 < j$. Thus $x_a \in (w_1, \ldots, w_{j-1}) : w_j$, as desired. $\square$

10.2 Powers of monomial ideals with linear resolution

We consider classes of monomial ideals for which all of its powers have a linear resolution.

10.2.1 Monomial ideals with 2-linear resolution

We begin with

Lemma 10.2.1. *Let $I \subset S$ be a squarefree monomial ideal with 2-linear resolution. Then, after suitable renumbering of the variables, one has the following property: if $x_i x_j \in I$ with $i \neq j$, $k > i$ and $k > j$, then either $x_i x_k$ or $x_j x_k$ belongs to I.*

Proof. Let G be the finite graph on $[n]$ with $I = I(G)$. Since I has a linear resolution, the complementary graph $\overline{G}$ is a chordal graph, see Theorem 9.2.3. Let Δ be the quasi-forest on $[n]$ whose 1-skeleton coincides with $\overline{G}$, see Section 9.2. Let $F_1, \ldots, F_m$ be a leaf order of Δ. Let i_1 be the number of free vertices of the leaf F_m. We label the free vertices of F_m by $n, n-1, \ldots, n-i_1+1$. Thus $\langle F_1, \ldots, F_{m-1} \rangle$ is a quasi-forest on $[n-i_1]$ and F_{m-1} is a leaf of $\langle F_1, \ldots, F_{m-1} \rangle$. Let i_2 be the number of free vertices of the leaf F_{m-1}. We label the free vertices of F_{m-1} by by $n - i_1, \ldots, n - (i_1 + i_2) + 1$. Proceeding in this way we label all the vertices of Δ, that is, those of G, and then choose the numbering of the variables of S according to this labelling.

Suppose there exist $x_i x_j \in I$ and $k > i, j$ such that $x_i x_k \notin I$ and $x_j x_k \notin I$. Let r be the smallest number such that $\Gamma = \langle F_1, \ldots, F_r \rangle$ contains the vertices $1, \ldots, k$. Then k is a free vertex of F_r in Γ. Since $x_i x_k \notin I$ and $x_j x_k \notin I$, it follows that $\{i, k\}$ and $\{j, k\}$ must be edges of Γ. Since k is a free vertex of F_r in Γ it follows that i and j must be vertices of F_r. Hence $\{i, j\}$ is an edge of F_r. Thus $\{i, j\}$ is an edge of $\overline{G}$. In other words, $\{i, j\}$ cannot be an edge of G. However, this contradicts the assumption that $x_i x_j \in I$. $\qquad\square$

Lemma 10.2.2. *Let I be a monomial ideal generated in degree 2 and $J \subset I$ the ideal generated by all squarefree monomials belonging to I. Suppose that I has a linear resolution. Then J has a linear resolution.*

Proof. Let $\{x_{i_1}^2, \ldots, x_{i_k}^2\} = I \cap \{x_1^2, \ldots, x_n^2\}$. Then $I = (x_{i_1}^2, \ldots, x_{i_k}^2, J)$. Recall from Subsection 1.6 that the polarization of I is the squarefree ideal $I^* = (x_{i_1} y_1, \ldots, x_{i_k} y_k, J)$ of $K[x_1, \ldots, x_n, y_1, \ldots, y_k]$. We regard I^* to be the edge ideal of the finite graph G^* with the vertices $-k, \ldots, -1, 1, \ldots, n$, where the vertices $-j$ correspond to the variables y_j and the vertices i to the variables x_i. Let G be the restriction of G^* to $\{1, \ldots, n\}$. In other words, $\{i, j\}$ with $1 \leq i < j \leq n$ is an edge of G if and only if it is an edge of G^*. It is clear that J is the edge ideal of G.

Since I has a linear resolution, Corollary 1.6.3 guarantees that I^* has a linear resolution. Hence $\overline{G^*}$ is chordal. Obviously the restriction of a chordal graph to a subset of the vertices is again chordal. Hence $\overline{G}$ is chordal and J has a linear resolution, as desired. $\qquad\square$

Lemma 10.2.3. *Work with the situation as in the proof of Lemma 10.2.2. Let Δ be the quasi-forest whose 1-skeleton coincides with $\overline{G}$. Then*

(a) *the vertex i_j is a free vertex of Δ for $j = 1, \ldots, k$;*
(b) *no two of these vertices $i_1, \ldots, i_k$ belong to the same facet of Δ.*

Proof. Let Δ^* be the quasi-forest whose 1-skeleton is $\overline{G^*}$.

(a) Suppose that i_j is not a free vertex of Δ. Then there exist edges $\{i_j, r\}$ and $\{i_j, s\}$ of $\overline{G}$ such that $\{r, s\}$ is not an edge of $\overline{G}$. Then $\{i_j, r\}$ and $\{i_j, s\}$ are also edges of $\overline{G^*}$, and $\{r, s\}$ is not an edge of $\overline{G^*}$. Since $x_{i_j} y_j \in I^*$, it follows that $\{i_j, -j\}$ is not an edge in G^*. Since $x_r y_j$ and $x_s y_j$ do not belong to I^*, it follows that $\{-j, r\}$ and $\{-j, s\}$ are edges of $\overline{G^*}$. Thus $\{i_j, r\}, \{r, -j\}, \{-j, s\}, \{s, i_j\}$ is a cycle of $\overline{G^*}$ of length 4 with no chord; a contradiction.

(b) Suppose that i_j and i_ℓ are free vertices belonging to the same facet of Δ. Then $\{i_j, i_\ell\}$ is an edge of $\overline{G^*}$. Since $x_{i_j} y_\ell$, $x_{i_\ell} y_j$ and $y_j y_\ell$ do not belong to I^*, it follows that $\{i_j, -\ell\}$, $\{i_\ell, -j\}$ and $\{-j, -\ell\}$ are egdes of $\overline{G^*}$. On the other hand, since $x_{i_j} y_j$ and $x_{i_\ell} y_\ell$ belong to I^*, it follows that $\{i_j, -j\}$ and $\{i_l, -\ell\}$ are not edges of $\overline{G^*}$. Hence $\{i_j, i_\ell\}, \{i_\ell, -j\}, \{-j, -\ell\}, \{-\ell, i_j\}$ is a cycle of length 4 with no chord; a contradiction. $\square$

Corollary 10.2.4. *Let I be a monomial ideal of S generated in degree 2. Suppose that I has a linear resolution and that $x_i^2 \in I$. Then with the numbering of the variables as given in Lemma 10.2.1 one has the following property: for all $j > i$ for which there exists k such that $x_k x_j \in I$, one has $x_i x_j \in I$ or $x_i x_k \in I$.*

Proof. Suppose $x_i^2 \in I$ and there exists $j > i$ for which there exists k such that $x_k x_j \in I$, but neither $x_i x_j$ nor $x_i x_k$ belongs to I. Since $x_i^2 \in I$, one has $k \neq i$.

Let $k \neq j$. Then $\{k, j\}$ is not an edge of Δ, where Δ is the quasi-forest as defined in the proof of Lemma 10.2.1. Since both $\{i, j\}$ and $\{i, k\}$ are edges of Δ, it follows that i cannot be a free vertex of Δ, contradicting Lemma 10.2.3.

Let $k = j$. Then $x_j^2 \in I$ and j is a free vertex of Δ by Lemma 10.2.3. Since $x_i x_j \notin I$, the edge $\{i, j\}$ belongs to Δ. Hence both i and j belong to the same facet, contradicting Lemma 10.2.3. $\square$

10.2.2 Powers of monomial ideals with 2-linear resolution

Recall from Lemma 10.2.1 and Corollary 10.2.4 that if I is a monomial ideal of S generated in degree 2 which has a linear resolution then I satisfies the conditions $(*)$ and $(**)$ listed in the following

Theorem 10.2.5. *Let $I \subset S = K[x_1, \ldots, x_n]$ be an ideal which is generated by quadratic monomials and suppose that I possesses the following properties $(*)$ and $(**)$:*

$(*)$ *if $x_i x_j \in I$ with $i \neq j$, $k > i$ and $k > j$, then either $x_i x_k$ or $x_j x_k$ belongs to I;*

$(**)$ *if $x_i^2 \in I$ and $j > i$ for which there is k such that $x_k x_j \in I$, then either $x_i x_j \in I$ or $x_i x_k \in I$*

Let $\mathcal{R}(I) = R/P$ be the Rees algebra of I. Then there exists a lexicographic order $<_{\mathrm{lex}}$ on R such that the reduced Gröbner basis $\mathcal{G}$ of the defining ideal P with respect to $<_{\mathrm{lex}}$ consists of binomials $f \in R$ with $\deg_x(f) \leq 1$.

Proof. Let Ω denote the finite graph with the vertices $1, \ldots, n, n+1$ whose edge set $E(\Omega)$ consists of those edges and loops $\{i, j\}$, $1 \leq i \leq j \leq n$, with $x_i x_j \in I$ together with the edges $\{1, n+1\}, \{2, n+1\}, \ldots, \{n, n+1\}$.

Let $K[\Omega]$ denote the subring of $S[x_{n+1}]$ generated by those quadratic monomials $x_i x_j$, $1 \leq i \leq j \leq n+1$, with $\{i, j\} \in E(\Omega)$. Let $R = K[x_1, \ldots, x_n, \{y_{\{i,j\}}\}_{\substack{1 \leq i \leq n, 1 \leq j \leq n \\ \{i,j\} \in E(\Omega)}}]$ be the polynomial ring and define the surjective homomorphism $\pi : R \to K[\Omega]$ by setting $\pi(x_i) = x_i x_{n+1}$ and $\pi(y_{\{i,j\}}) = x_i x_j$. Since the Rees algebra $\mathcal{R}(I)$ of I is isomorphic to $K[\Omega]$ in the obvious way, we will identify the defining ideal P of the Rees algebra with the kernel of π.

We introduce the lexicographic order $<_{\mathrm{lex}}$ on R induced by the ordering of the variables as follows: (i) $y_{\{i,j\}} > y_{\{p,q\}}$ if either $\min\{i, j\} < \min\{p, q\}$ or $(\min\{i, j\} = \min\{p, q\}$ and $\max\{i, j\} < \max\{p, q\})$ and (ii) $y_{\{i,j\}} > x_1 > x_2 > \cdots > x_n$ for all $y_{\{i,j\}}$. Let $\mathcal{G}$ denote the reduced Gröbner basis of P with respect to $<_{\mathrm{lex}}$.

It follows from Corollary 10.1.5 that the reduced Gröbner basis $\mathcal{G}$ consists of binomials of the form f_Γ, where Γ is a primitive even closed walk of Ω.

Now, let f be a binomial belonging to $\mathcal{G}$ and

$$\Gamma = (\{w_1, w_2\}, \{w_2, w_3\}, \ldots, \{w_{2m}, w_1\})$$

the primitive even closed walk of Ω associated with f. In other words, with setting $y_{\{i, n+1\}} = x_i$ and $w_{2m+1} = w_1$, one has

$$f = f_\Gamma = \prod_{k=1}^m y_{\{w_{2k-1}, w_{2k}\}} - \prod_{k=1}^m y_{\{w_{2k}, w_{2k+1}\}}.$$

What we must prove is that, among the vertices $w_1, w_2, \ldots, w_{2m}$, the vertex $n+1$ appears at most one time. Let $y_{\{w_1, w_2\}}$ be the biggest variable appearing in f with respect to $<_{\mathrm{lex}}$ with $w_1 \leq w_2$. Let $k_1, k_2, \ldots$ with $k_1 < k_2 < \cdots$ denote the integers $3 \leq k < 2m$ for which $w_k = n+1$.

Case I: Let k_1 be even. Since $\{n+1, w_1\} \in E(\Omega)$, the closed walk

$$\Gamma' = (\{w_1, w_2\}, \{w_2, w_3\}, \ldots, \{w_{k_1-1}, w_{k_1}\}, \{w_{k_1}, w_1\})$$

is an even closed walk in Ω with $\deg_x(f_{\Gamma'}) = 1$. Since the initial monomial $in_{<_{\mathrm{lex}}}(f_{\Gamma'}) = y_{\{w_1, w_2\}} y_{\{w_3, w_4\}} \cdots y_{\{w_{k_1-1}, w_{k_1}\}}$ of $f_{\Gamma'}$ divides $in_{<_{\mathrm{lex}}}(f_\Gamma) = \prod_{k=1}^m y_{\{w_{2k-1}, w_{2k}\}}$, it follows that $f_\Gamma \notin \mathcal{G}$ unless $\Gamma' = \Gamma$.

Case II: Let both k_1 and k_2 be odd. This is impossible since Γ is primitive and since the subwalk

$$\Gamma'' = (\{w_1, w_2\}, \ldots, \{w_{k_1-1}, w_{k_1}\}, \{w_{k_2}, w_{k_2+1}\}, \ldots, \{w_{2m}, w_1\})$$

of Γ is an even closed walk in Ω.

Case III: Let k_1 be odd and let k_2 be even. Let C be the odd closed walk

$$C = (\{w_{k_1}, w_{k_1+1}\}, \{w_{k_1+1}, w_{k_1+2}\}, \ldots, \{w_{k_2-1}, w_{k_2}\})$$

in Ω. Since both $\{w_2, w_{k_1}\}$ and $\{w_{k_2}, w_1\}$ are edges of Ω, the closed walk

$$\Gamma''' = (\{w_1, w_2\}, \{w_2, w_{k_1}\}, C, \{w_{k_2}, w_1\})$$

is an even closed walk in Ω and the initial monomial $in_{<_{\mathrm{lex}}}(f_{\Gamma'''})$ of $f_{\Gamma'''}$ divides $in_{<_{\mathrm{lex}}}(f_\Gamma)$. Thus we discuss Γ''' instead of Γ.

Since Γ''' is primitive and since C is of odd length, it follows that none of the vertices of C coincides with w_1 and that none of the vertices of C coincides with w_2.

(III – a) First, we study the case when there is $p \geq 0$ with $k_1 + p + 2 < k_2$ such that $w_{k_1+p+1} \neq w_{k_1+p+2}$. Let W and W' be the walks

$$W = (\{w_{k_1}, w_{k_1+1}\}, \{w_{k_1+1}, w_{k_1+2}\}, \ldots, \{w_{k_1+p+1}, w_{k_1+p+2}\}),$$
$$W' = (\{w_{k_2}, w_{k_2-1}\}, \{w_{k_2-1}, w_{k_2-2}\}, \ldots, \{w_{k_1+p+3}, w_{k_1+p+2}\})$$

in Ω.

(III – a – 1) Let $w_1 \neq w_2$. If either $\{w_2, w_{k_1+p+1}\}$ or $\{w_2, w_{k_1+p+2}\}$ is an edge of Ω, then it is possible to construct an even closed walk $\Gamma^\sharp$ in Ω such that $in_{<_{\mathrm{lex}}}(f_{\Gamma^\sharp})$ divides $in_{<_{\mathrm{lex}}}(f_{\Gamma'''})$ and $\deg_x(f_{\Gamma^\sharp}) = 1$. For example, if, say, $\{w_2, w_{k_1+p+2}\} \in E(\Omega)$ and if p is even, then

$$\Gamma^\sharp = (\{w_2, w_1\}, \{w_1, w_{k_2}\}, W', \{w_{k_1+p+2}, w_2\})$$

is a desired even closed walk.

(III – a – 2) Let $w_1 \neq w_2$. We assume that neither $\{w_2, w_{k_1+p+1}\}$ nor $\{w_2, w_{k_1+p+2}\}$ is an edge of Ω. Since $\{w_{k_1+p+1}, w_{k_1+p+2}\}$ is an edge of Ω, it follows from $(*)$ that either $w_2 < w_{k_1+p+1}$ or $w_2 < w_{k_1+p+2}$. Let, say, $w_2 < w_{k_1+p+2}$. Since $w_1 < w_2$ and $\{w_1, w_2\} \in E(\Omega)$, again by $(*)$ one has $\{w_1, w_{k_1+p+2}\} \in E(\Omega)$. If p is even, then consider the even closed walk

$$\Gamma^\flat = (\{w_1, w_2\}, \{w_2, w_{k_2}\}, W', \{w_{k_1+p+2}, w_1\})$$

in Ω. If p is odd, then consider the even closed walk

$$\Gamma^\flat = (\{w_1, w_2\}, \{w_2, w_{k_1}\}, W, \{w_{k_1+p+2}, w_1\})$$

in Ω. In each case, one has $\deg_x(f_{\Gamma^\flat}) = 1$. Since $y_{\{w_1, w_2\}} > y_{\{w_1, w_{k_1+p+2}\}}$, it follows that $in_{<_{\mathrm{lex}}}(f_{\Gamma^\flat})$ divides $in_{<_{\mathrm{lex}}}(f_{\Gamma'''})$.

(III − a − 3) Let $w_1 = w_2$. Since $w_1 < w_{k_1+p+1}$, it follows from (∗∗) that either $\{w_1, w_{k_1+p+1}\} \in E(\Omega)$ or $\{w_1, w_{k_1+p+2}\} \in E(\Omega)$. Thus the same technique as in (III − a − 2) can be applied.

(III − b) Second, if $C = (\{n+1, j\}, \{j, j\}, \{j, n+1\})$, then in each of the cases $w_1 < w_2 < j$, $w_1 < j < w_2$ and $w_1 = w_2 < j$, by either (∗) or (∗∗), one has either $\{w_1, j\} \in E(\Omega)$ or $\{w_2, j\} \in E(\Omega)$.   □

As the final conclusion of the present section we obtain

Theorem 10.2.6. *Let I be a monomial ideal of S generated in degree 2. Then the following conditions are equivalent:*

(a) *I has a linear resolution;*
(b) *I has linear quotients;*
(c) *Each power of I has a linear resolution.*

Proof. First of all, (c) ⇒ (a) is trivial, and (b) ⇒ (a) is guaranteed by Proposition 8.2.1. We will show that (a) ⇒ (c) and (a) ⇒ (b).

(a) ⇒ (c): If I has a linear resolution, then it follows from Proposition 10.2.1 and Corollary 10.2.4 that, after a suitable renumbering of the variables, the conditions (∗) and (∗∗) of Theorem 10.2.5 are satisfied. Hence Corollary 10.1.7 guarantees that each power of I has a linear resolution.

(a) ⇒ (b): Again we may assume that the conditions (∗) and (∗∗) are satisfied. We show that the following condition (q) is satisfied: the elements of $G(I)$ can be ordered such that if $u, v \in G(I)$ with $u > v$, then there exists $w > v$ such that $w/\gcd(w, v)$ is of degree 1 and $w/\gcd(w, v)$ divides $u/\gcd(u, v)$. This condition (q) guarantees that I has linear quotients.

The squarefree monomials belonging to $G(I)$ will be ordered by the lexicographical order induced by $x_n > x_{n-1} > \cdots > x_1$, and if $x_i^2 \in G(I)$ then we let $u > x_i^2 > v$, where u is the smallest squarefree monomial of the form $x_k x_i$ with $k < i$ and where v is the largest squarefree monomial less than u.

Now, for any two monomials u and v belonging to $G(I)$ with $u > v$, we must show that property (q) is satisfied. There are three cases:

Case 1: $u = x_s x_t$ and $v = x_i x_j$ both are squarefree monomials with $s < t$ and $i < j$. Since $u > v$, we have $t \geq j$. If $t = j$, take $w = u$. If $t > j$, then by (∗), either $x_i x_t \in G(I)$ or $x_j x_t \in G(I)$. Accordingly, let $w = x_i x_t$ or $w = x_j x_t$.

Case 2: $u = x_t^2$ and $v = x_i x_j$ with $i < j$. Since $u > v$, we have $t > j$. Hence by (∗), either $x_i x_t \in G(I)$ or $x_j x_t \in G(I)$. Accordingly, let $w = x_i x_t$ or $w = x_j x_t$.

Case 3: $u = x_s x_t$ with $s \leq t$ and $v = x_i^2$. If $t = i$, then $s \neq t$ and take $w = u$. If $t > i$, then by (∗∗), we have either $x_i x_t \in G(I)$ or $x_i x_s \in G(I)$. Both elements are greater than v. Accordingly, let $w = x_i x_t$ or $w = x_i x_s$. Then again (q) holds.   □

Example 10.2.7. (a) (Sturmfels) The monomial ideal

$$I = (def, cef, cdf, cde, bef, bcd, acf, ade)$$

of $K[a, b, c, d, e, f]$ has linear quotients. However, I^2 does not have a linear resolution.

(b) (Conca) The ideal

$$I = (a^2, ab, ac, ad, b^2, ae + bd, d^2)$$

of $K[a, b, c, d, e]$ has a linear resolution (even linear quotients with respect to the generators in the given order), but, at least in characteristic 0, the ideal I^2 does not have a linear resolution.

10.2.3 Powers of vertex cover ideals of Cohen–Macaulay bipartite graphs

One of the typical examples of monomial ideals for which Theorem 10.1.9 can be applied is the vertex cover ideal of a Cohen–Macaulay bipartite graph. Let G be a Cohen–Macaulay bipartite graph. We have seen in Theorem 9.1.13 that $G = G_P$ for some finite poset $P = \{p_1, \ldots, p_n\}$, and that the vertex cover ideal I_G of G is equal the squarefree monomial ideal H_P, which is the ideal associated to the poset ideals of P. Hence in what follows we will discuss the powers of H_P.

Let

$$K[\mathbf{x}, \mathbf{y}] = K[x_1, \ldots, x_n, y_1, \ldots, y_n]$$

be the polynomial ring in $2n$ variables over a field K. The ideal H_P is generated by the monomials $u_\alpha = (\prod_{p_i \in \alpha} x_i)(\prod_{p_j \in P \setminus \alpha} y_j)$ of $K[\mathbf{x}, \mathbf{y}]$ with $\alpha \in \mathcal{J}(P)$.

Let J the defining ideal of $\mathcal{R}(H_P)$. In other words,

$$\mathcal{R}(H_P) = K[\mathbf{x}, \mathbf{y}, \{u_\alpha t\}_{\alpha \in \mathcal{J}(P)}] \subset K[\mathbf{x}, \mathbf{y}, t]$$

and J is the kernel of the canonical surjective K-algebra homomorphism $\varphi \colon K[\mathbf{x}, \mathbf{y}, \mathbf{z}] \to \mathcal{R}(H_P)$, where

$$K[\mathbf{x}, \mathbf{y}, \mathbf{z}] = K[\mathbf{x}, \mathbf{y}, \{z_\alpha\}_{\alpha \in \mathcal{J}(P)}]$$

is the polynomial ring over K and where φ is defined by setting $\varphi(x_i) = x_i$, $\varphi(y_j) = y_j$ and $\varphi(z_\alpha) = u_\alpha t$.

Let $<_{\mathrm{lex}}$ denote the lexicographic order on $K[\mathbf{x}, \mathbf{y}]$ induced by the ordering $x_1 > \cdots > x_n > y_1 > \cdots > y_n$ and $<^\sharp$ the reverse lexicographic order on $K[\{z_\alpha\}_{\alpha \in \mathcal{J}(P)}]$ induced by an ordering of the variables z_α satisfying $z_\alpha > z_\beta$ if $\beta \subset \alpha$ in $\mathcal{J}(P)$. Finally let $<^\sharp_{\mathrm{lex}}$ be the monomial order on $K[\mathbf{x}, \mathbf{y}, \mathbf{z}]$, as defined before Theorem 10.1.9.

Theorem 10.2.8. *The reduced Gröbner basis $\mathcal{G}_{<^\sharp_{\mathrm{lex}}}(J)$ of the defining ideal $J \subset K[\mathbf{x}, \mathbf{y}, \mathbf{z}]$ with respect to the monomial order $<^\sharp_{\mathrm{lex}}$ consists of quadratic binomials whose initial monomials are squarefree. In particular, H_P satisfies the x-condition with respect to $<^\sharp_{\mathrm{lex}}$.*

Proof. The reduced Gröbner basis of $J \cap K[\{z_\alpha\}_{\alpha \in \mathcal{J}(P)}]$ with respect to the reverse lexicographic order $<^\sharp$ coincides with $\mathcal{G}_{<^\sharp_{lex}}(J) \cap K[\{z_\alpha\}_{\alpha \in \mathcal{J}(P)}]$. Theorem 10.1.3 guarantees that $\mathcal{G}_{<^\sharp_{lex}}(J) \cap K[\{z_\alpha\}_{\alpha \in \mathcal{J}(P)}]$ consists of those binomials $z_\alpha z_\beta - z_{\alpha \cap \beta} z_{\alpha \cup \beta}$ with neither $\alpha \subset \beta$ nor $\beta \subset \alpha$.

It follows from Proposition 10.1.2 that the reduced Gröbner basis of J consists of primitive binomials of $K[\mathbf{x}, \mathbf{y}, \mathbf{z}]$. Let

$$f = (\prod_{i=1}^n x_i^{a_i} y_i^{b_i})(z_{\alpha_1} \cdots z_{\alpha_q}) - (\prod_{i=1}^n x_i^{a_i'} y_i^{b_i'})(z_{\alpha_1'} \cdots z_{\alpha_q'})$$

be a primitive binomial of $K[\mathbf{x}, \mathbf{y}, \mathbf{z}]$ belonging to $\mathcal{G}_{<^\sharp_{lex}}(J)$ with

$$(\prod_{i=1}^n x_i^{a_i} y_i^{b_i})(z_{\alpha_1} \cdots z_{\alpha_q})$$

its initial monomial, where $\alpha_1 \leq \cdots \leq \alpha_q$ and $\alpha_1' \leq \cdots \leq \alpha_q'$.

Let $f \notin K[\{z_\alpha\}_{\alpha \in \mathcal{J}(P)}]$, and let j denote an integer for which $\alpha_j' \not\subset \alpha_j$. Such an integer exists. In fact, if $\alpha_j' \subset \alpha_j$ for all j, then each $a_i = 0$ and each $b_i' = 0$. This is impossible since $(\prod_{i=1}^n x_i^{a_i} y_i^{b_i})(z_{\alpha_1} \cdots z_{\alpha_q})$ is the initial monomial of f.

Let $p_i \in \alpha_j' \setminus \alpha_j$. Then p_i belongs to each of $\alpha_j', \alpha_{j+1}', \ldots, \alpha_q'$, and does not belong to each of $\alpha_1, \alpha_2, \ldots, \alpha_j$. Hence $a_i > 0$.

Let $p_{i_0} \in P$ with $p_{i_0} \in \alpha_j' \setminus \alpha_j$ such that $\alpha_j \cup \{p_{i_0}\} \in \mathcal{J}(P)$.

Thus $a_{i_0} > 0$. Let $\beta = \alpha_j \cup \{p_{i_0}\}$. Then the binomial $g = x_{i_0} u_{\alpha_j} - y_{i_0} u_\beta$ belongs to J with $x_{i_0} u_{\alpha_j}$ its initial monomial. Since $x_{i_0} u_{\alpha_j}$ divides the initial monomial of f, it follows that the initial monomial of f coincides with $x_{i_0} u_\alpha$, as desired. $\square$

In view of Theorem 10.1.9 the preceding theorem yields

Corollary 10.2.9. *Let G be a Cohen–Macaulay bipartite graph. Then all powers of the vertex cover ideal I_G of G have linear quotients.*

10.2.4 Powers of vertex cover ideals of Cohen–Macaulay chordal graphs

In a similar way to bipartite graphs one has

Theorem 10.2.10. *Let G be a Cohen–Macaulay chordal graph. Then all powers of I_G have a linear resolution.*

Let G be a chordal graph on $[n]$. In Theorem 9.3.1 we have seen that G is Cohen–Macaulay, if and only if $[n]$ is the disjoint union of those facets of $\Delta(G)$ with a free vertex. Thus Theorem 10.2.10 follows from

Theorem 10.2.11. *Let G be a graph on $[n]$, and suppose that $[n]$ is the disjoint union of those facets of the clique complex of G with a free vertex. Then all powers of I_G have a linear resolution.*

Proof. Let $F_1, \ldots, F_m$ be the facets of $\Delta(G)$ which have a free vertex. Since $[n] = F_1 \cup F_2 \cup \cdots \cup F_s$ is a disjoint union, we may assume that if $i \in F_p$, $j \in F_q$ and $p < q$, then $i < j$. In particular, $1 \in F_1$ and $n \in F_s$. Moreover, we may assume that if $i_1, i_2 \in F_i$ where i_1 is a nonfree vertex and i_2 is a free vertex, then $i_1 < i_2$.

Observe that any minimal vertex cover of G is of the following form:

$$(F_1 \setminus \{a_1\}) \cup (F_2 \setminus \{a_2\}) \cup \cdots \cup (F_s \setminus \{a_s\}), \quad \text{where} \quad a_j \in F_j.$$

In particular, G is unmixed and all generators of I_G have degree $n - s$.

Now let $R(I_G)$ be the Rees algebra of vertex cover ideal of G. Suppose $u_1, \ldots, u_m$ is the minimal set of monomial generators of I_G. Then there is a surjective K-algebra homomorphism

$$K[x_1, \ldots, x_n, y_1, \ldots, y_m] \longrightarrow R(I_G) \quad x_i \mapsto x_i \quad \text{and} \quad y_j \mapsto u_j,$$

whose kernel J is a binomial ideal. Let $<$ be the lexicographic order induced by the ordering $x_1 > x_2 > \cdots > x_n > y_1 > \cdots > y_m$. We are going to show that I_G satisfies the x-condition with respect to this monomial order. Suppose that $x_{i_1} x_{i_2} \cdots x_{i_p} y_{j_1} y_{j_2} \cdots y_{j_q}$ with $i_1 \leq i_2 \leq \ldots \leq i_p$ is a minimal generator of $\text{in}_<(J)$. Then

$$x_{i_1} x_{i_2} \cdots x_{i_p} y_{j_1} y_{j_2} \cdots y_{j_q} - x_{k_1} x_{k_2} \cdots x_{k_p} y_{\ell_1} y_{\ell_2} \cdots y_{\ell_q} \in J. \qquad (10.3)$$

It follows that $i_1 < \min\{k_1, \ldots, k_p\}$, and there exists an index j_r such that x_{i_1} does not divide u_{j_r}. Say, $i_1 \in F_c$. Then (10.3) implies that there exists $d \in [p]$ with $k_d \in F_c$. In particular, $i_1 \neq \max\{i : i \in F_d\}$. Let $i_0 = \max\{i : i \in F_d\}$. Since i_0 is a free vertex, it follows that $x_{i_1}(u_{j_r}/x_{i_0})$ is a minimal generator of I_G, say, u_g. Therefore, $f = x_{i_1} y_{j_r} - x_{i_0} y_g \in J$ and $\text{in}(f) = x_{i_1} y_{j_r}$ divides $x_{i_1} x_{i_2} \cdots x_{i_p} y_{j_1} y_{j_2} \cdots y_{j_q}$, as desired. $\qquad\square$

10.3 Depth and normality of powers of monomial ideals

10.3.1 The limit depth of a graded ideal

What can be said about the numerical function $f(k) = \text{depth}\, S/I^k$ for a graded ideal $I \subset S = K[x_1, \ldots, x_n]$ and $k \gg 0$? We first show

Proposition 10.3.1. *Let $I \subset S$ be a graded ideal. Then $\text{depth}\, S/I^k$ is constant for all $k \gg 0$.*

Proof. We will show that depth I^k is constant for $k \gg 0$. This will yield the desired conclusion. In order to show this we consider the Koszul homology $H(\mathbf{x}; \mathcal{R}(I))$ of the Rees algebra $\mathcal{R}(I)$ of I with respect to $\mathbf{x} = x_1, \ldots, x_n$. Each $H_i(\mathbf{x}; \mathcal{R}(I))$ is a finitely generated graded S-module with homogeneous components

$$H_i(\mathbf{x}; \mathcal{R}(I))_k = H_i(\mathbf{x}; I^k). \tag{10.4}$$

Corollary A.4.2 implies that

$$\operatorname{depth} I^k = n - \max\{i \colon H_i(\mathbf{x}; I^k) \neq 0\}. \tag{10.5}$$

Now, according to (10.4), we have $H_i(\mathbf{x}; I^k) \neq 0$ for all $k \gg 0$, if and only if the Krull dimension of $H_i(\mathbf{x}; I^k)$ is not zero. From this it follows that

$$\operatorname{depth} I^k = n - \max\{i \colon \dim H_i(\mathbf{x}; \mathcal{R}(I)) > 0\} \quad \text{for all} \quad k \gg 0. \tag{10.6}$$

$\square$

Let $\mathfrak{m} = (x_1, \ldots, x_n)$ be the graded maximal of S. The K-algebra $\overline{\mathcal{R}}(I) = \mathcal{R}(I)/\mathfrak{m}\mathcal{R}(I)$ is a called the **fibre ring**, and its Krull dimension the **analytic spread** of I. This invariant is a measure for the growth of the number of generators of the powers of I. Indeed, for $k \gg 0$, the Hilbert function $H(\overline{\mathcal{R}}(I), k) = \dim_K I^k/\mathfrak{m}I^k$, which counts the number of generators of the powers of I, is a polynomial function of degree $\ell(I) - 1$; see Theorem 6.1.3.

As we have seen before, the limit of the numerical function depth S/I^k exists. The next result gives an upper bound for this limit.

Proposition 10.3.2. *Let $I \subset S$ be a nonzero graded ideal. Then*

$$\lim_{k \to \infty} \operatorname{depth} S/I^k \leq n - \ell(I).$$

Equality holds if $\mathcal{R}(I)$ is Cohen–Macaulay.

Proof. Let $r > 0$ be any integer. Then $\lim_{k \to \infty} S/I^{kr} = \lim_{k \to \infty} S/I^k$ and $\ell(I^r) = \ell(I)$. Moreover, $\mathcal{R}(I^r) = \mathcal{R}(I)^{(r)}$ which is the rth Veronese subalgebra of $\mathcal{R}(I)$. It is know that if $\mathcal{R}(I)$ is Cohen–Macaulay, then $\mathcal{R}(I)^{(r)}$ is Cohen–Macaulay as well. Thus in the proof of the proposition we may replace I by I^r for any $r > 0$.

Let $c = \max\{i \colon \dim H_i(\mathbf{x}; \mathcal{R}(I)) > 0\}$. Then there exists an integer k_0 such that $H_i(\mathbf{x}; \mathcal{R}(I))_k = 0$ for all $i > c$ and all $k \geq k_0$. Thus, if we choose some $r \geq k_0$, then $H_i(\mathbf{x}; \mathcal{R}(I^r)) = 0$ for all $i > c$, while $\dim H_i(\mathbf{x}; \mathcal{R}(I^r)) > 0$ for $i \leq c$. Replacing I by I^r we may as well assume that $H_i(\mathbf{x}; \mathcal{R}(I)) = 0$ for $i > c$. Therefore, (10.6) and [BH98, Theorem 1.6.17] imply that

$$\lim_{k \to \infty} \operatorname{depth} S/I^k = \lim_{k \to \infty} \operatorname{depth} I^k - 1 = n - \operatorname{grade}(\mathfrak{m}, \mathcal{R}(I)) - 1. \tag{10.7}$$

By the graded version of [BH98, Theorem 2.1.2] one has $\operatorname{grade}(\mathfrak{m}, \mathcal{R}(I))) = \dim \mathcal{R}(I) - \dim \overline{\mathcal{R}}(I)$, with equality if and only if $\mathcal{R}(I)$ is Cohen–Macaulay. Hence, observing that $\dim \mathcal{R}(I) = n + 1$, the desired conclusion follows. $\square$

Example 10.3.3. The function $f(k) = \operatorname{depth} S/I^k$ may not be monotonic. Consider for example the monomial ideal

$$I = (a^6, a^5b, ab^5, b^6, a^4b^4c, a^4b^4d, a^4e^2f^3, b^4e^3f^2)$$

in $S = K[a, b, c, d, e, f]$. Then $\operatorname{depth} S/I = 0$, $\operatorname{depth} S/I^2 = 1$, $\operatorname{depth} S/I^3 = 0$, $\operatorname{depth} S/I^4 = 2$ and $\operatorname{depth} S/I^5 = 2$.

However, one has

Proposition 10.3.4. *Let I be a graded ideal all of whose powers have a linear resolution. Then $\operatorname{depth} S/I^k$ is a nonincreasing function of k.*

The proposition is a consequence of Corollary 10.3.5 stated below. We call the least degree of a homogeneous generator of a graded S-module M, the **initial degree** of M.

Lemma 10.3.5. *Let $J \subset I$ be graded ideals, and let d be the initial degree of I. Then*

$$\beta_{i,i+d}(J) \leq \beta_{i,i+d}(I)$$

for all i.

Proof. The short exact sequence

$$0 \longrightarrow J \longrightarrow I \longrightarrow I/J \longrightarrow 0$$

yields the long exact sequence

$$\cdots \longrightarrow \operatorname{Tor}_{i+1}(K, I/J)_{i+1+(d-1)} \longrightarrow \operatorname{Tor}_i(K, J)_{i+d} \longrightarrow \operatorname{Tor}_i(K, I)_{i+d} \longrightarrow \cdots$$

Since the initial degree of I/J is greater than or equal to d, it follows that $\operatorname{Tor}_{i+1}(K, I/J)_{i+1+(d-1)} = 0$. Hence $\operatorname{Tor}_i(K, J)_{i+d} \to \operatorname{Tor}_i(K, I)_{i+d}$ is injective. $\qquad\square$

10.3.2 The depth of powers of certain classes of monomial ideals

In this subsection we study the function $f(k) = \operatorname{depth} S/I^k$ for special classes of monomial ideals. We consider the case that I is generated in a single degree and that I has linear quotients. Say I is minimally generated by $f_1, \ldots, f_m$ and that the colon ideal $L_k = (f_1, \ldots, f_{k-1}) : f_k$ is generated by r_k elements. Then according to Corollary 8.2.2 one has $\operatorname{proj dim}(I) = \max\{r_1, r_2, \ldots, r_n\}$, so that by the Auslander–Buchsbaum formula,

$$\operatorname{depth} S/I = n - r(I) - 1 \quad \text{where} \quad r(I) = \max\{r_1, r_2, \ldots, r_n\}. \quad (10.8)$$

Fix positive integers d and $e_1, \ldots, e_n$ with $1 \leq e_1 \leq \cdots \leq e_n \leq d$. As a first application of (10.8) we consider the **ideal of Veronese type**

$$I_{(d;e_1,\ldots,e_n)}$$

of S indexed by d and $(e_1, \ldots, e_n)$ which is generated by those monomials $u = x_1^{a_1} \cdots x_n^{a_n}$ of S of degree d with $a_i \leq e_i$ for each $1 \leq i \leq n$. This class of ideals is a special class of polymatroidal ideals, introduced in Chapter 12.

Theorem 10.3.6. *Let* $t = d + n - 1 - \sum_{i=1}^{n} e_i$. *Then* $\operatorname{depth} S/I_{(d;e_1,\ldots,e_n)} = t$.

Proof. Let $u_0 = x_1^{e_1-1} \cdots x_{n-1}^{e_{n-1}-1} x_n^{e_n}$ and $u = x_{n-t} x_{n-t+1} \cdots x_{n-1} u_0 \in G(I)$. Let $J = (\{w \in G(I) : u <_{\mathrm{rev}} w\})$. For each $1 \leq i \leq n - t - 1$, one has $x_i u / x_n \in G(I)$ with $u <_{\mathrm{rev}} x_i u / x_n$. Hence $x_i \in J : u$ for all $1 \leq i \leq n - t - 1$. Moreover, one has $x_j u / x_{j_0} \notin G(I)$ for all $n - t \leq j \leq n$ and for all $j_0 \neq j$. Hence $x_j \notin J : u$ for all $n - t \leq j \leq n$. Thus $J : u = (x_1, \ldots, x_{n-t-1})$. On the other hand, for each $v = x_1^{a_1} \cdots x_n^{a_n} \in G(I)$ with $m(u) = \max\{i : a_i \neq 0\}$, the number of $i < m(v)$ with $a_i < e_i$ is at most $n - t - 1$. Thus the number of variables required to generate the colon ideal $(\{w \in G(I) : v <_{\mathrm{rev}} w\}) : v$ is at most $n - t - 1$. Hence $r(I) = n - t - 1$. Thus $\operatorname{depth} S/I = t$. □

The **squarefree Veronese ideal** of degree d in the variables $x_{i_1}, \ldots, x_{i_t}$ is the ideal of S which is generated by all squarefree monomials in $x_{i_1}, \ldots, x_{i_t}$ of degree d. A squarefree Veronese ideal is matroidal and Cohen–Macaulay; see Corollary 12.6.5 and Theorem 12.6.7.

Let $2 \leq d < n$ and $I = I_{n,d}$ be the squarefree Veronese ideal of degree d in the variables $x_1, \ldots, x_n$. Since for each k, the ideal I^k is the ideal of Veronese type indexed by kd and $(k, k, \ldots, k)$, Theorem 10.3.6 implies

Corollary 10.3.7. *Let* $2 \leq d < n$. *Then*

$$\operatorname{depth} S/I_{n,d}^k = \max\{0, n - k(n - d) - 1\}.$$

Let $P = \{p_1, \ldots, p_n\}$ be a finite poset. As a second application of formula (10.8) we study the powers of the ideal H_P introduced in Subsection 12.6.5.

Theorem 10.3.8. *Each power* H_P^k *has linear quotients.*

Proof. If α and β are poset ideals of P, then both $\alpha \cap \beta$ and $\alpha \cup \beta$ are poset ideals of P with $u_\alpha u_\beta = u_{\alpha \cap \beta} u_{\alpha \cup \beta}$. This fact guarantees that each monomial belonging to $G(H_P^k)$ possesses an expression of the form $u_{\alpha_1} u_{\alpha_2} \cdots u_{\alpha_k}$, where each α_j is a poset ideal of P, with $\alpha_1 \subset \alpha_2 \subset \cdots \subset \alpha_k$. We claim that such an expression is unique. In fact, suppose that $u_{\alpha_1} u_{\alpha_2} \cdots u_{\alpha_k}$ coincides with $u_{\beta_1} u_{\beta_2} \cdots u_{\beta_k}$, where each β_j is a poset ideal of P, with $\beta_1 \subset \beta_2 \subset \cdots \subset \beta_k$. Let $1 \leq \ell \leq k$ be the smallest integer for which $\alpha_\ell \neq \beta_\ell$. Let $p_i \in \alpha_\ell \setminus \beta_\ell$. Then $u_{\alpha_1} u_{\alpha_2} \cdots u_{\alpha_k}$ is divided by $x_i^{k-\ell+1}$. However, $u_{\beta_1} u_{\beta_2} \cdots u_{\beta_k}$ cannot be divided by $x_i^{k-\ell+1}$. This contradiction says that $\alpha_j = \beta_j$ for all $1 \leq j \leq k$. Thus such an expression is unique, as desired.

We fix an ordering $<$ of the monomials u_α, where α is a poset ideal of P, with the property that one has $u_\alpha < u_\beta$ if $\beta \subset \alpha$. We then introduce the lexicographic order $<_{\mathrm{lex}}$ of the monomials belonging to $G(H_P^k)$ induced by the ordering $<$ of the monomials u_α. We claim that H_P^k has linear quotients. More precisely, we show that, for each monomial $w = u_{\alpha_1} u_{\alpha_2} \cdots u_{\alpha_k} \in G(H_P^k)$, the colon ideal $(\{v \in G(H_P^k) : w <_{\mathrm{lex}} v\}) : w$ is generated by those variables y_i for which there is $1 \leq j \leq k$ with $p_i \in \alpha_j$ such that $\alpha_j \setminus \{p_i\}$ is a poset ideal of P.

First, let y_i be a variable with $p_i \in \alpha_j$ and suppose that $\beta = \alpha_j \setminus \{p_i\}$ is a poset ideal of P. One has $y_i u_{\alpha_j} = x_i u_\beta$. Hence

$$y_i w = x_i u_{\alpha_1} \cdots u_{\alpha_{j-1}} u_\beta u_{\alpha_{j+1}} \cdots u_{\alpha_k}.$$

Since each of the poset ideals $\alpha_1, \ldots, \alpha_{j-1}$ and β is a subset of α_j, it then follows that the monomial $u_{\alpha_1} \cdots u_{\alpha_{j-1}} u_\beta$ can be expressed uniquely in the form $u_{\alpha'_1} \cdots u_{\alpha'_{j-1}} u_{\alpha'_j}$ such that $\alpha'_1 \subset \cdots \subset \alpha'_{j-1} \subset \alpha'_j \subset \alpha_j$. Moreover, one has $u_{\alpha_1} \cdots u_{\alpha_{j-1}} u_\beta <_{\text{lex}} u_{\alpha'_1} \cdots u_{\alpha'_{j-1}} u_{\alpha'_j}$. Thus $w <_{\text{lex}} u_{\alpha'_1} \cdots u_{\alpha'_{j-1}} u_{\alpha'_j} u_{\alpha_{j+1}} \cdots u_{\alpha_k}$. Hence y_i belongs to the colon ideal $(\{v \in G(H_P^k) : w <_{\text{lex}} v\}) : w$.

Second, let δ be a monomial belonging to the colon ideal

$$(\{v \in G(H_P^k) : w <_{lex} v\}) : w.$$

Thus one has $\delta w = \mu v$ for monomials μ and v with $w <_{\text{lex}} v$. Say, $v = u_{\alpha'_1} \cdots u_{\alpha'_k}$ with $\alpha'_1 \subset \cdots \subset \alpha'_k$. What we must prove is that the monomial δ is divided by a variable y_i for which there is $1 \le j \le k$ such that $\alpha_j \setminus \{p_i\}$ is a poset ideal of P. Since $w <_{\text{lex}} v$, it follows that there is j_0 for which $\alpha_{j_0} < \alpha'_{j_0}$. In particular $\alpha_{j_0} \not\subset \alpha'_{j_0}$. Thus there is a maximal element p_{i_0} of α_{j_0} with $p_{i_0} \not\subset \alpha'_{j_0}$. Then p_{i_0} belongs to each of the poset ideals $\alpha_{j_0}, \alpha_{j_0+1}, \ldots, I_k$ and belongs to none of the poset ideals $\alpha'_1, \ldots, \alpha'_{j_0}$. Hence the power of y_{i_0} in the monomial v is at least j_0, but that in w is at most $j_0 - 1$. Hence y_0 must divide δ. Since p_{i_0} is a maximal element of α_{j_0}, the subset $\alpha_{j_0} \setminus \{p_{i_0}\}$ of P is a poset ideal of P, as desired. $\square$

By using Theorem 10.3.8 we can now compute depth S/H_P^k in terms of the combinatorics on P. Recall that an **antichain** of P is a subset $A \subset P$ any two of whose elements are incomparable in P. Given an antichain A of P, we write $\langle A \rangle$ for the poset ideal of P generated by A, which consists of those elements $p \in P$ such that there is $a \in A$ with $p \le a$. For each $k = 1, 2, \ldots$, we write $\delta(P; k)$ for the largest integer N for which there is a sequence $(A_1, A_2, \ldots, A_r)$ of antichains of P with $r \le k$ such that

(i) $A_i \cap A_j = \emptyset$ if $i \ne j$;
(ii) $\langle A_1 \rangle \subset \langle A_2 \rangle \subset \cdots \subset \langle A_r \rangle$;
(iii) $N = |A_1| + |A_2| + \cdots + |A_r|$.

We call such a sequence of antichains a k-**acceptable sequence**.

It follows from the definition that $\delta(P; 1)$ is the maximal cardinality of antichains of P and $\delta(P; 1) < \delta(P; 2) < \cdots < \delta(P; \text{rank}(P) + 1)$. Moreover, $\delta(P; k) = n$ for all $k \ge \text{rank}(P) + 1$. Here $\text{rank}(P)$ is the **rank** of P, that is, $\text{rank}(P) + 1$ is the maximal cardinality of chains contained in P. A **chain** is a totally ordered subset of P.

Corollary 10.3.9. *Let P be an arbitrary finite poset with $|P| = n$. Then*

$$\text{depth}\, S/H_P^k = 2n - \delta(P; k) - 1$$

for all $k \ge 1$.

Proof. We work with the same notation as in the proof of Theorem 10.3.8. Recall that, for a monomial $w = u_{\alpha_1} u_{\alpha_2} \cdots u_{\alpha_k} \in G(H_P^k)$, the colon ideal $(\{v \in G(H_P^k) : w <_{lex} v\}) : w$ is generated by those variables y_i for which there is $1 \leq j \leq k$ with $p_i \in \alpha_j$ such that $\alpha_j \setminus \{p_i\}$ is a poset ideal of P. Note that $\alpha_j \setminus \{p_i\}$ is a poset ideal of P if and only if p_i is a maximal element of α_j. Let B_j denote the set of maximal elements of α_j. Then the number of variables required to generate the colon ideal $(\{v \in G(H_P^k) : w <_{lex} v\}) : w$ is $|\bigcup_{j=1}^k B_j|$. Let $Q_w = \bigcup_{j=1}^k B_j$. One has $r = \text{rank}(Q_w) + 1 \leq k$. We then define a sequence $A_1, A_2, \ldots, A_r$ of subset of B_w as follows: A_1 is the set of minimal elements of Q_w and, for $2 \leq j \leq r$, A_j is the set of minimal element of $Q_w \setminus (A_1 \cup \cdots \cup A_{j-1})$. Then $(A_1, \ldots, A_r)$ is k-acceptable with $|Q_w| = \sum_{j=1}^r |A_j|$. Hence $|Q_w| \leq \delta(P; k)$.

On the other hand, there is a k-acceptable sequence $(A_1, A_2, \ldots, A_r)$ with $\delta(P; k) = \sum_{j=1}^r |A_j|$. Let $w = u_\emptyset^{k-r} u_{\langle A_1 \rangle} \cdots u_{\langle A_r \rangle} \in G(H_P^k)$. Then the number of variables required to generate the colon ideal $(\{v \in G(H_P^k) : w <_{lex} v\}) : w$ is $\delta(P; k)$.

Consequently, one has $r(H_P^k) = \delta(P; k)$. Thus depth $S/H_P^k = 2n - \delta(P; k) - 1$, as required. $\square$

Since $\{x_i, y_i\}$ is a minimal prime ideal of H_P for each $1 \leq i \leq n$, it follows that $\dim S/H_P = 2n - 2$. Hence H_P is Cohen–Macaulay if and only if $\delta(P; 1) = 1$. In other words, H_P is Cohen–Macaulay if and only if P is a chain.

Corollary 10.3.10. *Let P be an arbitrary finite poset with $|P| = n$ and* $\text{rank}(P) = r$. *Then*

(i) depth $S/H_P > $ depth $S/H_P^2 > \cdots > $ depth $S/H_P^r > $ depth S/H_P^{r+1};
(ii) depth $S/H_P^k = n - 1$ *for all* $k > \text{rank}(P)$;
(iii) $\lim_{k \to \infty}$ depth $S/H_P^k = n - 1$.

Corollary 10.3.11. *Given an integer $n > 0$ and given a finite sequence* $(a_1, a_2, \ldots, a_r)$ *of positive integers with* $a_1 \geq a_2 \geq \cdots \geq a_r$ *and with* $a_1 + \cdots + a_r = n$, *there exists a squarefree monomial ideal* $I \subset S = K[x_1, \ldots, x_n, y_1, \ldots, y_n]$ *such that*

(i) depth $S/I^k = 2n - (a_1 + \cdots + a_k) - 1$, $k = 1, 2, \ldots, r - 1$;
(ii) depth $S/I^k = n - 1$ *for all* $k \geq r$;
(iii) $\lim_{k \to \infty}$ depth $S/I^k = n - 1$.

Proof. Let $A(a_i)$ denote the antichain with $|A(a_i)| = a_i$ and P the ordinal sum of the antichains $A(a_1), A(a_2), \ldots, A(a_r)$. In other words, P is the poset whose underlying set is the disjoint union of the sets $A(a_1), \ldots, A(a_r)$ with the property that $\alpha < \beta$ for $\alpha, \beta \in P$ if and only if $\alpha \in A(a_i)$, $\beta \in A(a_j)$ with $i < j$. Thus $\text{rank}(P) = r - 1$. Since $a_1 \geq a_2 \geq \cdots \geq a_r$ and $a_1 + \cdots + a_r = n$, it follows that $\delta(P; k) = a_1 + a_2 + \cdots + a_k$ if $1 \leq k \leq r - 1$ and that $\delta(P; k) = n$ for all $k \geq r$. $\square$

In general, given a function $f : \mathbb{N} \to \mathbb{N}$, we introduce the function Δf by setting $(\Delta f)(k) = f(k) - f(k+1)$ for all $k \in \mathbb{N}$.

Corollary 10.3.12. *Given a nonincreasing function $f : \mathbb{N} \to \mathbb{N}$ with*

$$f(0) = 2 \lim_{k \to \infty} f(k) + 1$$

for which Δf is nonincreasing, there exists a monomial ideal $I \subset S$ such that depth $S/I^k = f(k)$ *for all $k \geq 1$.*

Proof. Let $\lim_{k \to \infty} f(k) = n-1$ and $f(0) = 2n-1$. Let $a_k = (\Delta f)(k-1)$ for all $k \geq 1$. Thus $f(k) = 2n - (a_1 + \cdots a_k) - 1$ for all $k \geq 1$. Since f is nonincreasing, one has $a_k \geq 0$ for all k. Since Δf is nonincreasing, one has $a_1 \geq a_2 \geq \cdots$. Let $r \geq 1$ denote the smallest integer for which $a_1 + a_2 + \cdots + a_r = n$. Thus $a_i > 0$ for $1 \leq i \leq r$ and $a_i = 0$ for all $i > r$. It then follows from Corollary 10.3.11 that there exists a monomial ideal $I \subset S$ for which depth $S/I^k = f(k)$ for all $k \geq 1$. $\square$

To complete the picture we quote without proof the following fact [HH06, Theorem 4.1]: given a bounded nondecreasing function $f : \mathbb{N} \setminus \{0\} \to \mathbb{N}$ there exists a polynomial ring S (with a suitable number of variables) and monomial ideal $I \subset S$ such that depth $S/I^k = f(k)$ for all k.

It is an open question whether any eventually constant numerical function can be the depth function of the powers of a monomial ideal.

10.3.3 Normally torsionfree squarefree monomial ideals and Mengerian simplicial complexes

In view of Proposition 10.3.2 it is of interest to know when the Rees algebra of an ideal is Cohen–Macaulay in order to compute the limit depth of an ideal.

Let $I \subset S$ be a monomial ideal. Then the Rees algebra $\mathcal{R}(I)$ of I is a toric ring. By a famous theorem of Hochster , a toric ring is Cohen–Macaulay if it is normal; see Theorem B.6.2. It is a well-known fact (see for example [HSV91]) that the Rees algebra $\mathcal{R}(I)$ is normal if and only if all powers of I are integrally closed, that is, if I is normal. Combining this fact with Theorem 1.4.6 we obtain

Theorem 10.3.13. *Let I be a squarefree normally torsionfree monomial ideal. Then $\mathcal{R}(I)$ is Cohen–Macaulay.*

Now we want to give a combinatorial interpretation of the condition on squarefree monomial ideal I to be normally torsionfree. To this end, we may view I as the facet ideal of a simplicial complex.

Let Δ be a simplicial complex on $[n]$. Generalizing the concept of vertex covers of a graph introduced in Subsection 9.1.1, we call a subset $C \subset [n]$ a **vertex cover** of the simplicial complex Δ, if $C \cap F \neq \emptyset$ for all $F \in \mathcal{F}(\Delta)$.

The vertex cover C is called a **minimal vertex cover**, if no proper subset of C is a vertex cover. We denote the set of monomial vertex covers of Δ by $\mathcal{C}(\Delta)$.

Obviously the minimal vertex covers of Δ correspond to the minimal prime ideals of the facet ideal $I(\Delta) = (\mathbf{x}_F \colon F \in \mathcal{F}(\Delta))$ of Δ, so that, according to Corollary 1.3.6,

$$I(\Delta) = \bigcap_{C \in \mathcal{C}(\Delta)} P_C.$$

By Theorem 1.4.6, $I(\Delta)$ is normally torsionfree if and only if $I(\Delta)^{(k)} = I(\Delta)^k$ for all k. In other word, $I(\Delta)$ is normally torsion free, if the **symbolic Rees algebra**

$$\mathcal{R}^s(I(\Delta)) = \bigoplus_{k \geq 0} I(\Delta)^{(k)} t^k$$

of $I(\Delta)$ is standard graded.

In order to analyze when $\mathcal{R}^s(I(\Delta))$ is standard graded, we introduce two invariants attached to an integer vector $\mathbf{a} = (a_1, \ldots, a_n) \in \mathbb{Z}_+^n$.

We let $o(\mathbf{a})$ be the largest integer k such that $\mathbf{x}^{\mathbf{a}} \in I(\Delta)^{(k)}$, and let $\sigma(\mathbf{a})$ the largest integer r such that $\mathbf{x}^{\mathbf{a}} \in I(\Delta)^r$.

Obviously, $\sigma(\mathbf{a}) \leq o(\mathbf{a})$, and equality holds for all $\mathbf{a} \in \mathbb{Z}_+^n$, if and only if $\mathcal{R}^s(I(\Delta))$ is standard graded.

Let $\mathcal{F}(\Delta) = \{F_1, \ldots, F_m\}$, and let M be the **incidence matrix** of Δ, that is, the $m \times n$-matrix $M = (e_{ij})$ with $e_{ij} = 1$ if $j \in F_i$ and $e_{ij} = 0$ if $j \notin F_i$.

For two vectors $\mathbf{a}$ and $\mathbf{b}$ in $\mathbb{Z}^n$ we write $\mathbf{a} \geq \mathbf{b}$ if $a_i \geq b_i$ for all i, and we set $\mathbf{1} = (1, \ldots, 1)$. Then the invariants $o(\mathbf{a})$ and $\sigma(\mathbf{a})$ can be expressed as follows.

Proposition 10.3.14. *Let* $\mathbf{a} \in \mathbb{Z}_+^n$. *Then*

(a) $o(\mathbf{a}) = \min\{\langle \mathbf{c}, \mathbf{a} \rangle \colon \mathbf{c} \in \mathbb{Z}_+^n, \ M \cdot \mathbf{c} \geq \mathbf{1}\}$;
(b) $\sigma(\mathbf{a}) = \max\{\langle \mathbf{b}, \mathbf{1} \rangle \colon \mathbf{b} \in \mathbb{Z}_+^m, \ M^t \cdot \mathbf{b} \leq \mathbf{a}\}$.

Here M^t denotes the transpose of M and $\langle \ , \ \rangle$ the standard scalar product.

Proof. (a) To say that $\mathbf{x}^{\mathbf{a}} \in I(\Delta)^{(k)} = \bigcap_{C \in \mathcal{C}(\Delta)} P_C^k$ is equivalent to saying that $\sum_{i \in C} a_i \geq k$ for all $C \in \mathcal{C}(\Delta)$. This implies that

$$o(\mathbf{a}) = \min\{\sum_{i \in C} a_i \colon C \in \mathcal{C}(\Delta)\}. \tag{10.9}$$

Let $\mathbf{c} \in \{0, 1\}^n$, and set $\mathrm{supp}(\mathbf{c}) = \{i \in [n] \colon c_i \neq 0\}$. Then $C \subset [n]$ is a vertex cover of Δ if and only if $C = \mathrm{supp}(\mathbf{c})$ with $M \cdot \mathbf{c} \geq \mathbf{1}$. Thus, since $\langle \mathbf{a}, \mathbf{c} \rangle = \sum_{i \in C} a_i$, it follows from (10.9) that

$$o(\mathbf{a}) = \min\{\langle \mathbf{a}, \mathbf{c} \rangle \colon \mathbf{c} \in \{0, 1\}^n, \ M \cdot \mathbf{c} \geq \mathbf{1}\}$$
$$= \min\{\langle \mathbf{a}, \mathbf{c} \rangle \colon \mathbf{c} \in \mathbb{Z}_+^n, \ M \cdot \mathbf{c} \geq \mathbf{1}\}.$$

(b) Let $\mathbf{a}_1, \ldots, \mathbf{a}_m$ be the $(0, 1)$-vectors with $\mathrm{supp}(a_i) = F_i$ for $i = 1, \ldots, m$. Then $\mathbf{x}^{\mathbf{a}} \in I(\Delta)^k$ if and only if there exist nonnegative integers $b_1, \ldots, b_m$ such that $k = \sum_{i=1}^m b_i$ and $\sum_{i=1}^m b_i \mathbf{a}_i \leq \mathbf{a}$. Let $\mathbf{b} = (b_1, \ldots, b_m)$. Then $\langle \mathbf{b}, \mathbf{1} \rangle = k$ and $M^t \cdot \mathbf{b} = \sum_{i=1}^m b_i \mathbf{a}_i$. This yields the desired formula for $\sigma(\mathbf{a})$.   □

A simplicial complex Δ on $[n]$ with incidence matrix M is called a **Mengerian simplicial complex** if for all $\mathbf{a} \in \mathbb{Z}_+^n$,

$$\min\{\langle \mathbf{c}, \mathbf{a} \rangle \colon \mathbf{c} \in \mathbb{Z}_+^n, \ M \cdot \mathbf{c} \geq \mathbf{1}\} = \max\{\langle \mathbf{b}, \mathbf{1} \rangle \colon \mathbf{b} \in \mathbb{Z}_+^m, \ M^t \cdot \mathbf{b} \leq \mathbf{c}\}.$$

Our discussion so far combined with Proposition 10.3.14 and Theorem 10.3.13 yields the following conclusion.

Corollary 10.3.15. *Let Δ be a simplicial complex. Then the following conditions are equivalent:*

(a) $I(\Delta)$ *is normally torsionfree;*
(b) $\mathcal{R}(I(\Delta)) = \mathcal{R}^s(I(\Delta))$;
(c) $\mathcal{R}^s(I(\Delta))$ *is standard graded;*
(d) Δ *is a Mengerian simplicial complex.*

If the equivalent conditions hold, then $\mathcal{R}(I(\Delta))$ is Cohen–Macaulay.

10.3.4 Classes of Mengerian simplicial complexes

Let Δ be a simplical complex on the vertex set $[n]$. A **cycle** or, more precisely, an s-cycle of Δ ($s \geq 2$) is an alternating sequence of distinct vertices and facets $v_1, F_1, \ldots, v_s, F_s, v_{s+1} = v_1$ such that $v_i, v_{i+1} \in F_i$ for $i = 1, \ldots, s$. A cycle is **special** if it has no facet containing more than two vertices of the cycle. Observe that a cycle of a graph is always special.

By a result of [FHO74, Theorem 5.1], a simplicial complex which has no special odd cycles is Mengerian. Thus from Corollary 10.3.15 we obtain

Theorem 10.3.16. *Let Δ be a simplicial complex which has no special odd cycles. Then $I(\Delta)$ is normally torsionfree.*

Since the bipartite graphs are exactly those which have no odd cycles, we obtain

Corollary 10.3.17. *Let G be a bipartite graph. Then $I(G)$ is normally torsionfree.*

Corollary 10.3.17 has an interesting consequence

Corollary 10.3.18. *Let G be a bipartite graph with c connected components. Then $\lim_{k \to \infty} \mathrm{depth}\, S/I(G)^k = c$.*

Proof. Since $I(G)$ is normally torsionfree, it follows from Corollary 10.3.15 that $\mathcal{R}(I(\Delta))$ is Cohen–Macaulay. Thus Proposition 10.3.2 implies that

$$\lim_{k\to\infty} \text{depth } S/I(G)^k = n - \ell(I(\Delta)).$$

It follows from the subsequent lemma that $\ell(I(\Delta))$ is the rank of the incidence matrix of G. By a result of [GKS95, Theorem 2.5], the rank of the incidence matrix of a graph G with n vertices is equal $n - c_0$, where c_0 is the number of components of G which do not contain an odd cycle. Since G is bipartite, we have $c = c_0$, and the assertion follows.   □

Lemma 10.3.19. *Let $I \subset S = K[x_1, \ldots, x_n]$ be a monomial ideal generated in a single degree with $G(I) = \{\mathbf{x}^{\mathbf{a}_1}, \ldots, \mathbf{x}^{\mathbf{a}_m}\}$, and let A be the $m \times n$ matrix whose columns are the vectors $\mathbf{a}_1, \ldots, \mathbf{a}_m$. Then $\ell(I) = \text{rank}_{\mathbb{Q}} A$.*

Proof. Since by assumption all generators of I have the same degree, it follows that $\mathcal{R}(I)/\mathfrak{m}\mathcal{R}(I) \cong R$ where $R = K[\mathbf{x}^{\mathbf{a}_1}, \ldots, \mathbf{x}^{\mathbf{a}_m}]$. Therefore, $\ell(I) = \dim R$. Since R is an affine K algebra, its Krull dimension is given by the transcendence degree $\text{tr deg } Q(R)/K$ of the quotient field of R over K; see [Mat80, (14G) Corollary1] or [Kun08, Corollary 3.6]. Let $\{\mathbf{a}_{i_1}, \ldots \mathbf{a}_{i_\ell}\}$ be a maximal set of linearly independent row vectors of A over $\mathbb{Q}$. Then $\text{rank}_{\mathbb{Q}} A = \ell$. On the other hand, the K subalgebra $T = K[\mathbf{x}^{\mathbf{a}_{i_1}}, \ldots, \mathbf{x}^{\mathbf{a}_{i_\ell}}]$ is a polynomial ring and $Q(R)$ is algebraic over $Q(T)$. Therefore, $\text{tr deg } Q(R) = \text{tr deg } Q(T) = \ell$, as desired.   □

Observe that if I is the facet ideal of a simplicial complex Δ, then A is nothing but the incidence matrix of Δ.

A simplicial complex without a special odd cycle can be also characterized in terms of its incidence matrix. In fact, a special cycle corresponds to an $s \times s$ submatrix of the form

$$\begin{pmatrix} 1 & 0 & 0 & \cdot & \cdot & 0 & 1 \\ 1 & 1 & 0 & \cdot & \cdot & 0 & 0 \\ 0 & 1 & 1 & \cdot & \cdot & 0 & 0 \\ \cdot & & & & & & \cdot \\ \cdot & & & & & & \cdot \\ \cdot & & & & & 1 & 0 \\ 0 & \cdot & \cdot & \cdot & 0 & 1 & 1 \end{pmatrix}.$$

with $s \geq 2$. Therefore Δ has no special odd cycle if and only if its incidence matrix has no such $s \times s$ submatrix with odd s, even after a permutation of rows and columns.

We say that Δ is an **unimodular** simplicial complex if every square submatrix of its incidence matrix has determinant equal to $0, \pm 1$.

The above matrix has determinant equal to 2 if s is odd. Therefore a unimodular simplicial complex has no special odd cycle. In particular it is Mengerian and consequently we have

Corollary 10.3.20. *Let Δ be a unimodular simplicial complex. Then $I(\Delta)$ is normally torsionfree.*

As a last example of Mengerian simplicial complexes we consider simplicial forests. A simplicial complex Δ with $\mathcal{F}(\Delta) = \{F_1, \ldots, F_m\}$ is called a **forest** if for each nonempty subset $\{F_{i_1}, \ldots, F_{i_k}\} \subset \mathcal{F}(\Delta)$, the simplical subcomplex Γ with $\mathcal{F}(\Gamma) = \{F_{i_1}, \ldots, F_{i_k}\}$ has a leaf. Recall from Chapter 9 that a facet F of Δ is called a leaf, if either F is the only facet of Δ, or there exists $G \in \mathcal{F}(\Delta)$, $G \neq F$ such that $H \cap F \subset G \cap F$ for each $H \in \mathcal{F}(\Delta)$ with $H \neq F$.

A graph which is a forest may also be viewed as a simplicial forest, and any forest is a quasi-forest.

Proposition 10.3.21. *Let Δ be a forest. Then Δ has no special cycles of length ≥ 3. In particular, $I(\Delta)$ is normally torsionfree and $\mathcal{R}(I(\Delta))$ is Cohen–Macaulay.*

Proof. Assume that Δ has a special cycle $v_1, F_1, \ldots, v_s, F_s, v_{s+1} = v_1$ with $s \geq 3$. Let Γ be the subcomplex with the facets $F_1, \ldots, F_s$ and F_1 a leaf of Γ. Then there exists a facet $F_i \neq F_1$ such that $F_i \cap F_1 \neq \emptyset$ and $F_j \cap F_1 \subseteq F_i \cap F_1$ for all $j \neq 1$. Therefore, $v_1, v_2 \in F_i$. Since F_1 is the only facet of the cycle which contains v_1, v_2, we get $F_i = F_1$; a contradiction. $\quad\square$

Corollary 10.3.22. *Let Δ be a forest with vertex set $[n]$. Assume that Δ is pure and has m facets. Then $\lim_{k \to \infty} \operatorname{depth} S/I(\Delta)^k = n - m$.*

Proof. By Proposition 10.3.21, $\mathcal{R}(I(\Delta))$ is Cohen–Macaulay, so that we may apply Proposition 10.3.2 to conclude that $\lim_{k \to \infty} S/I(\Delta)^k = n - \ell(I(\Delta))$. Since all generators of $I(\Delta)$ have the same degree, Lemma 10.3.19 implies that $\ell(I(\Delta)) = \operatorname{rank}_Q A$, where A is the incidence matrix of Δ. Any forest is a quasi-forest. Thus we may choose a leaf order $F_1, \ldots, F_m$. Since for each i, the facet F_i has a vertex which does not appear as a vertex of any F_j for $j < i$, one sees that $\operatorname{rank}_Q A = m$. The desired result follows. $\quad\square$

Problems

10.1. Compute the toric ideal of each of the following monomial configurations:

- $\{x_1^2, x_1 x_2, x_2^2\}$;
- $\{x_1 x_2, x_4 x_5, x_1 x_3, x_2 x_3, x_3 x_4, x_3 x_5\}$;
- $\{x_1 x_2, x_2 x_3, \ldots, x_{2q-1} x_{2q}, x_{2q} x_1\}$;
- $\{x_1^i x_2^{q-i} : i = 0, 1, \ldots, q\}$;
- $\{x_1 x_3 x_5, x_1 x_3 x_6, x_1 x_4 x_5, x_1 x_4 x_6, x_2 x_3 x_5, x_2 x_3 x_6, x_2 x_4 x_5, x_2 x_4 x_6\}$;
- $\{x_i x_j x_k : 1 \leq i < j < k \leq 5\}$.

10.2. (a) Compute the toric ideal I_{A_P}, where P is the poset of Problem 9.6.
(b) Compute the toric ideal I_{A_G}, where G is the finite graph of Problem 9.10.
(c) Compute the toric ideal I_{A_G}, where G is the complete graph with 4 vertices.

10.3. Show that every primitive binomial belonging to a toric ideal is irreducible.

10.4. Let G be the complete graph with 6 vertices.
(a) How many primitive even closed walks of length 4 does G possess?
(b) How many primitive even closed walks of length 6 does G possess?
(c) How many primitive even closed walks of length 8 does G possess?

10.5. Let G be a bipartite graph. Show that the toric ideal I_{A_G} is generated by quadratic binomials if and only if every cycle of length > 4 has a chord.

10.6. Among the following monomial ideals, which of them satisfy the x-condition?

- (x_1^2, x_1x_2, x_2^2);
- $(x_1x_2, x_4x_5, x_1x_3, x_2x_3, x_3x_4, x_3x_5)$;
- $(x_1x_3x_5, x_1x_3x_6, x_1x_4x_5, x_1x_4x_6, x_2x_3x_5, x_2x_3x_6, x_2x_4x_5, x_2x_4x_6)$;

10.7. Let $I \subset S = K[x_1, \ldots, x_6]$ be the monomial ideals generated by

$$x_1x_4, x_2x_5, x_3x_6, x_4x_5, x_4x_6, x_5x_6.$$

(a) Show that I has a linear resolution.
(b) Find an ordering of the monomials belonging to $G(I)$ for which I has linear quotients.
(c) Does I^2 have linear quotients?

10.8. Let I be a monomial ideal generated by quadratic monomials $u_1, \ldots, u_s$ and suppose that I has linear quotients with respect to this given ordering. Is it true or false that I^2 has linear quotients with respect to the lexicographic ordering

$$u_1^2, u_1u_2, \ldots, u_1u_s, u_2^2, u_2u_3, \ldots, u_s^2 ?$$

10.9. Let G be the Cohen–Macaulay tree on $\{1, \ldots, 6\}$ with the edges

$$\{1,2\}, \{2,3\}, \{3,4\}, \{3,5\}, \{5,6\}.$$

(a) Find the vertex cover ideal I_G of G.
(b) Show that I_G has linear quotients.
(c) Does $(I_G)^2$ have linear quotients?

10.10. Let G be the Cohen–Macauly chordal graph on $\{1, \ldots, 6\}$ with the edges

$$\{1,2\}, \{1,3\}, \{1,4\}, \{2,3\}, \{3,4\}, \{3,5\}, \{4,5\}, \{4,6\}, \{5,6\}.$$

(a) Find the vertex cover ideal I_G of G.
(b) Show that I_G has linear quotients.
(c) Does $(I_G)^2$ have linear quotients?

10.11. Find a Cohen–Macaulay finite graph G which is neither bipartite nor chordal such that all powers of the vertex cover ideal I_G of G have a linear resolution.

10.12. Let P be the poset of the positive integers dividing 12, ordered by divisibility. Compute the depth of S/H_P^k for all k.

10.13. Let Δ be a simplicial complex on $[n]$ and Δ^* the simplicial complex with $\mathcal{F}(\Delta^*) = \mathcal{C}(\Delta)$. Show:
(a) $\Delta^{**} = \Delta$.
(b) $I(\Delta^*) = \bigcap_{F \in \mathcal{F}(\Delta)} P_F$.

10.14. Let Δ be a simplicial complex on $[n]$. A vector $\mathbf{a} = (a_1, \ldots, a_n)$ with nonnegative integer coefficients is called a **vertex cover of order** k of Δ if $\sum_{i \in F} \geq k$ for all $F \in \mathcal{F}(\Delta)$.
(a) Let $\mathbf{a}$ be a vertex cover of order k and $\mathbf{b}$ a vertex cover of order ℓ. Show that $\mathbf{a} + \mathbf{b}$ is a vertex cover of order $k + \ell$.
(b) Let $S = K[x_1, \ldots, x_n]$ and $S[t]$ be the polynomial ring over S in the indeterminate t. For each $k \geq 0$, let $A_k(\Delta) \subset S[t]$ be the K-subspace of $S[t]$ spanned by the monomials $\mathbf{x}^{\mathbf{a}} t^k$ where $\mathbf{a}$ is a vertex cover of order k of Δ. Use (a) to prove that $A(\Delta) = \bigoplus_{k \geq 0} A_k(\Delta)$ has the natural structure of a graded S-algebra.
(c) Show that $A(\Delta) = \mathcal{R}^s(I(\Delta^*))$.
(d) Let G be a bipartite graph. Then show that $A(G)$ is standard graded.

Notes

The study of homological properties of powers of a graded ideal has been one of the main topics of commutative algebra in recent years. Important invariants, like the regularity, the depth or the set of associated prime ideals become stable for high powers; see [CHT99], [Kod00] and [Bro79]. The powers of an ideal with linear resolution or linear quotients need not have a linear resolution. On the other hand, Römer [Roe01b] gives an upper bound for the regularity of all powers of a graded ideal in terms of the so-called x-regularity of the corresponding Rees ring. This implies that if the x-regularity is zero, then all powers of the ideal do have a linear resolution. By using Römer's result another criterion in terms of the initial ideal of the defining ideal of the Rees algebra was obtained in [HHZ04a]. This criterion in combination with Fröberg's [Fro90], Dirac's theorem [Dir61] and [OH99] is used to prove Theorem 10.2.6. The result on quadratic Gröbner bases arising finite posets (Theorem 10.1.3) is taken from [Hib87].

For certain classes of monomial ideals which naturally arise in combinatorial contexts the depth of the powers can be computed by the method of linear quotients. The examples given here are taken from [HH06]. Theorem 10.3.13

is a direct of consequence of a result in the paper [SVV94] by Simis, Vasconcleos and Villarreal. Also the fact that edge ideal of a bipartite graph is normally torsionfree is shown in [SVV94]. Most of the results of Subsections 10.3.3 and 10.3.4 are taken from [HHTZ08]. There the symbolic Rees algebra is interpreted as a vertex cover algebra. Higher order vertex covers and vertex cover algebras were first introduced in [HHT07]. The relationship between facet ideals and Mengerian simplicial complexes was first studied in the paper [GVV07] of Gitler, Valencia and Villarreal. Combining the results of this paper with results of Escobar, Villarreal and Yoshino [EVY06], one obtains another proof of Corollary 10.3.15. Important classes of Mengerian simplicial complexes are the unimodular simplicial complexes. In [HHT09] it is shown that the vertex cover algebra of a weighted simplicial complex is standard graded for all weight functions if and only if the simplical complex unimodular.

Simplicial forests were introduced by Faridi [Far02]. It turns out that they are just the hypergraphs which have no special odd cycle of length ≥ 3. In hypergraph theory such hypergraphs are called totally balanced. In her paper [Far02], Faridi showed that the Rees algebra of the facet ideal of a simplicial tree is a normal Cohen–Macaulay domain. This result is related to Proposition 10.3.21. In [Far04] it is shown that simplicial trees are sequentially Cohen–Macaulay.

11

Shifting theory

Algebraic shifting, introduced by Gil Kalai, is one of the most powerful techniques to develop the extremal combinatorics of simplicial complexes. We will make a self-contained and systematic study of the algebraic aspects of shifting theory from a viewpoint of generic initial ideals and graded Betti numbers.

11.1 Combinatorial shifting

First of all, we discuss combinatorial shifting, which played an important role in the classical extremal combinatorics of finite sets.

11.1.1 Shifting operations

A simplicial complex Δ on $[n]$ is **shifted** if, for $F \in \Delta$, $i \in F$ and $j \in [n]$ with $j > i$, one has $(F \setminus \{i\}) \cup \{j\} \in \Delta$.

Note that Δ is shifted if and only if I_Δ is squarefree strongly stable.

A **shifting operation** on $[n]$ is a map which associates each simplicial complex Δ on $[n]$ with a simplicial complex $\mathrm{Shift}(\Delta)$ on $[n]$ and which satisfies the following conditions:

(S_1) $\mathrm{Shift}(\Delta)$ is shifted;
(S_2) $\mathrm{Shift}(\Delta) = \Delta$ if Δ is shifted;
(S_3) $f(\Delta) = f(\mathrm{Shift}(\Delta))$;
(S_4) $\mathrm{Shift}(\Delta') \subset \mathrm{Shift}(\Delta)$ if $\Delta' \subset \Delta$.

11.1.2 Combinatorial shifting

In classical combinatorics of finite sets, Erdös, Ko and Rado introduced combinatorial shifting.

Let Δ be a simplicial complex on $[n]$. Let $1 \leq i < j \leq n$. Write $\mathrm{Shift}_{ij}(\Delta)$ for the collection of subsets of $[n]$ consisting of the sets $C_{ij}(F) \subset [n]$, where $F \in \Delta$ and where

J. Herzog, T. Hibi, *Monomial Ideals*, Graduate Texts in Mathematics 260,   211
DOI 10.1007/978-0-85729-106-6_11, © Springer-Verlag London Limited 2011

$$C_{ij}(F) = \begin{cases} (F \setminus \{i\}) \cup \{j\}, & \text{if } i \in F, \ j \notin F \text{ and } (F \setminus \{i\}) \cup \{j\} \notin \Delta, \\ F, & \text{otherwise.} \end{cases}$$

It follows easily that $\text{Shift}_{ij}(\Delta)$ is a simplicial complex on $[n]$.

Lemma 11.1.1. *The operation* $\Delta \to \text{Shift}_{ij}(\Delta)$ *satisfies the conditions* (S_2), (S_3) *and* (S_4).

Proof. Since $|C_{ij}(F)| = |F|$ for all faces F of Δ and since $C_{ij}(F) \neq C_{ij}(G)$ if $F \neq G$, it follows that Δ and $\text{Shift}_{ij}(\Delta)$ have the same f-vector. Hence (S_3) is satisfied. The condition (S_4) is clearly satisfied.

Let Δ be shifted and F a face of Δ. Let $i \in F$ and $i < j$ with $j \notin F$. Since Δ is shifted, it follows that $(F \setminus \{i\}) \cup \{j\}$ must be a face of Δ. Thus $C_{ij}(F) = F$ for all faces F of Δ. Hence (S_2) is satisfied. $\square$

Lemma 11.1.2. *There exists a finite sequence of pairs of integers*

$$(i_1, j_1), (i_2, j_2), \ldots, (i_q, j_q)$$

with each $1 \leq i_k < j_k \leq n$ *such that*

$$\text{Shift}_{i_q j_q}(\text{Shift}_{i_{q-1} j_{q-1}}(\cdots(\text{Shift}_{i_1 j_1}(\Delta))\cdots))$$

is shifted.

Proof. For each face $F = \{j_1, \ldots, j_d\}$ of Δ, we set $c(F) = j_1 + \cdots + j_d$. Let $c(\Delta) = \sum_{F \in \Delta} c(F)$. Obviously one has $c(\Delta) \leq c(\text{Shift}_{ij}(\Delta))$. If Δ is not shifted, then there exists a face F together with i and j with $i < j$ such that $i \in F$, $j \notin F$ and $(F \setminus \{i\}) \cup \{j\} \notin \Delta$. Since $c(F) < c(C_{ij}(F))$, it follows that $c(\Delta) < c(\text{Shift}_{ij}(\Delta))$. This simple observation yields the desired result. $\square$

An arbitrary shifted complex which is obtained by a finite number of sequences of operations as described in Lemma 11.1.2 will be denoted by Δ^c and will be called a **combinatorial shifted complex** of Δ. It follows from Lemma 11.1.1 that the operation $\Delta \to \Delta^c$ is a shifting operation. Such an operation is called **combinatorial shifting**. A combinatorial shifted complex Δ^c of Δ is, however, not necessarily uniquely determined by Δ. Later, we will see some extremely bad behaviour of combinatorial shifting. The only advantage of combinatorial shifting is that it is easily computable.

11.2 Exterior and symmetric shifting

We now introduce exterior algebraic shifting and symmetric algebraic shifting.

11.2.1 Exterior algebraic shifting

Let K be an infinite field and $E = \bigoplus_{d=0}^n \bigwedge^d V$ the exterior algebra of a vector space V over K of dimension n with basis $e_1, \ldots, e_n$. Let Δ be a simplicial complex on $[n]$ and $J_\Delta \subset E$ the exterior face ideal of Δ. Let $<_{\text{rev}}$ denote the reverse lexicographic order on E induced by the ordering $e_1 > \cdots > e_n$. Let

$$\Delta^e = \text{gin}_{<_{\text{rev}}}(J_\Delta).$$

We know by Proposition 5.2.10 that the exterior face ideal J_{Δ^e} of Δ^e is strongly stable. Thus Δ^e is shifted. We call Δ^e the **exterior algebraic shifted complex** of Δ.

Proposition 11.2.1. *The operation $\Delta \to \Delta^e$ is a shifting operation.*

Proof. Since Δ^e is shifted, the condition (S_1) is satisfied. Since J_Δ is strongly stable, it follows that $\text{gin}_{<_{\text{rev}}}(J_\Delta) = J_\Delta$, see Theorem 5.2.9. Thus (S_2) is satisfied. On the other hand, since $\text{gin}_{<_{\text{rev}}}(J_\Delta)$ and J_Δ have the same Hilbert function, one has $f(\Delta^e) = f(\Delta)$. Thus (S_3) is satisfied. Finally, if Γ is a subcomplex of Δ, then $J_\Delta \subset J_\Gamma$. Thus $J_{\Delta^e} \subset J_{\Gamma^e}$. Hence $\Gamma^e \subset \Delta^e$. Thus (S_4) is satisfied. $\square$

The shifting operation $\Delta \to \Delta^e$ is called the **exterior algebraic shifting**.

11.2.2 Symmetric algebraic shifting

Let K be a field of characteristic 0 or of characteristic $> n$ and $S = K[x_1, \ldots, x_n]$ the polynomial ring in n variables over K. We work with the reverse lexicographic order $<_{\text{rev}}$ on S induced by the ordering $x_1 > \cdots > x_n$.

Let $I \subset S$ be a squarefree monomial ideal and $\text{gin}_{<_{\text{rev}}}(I)$ its generic initial ideal with respect to $<_{\text{rev}}$. Since K is of characteristic 0, it follows that $\text{gin}_{<_{\text{rev}}}(I)$ is strongly stable. However, $\text{gin}_{<_{\text{rev}}}(I)$ is no longer squarefree.

Lemma 11.2.2. *Let $I \subset S$ be a squarefree monomial ideal. Then*

$$m(u) + \deg u \le n + 1$$

for all monomial u belonging to $G(\text{gin}_{<_{\text{rev}}}(I))$.

Proof. Since $\text{gin}_{<_{\text{rev}}}(I)$ is strongly stable, it follows from Corollary 7.2.3 that

$$\beta_{i\,i+j}(I) = \sum_{u \in G(\text{gin}_{<_{\text{rev}}}(I))_j} \binom{m(u) - 1}{i},$$

where $G(\text{gin}_{<_{\text{rev}}}(I))_j$ is the set of monomials $u \in G(\text{gin}_{<_{\text{rev}}}(I))$ of degree j. Thus in particular

$$\max\{m(u) + \deg u - 1 : u \in G(\mathrm{gin}_{<_{\mathrm{rev}}}(I))\}$$

is the highest shift in the resolution of $\mathrm{gin}_{<_{\mathrm{rev}}}(I)$. Since I is squarefree ideal, Hochster's formula (Theorem 8.1.1) guarantees that the highest shift in the resolution of I is at most n. Since the Betti number with the highest shift in the resolution on I is extremal, it follows from Theorem 4.3.17 that the highest shift in the resolution of I and that of $\mathrm{gin}_{<_{\mathrm{rev}}}(I)$ coincides. Hence $m(u) + \deg u - 1 \leq n$ for all $u \in G(\mathrm{gin}_{<_{\mathrm{rev}}}(I))$, as desired. $\square$

In order to define symmetric algebraic shifting, we must introduce a certain operator, called the squarefree operator, which transfers $\mathrm{gin}_{<_{\mathrm{rev}}}(I)$ into a squarefree strongly stable ideal.

Let $u = x_{i_1} x_{i_2} \cdots x_{i_d}$ be a monomial of S, where $i_1 \leq i_2 \leq \cdots \leq i_d$, we set

$$u^\sigma = x_{i_1} x_{i_2+1} \cdots x_{i_j+(j-1)} \cdots x_{i_d+(d-1)}.$$

One has

$$m(u^\sigma) - \deg u^\sigma = m(u) - 1. \tag{11.1}$$

Thus u^σ belongs to S if and only if $m(u) + \deg u \leq n+1$. The operator $u \to u^\sigma$ will be called **squarefree operator**.

Corollary 11.2.3. *Let I be a squarefree ideal of S. Then u^σ belongs to S for all $u \in G(\mathrm{gin}_{<_{\mathrm{rev}}}(I)))$.*

The squarefree operator $u \to u^\sigma$ naturally arises in the very elementary stage of enumerative combinatorics.

Example 11.2.4. Let $A_{n,d}$ be the set of monomials in the variables $x_1, \ldots, x_n$ of degree d and $B_{n,d}$ the set of squarefree monomials in the variables $x_1, \ldots, x_n$ of degree d. In high school mathematics we learn $|B_{n,d}| = \binom{n}{d}$. What is $|A_{n,d}|$? We associate each monomial $u \in A_{n,d}$ with $u^\sigma \in B_{n+d-1,d}$. The map $u \to u^\sigma$ gives a bijection between $A_{n,d}$ and $B_{n+d-1,d}$. Its inverse is the map which associate each squarefree monomial $v = x_{i_1} \cdots x_{i_d}$ of $B_{n+d-1,d}$, where $1 \leq i_1 < \cdots < i_d \leq n + d - 1$, with the monomial

$$v^\tau = x_{i_1} x_{i_2-1} \cdots x_{i_j-(j-1)} \cdots x_{i_d-(d-1)}$$

belonging to $A_{n,d}$. Thus $|A_{n,d}| = |B_{n+d-1,d}|$. Hence $|A_{n,d}| = \binom{n+d-1}{d} = \binom{n+d-1}{n-1}$.

Let $I \subset S$ be strongly stable ideal. We write I^σ for the squarefree monomial ideal generated by the monomials $u_1^\sigma, \ldots, u_s^\sigma$.

Lemma 11.2.5. *If $I \subset S$ is strongly stable with $G(I) = \{u_1, \ldots, u_s\}$, then I^σ is squarefree strongly stable with $G(I^\sigma) = \{u_1^\sigma, \ldots, u_s^\sigma\}$.*

Proof. First, suppose that, for some $u_j \in G(I)$, one has $u_j^\sigma \notin G(I^\sigma)$. Then there exists an integer $i \neq j$ such that $u_i^\sigma | u_j \sigma$. If $u_i^\sigma = u_j^\sigma$, then $u_i = u_j$, a contradiction. Hence we may assume that u_i^σ is a proper divisor of u_j^σ.

Let $u_i = x_{k_1} x_{k_2} \cdots x_{k_t}$ and $u_j = x_{l_1} x_{l_2} \cdots x_{l_d}$. Then we have $u_i^\sigma = x_{k_1} x_{k_2+1} \cdots x_{k_t+(t-1)}$ and $u_j^\sigma = u_j = x_{l_1} x_{l_2+1} \cdots x_{l_d+(d-1)}$. Since u_i^σ divides u_j^σ properly, there exist $p_1 \leq p_2 \leq \cdots \leq p_t$ such that $k_1 = l_{p_1}+(p_1-1), k_2+1 = l_{p_2} + (p_2 - 1), \ldots, k_t + (t - 1) = l_{p_t} + (p_t - 1)$ with $t < d$. It follows that $k_r = l_{p_r} + (p_r - r)$ for $r = 1, \ldots, t$. Thus $u_i = \prod_{r=1}^{t} x_{l_{p_r}+(p_r-r)}$. Since I is strongly stable and $u_i \in I$ it follows that $\prod_{r=1}^{t} x_{l_{p_r}} \in I$, contradicting the fact that $u_j \in G(I)$.

Second, to see why I^σ is squarefree strongly stable, we take a monomial $u = x_{i_1} \cdots x_{i_d} \in G(I)$ together with $u_0 = (x_b u^\sigma)/x_{i_a+(a-1)}$, where x_b does not divide u^σ and where $b < i_a + (a - 1)$ and $a \in [d]$. We claim $u_0 \in I^\sigma$. Choose $p < a$ such that $i_p + (p - 1) < b < i_{p+1} + p$. (Here $i_0 = 1$). Let

$$v = (\prod_{j=1}^{p} x_{i_j}) x_{b-p} (\prod_{j=p+1}^{a-1} x_{i_j-1})(\prod_{j=a+1}^{d} x_{i_j}).$$

Since $b - p < i_{p+1} \leq i_a$ and since I is strongly stable, the monomial v belongs to I. One has $v^\sigma = (x_b u^\sigma)/x_{i_a+(a-1)} = u_0$. Let, say, $v = x_{\ell_1} \cdots x_{\ell_d}$ with $\ell_1 \leq \cdots \leq \ell_d$. Again, since I is strongly stable, it follows that $w = x_{\ell_1} \cdots x_{\ell_c} \in G(I)$ for some $c \leq d$. Since w^σ divides $v^\sigma = u_0$, one has $u_0 \in I^\sigma$, as desired. $\square$

Let Δ be a simplicial complex on $[n]$. Since the base field K is of characteristic 0 or of characteristic $> n$, Proposition 4.2.4 implies that $\mathrm{gin}_{<_{\mathrm{rev}}}(I_\Delta)$ is strongly stable. Thus $(\mathrm{gin}_{<_{\mathrm{rev}}}(I_\Delta))^\sigma$ is a squarefree strongly stable ideal of S. Now, the **symmetric algebraic shifted complex** of Δ is defined to be the shifted complex Δ^s on $[n]$ with

$$I_{\Delta^s} = (\mathrm{gin}_{<_{\mathrm{rev}}}(I_\Delta))^\sigma.$$

Lemma 11.2.6. *If $I \subset S$ is a strongly stable ideal, then $\beta_{ii+j}(I) = \beta_{ii+j}(I^\sigma)$ for all i and j.*

Proof. The desired formula follows from (11.1) together with Corollary 7.2.3 and Corollary 7.4.2 $\square$

Lemma 11.2.6 implies in particular that the operation $\Delta \to \Delta^s$ satisfies the condition (S_3). On the other hand, as in the case of exterior shifting one shows that the operation $\Delta \to \Delta^s$ satisfies the condition (S_4).

Finally, the fact that the operation $\Delta \to \Delta^s$ satisfies the condition (S_2) follows from Theorem 11.2.7 stated below. We call the shifting operation $\Delta \to \Delta^s$ **symmetric algebraic shifting**.

Theorem 11.2.7. *Let $I \subset S$ be a squarefree strongly stable ideal. Then*

$$I = \mathrm{gin}_{<_{\mathrm{rev}}}(I)^\sigma.$$

Proof. Working with induction on the largest integer $m(u)$ for which $u \in G(I)$, by using Lemma 11.2.8 we may suppose that there is $u \in G(I)$ with $m(u) = n$.

Let $I' = I : (x_n)$ and I'' the squarefree ideal of S generated by those squarefree monomials $u \in G(I)$ with $m(u) < n$. Each of I' and I'' is squarefree strongly stable and $I'' \subset I \subset I'$. Our assumption of induction guarantees that $I' = \mathrm{gin}_{<_{\mathrm{rev}}}(I')^\sigma$ and $I'' = \mathrm{gin}_{<_{\mathrm{rev}}}(I'')^\sigma$. Hence

$$I'' \subset \mathrm{gin}_{<_{\mathrm{rev}}}(I)^\sigma \subset I'.$$

We claim that $I \subset \mathrm{gin}_{<_{\mathrm{rev}}}(I)^\sigma$. Since $I'' \subset \mathrm{gin}_{<_{\mathrm{rev}}}(I)^\sigma$, each $u \in G(I)$ with $m(u) < n$ belongs to $\mathrm{gin}_{<_{\mathrm{rev}}}(I)^\sigma$.

Now, let $w_1, \ldots, w_q$ be the monomials belonging to $G(\mathrm{gin}_{<_{\mathrm{rev}}}(I)^\sigma)$ with each $m(w_j) = n$, where $\deg w_1 \leq \cdots \leq \deg w_q$. Since $\mathrm{gin}_{<_{\mathrm{rev}}}(I)^\sigma \subset I'$, each $x_n w_j$ belongs to I. However, since $m(u) = n$ and since I is squarefree, one has $w_j \in I$ for each $1 \leq i \leq q$. Thus each w_j must be divided by a monomial $u_j \in G(I)$. If $m(u_j) < n$, then $u_j \in I'' \subset \mathrm{gin}_{<_{\mathrm{rev}}}(I)^\sigma$. This is impossible because $w_j \in G(\mathrm{gin}_{<_{\mathrm{rev}}}(I)^\sigma)$ and $u_j \neq w_j$. Hence $m(u_j) = n$.

Recall from Corollary 7.2.3 that, for a squarefree strongly stable ideal I of S, one has

$$\beta_{n-i,n}(I) = |\{u \in G(I) : \deg u = i, \ m(u) = n\}|. \tag{11.2}$$

Therefore $\beta_{n-\deg u_1,n}(I) \neq 0$. Assume $\deg u_1 < \deg w_1$. Then

$$\beta_{n-\deg u_1,n}(\mathrm{gin}_{<_{\mathrm{rev}}}(I)^\sigma) = 0.$$

However, in general one has

$$\beta_{ii+j}(I) \leq \beta_{ii+j}(\mathrm{gin}_{<_{\mathrm{rev}}}(I)^\sigma) \tag{11.3}$$

for all i and j. Thus $\deg u_1$ cannot be less than $\deg w_1$. Hence $\deg u_1 = \deg w_1$ and $u_1 = w_1$. In particular, w_1 belongs to $G(I)$. Suppose now that $u_1 = w_1, \ldots, u_k = w_k$. The same argument that shows that $u_1 = w_1$ yields $u_{k+1} = w_{k+1}$. Hence each w_i belongs to $G(I)$. Thus in particular

$$\{w \in G(\mathrm{gin}_{<_{\mathrm{rev}}}(I)^\sigma) : m(u) = n\} \subset \{u \in G(I) : m(u) = n\}.$$

However, the inequalities (11.3) together with (11.2) guarantee that

$$|\{u \in G(I) : m(u) = n\}| \leq |\{w \in G(\mathrm{gin}_{<_{\mathrm{rev}}}(I)^\sigma) : m(u) = n\}|.$$

Hence

$$\{u \in G(I) : m(u) = n\} = \{w \in G(\mathrm{gin}_{<_{\mathrm{rev}}}(I)^\sigma) : m(u) = n\}.$$

Thus each $u \in G(I)$ with $m(u) = n$ belongs to $\mathrm{gin}_{<_{\mathrm{rev}}}(I)^\sigma$. This completes the proof of our claim that $I \subset \mathrm{gin}_{<_{\mathrm{rev}}}(I)^\sigma$.

Finally, the Hilbert function of I and that of $\mathrm{gin}_{<_{\mathrm{rev}}}(I)^\sigma$ coincides, see Corollary 6.1.5. Hence, $I = \mathrm{gin}_{<_{\mathrm{rev}}}(I)^\sigma$, as desired. □

Lemma 11.2.8. *Let $J \subset K[x_1, \ldots, x_m]$ be a graded ideal, where $m \leq n$. Then* $\mathrm{gin}_{<_{\mathrm{rev}}}(J)S = \mathrm{gin}_{<_{\mathrm{rev}}}(JS)$.

Proof. We may assume $m < n$. Let $I = JS$. There exists a nonempty Zariski open set $U \subset \mathrm{GL}(n; K)$ such that $\mathrm{in}_{<_{\mathrm{rev}}}(\alpha(I)) = \mathrm{gin} <_{\mathrm{rev}}(I)$ for all $\alpha = (a_{ij}) \in U$.

Let $\mathrm{M}_n(K)$ be the set of all $n \times n$-matrices with entries in K. Note that the restriction map $\mathrm{M}_n(K) \to \mathrm{M}_m(K)$ given by $(a_{ij})_{i,j=1,\cdots,n} \mapsto (a_{ij})_{i,j=1,\cdots,m}$ is an open map. Let $V \subset \mathrm{GL}(m; K)$ be the image of U under this restriction. Then V is a nonempty Zariski open subset of $\mathrm{GL}(m; K)$. Hence there exists $\alpha \in U$ whose restriction β satisfies $\mathrm{in}_{<_{\mathrm{rev}}}(\beta(J)) = \mathrm{gin} <_{\mathrm{rev}}(J)$.

By Lemma 4.3.7 and by the definition of the restriction we obtain

$$(\mathrm{in}_{<_{\mathrm{rev}}}(\alpha(I)), x_{m+1}, \cdots, x_n) = \mathrm{in}_{<_{\mathrm{rov}}}(\alpha(I), x_{m+1}, \cdots, x_n) \qquad (11.4)$$
$$= \mathrm{in}_{<_{\mathrm{rev}}}(\beta(J), x_{m+1}, \cdots, x_n) = (\mathrm{in}_{<_{\mathrm{rev}}}(\beta(J)), x_{m+1}, \cdots, x_n).$$

By using Corollary 4.3.18 we have $\mathrm{proj\,dim}\, S/\mathrm{gin}_{<_{\mathrm{rev}}}(I) = \mathrm{proj\,dim}\, S/I = \mathrm{proj\,dim}\, K[x_1, \ldots, x_m]/J \leq m$. Thus by using Corollary 7.2.3 we see that $m(v) \leq m$ for all $v \in G(\mathrm{gin}_{<_{\mathrm{rev}}}(I)) = G(\mathrm{in}_{<_{\mathrm{rev}}}(\alpha(I)))$. Hence by (11.4) the desired result follows. $\qquad\square$

We conclude this section with the following observation

Proposition 11.2.9. *Let I be a strongly stable monomial ideal. Then one has* $\mathrm{gin}_{<_{\mathrm{rev}}}(I^\sigma) = I$. *In particular, the squarefree operator establishes a bijection between the strongly stable ideals and the squarefree strongly stable ideals.*

Proof. Let $J = \mathrm{gin}_{<_{\mathrm{rev}}}(I^\sigma)$. Then J is strongly stable and by Theorem 11.2.7 one has $J^\sigma = I^\sigma$. Therefore $G(J^\sigma) = G(I^\sigma)$. By Lemma 11.2.5 it follows that $G(J) = G(I)$. $\qquad\square$

11.3 Comparison of Betti numbers

We now study the comparison of graded Betti numbers for the different shifting operations. We expect the following inequalities

$$\beta_{ij}(I_\Delta) \leq \beta_{ij}(I_{\Delta^s}) \leq \beta_{ij}(I_{\Delta^e}) \leq \beta_{ij}(I_{\Delta^c}) \leq \beta_{ij}(I_{\Delta^{\mathrm{lex}}}),$$

where Δ^{lex} is the simplicial complex whose Stanley–Reisner ideal is the unique squarefree lexsegment ideal with the same Hilbert function as I_Δ.

In this chain of inequalities, the inequality $\beta_{ij}(I_{\Delta^s}) \leq \beta_{ij}(I_{\Delta^e})$ and even the inequality $\beta_{ij}(I_\Delta) \leq \beta_{ij}(I_{\Delta^e})$ is not known. All other inequalities will be proved in the following subsections.

11.3.1 Graded Betti numbers of I_Δ and I_{Δ^s}

In this subsection K will be an infinite field of characteristic 0 or of characteristic $> n$. We will prove the following

Theorem 11.3.1. *Let Δ be a simplicial complex on the vertex set $[n]$. Then*

$$\beta_{ij}(I_\Delta) \leq \beta_{ij}(I_{\Delta^s}) \quad \text{for all } i \text{ and } j$$

Proof. By Corollary 3.3.3 we know that $\beta_{ij}(I) \leq \beta_{ij}(\mathrm{gin}_{<_{\mathrm{rev}}}(I))$. Hence the theorem follows from Lemma 11.2.6.   □

11.3.2 Graded Betti numbers of I_{Δ^e} and I_{Δ^c}

Our goal is to show the inequalities $\beta_{ii+j}(I_{\Delta^e}) \leq \beta_{ii+j}(I_{\Delta^c})$ for all i and j.

Let K be an infinite field, $S = K[x_1, \ldots, x_n]$ the polynomial ring in n variables over K and $E = \bigoplus_{d=0}^n E_d$ with $E_d = \bigwedge^d V$ the exterior algebra of a vector space V over K of dimension n with basis $e_1, \ldots, e_n$. Let the general linear group $\mathrm{GL}(n; K)$ act linearly on E. Let $<_{\mathrm{rev}}$ be the reverse lexicographic order on E induced by the ordering $e_1 > \cdots > e_n$.

Given an arbitrary graded ideal $I = \bigoplus_{d=0}^n I_d$ of E with each $I_d \subset E_d$, fix $\varphi \in \mathrm{GL}(n; K)$ for which $\mathrm{in}_{<_{\mathrm{rev}}}(\varphi(I))$ is the generic initial ideal $\mathrm{gin}_{<_{\mathrm{rev}}}(I)$ of I. Recall that the subspace $E_d = \bigwedge^d V$ is of dimension $\binom{n}{d}$ with a canonical K-basis $\mathbf{e}_F$, $F \in \binom{[n]}{d}$, where $\binom{[n]}{d}$ denotes the set of all d-element subsets of $[n]$. Fix a K-basis $f_1, \ldots, f_s$ of I_d, where $s = \dim_K I_d$. Write each $\varphi(f_i)$, $1 \leq i \leq s$, of the form

$$\varphi(f_i) = \sum_{F \in \binom{[n]}{d}} \alpha_i^F \, \mathbf{e}_F$$

with each $\alpha_i^F \in K$. Let $M(I, d)$ denote the $s \times \binom{n}{d}$ matrix

$$M(I, d) = (\alpha_i^F)_{1 \leq i \leq s, \, F \in \binom{[n]}{d}}$$

whose columns are indexed by $F \in \binom{[n]}{d}$. Moreover, for each $G \in \binom{[n]}{d}$, write $M_G(I, d)$ for the submatrix of $M(I, d)$ which consists of the columns of $M(I, d)$ indexed by those $F \in \binom{[n]}{d}$ with $\mathbf{e}_G \leq_{\mathrm{rev}} \mathbf{e}_F$ and write $M'_G(I, d)$ for the submatrix of $M_G(I, d)$ which is obtained by removing the column of $M_G(I, d)$ indexed by G.

Lemma 11.3.2. *Let $\mathbf{e}_G \in E_d$ with $G \in \binom{[n]}{d}$. Then one has $\mathbf{e}_G \in (\mathrm{gin}_{<_{\mathrm{rev}}}(I))_d$ if and only if $\mathrm{rank}(M'_G(I, d)) < \mathrm{rank}(M_G(I, d))$.*

Proof. One has $\mathrm{rank}(M'_G(I, d)) < \mathrm{rank}(M_G(I, d))$ if and only if the row vector $(0, \ldots, 0, 1)$ with "1" lying on the column indexed by G belongs to the vector space spanned by the row vectors of $M_G(I, d)$. This is equivalent to saying that there exist $c_1, \ldots, c_s \in K$ such that $\mathbf{e}_G = \mathrm{in}_{<_{\mathrm{rev}}}(\sum_{i=1}^s c_i \varphi(f_i))$.   □

Corollary 11.3.3. *The rank of a matrix* $M_G(I,d)$, $G \in \binom{[n]}{d}$, *is independent of the choice of* $\varphi \in \mathrm{GL}(n;K)$ *for which* $\mathrm{gin}_{<_{\mathrm{rev}}}(I) = \mathrm{in}_{<_{\mathrm{rev}}}(\varphi(I))$ *and of the choice of the K-basis* $f_1, \ldots, f_s$ *of* I_d. *More precisely,* $\mathrm{rank}(M_G(I,d))$ *is equal to the number of* $F \in \binom{[n]}{d}$ *for which* $\mathbf{e}_F \in (\mathrm{gin}_{<_{\mathrm{rev}}}(I))_d$ *and* $\mathbf{e}_G \leq_{rev} \mathbf{e}_F$.

Proof. Note that $\mathrm{rank}(M'_G(I,d)) = \mathrm{rank}(M_G(I,d)) - 1$ if $\mathrm{rank}(M'_G(I,d)) < \mathrm{rank}(M_G(I,d))$. Therefore, the assertion follows from Lemma 11.3.2. $\qquad \square$

Corollary 11.3.4. *Let* $I \subset E$ *be a homogeneous ideal and* $\psi \in \mathrm{GL}(n;K)$. *Then one has* $\mathrm{rank}(M_G(I,d)) = \mathrm{rank}(M_G(\psi(I),d))$ *for all* $G \in \binom{[n]}{d}$.

Proof. Recall that there is a nonempty Zariski open subset $U \subset \mathrm{GL}(n;K)$ such that $\mathrm{gin}_{<_{\mathrm{rev}}}(I) = \mathrm{in}_{<_{\mathrm{rev}}}(\varphi(I))$ for all $\varphi \in U$. Similarly, there is a nonempty Zariski open subset $V \subset \mathrm{GL}(n;K)$ such that $\mathrm{gin}_{<_{\mathrm{rev}}}(\psi(I)) = \mathrm{in}_{<_{\mathrm{rev}}}(\varphi'(\psi(I)))$ for all $\varphi' \in V$. Since $U\psi^{-1} = \{\varphi\psi : \varphi \in U\}$ is again a nonempty Zariski open subset of $GL_n(K)$ it follows that $U\psi^{-1} \cap V \neq \emptyset$. If $\rho \in U\psi^{-1} \cap V$, then $\mathrm{gin}_{<_{\mathrm{rev}}}(I) = \mathrm{in}_{<_{\mathrm{rev}}}(\rho(\psi(I)) = \mathrm{gin}_{<_{\mathrm{rev}}}(\psi(I))$, and the matrix $M(I,d)$ defined by using $\rho\psi \in U$ and the K-basis $f_1, \ldots, f_s$ of I_d coincides with the matrix $M(\psi(I),d)$ defined by using $\rho \in V$ and the K-basis $\psi(f_1), \ldots, \psi(f_s)$ of $\psi(I)_d$. $\qquad \square$

If $u = \mathbf{e}_F$ is a monomial of E, then we set $m(u) = \max\{j : j \in F\}$. Given a monomial ideal $I \subset E$, one defines $m_{\leq i}(I,d)$, where $1 \leq i \leq n$ and $1 \leq d \leq n$, by

$$m_{\leq i}(I,d) = |\{u = \mathbf{e}_F \in I : \deg(u) = d,\ m(u) \leq i\}|.$$

Corollary 11.3.5. *Let* $i \geq d$ *and set* $F_{(i,d)} = \{i-d+1, i-d+2, \ldots, i\} \in \binom{[n]}{d}$. *Then given a homogeneous ideal* $I \subset E$ *one has*

$$m_{\leq i}(\mathrm{gin}_{<_{\mathrm{rev}}}(I), d) = \mathrm{rank}(M_{F_{(i,d)}}(I,d)).$$

Proof. Let $G \in \binom{[n]}{d}$. Then $m(\mathbf{e}_G) \leq i$ if and only if $\mathbf{e}_{F_{(i,d)}} \leq_{\mathrm{rev}} \mathbf{e}_G$. On the other hand, Corollary 11.3.3 says that $\mathrm{rank}(M_{F_{(i,d)}}(I,d))$ coincides with the number of monomials $\mathbf{e}_G \in (\mathrm{gin}_{<_{\mathrm{rev}}}(I))_d$ with $\mathbf{e}_{F_{(i,d)}} \leq_{\mathrm{rev}} \mathbf{e}_G$. Thus $m_{\leq i}(\mathrm{gin}_{<_{\mathrm{rev}}}(I), d) = \mathrm{rank}(M_{F_{(i,d)}}(I,d))$, as required. $\qquad \square$

Let $I \subset E$ be a monomial ideal. Fix $1 \leq i < j \leq n$. Let $t \in K$ and introduce the K-linear injective map $S_{ij}^t : I \to E$ satisfying

$$S_{ij}^t(\mathbf{e}_F) = \begin{cases} \mathbf{e}_{(F\setminus\{j\})\cup\{i\}} + t\mathbf{e}_F, & \text{if } j \in F,\ i \notin F \text{ and } \mathbf{e}_{(F\setminus\{j\})\cup\{i\}} \notin I, \\ \mathbf{e}_F, & \text{otherwise,} \end{cases}$$

where $\mathbf{e}_F \in I$ is a monomial. Let $I_{ij}(t) \subset E$ denote the image of I by S_{ij}^t.

Lemma 11.3.6. (a) *If* $t \neq 0$, *then there is* $\lambda_{ij}^t \in \mathrm{GL}(n;K)$ *with* $I_{ij}(t) = \lambda_{ij}^t(I)$. *In particular, the subspace* $I_{ij}(t)$ *is an ideal of* E.

(b) *Let* Δ *denote a simplicial complex on* $[n]$ *and* J_Δ *its exterior face ideal. Then* $(J_\Delta)_{ij}(0) = J_{\mathrm{Shift}_{ij}(\Delta)}$.

Proof. (a) Let $\lambda_{ij}^t \in \mathrm{GL}(n; K)$ defined by

$$\lambda_{ij}^t(e_k) = \begin{cases} e_k, & \text{if } k \neq j, \\ e_i + te_j, & \text{if } k = j. \end{cases}$$

We claim $I_{ij}(t) = \lambda_{ij}^t(I)$. Let $\mathbf{e}_F \in I$. We distinguish several cases:

(i) If $j \notin F$, then $\lambda_{ij}^t(\mathbf{e}_F) = \mathbf{e}_F = S_{ij}^t(\mathbf{e}_F)$. Thus $\lambda_{ij}^t(\mathbf{e}_F) \in I_{ij}(t)$.

(ii) If $j \in F$ and $i \in F$, then $\lambda_{ij}^t(\mathbf{e}_F) = t\mathbf{e}_F = tS_{ij}^t(\mathbf{e}_F)$. Thus $\lambda_{ij}^t(\mathbf{e}_F) \in I_{ij}(t)$.

(iii) Let $j \in F$ and $i \notin F$ with $\mathbf{e}_{(F\setminus\{j\})\cup\{i\}} \in I$. Then $\lambda_{ij}^t(\mathbf{e}_F) = \mathbf{e}_{(F\setminus\{j\})\cup\{i\}} + t\mathbf{e}_F$ and $S_{ij}^t(\mathbf{e}_F) = \mathbf{e}_F$. Since $\mathbf{e}_{(F\setminus\{j\})\cup\{i\}} \in I$, $S_{ij}^t(\mathbf{e}_{(F\setminus\{j\})\cup\{i\}}) = \mathbf{e}_{(F\setminus\{j\})\cup\{i\}} \in I_{ij}(t)$. Thus $\lambda_{ij}^t(\mathbf{e}_F) \in I_{ij}(t)$.

(iv) Let $j \in F$ and $i \notin F$ with $\mathbf{e}_{(F\setminus\{j\})\cup\{i\}} \notin I$. Then $\lambda_{ij}^t(\mathbf{e}_F) = \mathbf{e}_{(F\setminus\{j\})\cup\{i\}} + t\mathbf{e}_F$ and $S_{ij}^t(\mathbf{e}_F) = \mathbf{e}_{(F\setminus\{j\})\cup\{i\}} + t\mathbf{e}_F$. Thus $\lambda_{ij}^t(\mathbf{e}_F) \in I_{ij}(t)$.

Hence $\lambda_{ij}^t(I) \subset I_{ij}(t)$. Since each of λ_{ij}^t and S_{ij}^t is injective, one has $I_{ij}(t) = \lambda_{ij}^t(I)$, as desired.

(b) We claim $\{F \subset [n] : \mathbf{e}_F \in (J_\Delta)_{ij}(0)\} \cap \mathrm{Shift}_{ij}(\Delta) = \emptyset$.

(i) If $\mathbf{e}_F \in (J_\Delta)_{ij}(0)$ with $\mathbf{e}_F \notin J_\Delta$, then there is $\mathbf{e}_G \in J_\Delta$ with $F = (G \setminus \{j\}) \cup \{i\}$. Since $F \in \Delta$, $G \notin \Delta$ and $G = (F \setminus \{i\}) \cup \{j\}$, one has $G = C_{ij}(F) \in \mathrm{Shift}_{ij}(\Delta)$. Thus $F \notin \mathrm{Shift}_{ij}(\Delta)$.

(ii) Let $\mathbf{e}_F \in (J_\Delta)_{ij}(0)$ with $\mathbf{e}_F \in J_\Delta$. Suppose $F \in \mathrm{Shift}_{ij}(\Delta)$. Since $F \notin \Delta$, there is $G \subset [n]$ with $G \in \Delta$ such that $F = (G \setminus \{i\}) \cup \{j\}$. Hence $j \in F$, $i \notin F$ and $\mathbf{e}_G = \mathbf{e}_{(F\setminus\{j\})\cup\{i\}} \notin J_\Delta$. Thus $\mathbf{e}_G \in (J_\Delta)_{ij}(0)$ and $\mathbf{e}_F \notin (J_\Delta)_{ij}(0)$, contradiction.

Hence $(J_\Delta)_{ij}(0) \subset J_{\mathrm{Shift}_{ij}(\Delta)}$. Since

$$\dim_K(J_\Delta)_{ij}(0) = \dim_K J_\Delta = \dim_K J_{\mathrm{Shift}_{ij}(\Delta)},$$

it follows that $(J_\Delta)_{ij}(0) = J_{\mathrm{Shift}_{ij}(\Delta)}$. $\qquad\square$

Corollary 11.3.7. *With the same notation as in Corollary 11.3.5 one has*

$$\mathrm{rank}(M_{F_{(i,d)}}(J_{\mathrm{Shift}_{ij}(\Delta)}, d)) \leq \mathrm{rank}(M_{F_{(i,d)}}(J_\Delta, d)).$$

Proof. Let $r(t)$ be the rank of the matrix $M_{F_{(i,d)}}((J_\Delta)_{ij}(t), d)$. By Corollary 11.3.4 we have $r(t) = \mathrm{rank}(M_{F_{(i,d)}}(J_\Delta, d))$ for all $t \neq 0$. In particular $r(t)$ is constant for $t \neq 0$. Suppose $r(0) > r(t)$. Then there exists a minor of size $r(0)$ of $M_{F_{(i,d)}}(J_\Delta)_{ij}(t), d))$ which we denote by $M(t)$ such that $M(0) \neq 0$. Since $M(t)$ is a polynomial in t and since K is infinite, there exists $t \neq 0$ with $M(t) \neq 0$, as well. But this contradicts the assumption that $r(0) > r(t)$. $\qquad\square$

Corollary 11.3.8. *Let Δ be a simplicial complex on $[n]$. Then for all i and d one has*

$$m_{\leq i}(J_{\Delta^e}, d) \geq m_{\leq i}(J_{\Delta^c}, d).$$

Proof. Corollary 11.3.5 together with Corollary 11.3.7 guarantees that

$$m_{\leq i}(\text{gin}(J_\Delta), d) \geq m_{\leq i}(\text{gin}(J_{\text{Shift}_{ij}(\Delta)}), d). \tag{11.5}$$

Hence $m_{\leq i}(\text{gin}(J_\Delta), d) \geq m_{\leq i}(\text{gin}(J_{\Delta^c}), d)$. In other words, $m_{\leq i}(J_{\Delta^e}, d) \geq m_{\leq i}(J_{(\Delta^c)^e}, d)$. However, since Δ^c is shifted, it follows that $(\Delta^c)^e = \Delta^c$. Thus $m_{\leq i}(J_{\Delta^e}, d) \geq m_{\leq i}(J_{\Delta^c}, d)$, as desired. □

We now approach to the final step to prove the inequalities $\beta_{ii+j}(I_{\Delta^e}) \leq \beta_{ii+j}(I_{\Delta^c})$. We first show

Proposition 11.3.9. *Let Δ and Δ' be shifted simplicial complexes on $[n]$ with $f(\Delta) = f(\Delta')$ and suppose that*

$$m_{\leq i}(J_\Delta, j) \geq m_{\leq i}(J_{\Delta'}, j)$$

for all i and j. Then for all i and j one has

$$\beta_{ii+j}(I_\Delta) \leq \beta_{ii+j}(I_{\Delta'}).$$

Proof. Since $f(\Delta) = f(\Delta')$, one has $m_{\leq n}(I_\Delta, j) = m_{\leq n}(I_{\Delta'}, j)$ for all j, see Subsection 6.2. Proposition 7.4.3 then yields the inequalities $\beta_{ii+j}(I_\Delta) \leq \beta_{ii+j}(I_{\Delta'})$ for all i and j, as desired. □

Theorem 11.3.10. *Let Δ be a simplicial complex, Δ^e the exterior algebraic shifted complex of Δ and Δ^c a combinatorial shifted complex of Δ. Then*

$$\beta_{ii+j}(I_{\Delta^e}) \leq \beta_{ii+j}(I_{\Delta^c})$$

for all i and j.

Proof. Corollary 11.3.8 guarantees $m_{\leq i}(J_{\Delta^c}, j) \leq m_{\leq i}(J_{\Delta^e}, j)$ for all i and j. Thus by virtue of Proposition 11.3.9 the required inequalities $\beta_{ii+j}(I_{\Delta^e}) \leq \beta_{ii+j}(I_{\Delta^c})$ follow immediately. □

11.3.3 Graded Betti numbers of I_Δ and I_{Δ^c}

Our goal is to show the inequalities $\beta_{ii+j}(I_\Delta) \leq \beta_{ii+j}(I_{\Delta^c})$ for all i and j.

Let Δ be a simplicial complex on $[n]$ and $I_\Delta \subset S$ its Stanley–Reisner ideal. Let $\tilde{H}_k(\Delta; K)$ denote the kth reduced homology group of Δ with coefficients K. If $W \subset [n]$, then Δ_W stands for the simplicial complex on W whose faces are those faces F of Δ with $F \subset W$.

Recall that Hochster's formula (Theorem 8.1.1) to compute the graded Betti numbers of I_Δ says that

$$\beta_{ii+j}(I_\Delta) = \sum_{W \subset [n], |W| = i+j} \dim_K \tilde{H}_{j-2}(\Delta_W; K) \tag{11.6}$$

for all i and j.

Fix $1 \leq i < j \leq n$ and set $\Gamma = \text{Shift}_{ij}(\Delta)$.

Lemma 11.3.11. *For all k, $\dim_K \tilde{H}_k(\Delta; K) \leq \dim_K \tilde{H}_k(\Gamma; K)$.*

Proof. By considering an extension field of K if necessary, we may assume that K is infinite. Let Δ^e denote the exterior algebraic shifted complex of Δ. By Proposition 11.4.7 we have $\tilde{H}_k(\Delta; K) \cong \tilde{H}_k(\Delta^e; K)$. Thus we need to show that $\dim_K \tilde{H}_k(\Delta^e; K) \leq \dim_K \tilde{H}_k(\Gamma^e; K)$ for all k. By using (11.6) one has $\beta_{in}(I_\Delta) = \dim_K \tilde{H}_{n-i-2}(\Delta; K)$. Hence it remains to show that $\beta_{in}(I_{\Delta^e}) \leq \beta_{in}(I_{\Gamma^e})$ for all i. Inequality (11.5) says that $m_{\leq i}(J_{\Delta^e}, j) \geq m_{\leq i}(J_{\Gamma^e}, j)$ for all i and j. It then follows from Corollary 11.3.9 that $\beta_{ii+j}(I_{\Delta^e}) \leq \beta_{ii+j}(I_{\Gamma^e})$ for all i and j. Thus in particular $\beta_{in}(I_{\Delta^e}) \leq \beta_{in}(I_{\Gamma^e})$ for all i.   □

Let $W \subset [n] \setminus \{i, j\}$, and let

$$\Delta_1 = \Delta_{W \cup \{i\}}, \quad \Delta_2 = \Delta_{W \cup \{j\}}, \quad \Gamma_1 = \Gamma_{W \cup \{i\}} \quad \text{and} \quad \Gamma_2 = \Gamma_{W \cup \{j\}}.$$

Then

$$\Delta_1 \cap \Delta_2 = \Gamma_1 \cap \Gamma_2 = \Delta_W = \Gamma_W, \quad \text{and} \quad \Gamma_1 \cup \Gamma_2 = \text{Shift}_{ij}(\Delta_1 \cup \Delta_2). \,(11.7)$$

The reduced Mayer–Vietoris exact sequence of Δ_1 and Δ_2 and that of Γ_1 and Γ_2 (see Proposition 5.1.8) is given by

$$\cdots \xrightarrow{\quad\quad} \tilde{H}_k(\Delta_W; K) \xrightarrow{\;\partial_{1,k}\;} \tilde{H}_k(\Delta_1; K) \oplus \tilde{H}_k(\Delta_2; K)$$

$$\xrightarrow{\;\partial_{2,k}\;} \tilde{H}_k(\Delta_1 \cup \Delta_2; K) \xrightarrow{\;\partial_{3,k}\;} \tilde{H}_{k-1}(\Delta_W; K) \xrightarrow{\;\partial_{1,k-1}\;} \cdots.$$

and

$$\cdots \xrightarrow{\quad\quad} \tilde{H}_k(\Gamma_W; K) \xrightarrow{\;\partial'_{1,k}\;} \tilde{H}_k(\Gamma_1; K) \oplus \tilde{H}_k(\Gamma_2; K)$$

$$\xrightarrow{\;\partial'_{2,k}\;} \tilde{H}_k(\Gamma_1 \cup \Gamma_2; K) \xrightarrow{\;\partial'_{3,k}\;} \tilde{H}_{k-1}(\Gamma_W; K) \xrightarrow{\;\partial'_{1,k-1}\;} \cdots.$$

Since $\Delta_W = \Gamma_W$ we can compare $\text{Ker}(\partial'_{1,k})$ and $\text{Ker}(\partial_{1,k})$.

Lemma 11.3.12. *Suppose that $j = i + 1$. Then one has*

$$\text{Ker}(\partial'_{1,k}) \subset \text{Ker}(\partial_{1,k}).$$

for all k.

Proof. Let $[a] \in \text{Ker}(\partial'_{1,k})$, where $a \in \tilde{C}_k(\Gamma_W)$. Since $([a], [a]) \in \tilde{H}_k(\Gamma_1; K) \oplus \tilde{H}_k(\Gamma_2; K)$ vanishes (in particular, $[a] \in \tilde{H}_k(\Gamma_1; K)$ vanishes), there exists $u \in \tilde{C}_{k+1}(\Gamma_1)$ with $\partial(u) = a$. Say,

$$u = \sum_{|F|=k+1,\, i\notin F,\, F\cup\{i\}\in\Gamma_1} a_{F\cup\{i\}}\mathbf{e}_{F\cup\{i\}} + \sum_{|G|=k+2,\, G\in\Delta_W} b_G\mathbf{e}_G, \quad (11.8)$$

where $a_{F\cup\{i\}}, b_G \in K$.

Let $F \subset W$ with $F \cup \{i\} \in \Gamma_1$. By the definition of Shift_{ij} it follows immediately that $F \cup \{i\} \in \Delta_1$ and $F \cup \{j\} \in \Delta_2$. Thus $F \cup \{j\} \in \Gamma_2$. In particular, $u \in \tilde{C}_{k+1}(\Delta_1)$ with $\partial(u) = a$. Hence $[a] \in \tilde{H}_k(\Delta_1; K)$ vanishes.

Since $a \in \tilde{C}_k(\Gamma_W)$ is a linear combination of those basis elements $\mathbf{e}_F$ with $F \in \Gamma$, $F \subset W$ and $|F| = k + 1$ and since $j = i + 1$, it follows that $\partial(v) = a$, where $v \in \tilde{C}_{k+1}(\Delta_2)$ is the element

$$v = \sum_{|F|=k+1,\, i \notin F,\, F \cup \{i\} \in \Gamma_1} a_{F \cup \{i\}} \mathbf{e}_{F \cup \{j\}} + \sum_{|G|=k+2,\, G \in \Delta_W} b_G \mathbf{e}_G.$$

Thus $[a] \in \tilde{H}_k(\Delta_2; K)$ vanishes.

These calculations now show that $([a],[a]) \in \tilde{H}_k(\Delta_1; K) \oplus \tilde{H}_k(\Delta_2; K)$ vanishes, as required. $\qquad\square$

Suppose again that $j = i + 1$. It then follows that

$$\dim_K(\text{Ker}(\partial_{1,k})) \geq \dim_K(\text{Ker}(\partial'_{1,k})),$$

$$\dim_K(\text{Im}(\partial_{1,k})) \leq \dim_K(\text{Im}(\partial'_{1,k})),$$

$$\dim_K(\text{Ker}(\partial_{2,k})) \leq \dim_K(\text{Ker}(\partial'_{2,k})). \tag{11.9}$$

On the other hand,

$$\dim_K(\tilde{H}_k(\Delta_1 \cup \Delta_2; K)) = \dim_K(\text{Ker}(\partial_{3,k})) + \dim_K(\text{Im}(\partial_{3,k})), \tag{11.10}$$

$$\dim_K(\tilde{H}_k(\Gamma_1 \cup \Gamma_2; K)) = \dim_K(\text{Ker}(\partial'_{3,k})) + \dim_K(\text{Im}(\partial'_{3,k})). \tag{11.11}$$

Lemma 11.3.11 together with (11.7) guarantees that

$$\dim_K(\tilde{H}_k(\Delta_1 \cup \Delta_2; K)) \leq \dim_K(\tilde{H}_k(\Gamma_1 \cup \Gamma_2; K)). \tag{11.12}$$

Since $\text{Im}(\partial_{3,k}) = \text{Ker}(\partial_{1,k-1})$ and $\text{Im}(\partial'_{3,k}) = \text{Ker}(\partial'_{1,k-1})$, Lemma 11.3.12 yields

$$\dim_K(\text{Im}(\partial_{3,k})) \geq \dim_K(\text{Im}(\partial'_{3,k})). \tag{11.13}$$

Since $\text{Im}(\partial_{2,k}) = \text{Ker}(\partial_{3,k})$ and $\text{Im}(\partial'_{2,k}) = \text{Ker}(\partial'_{3,k})$, it follows from formula (11.10) and (11.11) together with (11.12) and (11.13) that

$$\dim_K(\text{Im}(\partial_{2,k})) \leq \dim_K(\text{Im}(\partial'_{2,k})). \tag{11.14}$$

Finally, it follows from the reduced Mayer–Vietoris exact sequence of Δ_1 and Δ_2 and that of Γ_1 and Γ_2 together with (11.9) and (11.10) that

$$\dim_K(\tilde{H}_k(\Delta_1; K) \oplus \tilde{H}_k(\Delta_2; K)) \leq \dim_K(\tilde{H}_k(\Gamma_1; K) \oplus \tilde{H}_k(\Gamma_2; K)). \tag{11.15}$$

Now we are ready to prove the crucial

Lemma 11.3.13. *Fix $1 \leq p < q \leq n$. Let Δ be a simplicial complex on $[n]$ and $\Gamma = \mathrm{Shift}_{pq}(\Delta)$. Then*

$$\beta_{ii+j}(I_\Delta) \leq \beta_{ii+j}(I_\Gamma)$$

for all i and j.

Proof. Let π be a permutation on $[n]$ with $\pi(p) < \pi(q)$. Then π naturally induce the automorphism of $S = K[x_1, \ldots, x_n]$ by setting $x_i \mapsto x_{\pi(i)}$. Write $\pi(\Delta)$ for the simplicial complex $\{\pi(F) : F \in \Delta\}$ on $[n]$. Then

$$\pi(I_{\mathrm{Shift}_{pq}(\Delta)}) = I_{\mathrm{Shift}_{\pi(p)\pi(q)}(\pi(\Delta))}.$$

Thus in particular

$$\beta_{ii+j}(I_{\mathrm{Shift}_{pq}(\Delta)}) = \beta_{ii+j}(I_{\mathrm{Shift}_{\pi(p)\pi(q)}(\pi(\Delta))}).$$

Consequently, we may assume that $q = p + 1$.

The right-hand side of Hochster's formula (11.6) can be rewritten as

$$\beta_{ii+j}(I_\Delta) = \alpha_{ij}(\Delta) + \gamma_{ij}(\Delta) + \delta_{ij}(\Delta),$$

where

$$\alpha_{ij}(\Delta) = \sum_{W \subset [n] \setminus \{p,q\},\, |W|=i+j} \dim_K(\tilde{H}_{j-2}(\Delta_W; K)),$$

$$\gamma_{ij}(\Delta) = \sum_{W \subset [n] \setminus \{p,q\},\, |W|=i+j-1} \dim_K(\tilde{H}_{j-2}(\Delta_{W \cup \{p\}}; K))$$

$$+ \sum_{W \subset [n] \setminus \{p,q\},\, |W|=i+j-1} \dim_K(\tilde{H}_{j-2}(\Delta_{W \cup \{q\}}; K)),$$

$$\delta_{ij}(\Delta) = \sum_{W \subset [n] \setminus \{p,q\},\, |W|=i+j-2} \dim_K(\tilde{H}_{j-2}(\Delta_{W \cup \{p,q\}}; K)).$$

Let $W \subset [n] \setminus \{p,q\}$. Then $\Delta_W = \Gamma_W$. Thus $\alpha_{ij}(\Delta) = \alpha_{ij}(\Gamma)$. Since $\Gamma_{W \cup \{p,q\}} = \mathrm{Shift}(\Delta_{W \cup \{p,q\}})$, Lemma 11.3.11 says that $\delta_{ij}(\Delta) \leq \delta_{ij}(\Gamma)$. Finally, it follows from (10) that $\gamma_{ij}(\Delta) \leq \gamma_{ij}(\Gamma)$. Hence $\beta_{ii+j}(I_\Delta) \leq \beta_{ii+j}(I_\Gamma)$, as desired. □

Lemma 11.3.13 together with the definition of combinatorial shifting now implies

Theorem 11.3.14. *Let the base field be arbitrary. Let Δ be a simplicial complex and Δ^c a combinatorial shifted complex of Δ. Then*

$$\beta_{ii+j}(I_\Delta) \leq \beta_{ii+j}(I_{\Delta^c})$$

for all i and j.

Let Δ' be a shifted simplicial complex with the same f-vector as Δ and Δ^{lex} the unique lexsegment simplicial complex with the same f-vector as Δ. By Theorem 7.4.3 we have that $\beta_{ii+j}(I_{\Delta'}) \leq \beta_{ii+j}(I_{\Delta^{\mathrm{lex}}})$ for all i and j. Since Δ^c is shifted with $f(\Delta^c) = f(\Delta)$, it follows that $\beta_{ii+j}(I_{\Delta^c}) \leq \beta_{ii+j}(I_{\Delta^{\mathrm{lex}}})$ for all i and j. Hence

Corollary 11.3.15. *Let the base field be arbitrary. Let Δ be a simplicial complex and Δ^{lex} the unique lexsegment simplicial complex with the same f-vector as Δ. Then*

$$\beta_{ii+j}(I_\Delta) \leq \beta_{ii+j}(I_{\Delta^{\mathrm{lex}}})$$

for all i and j.

11.4 Extremal Betti numbers and algebraic shifting

In the previous sections we have discussed the comparison of Betti numbers between a simplicial complex and its shifted complex. For the extremal Betti numbers this comparison yields the following result.

Theorem 11.4.1. *Let Δ be a simplicial complex and $I_\Delta \subset K[x_1, \ldots, x_n]$ its Stanley–Reisner ideal, where K is an infinite field which we assume to be of characteristic 0 in the statements concerning Δ^s.*

(a) *For all i and j, the following conditions are equivalent:*
  (i) *the ijth Betti number of I_Δ is extremal;*
  (ii) *the ijth Betti number of I_{Δ^e} is extremal;*
  (iii) *the ijth Betti number of I_{Δ^s} is extremal.*
(b) *The corresponding extremal Betti numbers of I_Δ, I_{Δ^e} and I_{Δ^s} are equal.*

Proof. In the case of symmetric algebraic shifting the statements in (a) and (b) are direct consequences of Theorem 4.3.17 and Lemma 11.2.6. For exterior algebraic shifting they follow from the subsequent considerations. □

Our aim is to relate the extremal Betti numbers of I_Δ with with certain numerical data of the E-resolution of $J_\Delta \subset E$. In order to simplify notation we set $J = J_\Delta$ $I = I_\Delta$. We let

$$P_j(t) = \sum_{i \geq 0} \beta^E_{ii+j}(E/J)t^i$$

Then Corollary 7.5.2 yields

$$P_j(t) = \sum_{i \geq 0} (\sum_{k=0}^{i} \binom{i+j-1}{j+k-1} \beta^S_{kk+j}(S/I))t^i.$$

Setting $k(j) = \max\{k \colon \beta^S_{kk+j}(S/I) \neq 0\}$, we see that

$$P_j(t) = \frac{\sum_{k=0}^{k(j)} \beta_{kk+j}^S(S/I)t^k(1-t)^{k(j)-k} + R(t)(1-t)^{k(j)+j}}{(1-t)^{k(j)+j}}, \quad (11.16)$$

with a certain polynomial $R(t)$.

We set $d_j(E/J) = k(j) + j$ and $e_j(E/J) = \beta_{k(j),k(j)+j}^S(S/I)$.

Corollary 11.4.2. *The following conditions are equivalent:*

(a) $\beta_{ii+j}^S(S/I)$ *is an extremal Betti number of S/I;*
(b) $i = k(j)$, *and* $d_{j'}(E/J) - d_j(E/J) < j' - j$ *for all* $j' > j$.

For the further discussion we need a different interpretation of the numbers d_j and e_j. To this end we consider Cartan homology. We will use the exact sequence

$$\rightarrow H_i(\mathbf{v}'; M) \rightarrow H_i(\mathbf{v}; M) \rightarrow H_{i-1}(\mathbf{v}; M)(-1) \rightarrow H_{i-1}(\mathbf{v}'; M) \rightarrow \quad (11.17)$$

of graded E-modules, which by Corollary A.8.4 is attached to any sequence $\mathbf{v} = v_1, \ldots, v_m$ of elements in E_1. Here $\mathbf{v}' = v_1, \ldots, v_{m-1}$.

Proposition 11.4.3. *Let $M \in \mathcal{G}$, and let $v_1, \ldots, v_n$ be a generic basis of E_1. Then the natural maps*

$$H_i(v_1, \ldots, v_j; M) \longrightarrow H_{i-1}(v_1, \ldots, v_j; M)(-1)$$

arising in the long exact sequence (11.17) of Cartan homology attached to the sequence $v_1, \ldots, v_n$ are surjective for all $j = 1, \ldots, n$ and all $i \gg 0$.

Proof. Applying the exact functor $^*\mathrm{Hom}_E(-, E)$, Proposition A.8.5 yields the isomorphisms

$$H_i(v_1, \ldots, v_j; M)^\vee \cong H^i(v_1, \ldots, v_j; M^\vee),$$

and the natural maps $H_i(v_1, \ldots, v_j; M) \rightarrow H_{i-1}(v_1, \ldots, v_j; M)(-1)$ induce maps

$$H^{i-1}(v_1, \ldots, v_j; M^\vee) \rightarrow H^i(v_1, \ldots, v_j; M^\vee)(-1) \quad (11.18)$$

in Cartan cohomology. Thus $H_i(v_1, \ldots, v_j; M) \rightarrow H_{i-1}(v_1, \ldots, v_j; M)(-1)$ is surjective for $i \gg 0$ if and only if

$$H^{i-1}(v_1, \ldots, v_j; M^\vee) \rightarrow H^i(v_1, \ldots, v_j; M^\vee)(-1)$$

is injective for $i \gg 0$.

Now we use that $H^\bullet(\mathbf{v}; M^\vee) = H^\bullet(v_1, \ldots, v_n; M^\vee)$ is a finitely generated graded module over the polynomial ring $K[y_1, \ldots, y_n]$ with ith homogeneous components $H^i(\mathbf{v}; M^\vee)$, and that the natural map (11.18) is just multiplication by y_j; see Proposition A.8.6. Each of the graded components $H^i(\mathbf{v}; M^\vee)$

itself is a graded K-vector space. In what follows we disregard this internal grading since it is of no relevance for the next arguments.

Since $\mathbf{v} = v_1, \ldots, v_n$ a generic basis of E_1 it follows from Proposition A.8.7 that $y_n, \ldots, y_1$ is an almost regular sequence on $H^\bullet(v_1, \ldots, v_n; M^\vee)$. Therefore the multiplication map

$$H^{i-1}(\mathbf{v}; M^\vee)/(y_n, \ldots, y_{n-j+1})H^{i-2}(\mathbf{v}; M^\vee) \xrightarrow{y_{n-j}}$$
$$H^i(\mathbf{v}; M^\vee)/(y_n, \ldots, y_{-j+1})H^{i-1}(\mathbf{v}; M^\vee) \tag{11.19}$$

is injective for all $j = 0, \ldots, n-1$ and all $i \gg 0$. In particular it follows from (11.17) that

$$0 \longrightarrow H^{i-1}(\mathbf{v}; M^\vee) \xrightarrow{y_n} H^i(\mathbf{v}; M^\vee) \longrightarrow H^i(v_1, \ldots, v_{n-1}; M^\vee) \to 0$$

is exact for $i \gg 0$. Thus we see that the ith component of $H^\bullet(v_1, \ldots, v_{n-1}; M^\vee)$ and of $H^\bullet(\mathbf{v}; M^\vee)/(y_n)H^\bullet(\mathbf{v}; M^\vee)$ coincide for $i \gg 0$. By using the long exact sequences (11.17) and the injectivity of the multiplication map in (11.19), induction on j yields that the ith component of $H^\bullet(v_1, \ldots, v_{n-j}; M^\vee)$ and of $H^\bullet(\mathbf{v}; M^\vee)/(y_n, \ldots, y_{n-j+1})H^\bullet(\mathbf{v}; M^\vee)$ coincide for $i \gg 0$. This, together with (11.19), completes the proof. □

We now fix $M \in \mathcal{G}$ and a generic sequence $\mathbf{v} = v_1, \ldots, v_n$ in E_1. In order to simplify notation we set $H_i(k) = H_i(v_1, \ldots, v_k; M)$ for $i > 0$ and $H_0(k) = H^\bullet(M/(v_1, \ldots, v_{k-1})M, v_k)$ for $k = 1, \ldots, n$. Furthermore we set $H_i(0) = 0$ for all i. Notice that $H_0(k)$ is *not* the 0th Cartan homology of M with respect to $v_1, \ldots, v_k$, but is the cohomology of $M/(v_1, \ldots, v_{k-1})M$ with respect to v_k as defined in Section 5.1.4. From A.8.4 we obtain immediately the following long exact sequence of graded E-modules

$$H_2(k) \to H_1(k)(-1) \to H_1(k-1) \to H_1(k) \to H_0(k)(-1) \to 0 \tag{11.20}$$
$$\cdots \to H_i(k-1) \to H_i(k) \to H_{i-1}(k)(-1) \to H_{i-1}(k-1) \to \cdots$$

We fix an integer j. By Proposition 11.4.3 there exists an integer i_0 such that for all $i \geq i_0$ and all $k = 1, \ldots, n$ the sequences

$$0 \longrightarrow H_{i+1}(k-1)_{(i+1)+j} \longrightarrow H_{i+1}(k)_{(i+1)+j} \longrightarrow H_i(k)_{i+j} \longrightarrow 0 \tag{11.21}$$

are exact.

Set $h_i^k = \dim_K H_i(k)_{i+j}$, and $c_k = h_{i_0}^k$ for $k = 1, \ldots, n$. The exact sequences (11.21) yield the equations

$$h_{i+1}^k = h_{i+1}^{k-1} + h_i^k \tag{11.22}$$

for all $i \geq i_0$, and $k = 1, \ldots, n$. It follows from (11.22) that

$$h_{i_0+i}^n = \binom{i+n-2}{n-1}c_1 + \binom{i+n-3}{n-2}c_2 + \cdots + \binom{i}{1}c_{n-1} + c_n \quad \text{for all} \quad i \geq 0.$$

Since $\beta_{ii+j}^E(M) = h_i^n$ for all i, we see that

$$\sum_{i \geq 0} \beta_{ii+j}^E(M)t^i = t^{i_0+1} \sum_{k=1}^{n} \frac{c_k}{(1-t)^{n-k+1}} + Q(t),$$

where $Q(t)$ is a polynomial. Thus a comparison with formula 11.16 yields

Proposition 11.4.4. Let $d_j = d_j(E/J)$ and $e_j = e_j(E/J)$ be defined as above. Then

$$d_j = n + 1 - \min\{k \colon c_k \neq 0\} \quad \text{and} \quad e_j = c_{n-d_j+1} = \min\{k \colon c_k \neq 0\},$$

where $c_k = \dim_K H_{i_0}(k)_{i_0+j}$.

In order to relate the invariants d_j and e_j to the generalized simplicial homology modules $H_0(k)$ we need the following

Lemma 11.4.5. Let $1 \leq l \leq n$ and j be integers. With the notation introduced the following conditions are equivalent:

(a) (i) $H_0(k)_j = 0$ for $k < l$, and $H_0(l)_j \neq 0$
     (ii) $H_0(k)_{j'} = 0$ for all $j' > j$ and all $k \leq l + j - j'$.
(b) For all $i \geq 0$ we have
     (i) $H_i(k)_{i+j} = 0$ for $k < l$, and $H_i(l)_{i+j} \neq 0$
     (ii) $H_i(k)_{i+j'} = 0$ for all $j' > j$ and all $k \leq l + j - j'$.
(c) Condition (b) is satisfied for some i.

Moreover, if the equivalent conditions hold, then $H_i(l)_{i+j} \cong H_0(l)_j$ for all $i \geq 0$.

Proof. In our proof we will use the following exact sequence

$$H_i(k-1)_{i+j'} \to H_i(k)_{i+j'} \to H_{i-1}(k)_{(i-1)+j'} \tag{11.23}$$
$$\to H_{i-1}(k-1)_{(i-1)+(j'+1)}$$

(a)$\Rightarrow$ (b): We prove (b) by induction on i. For $i = 0$, there is nothing to show. So now let $i > 0$ and assume that (i) and (ii) hold for $i - 1$. By (11.23) we have the exact sequence

$$H_i(l)_{i+j} \longrightarrow H_{i-1}(l)_{(i-1)+j} \longrightarrow H_{i-1}(l-1)_{(i-1)+(j+1)}.$$

Since $l - 1 \leq l + j - (j + 1)$, we have $H_{i-1}(l-1)_{(i-1)+(j+1)} = 0$ by induction hypothesis. Also by induction hypothesis, $H_{i-1}(l)_{(i-1)+j} \neq 0$; therefore $H_i(l)_{i+j} \neq 0$.

Now let $k < l$. Then (11.23) yields the exact sequence

$$H_i(k-1)_{i+j} \longrightarrow H_i(k)_{i+j} \longrightarrow H_{i-1}(k)_{(i-1)+j}.$$

By induction hypothesis we have $H_{i-1}(k)_{(i-1)+j} = 0$. Now by induction on k we may assume that $H_i(k-1)_{i+j} = 0$. Therefore $H_i(k)_{i+j} = 0$, and this shows (1).

In order to prove (b)(ii), we let $j' > j$ and $k \leq l + (j - j')$, and consider the exact sequence

$$H_i(k-1)_{i+j'} \longrightarrow H_i(k)_{i+j'} \longrightarrow H_{i-1}(k-1)_{(i-1)+j'},$$

from which the assertion follows by induction on i and k.

(c)$\Rightarrow$ (a). We show that if the conditions (i) and (ii) hold for $i > 0$, then they also hold for $i - 1$. Therefore backwards induction yields the desired conclusion.

We begin with the proof of (ii) for $i-1$ by induction on k. For $k = 0$, there is nothing to show. Now let $j' > j$, and $0 < k \leq l + (j - j')$, and consider the exact sequence

$$H_i(k)_{i+j'} \longrightarrow H_{i-1}(k)_{(i-1)+j'} \longrightarrow H_{i-1}(k-1)_{(i-1)+(j'+1)}.$$

Since $k-1 \leq l+j-(j'+1)$ it follows by our induction hypothesis that $H_{i-1}(k-1)_{(i-1)+(j'+1)} = 0$. On the other hand, by assumption we have $H_i(k)_{i+j'} = 0$, and hence $H_{i-1}(k)_{(i-1)+j'} = 0$.

In order to prove (i) for $i - 1$ we consider the exact sequence

$$H_i(l-1)_{i+j} \longrightarrow H_i(l)_{i+j} \longrightarrow H_{i-1}(l)_{(i-1)+j} \longrightarrow H_{i-1}(l-1)_{(i-1)+(j+1)}.$$

Since $l - 1 \leq l + j - (j + 1)$, we know from (ii) (which we have already shown for $i - 1$) that $H_{i-1}(l-1)_{(i-1)+(j+1)} = 0$. By our assumption we have $H_i(l-1)_{i+j} = 0$, and hence

$$H_{i-1}(l)_{(i-1)+j} \cong H_i(l)_{i+j} \neq 0.$$

That $H_{i-1}(k)_{(i-1)+j} = 0$ for $k < l$ is proved similarly. This concludes the proof of the implication (c)$\Rightarrow$ (a).

In the proof of this implication we have just seen that $H_i(l)_{i+j} \cong H_{i-1}(l)_{(i-1)+j}$. By induction hypothesis we may assume that $H_{i-1}(l)_{(i-1)+j} \cong H_0(l)_j$, and hence $H_i(l)_{i+j} \cong H_0(l)_j$, as desired. $\square$

A pair of numbers (l, j) satisfying the equivalent conditions of Lemma 11.4.5 will be a called a **distinguished pair** (for M).

Now we may characterize the extremal Betti numbers of S/I as follows:

Corollary 11.4.6. *The Betti number $\beta_{ii+j}(S/I)$ is extremal if and only if $(n + 1 - i - j, j)$ is a distinguished pair for E/J. Moreover, if the equivalent conditions of Lemma 11.4.5 hold, then $\beta_{ii+j}(S/I) = \dim_K H_0(n+1-i-j)_j$.*

Proof. We know from Corollary 11.4.2 that $\beta_{ii+j}(S/I)$ is an extremal Betti number if and only if $d_{j'}(E/J) - d_j(E/J) < j' - j$ for all $j' > j$. By Proposition 11.4.4 this condition is equivalent to

$$\min\{k: H_{i_0}(k)_{i_0+j'} \neq 0\} > l + (j - j'),$$

where $l = \min\{k: H_{i_0}(k)_{i_0+j} \neq 0\}$. In particular, we have

$$H_{i_0}(k)_{i_0+j'} = 0 \quad \text{for} \quad k \leq l + (j - j').$$

Thus $\beta_{ii+j}(S/I)$ is an extremal Betti number if and only if (l, j) is a distinguished pair.

It follows from the definition of $d_j(E/J)$ and Proposition 11.4.4 that $l = n + 1 - i - j$. Finally, applying Corollary 11.4.2, Proposition 11.4.4 and Lemma 11.4.5 we see that

$$\beta_{ii+j}(S/I) = e_j(E/J) = c_l = \dim_K H_0(l)_j.$$

$\square$

We are ready for

Proof (of Theorem 11.4.1(a) (i) $\Leftrightarrow$ (ii) and (b)). Let $v_1, \ldots, v_n$ be a generic basis of E_1. By Theorem 5.2.11 we have

$$\dim_K H^i(K\{\Delta\}/(v_1, \ldots, v_{k-1})K\{\Delta\}, v_k)$$
$$= \dim_K H^i(K\{\Delta^e\}/(e_{n-k+2}, \ldots, e_n)K\{\Delta^e\}, e_{n-k+1}).$$

for all i and k. The same holds true for $K\{\Delta^e\}$. Therefore, since $\Delta^e = (\Delta^e)^e$, it follows that

$$\dim_K H^i(K\{\Delta\}/(v_1, \ldots, v_{k-1})K\{\Delta\}, v_k)$$
$$= \dim_K H^i(K\{\Delta^e\}/(v_1, \ldots, v_{k-1})K\{\Delta^e\}, v_k).$$

Since by Lemma 11.4.5 these dimensions determine the distinguished pairs, all assertions follow from Corollary 11.4.6 $\square$

As consequence of Theorem 11.4.1 we obtain (see also Corollary 5.2.12) the following two results of Kalai:

Corollary 11.4.7. *Let Δ be a simplicial complex and let K be a field. Then*

$$\tilde{H}_i(\Delta; K) \cong \tilde{H}_i(\Delta^e; K) \quad \text{for all} \quad i.$$

Moreover if char $K = 0$, *then we also have* $\tilde{H}_i(\Delta; K) \cong \tilde{H}_i(\Delta^s; K)$.

Proof. Hochster's formula (Theorem 8.1.1) implies that

$$\beta_{n-i-1,n}(S/I_\Gamma) = \dim_K \tilde{H}_i(\Gamma; K) \tag{11.24}$$

for all i, for a simplicial complex Γ on the vertex set $[n]$. Thus the assertion follows from Theorem 11.4.1 together with Remark 4.3.14. $\square$

The simplicial homology of a shifted complex is easy to compute because of the following

Proposition 11.4.8. *Let Δ be a simplicial complex on the vertex set $[n]$ such that I_Δ is squarefree strongly stable. Then*

$$\dim_K \tilde{H}_i(\Delta; K) = |\{u \in G(I_\Delta)_{i+2}:\ m(u) = n\}|$$
$$= |\{\sigma \in \Delta:\ \dim \sigma = i,\quad \sigma \cup \{n\} \notin \Delta\}|$$

Proof. The first equation follows from (11.24) and Corollary 7.4.2, while the second equation follows trivially from the definitions. □

Corollary 11.4.9. *Let Δ be a simplicial complex and let K be a field as in Theorem 11.4.1. Then the following conditions are equivalent:*

(a) Δ *is Cohen–Macaulay over K;*
(b) Δ^e *(resp. Δ^s) is Cohen–Macaulay;*
(c) Δ^e *(resp. Δ^s) is pure.*

Proof. (a) $\Leftrightarrow$ (b): Since shifting operators preserve f-vectors, it follows that $\dim K[\Delta] = \dim K[\Delta^e] = \dim K[\Delta^s]$. Now Theorem 11.4.1 implies that $\operatorname{proj\,dim} K[\Delta] = \operatorname{proj\,dim} K[\Delta^e] = \operatorname{proj\,dim} K[\Delta^s]$. Thus $\operatorname{depth} K[\Delta] = \operatorname{depth} K[\Delta^e] = \operatorname{depth} K[\Delta^s]$ by the Auslander–Buchbaum theorem; cf. Corollary A.4.3. This shows the equivalence of statements (a) and (b).

(b) $\Leftrightarrow$ (c): We first observe that I_{Δ^e} as well as I_{Δ^s} is squarefree strongly stable. This follows from Proposition 5.2.10 and Proposition 4.2.4 together with Lemma 11.2.5. Thus we have to show that a squarefree strongly stable ideal I is Cohen–Macaulay if and only if all minimal prime ideals of I have the same height.

The ideal I is the Stanley–Reisner ideal of a simplicial complex Γ. We denote by $I^\vee$ the Stanley–Reisner ideal of Alexander dual $\Gamma^\vee$ of Γ. It is easily seen that $I^\vee$ is again squarefree strongly stable. Hence it follows from Corollary 1.5.5 that all minimal prime ideals of I have the same height if and only if $I^\vee$ is generated in one degree. Corollary 7.4.2 implies that this is the case if and only if $I^\vee$ has linear resolution, which by Theorem 8.1.9 is equivalent to saying that I is a Cohen–Macaulay ideal. □

11.5 Superextremal Betti numbers

In this section we give an algebraic proof of a theorem of Björner and Kalai [BK88]. The version presented here is slightly more general than the original theorem, as it applies to any graded ideal in the exterior algebra (not just to monomial ideals). Nevertheless the proof follows closely the arguments given by Björner and Kalai in their paper. A non-squarefree version of their theorem will also be presented.

Let $J \subset E$ be a graded ideal. We set $f_{i-1} = \dim_K(E/J)_i$ for all $i \geq 0$, and call $f = (f_0, f_1, \ldots)$ the f-vector of E/J. As in Chapter 5 we denote by $H^i(E/J)$ the generalized simplicial cohomology of E/J. We let $\beta_{i-1} = \dim_K H^i(E/J)$, and call $\beta = (\beta_{-1}, \beta_0, \beta_1 \ldots)$ the (topological) Betti sequence of E/J. In case $J = J_\Delta$ for some simplicial complex Δ, the β_i are the ordinary (topological) Betti numbers of Δ.

A pair of sequences $(f, \beta) \in \mathbb{N}_0^\infty$ is called **compatible** if there exists a graded K-algebra E/J such that f is the f-sequence and β the Betti sequence of E/J.

Theorem 11.5.1 (Björner and Kalai). *Let K be a field. The following conditions are equivalent:*

(a) *The pair of sequences (f, β) is compatible.*
(b) *Set $\chi_i = (-1)^i \sum_{j=-1}^i (-1)^j (f_j - \beta_j)$ for all i. Then*
   (i) *$\chi_{-1} = 1$ and $\chi_i \geq 0$ for all i,*
   (ii) *$\beta_i \leq \chi_{i-1}^{(i)} - \chi_i$ for all i.*

Proof. Choosing a suitable field extension of K, we may as well assume that K is infinite. We fix a monomial order on E with $e_1 > e_2 > \cdots > e_n$.

(a) $\Rightarrow$ (b): The f-vectors of E/J and $E/\operatorname{gin}_<(J)$ coincide; see Corollary 6.1.5, where a similar statement is made for graded ideals in the polynomial ring. By Corollary 5.2.12 we have $H^i(E/J) \cong H^i(E/\operatorname{gin}_<(J))$ for all i. Hence also the Betti sequences of E/J and $E/\operatorname{gin}_<(J)$ coincide. Thus we may replace J by $\operatorname{gin}_<(J)$, and hence may as well assume that J is strongly stable; see Proposition 5.2.10.

Let J' be the ideal generated by all $u \in G(J)$ with $m(u) < n$ and all monomials $u \in E$ such that $u \wedge e_n \in G(J)$. Then J' is again strongly stable and $E_1 J' \subset J$. By Proposition 11.4.8, the last property implies that

$$\dim_K(J'/J)_i = |\{u \in G(J)_{i+1} : m(u) = n\}| = \beta_{i-1}(E/J).$$

It follows that $\dim_K(E/J')_i = f_{i-1} - \beta_{i-1}$ for all i. Now we notice that e_n is regular on E/J', in the sense that the complex

$$E/J' \xrightarrow{e_n} E/J' \xrightarrow{e_n} E/J'$$

is exact. Therefore for each i we obtain an exact sequence of K-vector spaces

$$\rightarrow (E/J')_{i-1} \rightarrow (E/J')_i \rightarrow (E/J')_{i+1} \rightarrow (E/(J' + e_n E))_{i+1} \rightarrow 0, \quad (11.25)$$

and hence $\chi_i = \dim_K(E/(J' + e_n E))_{i+1}$.

Next we observe that $J'/J \cong (J' + e_n E)/(J + e_n E)$ and $E_1(J' + e_n E) \subset J + e_n E$, so that together with the algebraic version of the Kruskal–Katona theorem (cf. Theorem 6.4.4) we obtain

$$\chi_i + \beta_i = \dim_K E_{i+1} - \dim_K(J + e_n E)_{i+1}$$
$$\leq \dim_K E_{i+1} - \dim_K E_1(J' + e_n E)_i \leq \chi_{i-1}^{(i)},$$

as required.

(b) $\Rightarrow$ (a): The hypotheses imply that $\chi_i \leq \chi_{i-1}^{(i)}$ and $\chi_i + \beta_i \leq (\chi_{i-1} + \beta_{i-1})^{(i)}$. Thus the Kruskal–Katona theorem yields an integer m, and lexsegment ideals $L \subset N$ in the exterior algebra $E' = K\langle e_1, \ldots, e_{m-1}\rangle$ such that $\dim_K(E/N)_{i+1} = \chi_i$ and $\dim_K(E/L)_{i+1} = \chi_i + \beta_i$ that for all i.

Now let $J \subset E = K\langle e_1, \ldots, e_m\rangle$ be the ideal generated by the elements in $G(L)$ and all elements $u \wedge e_m$ with $u \in G(N)$. Moreover we set $J' = NE$. Then $J'/J \cong N/L$, so

$$\dim_K(E/J)_{i+1} = \dim_K(N/L)_{i+1} + \dim_K(E/J')_{i+1} \qquad (11.26)$$
$$= \beta_i + \dim_K(E/J')_{i+1}.$$

On the other hand, e_m is regular on E/J', so (11.25) yields

$$\dim_K(E/(J' + e_mE))_{i+1} = (-1)^{i+1}\sum_{j=0}^{i+1}(-1)^j \dim_K(E/J')_j \qquad (11.27)$$

for all i. Thus, since $E/(J' + e_mE) \cong E'/N$, it follows from (11.27) that

$$\dim_K(E/J')_{i+1} = \dim_K(E'/N)_{i+1} + \dim_K(E'/N)_i = \chi_i + \chi_{i-1} = f_i - \beta_i.$$

This, together with (11.26), implies that $\dim_K(E/J)_{i+1} = f_i$.

Finally, it is clear from the construction of J that $|\{u \in G(J)_{i+2} : m(u) = m\}|$ equals $\dim_K(N/L)_{i+1}$ which is β_i. Thus, by Proposition 11.4.8, the assertion follows. $\qquad\square$

The Björner–Kalai Theorem can be translated into a theorem on superextremal Betti numbers. Let $I \subset S$ be a graded ideal. We let m be the maximal integer j such that $\beta_{ij}(S/I) \neq 0$ for some i. In other words, m is the largest shift in the graded minimal free S-resolution of S/I. It is clear that $\beta_{im}(S/I)$ is an extremal Betti number for all i with $\beta_{im}(S/I) \neq 0$, and that there is at least one such i. These Betti numbers are distinguished by the fact that they are positioned on the diagonal $\{(i, m-i) : i = 0, \ldots, m\}$ in the Betti diagram, and that all Betti numbers on the right lower side of the diagonal are zero. The ring S/I may of course have other extremal Betti numbers not sitting on this diagonal. The Betti numbers β_{im}, $i = 0, \ldots, m$ are called **superextremal**, regardless of whether they are zero or not. We want to find out which sequences of numbers $(b_0, b_1, \ldots, b_m)$ appear as sequences of superextremal Betti numbers for graded rings with given Hilbert function.

Before answering this question we have to encode the Hilbert function $H_{S/I}(t)$ of S/I in a suitable way. Using the additivity of the Hilbert function, the graded minimal free resolution of S/I yields the following formula:

$$H_{S/I}(t) = \frac{a_0 + a_1 t + a_2 t^2 + \cdots + a_m t^m}{(1-t)^n}$$

with $a_i \in \mathbb{Z}$; see Section 6.1.3. It follows that

$$(1 - t)^{n-m} H_{S/I}(t) = \frac{a_0 + a_1 t + a_2 t^2 + \cdots + a_m t^m}{(1 - t)^m}$$

Notice that $n - m$ may take positive or negative values. At any rate, the rational function $(1 - t)^{n-m} H_{S/I}(t)$ has degree ≤ 0. One easily verifies that there is a unique expansion

$$(1 - t)^{n-m} H_{S/I}(t) = \sum_{i=0}^{m} f_{i-1} \frac{t^i}{(1 - t)^i}$$

with $f_i \in \mathbb{Z}$. It is clear that $f_{-1} = 1$, and we shall see later that all $f_i \geq 0$. We call $f = (f_{-1}, f_0, f_1, \ldots, f_{m-1})$ the **f-vector** of S/I. Given the highest shift in the resolution, the f-vector of S/I determines the Hilbert function of S/I, and vice versa.

We set $b_i = \beta_{m-i-1,m}(S/I)$ for $i = 0, \ldots, m$, and call $b = (b_{-1}, \ldots, b_{m-1})$ the **superextremal sequence** of S/I. Finally we set

$$\chi_i = (-1)^i \sum_{j=-1}^{i} (-1)^j (f_j - b_j) \quad \text{for} \quad i = -1, 0. \ldots, m - 1.$$

The Björner–Kalai theorem has the following counterpart.

Theorem 11.5.2. *Let K be a field of characteristic 0, and let $f = (f_{-1}, f_0, \ldots, f_{m-1})$ and $b = (b_{-1}, b_0, \ldots, b_{m-1})$ be sequences of non-negative integers. The following conditions are equivalent:*

(a) *there exists a homogeneous K-algebra S/I such that f is the f-vector, and b the superextremal sequence of S/I;*
(b) (i) $\chi_{-1} = 1$ *and* $\chi_i \geq 0$ *for all i,*
   (ii) $b_i \leq \chi_{i-1}^{(i)} - \chi_i$ *for all i.*

Proof. (a) $\Rightarrow$ (b) Let $<$ be the reverse lexicographic order. Since by Theorem 4.3.17 the extremal Betti numbers are preserved when we pass from I to $\mathrm{gin}_<(I)$, it follows that I and $\mathrm{gin}_<(I)$ have the same highest shift m, and hence the same b-vector. By Corollary 6.1.5, S/I and $S/\mathrm{gin}_<(I)$ have the same Hilbert function. Hence it follows that the f-vectors of S/I and $S/\mathrm{gin}_<(I)$ coincide. Thus, since $\mathrm{char}(K) = 0$, we may assume that I is a strongly stable monomial ideal; see Proposition 4.2.6.

The ideal I^σ is defined in $S' = K[x_1, \ldots, x_m]$ and by Lemma 11.2.6 we have $\beta_{ii+j}(I) = \beta_{ii+j}(I^\sigma)$. This implies that

$$H_{S'/I^\sigma}(t) = (1 - t)^{n-m} H_{S/I}(t).$$

Hence, if we let Δ be the simplicial complex with $I_\Delta = I^\sigma$, then Δ and S/I have the same f-vector, and the theorem of Hochster (Theorem 8.1.1) implies that $b_i = \dim_K \tilde{H}_i(\Delta; K)$. Therefore the conclusion follows from Theorem 11.5.1.

(b) $\Rightarrow$ (a): Given f- and b-sequences satisfying conditions (b), Theorem 11.5.1 guarantees the existence of an integer m and a simplicial complex Δ on the vertex set $[m]$ whose f-vector is f and whose β-sequence is b. Then $K[x_1, \ldots, x_m]/I_\Delta$ is a homogeneous K-algebra satisfying (a). $\square$

Problems

11.1. Show that Δ is shifted if and only if I_Δ is strongly stable.

11.2. Show that $\mathrm{Shift}_{ij}(\Delta)$ is a simplicial complex.

11.3. Suppose that $\mathrm{Shift}_{ij}(\Delta)$ is pure. Then is Δ pure?

11.4. Let $M_n(K)$ be the set of all $n \times n$-matrices with entries in K. Show that the restriction map $M_n(K) \to M_m(K)$ given by $(a_{ij})_{i,j=1,\cdots,n} \mapsto (a_{ij})_{i,j=1,\cdots,m}$ is an open map.

11.5. Show by an example that a simplical complex may have different combinatorial shiftings.

11.6. Let Δ be the simplicial complex on $[6]$ whose facets $\{i, j\}$ are, where $i = 1, 2, 3$ and $j = 4, 5, 6$. Show that $\Delta^s \neq \Delta^e$.

11.7. Give an example that in general $(\Delta^c)^\vee \neq (\Delta^\vee)^c$.

11.8. Let Δ be the cycle of length n, i.e. $\mathcal{F}(\Delta) = \{\{1,2\}, \{2,3\}, \{3,4\}, \{n-1,n\}, \{n,1\}\}$. Show that $\Delta^c = \Delta^v = \Delta^v$.

11.9. Let Δ be the simplicial complex on $[6]$ whose facets are $\{i, j\}$ with $i = 1, 2, 3$ and $j = 4, 5, 6$. Show that $\Delta^s \neq \Delta^e$.

11.10. Give an example of a simplicial complex Δ which is not lexsegment, but $\beta_{ii+j}(I_\Delta) = \beta_{ii+j}(I_{\Delta^{\mathrm{lex}}})$ for all i and j.

Notes

In classical combinatorics on finite sets, combinatorial shifting [EKR61], introduced by Erdös, Ko and Rado in 1961, was one of the most useful techniques for studying extremal properties of finite sets. One of the reasons why shifted complexes are important in combinatorics is that the f-vector of a simplicial complex and its shifted complex coincide, and in addition, the computation of f-vectors of a shifted complex is quite easy; see [Kal01]. While Kalai invented shifting theory for the development of f-vector theory, the algebraic aspects of the theory have been stressed more in the papers [HT99], [Her01], [AHH00a], [AH00] and [BNT06]. One of the basic algebraic problems in shifting theory is the comparison of the graded Betti numbers of a simplicial complex and its shifted complex. The comparison of I_Δ and $I_{\Delta^{\text{lex}}}$, given in Corollary 11.3.15, was first proved by [AHH00a] under the assumption of char$(K) = 0$. The proof of this result in all characteristics presented here uses combinatorial shifting [MH09].

The comparison of I_Δ and I_{Δ^s} for symmetric algebraic shifting is well understood. Unfortunately, it is not known whether symmetric algebraic shifting can be defined in all characteristics. Also, it is not known whether the graded Betti numbers of a simplicial complex are bounded by the corresponding graded Betti numbers of the exterior shifted complex. However, it has been shown [AHH00a] that their extremal Betti numbers coincide; see Theorem 11.4.1.

The Björner–Kalai theorem [BK88] characterizes the possible (f, β) pairs of a simplicial complex. This theorem can be interpreted as a statement about superextremal Betti numbers and also has a symmetric algebraic version, as presented in Theorem 11.5.2.

12

Discrete Polymatroids

Matroid theory is a very active and fascinating research area in combinatorics. The discrete polymatroid is a multiset analogue of the matroid. Based on the classical polyhedral theory on integral polymatroids developed in the late 1960s and early 1970s, the combinatorics and algebra of discrete polymatroids will be studied. In particular, base rings of polymatroids and polymatroidal ideals are considered.

12.1 Classical polyhedral theory on polymatroids

Fix an integer $n > 0$ and set $[n] = \{1, 2, \ldots, n\}$. The canonical basis vectors of $\mathbb{R}^n$ will be denoted by $\epsilon_1, \ldots, \epsilon_n$. Let $\mathbb{R}_+^n$ denote the set of those vectors $\mathbf{u} = (u_1, \ldots, u_n) \in \mathbb{R}^n$ with each $u_i \geq 0$, and $\mathbb{Z}_+^n = \mathbb{R}_+^n \cap \mathbb{Z}^n$. For a vector $\mathbf{u} = (u_1, \ldots, u_n) \in \mathbb{R}_+^n$ and for a subset $A \subset [n]$, we set

$$\mathbf{u}(A) = \sum_{i \in A} u_i.$$

Thus in particular $\mathbf{u}(\{i\})$, or simply $\mathbf{u}(i)$, is the ith component u_i of $\mathbf{u}$. The *modulus* of $\mathbf{u}$ is

$$|\mathbf{u}| = \mathbf{u}([n]) = \sum_{i=1}^{n} u_i.$$

Let $\mathbf{u} = (u_1, \ldots, u_n)$ and $\mathbf{v} = (v_1, \ldots, v_n)$ be two vectors belonging to $\mathbb{R}_+^n$. We write $\mathbf{u} \leq \mathbf{v}$ if all components $v_i - u_i$ of $\mathbf{v} - \mathbf{u}$ are nonnegative and, moreover, write $\mathbf{u} < \mathbf{v}$ if $\mathbf{u} \leq \mathbf{v}$ and $\mathbf{u} \neq \mathbf{v}$. We say that $\mathbf{u}$ is a **subvector** of $\mathbf{v}$ if $\mathbf{u} \leq \mathbf{v}$. In addition, we set

$$\mathbf{u} \vee \mathbf{v} = (\max\{u_1, v_1\}, \ldots, \max\{u_n, v_n\}),$$
$$\mathbf{u} \wedge \mathbf{v} = (\min\{u_1, v_1\}, \ldots, \min\{u_n, v_n\}).$$

Thus $\mathbf{u} \wedge \mathbf{v} \leq \mathbf{u} \leq \mathbf{u} \vee \mathbf{v}$ and $\mathbf{u} \wedge \mathbf{v} \leq \mathbf{v} \leq \mathbf{u} \vee \mathbf{v}$.

J. Herzog, T. Hibi, *Monomial Ideals*, Graduate Texts in Mathematics 260,   237
DOI 10.1007/978-0-85729-106-6_12, © Springer-Verlag London Limited 2011

Definition 12.1.1. A **polymatroid** on the **ground set** $[n]$ is a nonempty compact subset $\mathcal{P}$ in $\mathbb{R}_+^n$, the set of **independent vectors**, such that

(P1) every subvector of an independent vector is independent;
(P2) if $\mathbf{u}, \mathbf{v} \in \mathcal{P}$ with $|\mathbf{v}| > |\mathbf{u}|$, then there is a vector $\mathbf{w} \in P$ such that

$$\mathbf{u} < \mathbf{w} \leq \mathbf{u} \vee \mathbf{v}.$$

A **base** of a polymatroid $\mathcal{P} \subset \mathbb{R}_+^n$ is a maximal independent vector of $\mathcal{P}$, i.e. an independent vector $\mathbf{u} \in \mathcal{P}$ with $\mathbf{u} < \mathbf{v}$ for no $\mathbf{v} \in \mathcal{P}$. Every base of $\mathcal{P}$ has the same modulus $\mathrm{rank}(\mathcal{P})$, the **rank** of $\mathcal{P}$. In fact, if u and v are bases of $\mathcal{P}$ with $|\mathbf{u}| < |\mathbf{v}|$, then by (P2) there exists $\mathbf{w} \in P$ with $\mathbf{u} < \mathbf{w} \leq \mathbf{u} \vee \mathbf{v}$, contradicting the maximality of $\mathbf{u}$.

Let $\mathcal{P} \subset \mathbb{R}_+^n$ be a polymatroid on the ground set $[n]$. Let $2^{[n]}$ denote the set of all subsets of $[n]$. The **ground set rank function** of $\mathcal{P}$ is a function $\rho : 2^{[n]} \to \mathbb{R}_+$ defined by setting

$$\rho(A) = \max\{\mathbf{v}(A) : \mathbf{v} \in \mathcal{P}\}$$

for all $\emptyset \neq A \subset [n]$ together with $\rho(\emptyset) = 0$.

Given a vector $\mathbf{x} \in \mathbb{R}_+^n$, an independent vector $\mathbf{u} \in \mathcal{P}$ is called a **maximal independent subvector** of $\mathbf{x}$ if (i) $\mathbf{u} \leq \mathbf{x}$ and (ii) $\mathbf{u} < \mathbf{v} \leq \mathbf{x}$ for no $\mathbf{v} \in \mathcal{P}$. Since $\mathcal{P}$ is compact, a maximal independent subvector of $\mathbf{x} \in \mathbb{R}_+^n$ exists. Moreover, if $\mathbf{x} \in \mathbb{R}_+^n$ and $\mathbf{w} \in \mathcal{P}$ with $\mathbf{w} \leq \mathbf{x}$, then, since $\{\mathbf{y} \in \mathcal{P} : \mathbf{w} \leq \mathbf{y}\}$ is compact, there is a maximal independent subvector $\mathbf{u} \in \mathcal{P}$ with $\mathbf{w} \leq \mathbf{u} \leq \mathbf{x}$.

If each of $\mathbf{u}$ and $\mathbf{u}'$ is a maximal independent subvector of $\mathbf{x} \in \mathbb{R}_+^n$, then $|\mathbf{u}| = |\mathbf{u}'|$. In fact, if $|\mathbf{u}| < |\mathbf{u}'|$, then by (P2) there is $\mathbf{u}'' \in \mathcal{P}$ with $\mathbf{u} < \mathbf{u}'' \leq \mathbf{u} \vee \mathbf{u}' \leq \mathbf{x}$, contradicting the maximality of $\mathbf{u}$. For a vector $\mathbf{x} \in \mathbb{R}_+$, we define $\xi(\mathbf{x}) = |\mathbf{u}|$, where $\mathbf{u} \in \mathcal{P}$ is a maximal independent subvector of $\mathbf{x}$.

Lemma 12.1.2. *Let* $\mathbf{x}, \mathbf{y} \in \mathbb{R}_+^n$. *Then*

$$\xi(\mathbf{x}) + \xi(\mathbf{y}) \geq \xi(\mathbf{x} \vee \mathbf{y}) + \xi(\mathbf{x} \wedge \mathbf{y}).$$

Proof. Let $\mathbf{a} \in \mathcal{P}$ be a maximal independent subvector of $\mathbf{x} \wedge \mathbf{y}$. Since $\mathbf{a} \leq \mathbf{x} \vee \mathbf{y}$, there exists a maximal independent subvector $\mathbf{b} \in \mathcal{P}$ of $\mathbf{x} \vee \mathbf{y}$ with $\mathbf{a} \leq \mathbf{b} \leq \mathbf{x} \vee \mathbf{y}$. Since $\mathbf{b} \wedge (\mathbf{x} \wedge \mathbf{y}) \in \mathcal{P}$ and $\mathbf{a} \leq \mathbf{b} \wedge (\mathbf{x} \wedge \mathbf{y}) \leq \mathbf{x} \wedge \mathbf{y}$, one has $\mathbf{a} = \mathbf{b} \wedge (\mathbf{x} \wedge \mathbf{y})$. We claim

$$\mathbf{a} + \mathbf{b} = \mathbf{b} \wedge \mathbf{x} + \mathbf{b} \wedge \mathbf{y}.$$

In fact, since $\mathbf{b} \leq \mathbf{x} \vee \mathbf{y}$, one has $\mathbf{b}(i) \leq \max\{\mathbf{x}(i), \mathbf{y}(i)\}$ for each $i \in [n]$. Let $\mathbf{x}(i) \leq \mathbf{y}(i)$. Then $\mathbf{a}(i) = \min\{\mathbf{b}(i), \mathbf{x}(i)\}$ and $\mathbf{b}(i) = \min\{\mathbf{b}(i), \mathbf{y}(i)\}$. Thus $\mathbf{a}(i) + \mathbf{b}(i) = (\mathbf{b} \wedge \mathbf{x})(i) + (\mathbf{b} \wedge \mathbf{y})(i)$, as required.

Since $\mathbf{b} \wedge \mathbf{x} \in \mathcal{P}$ is a subvector of $\mathbf{x}$ and since $\mathbf{b} \wedge \mathbf{y} \in \mathcal{P}$ is a subvector of $\mathbf{y}$, it follows that $|\mathbf{b} \wedge \mathbf{x}| \leq \xi(\mathbf{x})$ and $|\mathbf{b} \wedge \mathbf{y}| \leq \xi(\mathbf{y})$. Thus

$$\xi(\mathbf{x} \wedge \mathbf{y}) + \xi(\mathbf{x} \vee \mathbf{y}) = |\mathbf{a}| + |\mathbf{b}|$$
$$= |\mathbf{b} \wedge \mathbf{x}| + |\mathbf{b} \wedge \mathbf{y}|$$
$$\leq \xi(\mathbf{x}) + \xi(\mathbf{y}),$$

as desired. □

We now come to an important result of polymatroids.

Theorem 12.1.3. (a) *Let $\mathcal{P} \subset \mathbb{R}^n_+$ be a polymatroid on the ground set $[n]$ and ρ its ground set rank function. Then ρ is nondecreasing, i.e. if $A \subset B \subset [n]$, then $\rho(A) \leq \rho(B)$, and is submodular, i.e.*

$$\rho(A) + \rho(B) \geq \rho(A \cup B) + \rho(A \cap B)$$

for all $A, B \subset [n]$. Moreover, $\mathcal{P}$ coincides with the compact set

$$\{\mathbf{x} \in \mathbb{R}^n_+ : \mathbf{x}(A) \leq \rho(A), A \subset [n]\}. \tag{12.1}$$

(b) *Conversely, given a nondecreasing and submodular function $\rho : 2^{[n]} \to \mathbb{R}_+$ with $\rho(\emptyset) = 0$. Then the compact set (12.1) is a polymatroid on the ground set $[n]$ with ρ its ground set rank function.*

Proof. (a) ⇒ (b): Clearly, ρ is nondecreasing.

In general, for $X \subset [n]$ and for $\mathbf{y} \in \mathbb{R}_+$, we define $\mathbf{y}_X \in \mathbb{R}_+$ by setting $\mathbf{y}_X(i) = \mathbf{y}(i)$, $i \in X$ and $\mathbf{y}_X(i) = 0$, $i \in [n] \setminus X$.

Let $r = \mathrm{rank}(\mathcal{P})$ and $\mathbf{a} = (r, r, \ldots, r) \in \mathbb{R}^n_+$. Thus $\mathbf{w} \leq \mathbf{a}$ for all $\mathbf{w} \in \mathcal{P}$. We claim that, for each subset $X \subset [n]$, one has

$$\rho(X) = \xi(\mathbf{a}_X).$$

To see why this is true, write $\mathbf{b}$ for a maximal independent subvector of $\mathbf{a}_X$. Then $\xi(\mathbf{a}_X) = |\mathbf{b}| = \mathbf{b}(X) \leq \rho(X)$. On the other hand, if $\rho(X) = \mathbf{w}(X)$, where $\mathbf{w} \in \mathcal{P}$, then, since $\mathbf{w}_X \leq \mathbf{a}_X$, one has $\rho(X) = \mathbf{w}(X) = |\mathbf{w}_X| \leq \xi(\mathbf{a}_X)$. Thus $\rho(X) = \xi(\mathbf{a}_X)$.

Let A and B be subsets of $[n]$. Then $\rho(A \cup B) = \xi(\mathbf{a}_{A \cup B}) = \xi(\mathbf{a}_A \vee \mathbf{a}_B)$ and $\rho(A \cap B) = \xi(v_{A \cap B}) = \xi(\mathbf{a}_A \wedge \mathbf{a}_B)$. Thus the submodularity of ρ follows from Lemma 12.1.2.

Let $\mathcal{Q}$ denote the compact set (12.1). It follows from the definition of ρ that $\mathcal{P} \subset \mathcal{Q}$. We will show $\mathcal{Q} \subset \mathcal{P}$.

Suppose that there exists $\mathbf{v} \in \mathcal{Q}$ with $\mathbf{v} \notin \mathcal{P}$. Let $\mathbf{u} \in \mathcal{P}$ be a maximal independent subvector of $\mathbf{v}$ which maximizes $|N(\mathbf{u})|$, where

$$N(\mathbf{u}) = \{i \in [n] : \mathbf{u}(i) < \mathbf{v}(i)\}.$$

Let $\mathbf{w} = (\mathbf{u} + \mathbf{v})/2 \in \mathbb{R}^n_+$ and $\mathbf{b} \in \mathcal{P}$ with $\mathbf{b}(N(\mathbf{u})) = \rho(N(\mathbf{u}))$. Since $\mathbf{w} \in \mathcal{Q}$, it follows that

$$\mathbf{u}(N(\mathbf{u})) < \mathbf{w}(N(\mathbf{u})) \leq \rho(N(\mathbf{u})) = \mathbf{b}(N(\mathbf{u})).$$

Since $|\mathbf{u}_{N(\mathbf{u})}| < |\mathbf{b}_{N(\mathbf{u})}|$, there is $\mathbf{u}' \in \mathcal{P}$, with $\mathbf{u}_{N(\mathbf{u})} < \mathbf{u}' \leq \mathbf{u}_{N(\mathbf{u})} \vee \mathbf{b}_{N(\mathbf{u})}$. Thus $\mathbf{u}_{N(\mathbf{u})} < \mathbf{u}' \wedge \mathbf{w}_{N(\mathbf{u})} \leq \mathbf{w}_{N(\mathbf{u})}$. Hence $\mathbf{u}_{N(\mathbf{u})}$ cannot be a maximal independent subvector of $\mathbf{w}_{N(\mathbf{u})}$. Let $\mathbf{u}'' \in \mathcal{P}$ with $\mathbf{u}_{N(\mathbf{u})} < \mathbf{u}''$ be a maximal independent subvector of $\mathbf{w}_{N(\mathbf{u})}$.

Let $\mathbf{u}^* \in \mathcal{P}$ be a maximal independent subvector of $\mathbf{w}$ with $\mathbf{u}'' \leq \mathbf{u}^*$. Since each of $\mathbf{u}$ and $\mathbf{u}^*$ is a maximal independent subvector of $\mathbf{w}$, one has $|\mathbf{u}| = |\mathbf{u}^*|$. However, since $\mathbf{u}(N(\mathbf{u})) < \mathbf{u}''(N(\mathbf{u})) \leq \mathbf{u}^*(N(\mathbf{u}))$, there is $j \in [n] \setminus N(\mathbf{u})$ with $\mathbf{u}^*(j) < \mathbf{u}(j) \ (= \mathbf{v}(j))$. Since $\mathbf{u}^*(i) \leq \mathbf{w}(i) < \mathbf{v}(i)$ for all $i \in N(\mathbf{u})$, one has $|N(\mathbf{u}^*)| > |N(\mathbf{u})|$. This contradicts the maximality of $|N(\mathbf{u})|$.

(b) $\Rightarrow$ (a): Let $\mathcal{P}$ denote the compact set (12.1). Suppose that there exist $\mathbf{u}, \mathbf{v} \in \mathcal{P}$ with $|\mathbf{v}| > |\mathbf{u}|$ for which (P2) fails. Let $V = \{i \in [n] : \mathbf{v}(i) > \mathbf{u}(i)\}$.

We claim that, for each $i \in V$, there is a subset $A_i \subset [n]$ with $i \in A_i$ such that $\mathbf{u}(A_i) = \rho(A_i)$. In fact, if there is $i \in V$ with $\mathbf{u}(A) < \rho(A)$ for all subsets $A \subset [n]$ with $i \in A$, then for $N \gg 0$ the vector $\mathbf{w} = \mathbf{u} + (1/N)\epsilon_i$ belongs to $\mathcal{P}$ and satisfies $\mathbf{u} < \mathbf{w} \leq \mathbf{u} \vee \mathbf{v}$, a contradiction.

Now, let A be a maximal subset of $[n]$ with $\mathbf{u}(A) = \rho(A)$. By using the submodularity of ρ together with $\mathbf{u} \in \mathcal{P}$, it follows that

$$
\begin{aligned}
\rho(A \cup A_i) + \rho(A \cap A_i) &\leq \rho(A) + \rho(A_i) \\
&= \mathbf{u}(A) + \mathbf{u}(A_i) \\
&= \mathbf{u}(A \cup A_i) + \mathbf{u}(A \cap A_i) \\
&\leq \rho(A \cup A_i) + \rho(A \cap A_i).
\end{aligned}
$$

Hence there must be equality throughtout, so that $\mathbf{u}(A \cup A_i) = \rho(A \cup A_i)$.

Then, by the maximality of A, one has $A_i \subset A$. In particular $i \in A$ for all $i \in V$. Thus $V \subset A$. Hence $\rho(A) = \mathbf{u}(A) < \mathbf{v}(A)$. This contradicts $\mathbf{v} \in \mathcal{P}$. Hence (P2) holds. $\qquad \square$

We refer the reader to Appendix B for basic material on convex polytopes. A sketch of the proofs of Theorems 12.1.4 and 12.1.5 is given in Appendix B.

It follows from Theorem 12.1.3 (a) that a polymatroid $\mathcal{P} \subset \mathbb{R}_+^n$ on the ground set $[n]$ is a convex polytope in $\mathbb{R}^n$. In addition, the set of bases of $\mathcal{P}$ is a face of $\mathcal{P}$ with supporting hyperplane

$$
\{\mathbf{x} = (x_1, \ldots, x_n) \in \mathbb{R}^n : \sum_{i=1}^n x_i = \text{rank}(\mathcal{P})\}.
$$

How can we find the vertices of a polymatroid? We will associate any permutation $\pi = (i_1, \ldots, i_n)$ of $[n]$ with $A_\pi^1 = \{i_1\}, A_\pi^2 = \{i_1, i_2\}, \ldots, A_\pi^n = \{i_1, \ldots, i_n\}$.

Theorem 12.1.4. *Let $\mathcal{P} \subset \mathbb{R}_+^n$ be a polymatroid on the ground set $[n]$ and ρ its ground set rank function. Then the vertices of $\mathcal{P}$ are all points $\mathbf{v} = \mathbf{v}(k, \pi) \in \mathbb{R}_+^n$, where $\mathbf{v} = (v_1, \ldots, v_n)$ and*

$$v_{i_1} = \rho(A_\pi^1),$$
$$v_{i_2} = \rho(A_\pi^2) - \rho(A_\pi^1),$$
$$v_{i_3} = \rho(A_\pi^3) - \rho(A_\pi^2),$$
$$\cdots$$
$$v_{i_k} = \rho(A_\pi^k) - \rho(A_\pi^{k-1}),$$
$$v_{i_{k+1}} = v_{i_{k+2}} = \cdots = v_{i_n} = 0,$$

and k ranges over the integers belonging to $[n]$, and $\pi = (i_1, \ldots, i_n)$ ranges over all permutations of $[n]$. In particular the vertices of the face of $\mathcal{P}$ consisting of all bases of $\mathcal{P}$ are all points $\mathbf{v} = \mathbf{v}(n, \pi) \in \mathbb{R}_+^n$, where π ranges over all permutations of $[n]$.

We say that a polymatroid is **integral** if all of its vertices have integer coordinates; in other words, a polymatroid is integral if and only if its ground set rank function is integer valued.

Let $\mathcal{P}_1, \ldots, \mathcal{P}_k$ be polymatroids on the ground set $[n]$. The **polymatroidal sum** $\mathcal{P}_1 \vee \cdots \vee \mathcal{P}_k$ of $\mathcal{P}_1, \ldots, \mathcal{P}_k$ is the compact subset in $\mathbb{R}_+^n$ consisting of all vectors $\mathbf{x} \in \mathbb{R}_+^n$ of the form

$$\mathbf{x} = \sum_{i=1}^k \mathbf{x}_i, \qquad \mathbf{x}_i \in \mathcal{P}_i.$$

Theorem 12.1.5. *Let $\mathcal{P}_1, \ldots, \mathcal{P}_k$ be polymatroids on the ground set $[n]$ and ρ_i the ground set rank function of $\mathcal{P}_i$, $1 \leq i \leq k$. Then the polymatroid sum $\mathcal{P}_1 \vee \cdots \vee \mathcal{P}_k$ is a polymatroid on $[n]$ and $\sum_{i=1}^k \rho_i$ is its ground set rank function. Moreover, if each $\mathcal{P}_i$ is integral, then $\mathcal{P}_1 \vee \cdots \vee \mathcal{P}_k$ is integral, and for each integer vector $\mathbf{x} \in \mathcal{P}_1 \vee \cdots \vee \mathcal{P}_k$ there exist integer vectors $\mathbf{x}_i \in \mathcal{P}_i$, $1 \leq i \leq k$, with $\mathbf{x} = \sum_{i=1}^k \mathbf{x}_i$.*

12.2 Matroids and discrete polymatroids

A few examples and basic properties of matroids and discrete polymatroids will be discussed. We begin with

Definition 12.2.1. A **discrete polymatroid** on the ground set $[n]$ is a nonempty finite set $P \subset \mathbb{Z}_+^n$ satisfying

(D1) if $\mathbf{u} \in P$ and $\mathbf{v} \in \mathbb{Z}_+^n$ with $\mathbf{v} \leq \mathbf{u}$, then $\mathbf{v} \in P$;
(D2) if $\mathbf{u} = (u_1, \ldots, u_n) \in P$ and $\mathbf{v} = (v_1, \ldots, v_n) \in P$ with $|\mathbf{u}| < |\mathbf{v}|$, then there is $i \in [n]$ with $u_i < v_i$ such that $\mathbf{u} + \epsilon_i \in P$.

A **base** of a discrete polymatroid $P \subset \mathbb{Z}_+^n$ is a vector $\mathbf{u} \in P$ such that $\mathbf{u} < \mathbf{v}$ for no $\mathbf{v} \in P$. Let $B(P)$ denote the set of bases of P. It follows from

(D2) that if $\mathbf{u}$ and $\mathbf{v}$ are bases of P, then $|\mathbf{u}| = |\mathbf{v}|$. The nonnegative integer $|\mathbf{u}|$ with $\mathbf{u} \in B(P)$ is called the **rank of** P.

Let $2^{[n]}$ denote the set of all subsets of $[n]$. We associate each $F \subset [n]$ with the $(0, 1)$-vector $\mathbf{w}_F = (w_1, \ldots, w_n)$, where $w_i = 1$ if $i \in F$ and where $w_i = 0$ if $i \notin F$. A **matroid** on the ground set $[n]$ is a subset $\mathcal{M} \subset 2^{[n]}$ such that $\{\mathbf{w}_F : F \in \mathcal{M}\}$ is a discrete polymatroid on $[n]$. Note that a matroid $\mathcal{M}$ is a simplicial complex with the property that for all faces F and G in $\mathcal{M}$ with $|F| < |G|$, there exists $i \in G \setminus F$ such that $F \cup \{i\} \in \mathcal{M}$.

Example 12.2.2. (a) Let $\mathbf{v}_1, \ldots, \mathbf{v}_n$ be vectors of a vector space. Let $\mathcal{M}$ denote the subset of $2^{[n]}$ consisting of those $F \subset [n]$ such that the vectors $\mathbf{v}_k$ with $k \in F$ are linearly independent. A fundamental fact on linear algebra guarantees that $\mathcal{M}$ is a matroid. Such a matroid is called a **linear matroid**.

(b) Let G be a finite graph with the edges $e_1, \ldots, e_n$. Let $\mathcal{M}$ denote the subset of $2^{[n]}$ consisting of those $F \subset [n]$ such that the subgraph of G whose edges are e_k with $k \in F$ is a forest. It follows that $\mathcal{M}$ is a matroid. Such a matroid is called a **graphical matroid**.

Lemma 12.2.3. *Let P be a discrete polymatroid.*

(a) *Let $d \le \operatorname{rank} P$. Then the set $P' = \{\mathbf{u} \in P : |\mathbf{u}| \le d\}$ is a discrete polymatroid of rank d with the set of bases $\{\mathbf{u} \in P : |\mathbf{u}| = d\}$.*
(b) *Suppose that $d = \operatorname{rank} P$. Then for each $\mathbf{x} \in P$ the set*

$$P_{\mathbf{x}} = \{\mathbf{v} - \mathbf{x} : \mathbf{v} \in P, \mathbf{v} \ge \mathbf{x}\}$$

is a discrete polymatroid of rank $d - |\mathbf{x}|$.

Proof. (a) Let $\mathbf{u}, \mathbf{v} \in P$ with $d \ge |\mathbf{v}| > |\mathbf{u}|$. There exists $\mathbf{w} \in P$ such that $\mathbf{u} < \mathbf{w} \le \mathbf{u} \vee \mathbf{v}$. Since $\mathbf{w} > \mathbf{u}$, and since P contains all subvectors of $\mathbf{w}$, there exists an integer i such that $\mathbf{u} + \epsilon_i \le \mathbf{w}$. Then $\mathbf{u} < \mathbf{u} + \epsilon_i \le \mathbf{u} \vee \mathbf{v}$, and since $\mathbf{u} + \epsilon_i \le d$, it belongs to P'. This proves that P' is a discrete polymatroid. It is clear that $\{\mathbf{u} \in P : |\mathbf{u}| = d\}$ is the set of bases of P'.

(b) Let $\mathbf{u}', \mathbf{v}' \in P_{\mathbf{x}}$ with $|\mathbf{v}'| > |\mathbf{u}'|$, and set $\mathbf{u} = \mathbf{u}' + \mathbf{x}$ and $\mathbf{v} = \mathbf{v}' + \mathbf{x}$. Then $\mathbf{u}, \mathbf{v} \in P$ and $|\mathbf{v}| > |\mathbf{u}|$. Hence there exists $\mathbf{w} \in P$ with $\mathbf{u} < \mathbf{w} \le \mathbf{u} \vee \mathbf{v}$. Set $\mathbf{w}' = \mathbf{w} - \mathbf{x}$. Then $\mathbf{w}' \in P_{\mathbf{x}}$ and $\mathbf{u}' < \mathbf{w}' \le \mathbf{u}' \vee \mathbf{v}'$. $\square$

Discrete polymatroids can be characterized in terms of their sets of bases.

Theorem 12.2.4. *Let P be a nonempty finite set of integer vectors in $\mathbb{R}^n_+$ which contains with each $\mathbf{u} \in P$ all its integral subvectors, and let $B(P)$ be the set of vectors $\mathbf{u} \in P$ with $\mathbf{u} < \mathbf{v}$ for no $\mathbf{v} \in P$. Then the following conditions are equivalent:*

(a) *P is a discrete polymatroid;*
(b) *if $\mathbf{u}, \mathbf{v} \in P$ with $|\mathbf{v}| > |\mathbf{u}|$, then there is an integer i such that $\mathbf{u} + \epsilon_i \in P$ and*

$$\mathbf{u} + \epsilon_i \le \mathbf{u} \vee \mathbf{v};$$

(c) (i) *all* $\mathbf{u} \in B(P)$ *have the same modulus,*

   (ii) **(The exchange property)** *if* $\mathbf{u}, \mathbf{v} \in B(P)$ *with* $\mathbf{u}(i) > \mathbf{v}(i)$, *then there exists* j *with* $\mathbf{u}(j) < \mathbf{v}(j)$ *such that* $\mathbf{u} - \epsilon_i + \epsilon_j \in B(P)$.

Proof. (a) $\Rightarrow$ (b): Already shown in the proof of Lemma 12.2.3.

(b) $\Rightarrow$ (a): Obvious.

(b) $\Rightarrow$ (c): We have already noted that (c)(i) holds. Thus it remains to prove (c)(ii). Let $\mathbf{u}, \mathbf{v} \in B(P)$ with $\mathbf{u}(i) > \mathbf{v}(i)$ for some i. Then $\mathbf{u}(i) - 1 \geq \mathbf{v}(i)$, and hence $|\mathbf{u} - \epsilon_i| = |\mathbf{v}| - 1 < |\mathbf{v}|$. Thus by (b) there exists an integer j such that $(\mathbf{u} - \epsilon_i) + \epsilon_j \leq (\mathbf{u} - \epsilon_i) \vee \mathbf{v}$. We have $j \neq i$, because otherwise $\mathbf{u}(i) = (\mathbf{u} - \epsilon_i + \epsilon_j)(i) \leq \max\{\mathbf{u}(i) - 1, \mathbf{v}(i)\} = \mathbf{u}(i) - 1$; a contradiction. Thus $\mathbf{u}(j) + 1 = (\mathbf{u} - \epsilon_i + \epsilon_j)(j) \leq \max\{\mathbf{u}(j), \mathbf{v}(j)\} \leq \mathbf{v}(j)$, that is, $\mathbf{v}(j) > \mathbf{u}(j)$.

(c) $\Rightarrow$ (b): Let $\mathbf{v}, \mathbf{u} \in P$ with $|\mathbf{v}| > |\mathbf{u}|$, and let $\mathbf{w}' \in B(P)$ with $\mathbf{u} < \mathbf{w}'$. Since all $\mathbf{w}' \in B(P)$ have the same modulus, it follows that $|\mathbf{v}| < |\mathbf{w}'|$. Thus we can choose a subvector $\mathbf{w}$ of $\mathbf{w}'$ in P with $\mathbf{u} \leq \mathbf{w}$ and $|\mathbf{w}| = |\mathbf{v}|$. Let $P' = \{\mathbf{x} \in P : |\mathbf{x}| \leq |\mathbf{v}|\}$. Later, in Lemma 12.2.5, we will show that P' possesses the property (c). In particular, $\mathbf{w}$ and $\mathbf{v}$ satisfy the exchange property (c)(ii).

Suppose that $\mathbf{w}(j) \leq \max\{\mathbf{u}(j), \mathbf{v}(j)\}$ for all j. Since $\mathbf{u}(j) \leq \mathbf{w}(j)$ for all j, it follows that $\mathbf{w}(j) \leq \mathbf{v}(j)$ for all j. However, since $|\mathbf{w}| = |\mathbf{v}|$, this implies $\mathbf{w} = \mathbf{v}$. Then $\mathbf{u} < \mathbf{v}$, and the assertion is trivial.

Now assume that there exists an integer j such that $\mathbf{w}(j) > \max\{\mathbf{u}(j), \mathbf{v}(j)\}$. Then by the exchange property, there exists an integer i with $\mathbf{v}(i) > \mathbf{w}(i)$ such that $\mathbf{w} - \epsilon_i + \epsilon_j \in P$. Since $\mathbf{u} + \epsilon_i$ is a subvector of $\mathbf{w} - \epsilon_j + \epsilon_i$, it follows that $\mathbf{u} + \epsilon_i \in P$. Furthermore we have $(\mathbf{u} + \epsilon_i)(i) = \mathbf{u}(i) + 1 \leq \mathbf{w}(i) + 1 \leq \mathbf{v}(i)$, so that $\mathbf{u} + \epsilon_i \leq \mathbf{u} \vee \mathbf{v}$. $\qquad\square$

Lemma 12.2.5. *Let P be a nonempty finite set of integer vectors in $\mathbb{R}_+^n$ which contains with each $\mathbf{u} \in P$ all its integral subvectors. Let $B(P)$ denote the set of vectors $\mathbf{u} \in P$ with $\mathbf{u} < \mathbf{v}$ for no $\mathbf{v} \in P$ and suppose that all $\mathbf{u} \in B(P)$ have the same modulus (say, $= r$) and that $B(P)$ satisfies the exchange property. Let $P' = \{\mathbf{x} \in P : |\mathbf{x}| \leq d\}$, where $d \leq r$, and $B(P')$ the set of vectors $\mathbf{u} \in P'$ with $\mathbf{u} < \mathbf{v}$ for no $\mathbf{v} \in P'$. Then all $\mathbf{u} \in B(P')$ have the same modulus and that $B(P)$ satisfies the exchange property.*

Proof. By using the inductive argument, we assume that $d = r - 1$. Since all $\mathbf{u} \in B(P)$ have the same modulus, it follows that all $\mathbf{u} \in B(P')$ have the same modulus. Let $\mathbf{u}, \mathbf{v} \in B(P')$. One has $\mathbf{u}' = \mathbf{u} + \epsilon_a \in B(P)$ and $\mathbf{v}' = \mathbf{v} + \epsilon_b \in B(P)$ for some a and b. If $a = b$, then clearly $\mathbf{u}$ and $\mathbf{v}$ satisfies the exchange property. We thus assume that $\mathbf{u} + \epsilon_b \notin B(P)$. Let $\mathbf{u}(i_0) > \mathbf{v}(i_0)$. What we must prove is that there is j_0 with $\mathbf{u}(j_0) < \mathbf{v}(j_0)$ such that $\mathbf{u} + \epsilon_{j_0} \in B(P)$. In fact, if $\mathbf{u} + \epsilon_{j_0} \in B(P)$, then $\mathbf{u} - \epsilon_{i_0} + \epsilon_{j_0} \in B(P')$.

Suppose that for each c with $\mathbf{u}(c) < \mathbf{v}(c)$ one has $\mathbf{u} + \epsilon_c \notin B(P)$. In particular, since $\mathbf{u} + \epsilon_a \in B(P)$, one has $\mathbf{u}(a) \geq \mathbf{v}(a)$. Hence $\mathbf{u}'(a) > \mathbf{v}'(a)$. Thus there is k with $\mathbf{u}'(k) < \mathbf{v}'(k)$ such that $\mathbf{u}' - \epsilon_a + \epsilon_k \in B(P)$. In other words, $\mathbf{u} + \epsilon_k \in B(P)$. Thus $\mathbf{u}(k) \geq \mathbf{v}(k)$. Since $\mathbf{u}'(k) < \mathbf{v}'(k)$, one has $\mathbf{u}(k) = \mathbf{v}(k)$ and $k = b$. Hence $\mathbf{u} + \epsilon_b \in B(P)$; a contradiction. $\qquad\square$

We define the **distance** between $\mathbf{u}$ and $\mathbf{v}$, where $\mathbf{u}, \mathbf{v} \in B(P)$, by

$$\operatorname{dis}(\mathbf{u}, \mathbf{v}) = \frac{1}{2} \sum_{i=1}^{n} |\mathbf{u}(i) - \mathbf{v}(i)|.$$

A crucial property of $\operatorname{dis}(\mathbf{u}, \mathbf{v})$ is that if $\mathbf{u}(i) > \mathbf{v}(i)$ and $\mathbf{u}(j) < \mathbf{v}(j)$ together with $\mathbf{u}' = \mathbf{u} - \epsilon_i + \epsilon_j \in B(P)$, then $\operatorname{dis}(\mathbf{u}, \mathbf{v}) > \operatorname{dis}(\mathbf{u}', \mathbf{v})$.

Proposition 12.2.6. *Work with the same situation as in Theorem 12.2.4 and suppose that the condition (c) is satisfied. Then, for $\mathbf{u}, \mathbf{v} \in B(P)$ with $\mathbf{u}(i) < \mathbf{v}(i)$, there exists j with $\mathbf{u}(j) > \mathbf{v}(j)$ such that $\mathbf{u} + \epsilon_i - \epsilon_j \in B(P)$.*

Proof. Fix i with $\mathbf{u}(i) < \mathbf{v}(j)$. If there is $k_1 \neq i$ with $\mathbf{u}(k_1) < \mathbf{v}(k_1)$, then there is ℓ_1 with $\mathbf{u}(\ell_1) > \mathbf{v}(\ell_1)$ such that $\mathbf{w}_1 = \mathbf{v} - \epsilon_{k_1} + \epsilon_{\ell_1} \in B(P)$. Then $\mathbf{w}(i) = \mathbf{v}(i)$ and $\operatorname{dist}(\mathbf{u}, \mathbf{w}_1) < \operatorname{dist}(\mathbf{u}, \mathbf{v})$. Again, if there is $k_2 \neq i$ with $\mathbf{u}(k_2) < \mathbf{w}_1(k_2)$, then there is ℓ_2 with $\mathbf{u}(\ell_2) > \mathbf{w}_1(\ell_2)$ such that $\mathbf{w}_2 = \mathbf{w}_1 - \epsilon_{k_2} + \epsilon_{\ell_2} \in B(P)$. Then $\mathbf{w}_2(i) = \mathbf{v}_2(i)$ and $\operatorname{dist}(\mathbf{u}, \mathbf{w}_2) < \operatorname{dist}(\mathbf{u}, \mathbf{w}_1)$. Repeating these procedures yields $\mathbf{w}^* \in B(P)$ with $\mathbf{w}^*(i) = \mathbf{v}(i) > \mathbf{u}(i)$ and $\mathbf{w}^*(j) \leq \mathbf{u}(j)$ for all $j \neq i$. One has $j_0 \neq i$ with $\mathbf{w}^*(j_0) < \mathbf{u}(j_0)$. Then by the exchange property it follows that $\mathbf{u} - \epsilon_{j_0} + \epsilon_i \in B(P)$, as desired.   □

Proposition 12.2.7. *Let P be a nonempty finite set of integer vectors in $\mathbb{R}_+^n$ which contains with each $\mathbf{u} \in P$ all its integral subvectors. Then the following conditions are equivalent:*

(a) P is a discrete polymatroid of rank d on the ground set $[n]$;
(b) $B = \{(\mathbf{u}, d - |\mathbf{u}|) : \mathbf{u} \in P\}$ is the set of bases of a discrete polymatroid of rank d on the ground set $[n+1]$.

Proof. (a)$\Rightarrow$(b): We will show that B satisfies condition (c) of Theorem 12.2.4. Let $\mathbf{u}, \mathbf{v} \in P$, $i = d - |\mathbf{u}|$, $j = d - |\mathbf{v}|$ and set $\mathbf{u}' = (\mathbf{u}, i)$ and $\mathbf{v}' = (\mathbf{v}, j)$. We may suppose that $|\mathbf{v}| \geq |\mathbf{u}|$, so that $j \leq i$. If $i = j$, then $\mathbf{u}$ and $\mathbf{v}$ are bases of $P' = \{\mathbf{w} \in P : |\mathbf{w}| \leq d - i\}$, see Lemma 12.2.3. Thus $\mathbf{u}'$ and $\mathbf{v}'$ satisfy the exchange property, and hence we may assume that $j < i$. We consider two cases.

In the first case assume that $\mathbf{v}'(k) > \mathbf{u}'(k)$ for some k. Then $k \leq n$, and $\mathbf{v} - \epsilon_k \in P$, since $\mathbf{v} - \epsilon_k$ is a subvector of $\mathbf{v}$, and so $\mathbf{v}' - \epsilon_k + \epsilon_{n+1} \in B$.

In the second case assume that $\mathbf{u}'(k) > \mathbf{v}'(k)$ for some k. Since $|\mathbf{v}| > |\mathbf{u}|$, Theorem 12.2.4 (b) guarantees that there exists an integer ℓ such that $\mathbf{u} + \epsilon_\ell \in P$ and $\mathbf{u} + \epsilon_\ell \leq \mathbf{u} \wedge \mathbf{v}$. It follows that $\mathbf{u}(\ell) < \mathbf{u}(\ell) + 1 \leq \mathbf{v}(\ell)$. If $k \leq n$, then $\mathbf{u} - \epsilon_k + \epsilon_\ell \in P$, because it is a subvector of $\mathbf{u} - \epsilon_\ell$. Hence $\mathbf{u}' - \epsilon_k + \epsilon_\ell = (\mathbf{u} - \epsilon_k + \epsilon_\ell, i) \in B$. On the other hand, if $k = n + 1$, that is, $\mathbf{u}'(k) = i$, then $\mathbf{u}' - \epsilon_k + \epsilon_\ell = (\mathbf{u} + \epsilon_\ell, i - 1) \in B$, because $\mathbf{u} + \epsilon_\ell \in P$.

((b) $\Rightarrow$ (a)) Let $\mathbf{u}, \mathbf{v} \in P$ with $|\mathbf{v}| > |\mathbf{u}|$. Then $d - |\mathbf{v}| < d - |\mathbf{u}|$, and so by the exchange property of B there exists an integer i with $\mathbf{v}(i) > \mathbf{u}(i)$ and such that $(\mathbf{u} + \epsilon_i, d - |\mathbf{u}|) \in B$. This implies that $\mathbf{u} + \epsilon_i \in P$, and since $\mathbf{v}(i) > \mathbf{u}(i)$ we also have that $\mathbf{u} + \epsilon_i \leq \mathbf{u} \wedge \mathbf{v}$.   □

As a consequence of Theorem 12.2.4, a matroid can be characterized by the following exchange property: a pure simplicial complex $\mathcal{M}$ is a matroid, if and only if for any two facets F and G of $\mathcal{M}$ with $F \neq G$, and for any $i \in F \setminus G$, there exists $j \in G \setminus F$ such that $F \setminus \{i\} \cup \{j\} \in \mathcal{M}$.

Example 12.2.8. Fix positive integers $d_1, \ldots, d_n$ and d with $d_1 + \cdots + d_n \geq d$. Let $P \subset \mathbb{Z}_+^n$ be the set of vectors $\mathbf{u} \in \mathbb{Z}_+^n$ with $\mathbf{u}(i) \leq d_i$ for all $1 \leq i \leq n$ and with $|\mathbf{u}| \leq d$. Then P is a discrete polymatroid on $[n]$. Such a discrete polymatroid is called a **discrete polymatroid of Veronese type**.

To see why P is a discrete polymatroid, we use Theorem 12.2.4. Let $B(P)$ be the set of vectors $\mathbf{u} \in P$ with $\mathbf{u} < \mathbf{v}$ for no $\mathbf{v} \in P$. Thus $\mathbf{u} \in P$ belongs to $B(P)$ if and only if $|\mathbf{u}| = d$. What we must prove is that $B(P)$ possesses the exchange property. Let $\mathbf{u}, \mathbf{v} \in B(P)$ with $\mathbf{u}(i) > \mathbf{v}(i)$. Since $|\mathbf{u}| = |\mathbf{v}| = d$, it follows that there is j with $\mathbf{u}(j) < \mathbf{v}(j)$. Clearly, $\mathbf{u} - \epsilon_i + \epsilon_j \in P$. Thus $\mathbf{u} - \epsilon_i + \epsilon_j \in B(P)$, as desired.

12.3 Integral polymatroids and discrete polymatroids

We will establish our first fundamental Theorem 12.3.4, which says that a nonempty finite set $P \subset \mathbb{Z}_+^n$ is a discrete polymatroid if and only if $\mathrm{Conv}(P) \subset \mathbb{R}_+^n$ is an integral polymatroid with $\mathrm{Conv}(P) \cap \mathbb{Z}^n = P$. Here $\mathrm{Conv}(P)$ is the convex hull of P in $\mathbb{R}^n$.

First of all, we collect a few basic lemmata on integral polymatroids and discrete polymatroids which will be required to prove Theorem 12.3.4.

Lemma 12.3.1 can be proved by imitating the proof of (b) $\Rightarrow$ (a) of Theorem 12.1.3. However, for the sake of completeness, we give its proof in detail.

Lemma 12.3.1. *If $\mathcal{P} \subset \mathbb{R}_+^n$ is an integral polymatroid and if $\mathbf{u}, \mathbf{v} \in \mathcal{P} \cap \mathbb{Z}^n$ with $|\mathbf{v}| > |\mathbf{u}|$, then there is $\mathbf{w} \in \mathcal{P} \cap \mathbb{Z}^n$ such that $\mathbf{u} < \mathbf{w} \leq \mathbf{u} \vee \mathbf{v}$.*

Proof. Suppose, on the contrary, that no $\mathbf{w} \in \mathcal{P} \cap \mathbb{Z}^n$ satisfies $\mathbf{u} < \mathbf{w} \leq \mathbf{u} \vee \mathbf{v}$. Let $V = \{i \in [n] : \mathbf{v}(i) > \mathbf{u}(i)\}$. We claim that, for each $i \in V$, there is a subset $A_i \subset [n]$ with $i \in A_i$ such that $\mathbf{u}(A_i) = \rho(A_i)$. In fact, if there is $i \in V$ with $\mathbf{u}(A) < \rho(A)$ for all subsets $A \subset [n]$ with $i \in A$, then the integer vector $\mathbf{w} = \mathbf{u} + \epsilon_i$ belongs to $\mathcal{P}$ and satisfies $\mathbf{u} < \mathbf{w} \leq \mathbf{u} \vee \mathbf{v}$, a contradiction.

Now, let A be a maximal subset of $[n]$ with $\mathbf{u}(A) = \rho(A)$. By using the submodularity of ρ together with $\mathbf{u} \in \mathcal{P}$, it follows that

$$\rho(A \cup A_i) + \rho(A \cap A_i) \leq \rho(A) + \rho(A_i)$$
$$= \mathbf{u}(A) + \mathbf{u}(A_i)$$
$$= \mathbf{u}(A \cup A_i) + \mathbf{u}(A \cap A_i)$$
$$\leq \rho(A \cup A_i) + \rho(A \cap A_i).$$

Hence there must be equality throughout, so that $\mathbf{u}(A \cup A_i) = \rho(A \cup A_i)$.

Then the maximality of A guarantees that $A_i \subset A$. Thus in particular $i \in A$ for all $i \in V$. In other words, $V \subset A$. Hence $\rho(A) = \mathbf{u}(A) < \mathbf{v}(A)$. This contradicts $\mathbf{v} \in \mathcal{P}$. $\square$

Let $P \subset \mathbb{Z}_+^n$ be a discrete polymatroid and $B(P)$ the set of bases of P. We define the nondecreasing function $\rho_P : 2^{[n]} \to \mathbb{R}_+$ associated with P by setting

$$\rho_P(X) = \max\{\mathbf{u}(X) : \mathbf{u} \in B(P)\}$$

for all $\emptyset \neq X \subset [n]$ together with $\rho_P(\emptyset) = 0$.

Lemma 12.3.2. *If $X_1 \subset X_2 \subset \cdots \subset X_s \subset [n]$ is a sequence of subsets of $[n]$, then there is $\mathbf{u} \in B(P)$ such that $\mathbf{u}(X_k) = \rho_P(X_k)$ for all $1 \leq k \leq s$.*

Proof. We work with induction on s and suppose that there is $\mathbf{u} \in B(P)$ such that $\mathbf{u}(X_k) = \rho_P(X_k)$ for all $1 \leq k < s$. Choose $\mathbf{v} \in B(P)$ with $\mathbf{v}(X_s) = \rho_P(X_s)$. If $\mathbf{u}(X_s) < \mathbf{v}(X_s)$, then there is $i \in [n]$ with $i \notin X_s$ such that $\mathbf{u}(\{i\}) > \mathbf{v}(\{i\})$. The exchange property 12.2.4 (c) (ii) says that there is $j \in [n]$ with $\mathbf{u}(j) < \mathbf{v}(j)$ such that $\mathbf{u}_1 = \mathbf{u} - \epsilon_i + \epsilon_j \in B(P)$. Since $\mathbf{u}(X_{s-1}) = \rho_P(X_{s-1})$, it follows that $j \notin X_{s-1}$. Hence $\mathbf{u}_1(X_k) = \rho_P(X_k)$ for all $1 \leq k < s$. Moreover, $\mathbf{u}_1(X_s) \geq \mathbf{u}(X_s)$ and $\mathrm{dis}(\mathbf{u}, \mathbf{v}) > \mathrm{dis}(\mathbf{u}_1, \mathbf{v})$.

If $\mathbf{u}_1(X_s) = \mathbf{v}(X_s)$, then $\mathbf{u}_1$ is a desired base of P. If $\mathbf{u}_1(X_s) < \mathbf{v}(X_s)$, then the above technique will yield a base $\mathbf{u}_2$ of P such that $\mathbf{u}_2(X_k) = \rho_P(X_k)$ for all $1 \leq k < s$, $\mathbf{u}_2(X_s) \geq \mathbf{u}_1(X_s)$ and $\mathrm{dis}(\mathbf{u}_1, \mathbf{v}) > \mathrm{dis}(\mathbf{u}_2, \mathbf{v})$. It is now clear that repeated applications of this argument guarantee the existence of a base $\mathbf{u}_q$ of P such that $\mathbf{u}_q(X_k) = \rho_P(X_k)$ for all $1 \leq k \leq s$. $\square$

Corollary 12.3.3. *The function $\rho_P : 2^{[n]} \to \mathbb{R}_+$ is submodular.*

Proof. Let $A, B \subset [n]$. By Lemma 12.3.2 there is $\mathbf{u} \in B(P)$ such that $\mathbf{u}(A \cap B) = \rho_P(A \cap B)$ and $\mathbf{u}(A \cup B) = \rho_P(A \cup B)$. Hence

$$\begin{aligned} \rho_P(A) + \rho_P(B) &\geq \mathbf{u}(A) + \mathbf{u}(B) \\ &= \mathbf{u}(A \cup B) + \mathbf{u}(A \cap B) \\ &= \rho_P(A \cup B) + \rho_P(A \cap B), \end{aligned}$$

as desired. $\square$

We now come to our first fundamental

Theorem 12.3.4. *A nonempty finite set $P \subset \mathbb{Z}_+^n$ is a discrete polymatroid if and only if $\mathrm{Conv}(P) \subset \mathbb{R}_+^n$ is an integral polymatroid with $\mathrm{Conv}(P) \cap \mathbb{Z}^n = P$.*

Proof. The "if" part follows from Lemma 12.3.1. To see why the "only if" part is true, let $P \subset \mathbb{Z}_+^n$ be a discrete polymatroid and $\rho_P : 2^{[n]} \to \mathbb{R}_+$ the nondecreasing and submodular function associated with P. Write $\mathcal{P} \subset \mathbb{R}_+^n$ for the integral polymatroid with ρ_P its ground set rank function, i.e.

$$\mathcal{P} = \{\mathbf{u} \in \mathbb{R}_+^n : \mathbf{u}(X) \leq \rho_P(X), X \subset [n]\}.$$

Since each base $\mathbf{u}$ of P satisfies $\mathbf{u}(X) \leq \rho_P(X)$ for all $X \subset [n]$, it follows that $P \subset \mathcal{P}$. Moreover, since $\mathcal{P}$ is convex, one has $\mathrm{Conv}(P) \subset \mathcal{P}$. Now, Lemma 12.3.2 together with Proposition 12.1.4 guarantees that all vertices of $\mathcal{P}$ belong to P. Thus $\mathcal{P} = \mathrm{Conv}(P)$.

To complete our proof we must show $\mathcal{P} \cap \mathbb{Z}^n = P$. For each $i \in [n]$, write $P_i \subset \mathbb{Z}_+^n$ for the discrete polymatroid P_{ϵ_i} in the notation of Lemma 12.2.3 (b) and $B_i = B(P_i)$, the set of bases of P_i. We compute the nondecreasing and submodular function $\rho_{P_i} : 2^{[n]} \to \mathbb{R}_+$ associated with P_i. We distinguish three cases:

(a) Let $i \notin X \subset [n]$ with $\rho_P(X \cup \{i\}) > \rho_P(X)$. Choose $\mathbf{u} \in B(P)$ with $\mathbf{u}(X) = \rho_P(X)$ and with $\mathbf{u}(X \cup \{i\}) = \rho_P(X \cup \{i\})$. Then $\mathbf{u}(i) = \mathbf{u}(X\cup\{i\}) - \mathbf{u}(X) \geq 1$ and $\mathbf{u} \in B(P)$. Since $i \notin X$, one has $(\mathbf{u} - \epsilon_i)(X) = \mathbf{u}(X)$. Since $(\mathbf{u} - \epsilon_i)(X) \leq \rho_{P_i}(X) \leq \rho_P(X) = \mathbf{u}(X)$, it follows that $\rho_{P_i}(X) = \rho_P(X)$.

(b) Let $i \notin X \subset [n]$ with $\rho_P(X \cup \{i\}) = \rho_P(X)$. If $\mathbf{u} \in B(P)$ with $\mathbf{u}(i) \geq 1$, then

$$(\mathbf{u} - \epsilon_i)(X) \leq (\mathbf{u} - \epsilon_i)(X \cup \{i\})$$
$$= \mathbf{u}(X \cup \{i\}) - 1$$
$$\leq \rho_P(X \cup \{i\}) - 1 = \rho_P(X) - 1.$$

Thus $\rho_{P_i}(X) \leq \rho_P(X) - 1$. Choose $\mathbf{v} \in B(P)$ with $\mathbf{v}(X) = \rho_P(X)$. Then $\mathbf{v}(i) = 0$. (Otherwise, since $i \notin X$, $\rho_P(X \cup \{i\}) \geq \mathbf{v}(X \cup \{i\}) = \mathbf{v}(X) + \mathbf{v}(i) > \mathbf{v}(X) = \rho_P(X)$; a contradiction.) Let $\mathbf{u}_0 \in B(P)$ with $\mathbf{u}_0(i) \geq 1$. Then the exchange property says that there is $j \in [n]$ with $\mathbf{u}_0(j) < \mathbf{v}(j)$ such that $\mathbf{u}_0 - \epsilon_i + \epsilon_j \in B(P)$. Thus we assume $\mathbf{u}_0(i) = 1$. If $\mathbf{u}_0(X) < \mathbf{v}(X) - 1$, then $\mathbf{u}_0(X \cup \{i\}) = \mathbf{u}_0(X) + 1 < \mathbf{v}(X) = \mathbf{v}(X \cup \{i\})$. Thus there is $j \notin X \cup \{i\}$ with $\mathbf{u}_0(j) > \mathbf{v}(j)$. Hence there is $i \neq k \in [n]$ with $\mathbf{u}_0(k) < \mathbf{v}(k)$ such that $\mathbf{u}_1 = \mathbf{u}_0 - \epsilon_j + \epsilon_k \in B(P)$. Then $\mathbf{u}_1(i) = 1$, $\mathbf{u}_1(X) \geq \mathbf{u}_0(X)$ and $\mathrm{dis}(\mathbf{u}_0, \mathbf{v}) > \mathrm{dis}(\mathbf{u}_1, \mathbf{v})$. Thus, as in the proof of Lemma 12.3.2, we can find $\mathbf{u} \in B(P)$ with $\mathbf{u}(i) = 1$ such that $\mathbf{u}(X) = (\mathbf{u} - \epsilon_i)(X) = \mathbf{v}(X) - 1$. Hence $\rho_{P_i}(X) = \rho_P(X) - 1$.

(c) Let $i \in X \subset [n]$. Then $\rho_{P_i}(X) = \rho_P(X) - 1$. In fact, since $i \in X$, by Lemma 12.3.2 there is $\mathbf{u} \in B(P)$ with $\mathbf{u}(i) = \rho_P(\{i\}) \geq 1$ and with $\mathbf{u}(X) = \rho_P(X)$. Then $(\mathbf{u} - \epsilon_i)(X) = \rho_P(X) - 1$.

Let $\mathcal{P}_i \subset \mathbb{R}_+^n$ denote the integral polymatroid with ρ_{P_i} its ground set rank function. Then $\mathcal{P}_i = \mathrm{Conv}(P_i)$ and, working with induction on the rank of P enables us to assume that $\mathcal{P}_i \cap \mathbb{Z}^n = P_i$. If $\mathbf{x} \in \mathcal{P} \cap \mathbb{Z}^n$ with $\mathbf{x}(i) \geq 1$, then $\mathbf{y} = \mathbf{x} - \epsilon_i$ belongs to $\mathcal{P}_i$. (In fact, if $i \notin X \subset [n]$ with $\rho_{P_i}(X) = \rho_P(X) - 1$, then $\rho_P(X \cup \{i\}) = \rho_P(X)$. Thus replacing $\mathbf{u}$ with $\mathbf{x}$ in the inequalities $(\mathbf{u} - \epsilon_i)(X) \leq \cdots = \rho_P(X) - 1$ appearing in the discussion (b) shows that $\mathbf{y}(X) \leq \rho_{P_i}(X)$.) Thus $\mathbf{y} \in \mathcal{P}_i \cap \mathbb{Z}^n = P_i$. Hence $\mathbf{y} \leq \mathbf{u} - \epsilon_i$ for some $\mathbf{u} \in B(P)$ with $\mathbf{u}(i) \geq 1$. Thus $\mathbf{x} \leq \mathbf{u} \in B(P)$ and $\mathbf{x} \in P$. Hence $\mathcal{P} \cap \mathbb{Z}^n = P$, as desired. $\square$

12.4 The symmetric exchange theorem

We now establish our second fundamental theorem on discrete polymatroids, which is the symmetric exchange theorem.

Theorem 12.4.1. *If* $\mathbf{u} = (a_1, \ldots, a_n)$ *and* $\mathbf{v} = (b_1, \ldots, b_n)$ *are bases of a discrete polymatroid* $P \subset \mathbb{Z}_+^n$, *then for each* $i \in [n]$ *with* $a_i > b_i$ *there is* $j \in [n]$ *with* $a_j < b_j$ *such that both* $\mathbf{u} - \epsilon_i + \epsilon_j$ *and* $\mathbf{v} - \epsilon_j + \epsilon_i$ *are bases of* P.

Proof. Let B' denote the set of those bases $\mathbf{w}$ of P with $\mathbf{u} \wedge \mathbf{v} \leq \mathbf{w} \leq \mathbf{u} \vee \mathbf{v}$. It then turns out that B' satisfies the exchange property 12.2.4 (c) (ii) for polymatroids. Thus B' is the set of bases of a discrete polymatroid $P' \subset \mathbb{Z}_+^n$. Considering $\mathbf{u}' = \mathbf{u} - \mathbf{u} \wedge \mathbf{v}$ and $\mathbf{v}' = \mathbf{v} - \mathbf{u} \wedge \mathbf{v}$ instead of $\mathbf{u}$ and $\mathbf{v}$, we will assume that $P' \subset \mathbb{Z}_+^s$ is a discrete polymatroid, where $s \leq n$, and

$$\mathbf{u} = (a_1, \ldots, a_r, 0, \ldots, 0) \in \mathbb{Z}_+^s, \quad \mathbf{v} = (0, \ldots, 0, b_{r+1}, \ldots, b_s) \in \mathbb{Z}_+^s,$$

where each $0 < a_i$ and each $0 < b_j$ and where $|\mathbf{u}| = |\mathbf{v}| = \mathrm{rank}(P')$. Our work is to show that for each $1 \leq i \leq r$ there is $r + 1 \leq j \leq s$ such that both $\mathbf{u} - \epsilon_i + \epsilon_j$ and $\mathbf{v} - \epsilon_j + \epsilon_i$ are bases of P'. Let, say, $i = 1$.

First case: Suppose that $\mathbf{u} - \epsilon_1 + \epsilon_j$ are bases of P' for all $r + 1 \leq j \leq s$. It follows from the exchange property that, given arbitrary r integers $a_1', \ldots, a_r'$ with each $0 \leq a_i' \leq a_i$, there is a base $\mathbf{w}'$ of P' of the form

$$\mathbf{w}' = (a_1', \ldots, a_r', b_{r+1}', \ldots, b_s'),$$

where each $b_j' \in \mathbb{Z}$ with $0 \leq b_j' \leq b_j$. Thus in particular there is $r + 1 \leq j_0 \leq s$ such that $\mathbf{v} - \epsilon_{j_0} + \epsilon_1$ is a base of P'. Since $\mathbf{u} - \epsilon_1 + \epsilon_j$ is a base of P' for each $r + 1 \leq j \leq s$, both $\mathbf{u} - \epsilon_1 + \epsilon_{j_0}$ and $\mathbf{v} - \epsilon_{j_0} + \epsilon_1$ are bases of P', as desired.

Second case: Let $r \geq 2$ and $r + 2 \leq s$. Suppose that there is $r + 1 \leq j \leq s$ with $\mathbf{u} - \epsilon_1 + \epsilon_j \notin P'$. Let $X \subset \{r+1, \ldots, s\}$ denote the set of those $r + 1 \leq j \leq s$ with $\mathbf{u} - \epsilon_1 + \epsilon_j \notin P'$. Recall that Theorem 12.3.4 guarantees that $\mathrm{Conv}(P') \subset \mathbb{R}_+^s$ is an integral polymatroid on the ground set $[s]$ with $\mathrm{Conv}(P') \cap \mathbb{Z}^s = P'$. Let $\rho = \rho_{P'}$ denote the ground set rank function of the integral polymatroid $\mathrm{Conv}(P') \subset \mathbb{R}_+^s$. Thus $\rho(Y) = \max\{\mathbf{w}(Y) : \mathbf{w} \in B'\}$ for $\emptyset \neq Y \subset [s]$ together with $\rho(\emptyset) = 0$. In particular $\rho(Y) = \mathbf{u}(Y)$ if $Y \subset \{1, \ldots, r\}$ and $\rho(Y) = \mathbf{v}(Y)$ if $Y \subset \{r+1, \ldots, s\}$.

For each $j \in X$, since $\mathbf{u} - \epsilon_1 + \epsilon_j \notin \mathrm{Conv}(P')$, there is a subset $A_j \subset \{2, 3, \ldots, r\}$ with

$$\rho(A_j \cup \{j\}) \leq \mathbf{u}(A_j).$$

Thus

$$\rho(\{2, 3, \ldots, r\} \cup \{j\}) \leq \rho(A_j \cup \{j\}) + \rho(\{2, 3, \ldots, r\} \setminus A_j)$$
$$\leq \mathbf{u}(A_j) + \mathbf{u}(\{2, 3, \ldots, r\} \setminus A_j)$$
$$= \mathbf{u}(\{2, 3, \ldots, r\}) = \rho(\{2, 3, \ldots, r\}).$$

Hence, for all $j \in X$,

$$\rho(\{2,3,\ldots,r\} \cup \{j\}) = \mathbf{u}(\{2,3,\ldots,r\}).$$

By using Lemma 12.4.2 below, it follows that

$$\rho(\{2,3,\ldots,r\} \cup X) = \mathbf{u}(\{2,3,\ldots,r\}).$$

Now, since ρ is submodular,

$$\rho(\{2,3,\ldots,r\} \cup X) + \rho(\{1\} \cup X) \geq \rho(X) + \rho(\{1,2,\ldots,r\} \cup X)$$
$$= \mathbf{v}(X) + \operatorname{rank}(P').$$

Thus

$$\mathbf{u}(\{2,3,\ldots,r\}) + \rho(\{1\} \cup X) \geq \mathbf{v}(X) + \operatorname{rank}(P').$$

Since

$$\operatorname{rank}(P') - \mathbf{u}(\{2,3,\ldots,r\}) = a_1, \quad \text{and}$$
$$\rho(\{1\} \cup X) \leq \rho(\{1\}) + \rho(X) = a_1 + \mathbf{v}(X),$$

it follows that

$$\rho(\{1\} \cup X) = a_1 + \mathbf{v}(X).$$

Hence, for all $X' \subset X$,

$$a_1 + \mathbf{v}(X) = a_1 + \mathbf{v}(X') + \mathbf{v}(X \setminus X')$$
$$= \rho(\{1\}) + \rho(X') + \rho(X \setminus X')$$
$$\geq \rho(\{1\} \cup X') + \rho(X \setminus X')$$
$$\geq \rho(\{1\} \cup X)$$
$$= a_1 + \mathbf{v}(X).$$

Thus, for all $X' \subset X$,

$$\rho(\{1\} \cup X') = a_1 + \mathbf{v}(X').$$

By virtue of Lemma 12.3.2 there is a base $\mathbf{w}$ of P' with $\mathbf{w}(1) = a_1$ and with $\mathbf{w}(j) = \mathbf{v}(j) \ (= \rho(\{j\}))$ for all $j \in X$. Again the exchange property (for $\mathbf{w}$ and $\mathbf{v}$) guarantees that for each $1 \leq i \leq r$ with $\mathbf{w}(i) > 0$ there is $j \in \{r+1,\ldots,s\} \setminus X$ such that $\mathbf{w} - \epsilon_i + \epsilon_j$ is a base of P'. Hence repeated applications of the exchange property yield a base of $\mathbf{w}'$ of P' of the form $\mathbf{w}' = \mathbf{v} - \epsilon_{j_0} + \epsilon_1$, where $j_0 \in \{r+1,\ldots,s\} \setminus X$. Hence both $\mathbf{u} - \epsilon_1 + \epsilon_{j_0}$ and $\mathbf{v} - \epsilon_{j_0} + \epsilon_1$ are bases of P', as required. $\square$

Lemma 12.4.2. *We work with the same situation as in the proof of Theorem 12.4.1. Suppose that $\rho(\{2,3,\ldots,r\} \cup \{j\}) = \rho(\{2,3,\ldots,r\})$ for all $j \in X$. Then $\rho(\{2,3,\ldots,r\} \cup X) = \rho(\{2,3,\ldots,r\})$.*

Proof. We proceed with induction on $|X|$. Clearly, the assertion is true if $|X| = 1$. Now, let $|X| > 1$ and fix $j_0 \in X$. Let $Z = \{2, 3, \ldots, r\}$. Then, by using the assumption of induction, one has

$$
\begin{aligned}
&\rho(Z) + \rho(Z) \\
&= \rho(Z \cup (X \setminus \{j_0\})) + \rho(Z \cup \{j_0\}) \\
&\geq \rho((Z \cup (X \setminus \{j_0\})) \cup (Z \cup \{j_0\})) + \rho((Z \cup (X \setminus \{j_0\})) \cap (Z \cup \{j_0\})) \\
&= \rho(Z \cup X) + \rho(Z) \\
&\geq \rho(Z) + \rho(Z).
\end{aligned}
$$

Since the first and last lines of the above are equal, it follows that equality must hold throughout. Hence $\rho(Z \cup X) = \rho(Z)$, as desired. □

12.5 The base ring of a discrete polymatroid

Let P be a discrete polymatroid of rank d on the ground set $[n]$ with set of bases $B = B(P)$. Let, as usual, $S = K[x_1, \ldots, x_n]$ denote the polynomial ring in n variables over a field K. For each basis $\mathbf{u} \in B$ we write $\mathbf{x}^{\mathbf{u}}$ for the monomial $x_1^{u(1)} \cdots x_n^{u(n)}$ of S. The **base ring** of P is the toric ring $K[B]$ which is generated by those monomials $\mathbf{x}^{\mathbf{u}}$ with $\mathbf{u} \in B$. Since all the bases of P have the same modulus, it follows that $K[B]$ possesses a standard grading with each $\deg \mathbf{x}^{\mathbf{u}} = 1$.

Theorem 12.5.1. *The base ring of a discrete polymatroid is normal.*

Proof. Let P be a discrete polymatroid on $[n]$ with set of bases $B = B(P)$. Let $\mathcal{P} = \mathrm{Conv}(P) \subset \mathbb{R}^n$ and $\mathcal{Q} = \mathrm{Conv}(B) \subset \mathbb{R}^n$. Then by using Theorem 12.3.4 together with the fact that the $\mathcal{Q}$ is the face of $\mathcal{P}$, it follows that $B = \mathcal{Q} \cap \mathbb{Z}^n$. Now, by virtue of Lemma B.6.1, in order to show that $K[B]$ is normal what we must prove is that $\mathcal{Q}$ possesses the **integer decomposition property**, i.e. if $\mathbf{w} \in \mathbb{Z}^n$ belongs to $q\mathcal{Q} = \{q\mathbf{v} : \mathbf{v} \in \mathcal{Q}\}$, where $q \in \mathbb{Z}_+$, then there are $\mathbf{u}_1, \ldots, \mathbf{u}_q$ belonging to B such that $\mathbf{w} = \mathbf{u}_1 + \cdots + \mathbf{u}_q$.

Since $q\mathcal{P} = \mathcal{P} \vee \cdots \vee \mathcal{P}$ (q times), Theorem 12.1.5 guarantees that $\mathcal{P}$ possesses the integer decomposition property. Since $\mathcal{Q}$ is a face of $\mathcal{P}$, it follows easily that $\mathcal{Q}$ possesses the integer decomposition property. □

As a consequence of Hochster's theorem (Theorem B.6.2) we obtain

Corollary 12.5.2. *The base ring of a discrete polymatroid is Cohen–Macaulay.*

Let, in general, $\mathcal{P} \subset \mathbb{R}^n$ be an integral convex polytope of dimension n which possesses the integer decomposition property. The toric ring $K[\mathcal{P}]$ of $\mathcal{P}$ is a subring of $K[x_1, \ldots, x_n, t]$ which is generated by those monomials $\mathbf{x}^{\mathbf{u}} t$ with $\mathbf{u} \in \mathcal{P} \cap \mathbb{Z}^n$. Since $K[\mathcal{P}]$ is normal and is Cohen–Macaulay (Theorem B.6.2), it would be of interest when $K[\mathcal{P}]$ is Gorenstein. Let $\delta > 0$ denote the smallest integer for which $\delta(\mathcal{P} \setminus \partial\mathcal{P}) \cap \mathbb{Z}^n \neq \emptyset$. Here $\partial\mathcal{P}$ is the boundary of $\mathcal{P}$ and $\mathcal{P} \setminus \partial\mathcal{P}$ is the interior of $\mathcal{P}$. The canonical module $\Omega(K[\mathcal{P}])$ of $K[\mathcal{P}]$ is isomorphic to the ideal of $K[\mathcal{P}]$ generated by those monomials $\mathbf{x}^{\mathbf{u}} t^q$ with $\mathbf{u} \in q(\mathcal{P} \setminus \partial\mathcal{P}) \cap \mathbb{Z}^n$ (Theorem B.6.3). Note that $K[\mathcal{P}]$ is Gorenstein if and only if $\Omega(K[\mathcal{P}])$ is a principal ideal; see Corollary A.6.7. In particular, if $K[\mathcal{P}]$ is Gorenstein, then $\delta(\mathcal{P} \setminus \partial\mathcal{P})$ must possess a unique integer vector. Suppose that $\delta(\mathcal{P} \setminus \partial\mathcal{P})$ possesses a unique integer vector, say $\mathbf{w}^* \in \mathbb{Z}^n$, and let $\mathcal{Q} = \delta\mathcal{P} - \mathbf{w}^* = \{\mathbf{w} - \mathbf{w}^* : \mathbf{w} \in \delta\mathcal{P}\}$. Thus $\mathcal{Q} \subset \mathbb{R}^n$ is an integral convex polytope of dimension n and the origin of $\mathbb{R}^n$ is a unique integer vector belonging to the interior $\mathcal{Q} \setminus \partial\mathcal{Q}$ of $\mathcal{Q}$. Then $K[\mathcal{P}]$ is Gorenstein if and only if the following condition is satisfied:

($\sharp$) If the hyperplane $\mathcal{H} \subset \mathbb{R}^n$ determined by a linear equation $\sum_{i=1}^n a_i z_i = b$, where each a_i and b are integers and where the greatest common divisor of $a_1, \ldots, a_n, b$ is equal to 1, is a supporting hyperplane of $\mathcal{Q}$ such that $\mathcal{H} \cap \mathcal{Q}$ is a facet of $\mathcal{Q}$, then b is either 1 or -1.

We give a sketch of a proof of the above criterion for $K[\mathcal{P}]$ to be Gorenstein. Since $K[\mathcal{P}]$ is Gorenstein if and only if $\Omega(K[\mathcal{P}])$ is generated by $\mathbf{x}^{\mathbf{w}^*} t^\delta$, it follows that $K[\mathcal{P}]$ is Gorenstein if and only if

$$\mathbf{w}^* + q\mathcal{P} \cap \mathbb{Z}^n = (q + \delta)(\mathcal{P} \setminus \partial\mathcal{P}) \cap \mathbb{Z}^n, \quad q \geq 1. \tag{12.2}$$

First, assuming ($\sharp$) yields that the linear equation of a supporting hyperplane which defines a facet of $\mathcal{P}$ is of the form

$$\delta \sum_{i=1}^n a_i z_i = 1 + \sum_{i=1}^n a_i w_i^*, \tag{12.3}$$

where $\mathbf{w}^* = (w_1^*, \ldots, w_n^*)$. Since $\mathcal{P}$ is integral, the equation (12.3) possesses an integer solution. Hence $1 + \sum_{i=1}^n a_i w_i^*$ is divided by δ. Thus if $\mathbf{u} = (u_1, \ldots, u_n) \in \mathbb{Z}^n$ satisfies

$$\delta \sum_{i=1}^n a_i u_i < (q + \delta)(1 + \sum_{i=1}^n a_i w_i^*),$$

then

$$\delta \sum_{i=1}^n a_i u_i \leq (q + \delta)(1 + \sum_{i=1}^n a_i w_i^*) - \delta.$$

Hence

$$\delta \sum_{i=1}^n a_i(u_i - w_i^*) \leq q\,(1 + \sum_{i=1}^n a_i w_i^*).$$

In other words,

$$(q + \delta)(\mathcal{P} \setminus \partial\mathcal{P}) \cap \mathbb{Z}^n \subset \mathbf{w}^* + q\mathcal{P} \cap \mathbb{Z}^n. \tag{12.4}$$

Since the opposite inclusion of (12.4) is obvious, one has (12.2) and $K[\mathcal{P}]$ is Gorenstein.

Second, if $K[\mathcal{P}]$ is Gorenstein, then $\Omega(K[\mathcal{P}])$ is generated by $\mathbf{w}^* t^\delta$. It then follows immediately that $\Omega(K[\delta \mathcal{P}])$ is generated by $\mathbf{w}^* t^\delta$. Hence $K[\delta \mathcal{P}]$ is Gorenstein. Since $K[\mathcal{Q}] \simeq K[\delta \mathcal{P}]$, the toric ring $K[\mathcal{Q}]$ is Gorenstein. Thus $\Omega(K[\mathcal{Q}])$ is generated by t. In other words,

$$q(\mathcal{Q} \setminus \partial \mathcal{Q}) \cap \mathbb{Z}^n = (q-1)\mathcal{Q} \cap \mathbb{Z}^n, \quad q \geq 1. \tag{12.5}$$

A geometric observation easily says that (12.5) is equivalent to ($\sharp$). Hence the condition ($\sharp$) is satisfied, if $K[\mathcal{P}]$ is Gorenstein, as desired.

We now turn to the problem when the base ring of a discrete polymatroid is Gorenstein. To obtain a perfect answer to this problem seems, however, quite difficult. In what follows we introduce the concept of "generic" discrete polymatroids and find a characterization for the base ring of a generic discrete polymatroid to be Gorenstein.

Let $P \subset \mathbb{Z}_+^n$ be a discrete polymatroid of rank d and $B = B(P)$ the set of bases of P. We will assume that the canonical basis vectors $\epsilon_1, \dots, \epsilon_n$ of $\mathbb{R}^n$ belong to P. Let $\mathcal{F} = \mathrm{Conv}(B)$, the set of bases of the integral polymatroid $\mathcal{P} = \mathrm{Conv}(P) \subset \mathbb{R}_+^n$. Recall that $\mathcal{F}$ is a face of $\mathcal{P}$ with the supporting hyperplane $\mathcal{H}_{[n]} \subset \mathbb{R}^n$, i.e. $\mathcal{F} = \mathcal{H}_{[n]} \cap \mathcal{P}$. Let $\rho: 2^{[n]} \to \mathbb{R}_+$ denote the ground set rank function of $\mathcal{P}$. Then

$$\mathcal{F} = \{\mathbf{u} \in \mathcal{H}_{[n]} \cap \mathbb{Z}_+^n : \mathbf{u}(A) \leq \rho(A), \emptyset \neq A \subset [n], A \neq [n]\}.$$

Let $\varphi: \mathcal{H}_{[n]} \to \mathbb{R}^{n-1}$ denote the affine transformation defined by

$$\varphi(u_1, \dots, u_n) = (u_1, \dots, u_{n-1}).$$

Thus φ is injective and $\varphi(\mathcal{H}_{[n]} \cap \mathbb{Z}^n) = \mathbb{Z}^{n-1}$. Since for all $A \subset [n]$ with $n \in A$ and $A \neq [n]$ the hyperplane $\varphi(\mathcal{H}_A \cap \mathcal{H}_{[n]}) \subset \mathbb{R}^{n-1}$ is determined by the linear equation $\sum_{i \in [n] \setminus A} x_i = d - \rho(A)$, it follows that

$$\varphi(\mathcal{F}) = \{\mathbf{u} \in \mathbb{R}_+^{n-1} : d - \rho([n] \setminus A) \leq \mathbf{u}(A) \leq \rho(A), \emptyset \neq A \subset [n-1]\}.$$

We say that P is **generic** if

(G1) each base $\mathbf{u}$ of P satisfies $\mathbf{u}(i) > 0$ for all $1 \leq i \leq n$;
(G2) $\mathcal{F} = \mathrm{Conv}(B)$ is a facet of $\mathcal{P} = \mathrm{Conv}(P)$;
(G3) $\mathcal{F} \cap \mathcal{H}_A$ is a facet of $\mathcal{F}$ for all $\emptyset \neq A \subset [n]$ with $A \neq [n]$.

It follows that P is generic if and only if

(i) ρ is strictly increasing;
(ii) $\dim \varphi(\mathcal{F}) = n - 1$;
(iii) the facets of $\varphi(\mathcal{F})$ are all $\{\mathbf{u} \in \varphi(\mathcal{F}) : \mathbf{u}(A) = \rho(A)\}$ together with all $\{\mathbf{u} \in \varphi(\mathcal{F}) : \mathbf{u}(A) = d - \rho([n] \setminus A)\}$, where A ranges over all nonempty subsets of $[n-1]$.

Example 12.5.3. (a) Let $n = 2$ and let $a_1, a_2 > 0$ be integers. Let $P \subset \mathbb{Z}_+^2$ denote the discrete polymatroid of rank d consisting of those $\mathbf{u} = (u_1, u_2) \in \mathbb{Z}_+^2$ such that $u_1 \leq a_1$, $u_2 \leq a_2$ and $u_1 + u_2 \leq d$. Then P is generic if and only if $a_1 < d$, $a_2 < d$, and $d < a_1 + a_2$. If P is generic, then the bases of P are $(a_1, d - a_1), (a_1 - 1, d - a_1 + 1), \ldots, (d - a_2, a_2)$. Thus the base ring $K[B]$ of P is Gorenstein if and only if either $a_1 + a_2 = d + 1$ or $a_1 + a_2 = d + 2$.

(b) Let $n = 3$. Let $P \subset \mathbb{Z}_+^3$ be a discrete polymatroid of rank d with B its set of bases, and ρ the ground set rank function of the integral polymatroid $\mathrm{Conv}(P) \subset \mathbb{R}_+^3$. Then $\varphi(\mathcal{F}) \subset \mathbb{Z}_+^2$, where $\mathcal{F} = \mathrm{Conv}(B)$, consists of those $\mathbf{u} = (u_1, u_2) \in \mathbb{R}_+^2$ such that

$$
\begin{aligned}
d - \rho(\{2, 3\}) \leq u_1 &\qquad \leq \rho(\{1\}), \\
d - \rho(\{1, 3\}) \leq &\qquad u_2 \leq \rho(\{2\}), \\
d - \rho(\{3\}) \leq u_1 &+ u_2 \leq \rho(\{1, 2\}).
\end{aligned}
$$

Hence P is generic if and only if

$$
\begin{aligned}
0 < \rho(\{i\}) < \rho(\{i, j\}) < d, \ 1 \leq i \neq j \leq 3, \\
\rho(\{i\}) + \rho(\{j\}) > \rho(\{i, j\}), \ 1 \leq i < j \leq 3, \\
\rho(\{i, j\}) + \rho(\{j, k\}) > d + \rho(\{j\}), \ \{i, j, k\} = [3].
\end{aligned}
$$

Moreover, if P is generic, then the base ring $K[B]$ is Gorenstein if and only if

$$
\rho(\{i\}) + \rho(\{j, k\}) = d + 2, \quad \{i, j, k\} = [3].
$$

Theorem 12.5.4. (a) *Let $n \geq 3$. Let $P \subset \mathbb{Z}_+^n$ be a discrete polymatroid of rank d and suppose that the canonical basis vectors $\epsilon_1, \ldots, \epsilon_n$ of $\mathbb{R}^n$ belong to P. Let $\rho \colon 2^{[n]} \to \mathbb{R}_+$ denote the ground set rank function of the integral polymatroid $\mathrm{Conv}(P) \subset \mathbb{R}_+^n$. If P is generic and if the base ring $K[B]$ of P is Gorenstein, then there is a vector $\alpha = (\alpha_1, \ldots, \alpha_{n-1}) \in \mathbb{Z}_+^{n-1}$ with each $\alpha_i > 1$ and with $d > |\alpha| + 1$ such that*

$$
\rho(A) = \begin{cases} \alpha(A) + 1, & \text{if } \ \emptyset \neq A \subset [n-1], \\ d - \alpha([n] \setminus A) + 1, & \text{if } \ n \in A \neq [n]. \end{cases}
$$

(b) *Conversely, given $\alpha = (\alpha_1, \ldots, \alpha_{n-1}) \in \mathbb{Z}_+^{n-1}$, where $n \geq 3$, with each $\alpha_i > 1$ and $d \in \mathbb{Z}$ with $d > |\alpha| + 1$, define the function $\rho \colon 2^{[n]} \to \mathbb{R}_+$ by (a) together with $\rho(\emptyset) = 0$ and $\rho([n]) = d$. Then*

(i) *ρ is strictly increasing and submodular;*
(ii) *the discrete polymatroid $P = \{\mathbf{u} \in \mathbb{Z}_+^n \colon \mathbf{u}(A) \leq \rho(A), \emptyset \neq A \subset [n]\} \subset \mathbb{Z}_+^n$ arising from ρ is generic;*
(iii) *the base ring $K[B]$ of P is Gorenstein.*

Proof. (a) Suppose that a discrete polymatroid $P \subset \mathbb{Z}_+^n$ of rank d is generic and that the base ring $K[B]$ of P is Gorenstein. Let $\mathcal{F} = \mathrm{Conv}(B)$. Since $K[B]$ is Gorenstein, there is an integer $\delta \geq 1$ such that

$$\delta(\rho(A) - (d - \rho([n] \setminus A))) = 2$$

for all $\emptyset \neq A \subset [n-1]$. Hence either $\delta = 1$ or $\delta = 2$.

If $\delta = 2$, then $(2\rho(\{1\}) - 1, \ldots, 2\rho(\{n-1\}) - 1) \in \mathbb{Z}_+^{n-1}$ must be a unique integer vector belonging to the interior of $2\varphi(\mathcal{F})$. Since $K[B]$ is Gorenstein it follows that

$$\sum_{i \in A}(2\rho(\{i\}) - 1) = 2\rho(A) - 1, \quad \emptyset \neq A \subset [n-1].$$

Thus

$$\rho(A) = \sum_{i \in A} \rho(\{i\}) - \frac{1}{2}(|A| - 1), \quad \emptyset \neq A \subset [n-1].$$

Since $n \geq 3$, it follows that $\rho(\{1,2\}) \notin \mathbb{Z}$, a contradiction.

Now let $\delta = 1$ and set

$$\alpha_i = \rho(\{i\}) - 1 = d - \rho([n] \setminus \{i\}) + 1 > 1, \quad 1 \leq i \leq n-1.$$

Then $\alpha = (\alpha_1, \ldots, \alpha_{n-1}) \in \mathbb{Z}_+^{n-1}$ is a unique integer vector belonging to the interior of $\varphi(\mathcal{F}) \subset \mathbb{R}_+^{n-1}$ and $\varphi(\mathcal{F}) - \alpha$ consists of those $\mathbf{u} = (u_1, \ldots, u_{n-1}) \in \mathbb{R}^{n-1}$ such that

$$d - \rho([n] \setminus A) - \alpha(A) \leq \sum_{i \in A} u_i \leq \rho(A) - \alpha(A), \quad \emptyset \neq A \subset [n-1].$$

Since P is generic, the desired equality on ρ follows immediately. Moreover, since $\rho([n-1]) = |\alpha| + 1 < \rho([n]) = d$, one has $d > |\alpha| + 1$, as required.

(b) Since each $\alpha_i > 1$ and since $d > |\alpha| + 1$, it follows that $0 < \rho(A) < \rho([n]) = d$ for all $\emptyset \neq A \subset [n]$ with $A \neq [n]$. Moreover, $\rho(\{i\}) > 2$ for all $1 \leq i \leq n$. If $\emptyset \neq A \subset B \subset [n]$ with $n \notin A$ and $n \in B$, then $\rho(B) - \rho(A) = d - (\alpha([n] \setminus B) + \alpha(A)) > d - |\alpha| > 1$. Hence ρ is strictly increasing.

To see why ρ is submodular, we distinguish three cases as follows. First, if $A, B \subset [n-1]$ with $A \neq \emptyset$ and $B \neq \emptyset$, then

$$\begin{aligned}
\rho(A) + \rho(B) &= \alpha(A) + \alpha(B) + 2 \\
&= \alpha(A \cup B) + \alpha(A \cap B) + 2 \\
&= \rho(A \cup B) + \rho(A \cap B)
\end{aligned}$$

unless $A \cap B \neq \emptyset$. Second, if $A, B \subset [n]$ with $n \in A \neq [n]$ and $n \in B \neq [n]$, then

$$\begin{aligned}
\rho(A) + \rho(B) &= 2d - (\alpha([n] \setminus A) + \alpha([n] \setminus B)) + 2 \\
&= 2d - (\alpha([n] \setminus (A \cap B)) + \alpha([n] \setminus (A \cup B))) + 2 \\
&= \rho(A \cup B) + \rho(A \cap B)
\end{aligned}$$

unless $A \cup B \neq [n]$. Third, if $n \in A$ and $B \subset [n-1]$, then assuming $A \cap B \neq \emptyset$ and $A \cup B \neq [n]$ one has

$$\begin{aligned}
\rho(A) + \rho(B) &= d - \alpha([n] \setminus A) + 1 + \alpha(B) + 1 \\
&= d - \alpha(([n] \setminus A) \setminus B) + 1 + \alpha(B \setminus ([n] \setminus A)) + 1 \\
&= d - \alpha([n] \setminus (A \cup B)) + 1 + \alpha(A \cap B) + 1 \\
&= \rho(A \cup B) + \rho(A \cap B),
\end{aligned}$$

as desired.

Let $\mathcal{F} = \mathrm{Conv}(B)$. Since

$$\varphi(\mathcal{F}) = \{\mathbf{u} \in \mathbb{R}_+^{n-1} : \alpha(A) - 1 \leq \mathbf{u}(A) \leq \alpha(A) + 1, \emptyset \neq A \subset [n-1]\},$$

it follows that $\varphi(\mathcal{F}) - \alpha \subset \mathbb{R}^{n-1}$ consists of those $\mathbf{u} = (u_1, \ldots, u_{n-1}) \in \mathbb{R}^{n-1}$ such that

$$-1 \leq u_{i_1} + \cdots + u_{i_k} \leq 1, \quad 1 \leq i_1 < \cdots < i_k \leq n - 1.$$

Hence $(\varphi(\mathcal{F}) - \alpha) \cap \mathbb{Z}^{n-1}$ consists of those $\mathbf{v} = (v_1, \ldots, v_{n-1}) \in \mathbb{Z}^{n-1}$ such that $-1 \leq v_i \leq 1$ for all $1 \leq i \leq n - 1$, $|\{i : v_i = 1\}| \leq 1$, and $|\{i : v_i = -1\}| \leq 1$. In particular, the canonical basis vectors $\epsilon_1, \ldots, \epsilon_{n-1}$ of $\mathbb{R}^{n-1}$ belong to $\varphi(\mathcal{F}) - \alpha$. Thus $\mathcal{F}$ is a facet of $\mathcal{P} = \mathrm{Conv}(P)$. For $1 \leq i_1 < \cdots < i_k \leq n - 1$ write $\mathcal{H}_{i_1 \cdots i_k} \subset \mathbb{R}^{n-1}$ (resp. $\mathcal{H}'_{i_1 \cdots i_k} \subset \mathbb{R}^{n-1}$) for the supporting hyperplane of $\mathcal{F}$ determined by the linear equation $x_{i_1} + \cdots + x_{i_k} = 1$ (resp. $x_{i_1} + \cdots + x_{i_k} = -1$). Then the vectors $\epsilon_{i_1}, \ldots, \epsilon_{i_k}$ (resp. $-\epsilon_{i_1}, \ldots, -\epsilon_{i_k}$) and $\epsilon_{i_1} - \epsilon_j$ (resp. $-\epsilon_{i_1} + \epsilon_j$), $j \in [n-1] \setminus \{i_1, \ldots, i_k\}$, belong to the face $(\varphi(\mathcal{F}) - \alpha) \cap \mathcal{H}_{i_1 \cdots i_k}$ (resp. $(\varphi(\mathcal{F}) - \alpha) \cap \mathcal{H}'_{i_1 \cdots i_k}$) of $\varphi(\mathcal{F}) - \alpha$. Thus $(\varphi(\mathcal{F}) - \alpha) \cap \mathcal{H}_{i_1 \cdots i_k}$ (resp. $(\varphi(\mathcal{F}) - \alpha) \cap \mathcal{H}'_{i_1 \cdots i_k}$) is a facet of $\varphi(\mathcal{F}) - \alpha$. Hence P is generic. Moreover, since the Ehrhart ring $K[\varphi(\mathcal{F}) - \alpha]$ is Gorenstein, the base ring $K[B]$ ($\cong K[\varphi(\mathcal{F}) - \alpha]$) is Gorenstein, as desired. $\qquad\square$

12.6 Polymatroidal ideals

We now turn to the study of monomial ideals arising from discrete polymatroids.

Definition 12.6.1. *A monomial ideal I of S with $G(I) = \{\mathbf{x}^{\mathbf{u}_1}, \ldots, \mathbf{x}^{\mathbf{u}_s}\}$ is called **polymatroidal** if $\{\mathbf{u}_1, \ldots, \mathbf{u}_s\}$ is the set of bases of a discrete polymatroid on $[n]$. In other words, all elements in $G(I)$ have the same degree, and if $\mathbf{x}^{\mathbf{u}_r} = x^{a_1} \cdots x^{a_n}$ and $\mathbf{x}^{\mathbf{u}_t} = x^{b_1} \cdots x^{b_n}$ belong to $G(I)$ with $a_i > b_i$, then there exists j with $a_j < b_j$ such that $x_j(\mathbf{x}^{\mathbf{u}_r}/x_i) \in G(I)$.*

A fundamental fact on polymatroidal ideals is

Theorem 12.6.2. *A polymatroidal ideal has linear quotients.*

Proof. Let I be a polymatroidal ideal with $G(I) = \{\mathbf{x}^{\mathbf{u}_1}, \ldots, \mathbf{x}^{\mathbf{u}_s}\}$, where $\mathbf{x}^{\mathbf{u}_1} > \cdots > \mathbf{x}^{\mathbf{u}_s}$ with respect to the reverse lexicographic order. Let $J = (\mathbf{x}^{\mathbf{u}_1}, \ldots, \mathbf{x}^{\mathbf{u}_{q-1}})$ with $q < s$. Then

$$J : \mathbf{x}^{\mathbf{u}_q} = (\mathbf{x}^{\mathbf{u}_1}/[\mathbf{x}^{\mathbf{u}_1}, \mathbf{x}^{\mathbf{u}_q}], \ldots, \mathbf{x}^{\mathbf{u}_{q-1}}/[\mathbf{x}^{\mathbf{u}_{q-1}}, \mathbf{x}^{\mathbf{u}_q}]).$$

Thus in order to show that $J : \mathbf{x}^{\mathbf{u}_q}$ is generated by variables, what we must prove is that, for each $1 \leq k < q$, there is $x_j \in J : \mathbf{x}^{\mathbf{u}_q}$ such that x_j divides $\mathbf{x}^{\mathbf{u}_k}/[\mathbf{x}^{\mathbf{u}_k}, \mathbf{x}^{\mathbf{u}_q}]$. Let $\mathbf{u}_k = (a_1, \ldots, a_n)$ and $\mathbf{u}_q = (b_1, \ldots, b_n)$. Since $\mathbf{x}^{\mathbf{u}_k} > \mathbf{x}^{\mathbf{u}_q}$, there is an integer $1 < i < n$ with $a_i < b_i$ and with $a_{i+1} = b_{i+1}, \ldots, a_n = b_n$. Hence by the exchange property 12.2.4 (c) (ii) there is an integer $1 \leq j < i$ with $b_j < a_j$ such that $x_j(\mathbf{x}^{\mathbf{u}_q}/x_i) \in G(I)$. Since $j < i$ it follows that $x_j(\mathbf{x}^{\mathbf{u}_q}/x_i) \in J$. Thus $x_j \in J : \mathbf{x}^{\mathbf{u}_q}$. Finally, since $b_j < a_j$, it follows that x_j divides $\mathbf{x}^{\mathbf{u}_k}/[\mathbf{x}^{\mathbf{u}_k}, \mathbf{x}^{\mathbf{u}_q}]$. $\square$

Theorem 12.6.3. *Let I and J be polymatroidal ideals. Then IJ is again polymatroidal.*

Proof. Let P and Q be discrete polymatroids, and let $B(P)$ and $B(Q)$ be their bases. Theorem 12.3.4 together with Theorem 12.1.5 says that $\{\mathbf{u} + \mathbf{v} : \mathbf{u} \in B(P), \mathbf{v} \in B(Q)\}$ is the set of a discrete polymatroid $\{\mathbf{u} + \mathbf{v} : \mathbf{u} \in P, \mathbf{v} \in Q\}$ on $[n]$. Hence if I and J are polymatroidal ideals, then IJ is a polymatroidal ideal, as desired. $\square$

Corollary 12.6.4. *All powers of a polymatroidal ideal have linear quotients.*

We now classify all Cohen–Macaulay polymatroidal ideals. Recall that a monomial ideal $I \subset S$ is Cohen–Macaulay if the quotient ring S/I is Cohen–Macaulay. Typical examples of Cohen–Macaulay polymatroidal ideals are:

Example 12.6.5. (a) The Veronese ideal of S of degree d is the ideal of S which is generated by all monomials of S of degree d. The Veronese ideal is polymatroidal and is Cohen–Macaulay.

(b) The squarefree Veronese ideal of S of degree d is the ideal of S which is generated by all squarefree monomials of S of degree d. The squarefree Veronese ideal is matroidal (i.e. polymatroidal and squarefree). Moreover, by using the fact that each skeleton of a Cohen–Macaulay complex is Cohen–Macauly, the squarefree Veronese ideal is Cohen–Macaulay.

It turns out that Cohen–Macaulay polymatroidal ideals are essentially either Veronese ideals or squarefree Veronese ideals.

In order to prove Theorem 12.6.7 stated below, a formula to compute the dimension and depth of a monomial ideal with linear quotient will be required.

Let I be a monomial ideal of $S = K[x_1, \ldots, x_n]$ with $G(I)$ its unique minimal set of monomial generators. According to Corollary 1.3.9 the minimal prime ideals of I are generated by subsets of the variables. Hence I is unmixed if all minimal prime ideals of I are generated by the same number of variables. For a monomial prime ideal P, let $\mu(P)$ denote the number of variables which generates P, and set $c(I) = \min\{\mu(P) : P \in \mathrm{Min}(I)\}$. Then

$$\dim S/I = n - c(I).$$

Now assume in addition that $I \subset S$ is generated in one degree and suppose that I has linear quotients with respect to the ordering $u_1, u_2, \ldots, u_s$ of the monomials belonging to $G(I)$. Thus the colon ideal $(u_1, u_2, \ldots, u_{j-1}) : u_j$ is generated by a subset of $\{x_1, \ldots, x_n\}$ for each $2 \leq j \leq s$. Let r_j denote the number of variables which is required to generate $(u_1, u_2, \ldots, u_{j-1}) : u_j$. Let $r(I) = \max_{2 \leq j \leq s} r_j$. It follows from Corollary 8.2.2 that

$$\text{depth } S/I = n - r(I) - 1.$$

Hence a monomial ideal I which is generated in one degree with linear quotients is Cohen–Macaulay if and only if $c(I) = r(I) + 1$.

Lemma 12.6.6. *If $I \subset S$ is a Cohen–Macaulay polymatroidal ideal, then its radical $\sqrt{I}$ is squarefree Veronese.*

Proof. Let $I \subset S$ be a Cohen–Macaulay polymatroidal ideal. We may assume that $\bigcup_{u \in G(I)} \text{supp}(u) = \{x_1, \ldots, x_n\}$. Let $u \in G(I)$ be a monomial for which $|\text{supp}(u)|$ is minimal. Let, say, $\text{supp}(u) = \{x_{n-d+1}, x_{n-d+2}, \ldots, x_n\}$. Let J denote the monomial ideal generated by those monomials $w \in G(I)$ such that w is bigger than u with respect to the reverse lexicographic order. We know that the colon ideal $J : u$ is generated by a subset M of $\{x_1, \ldots, x_n\}$. We claim that $\{x_1, \ldots, x_{n-d}\} \subset M$. For each $1 \leq i \leq n - d$, there is a monomial belonging to $G(I)$ which is divided by x_i. It follows from Proposition 12.2.6 that there is a variable x_j with $n - d + 1 \leq j \leq n$ such that $v = x_i u / x_j \in G(I)$. One has $v \in J$. Since $x_i u = x_j v \in J$, it follows that $x_i \in J : u$, as required. Consequently, $r(I) \geq n - d$. Since I is Cohen–Macaulay, it follows that $c(I) \geq n - d + 1$. It then turns out that I is not contained in the ideal $(\{x_1, \ldots, x_n\} \setminus W)$ for each subset $W \subset \{x_1, \ldots, x_n\}$ with $|W| = d$. Hence for each $W \subset \{x_1, \ldots, x_n\}$ with $|W| = d$ there is a monomial $w \in G(I)$ with $\text{supp}(w) \subset W$. Since $|\text{supp}(w)| \geq |\text{supp}(u)| = d$, one has $\text{supp}(w) = W$. Hence $\sqrt{I}$ is generated by all squarefree monomials of degree d in $x_1, \ldots, x_n$, as desired. $\qquad \square$

Theorem 12.6.7. *A polymatroidal ideal I is Cohen–Macaulay if and only if I is*

(i) *a principal ideal, or*
(ii) *a Veronese ideal, or*
(iii) *a squarefree Veronese ideal.*

Proof. By using Lemma 12.6.6 we assume that $\sqrt{I}$ is generated by all squarefree monomials of degree d in $x_1, \ldots, x_n$, where $2 \leq d < n$. One has $c(I) = c(\sqrt{I}) = n - d + 1$. Suppose that I is not squarefree (or, equivalently, each monomial belonging to $G(I)$ is of degree $> d$). Let $u = \prod_{i=n-d+1}^{n} x_i^{a_i} \in G(I)$ be a monomial with $\text{supp}(u) = \{x_{n-d+1}, x_{n-d+2}, \ldots, x_n\}$. For a while, we assume that $(*)$ there is a monomial $v = \prod_{i=1}^{n} x_i^{b_i} \in G(I)$ with $b_{n-d+1} > a_{n-d+1}$. Let J denote the monomial ideal generated by those monomials

$w \in G(I)$ such that w is bigger than u with respect to the reverse lexicographic order. As was shown in the proof of Lemma 12.6.6, the colon ideal $J : u$ is generated by a subset M of $\{x_1, \ldots, x_n\}$ with $\{x_1, \ldots, x_{n-d}\} \subset M$. We claim that $x_{n-d+1} \in J : u$. By using Proposition 12.2.6 our assumption $(*)$ guarantees that there is a variable x_j with $n - d + 1 < j \leq n$ such that $u_0 = x_{n-d+1}u/x_j \in G(I)$. Since $u_0 \in J$, one has $x_{n-d+1} \in M$. Hence $r(I) \geq n - d + 1$. Thus $c(I) < r(I) + 1$ and I cannot be Cohen–Macaulay.

To complete our proof, we must examine our assumption $(*)$. For each d-element subset $\sigma = \{x_{i_1}, x_{i_2}, \ldots, x_{i_d}\}$ of $\{x_1, \ldots, x_n\}$, there is a monomial $u_\sigma \in G(I)$ with $\mathrm{supp}(u_\sigma) = \sigma$. If there are d-element subsets σ and τ of $\{x_1, \ldots, x_n\}$ and a variable $x_{i_0} \in \sigma \cap \tau$ with $a_{i_0} < b_{i_0}$, where a_{i_0} (resp. b_{i_0}) is the power of x_{i_0} in u_σ (resp. u_τ), then after relabelling the variables if necessarily we may assume that $\sigma = \{x_{n-d+1}, x_{n-d+2}, \ldots, x_n\}$ with $i_0 = n - d + 1$. In other words, the condition $(*)$ is satisfied. Thus in the case that the condition $(*)$ fails to be satisfied, there is a positive integer $e \geq 2$ such that, for each d-element subset $\{x_{i_1}, x_{i_2}, \ldots, x_{i_d}\}$ of $\{x_1, \ldots, x_n\}$ one has $u = (x_{i_1}x_{i_2} \cdots x_{i_d})^e \in G(I)$. Let $w = x_{n-d}x_{n-d+1}^{e-1}(\prod_{i=n-d+2}^n x_i^e) \in G(I)$. Let J denote the monomial ideal generated by those monomials $v \in G(I)$ such that v is bigger than w with respect to the reverse lexicographic order. Since $\prod_{i=n-d}^{n-1} x_i^e \in G(I)$, by using Proposition 12.2.6 one has $w_0 = x_{n-d}w/x_n \in J$ and $w_1 = x_{n-d+1}w/x_n \in J$. Thus the colon ideal $J : w$ is generated by a subset M of $\{x_1, \ldots, x_n\}$ with $\{x_1, \ldots, x_{n-d}, x_{n-d+1}\} \subset M$. Hence $r(I) \geq n - d + 1$, and thus one has $c(I) < r(I) + 1$, a contradiction. $\square$

A Cohen–Macaulay ideal is always unmixed. The converse is in general false, even for matroid ideals. For example, let $I \subset K[x_1, \ldots, x_6]$ be the monomial ideal generated by

$$x_1x_3, x_1x_4, x_1x_5, x_1x_6, x_2x_3, x_2x_4, x_2x_5, x_2x_6, x_3x_5, x_3x_6, x_4x_5, x_4x_6.$$

Then I is matroidal and unmixed. However, I is not Cohen–Macaulay.

12.7 Weakly polymatroidal ideals

The purpose of this section is to extend the notion of polymatroidal ideals introduced in the previous section and to show that this class of ideals has again linear quotients.

Let $S = K[x_1, \ldots, x_n]$ be the polynomial ring over the field K. For any monomial u, $m(u)$ denotes the greatest integer i for which x_i divides u. For $u = x_1^{a_1} \cdots x_n^{a_n}$ we set $\deg_{x_i} u = a_i$, and call it the x_i-**degree** of u.

Definition 12.7.1. A monomial ideal I is called **weakly polymatroidal** if for every two monomials $u = x_1^{a_1} \cdots x_n^{a_n}$ and $v = x_1^{b_1} \cdots x_n^{b_n}$ in $G(I)$ such that $a_1 = b_1, \cdots, a_{t-1} = b_{t-1}$ and $a_t > b_t$ for some t, there exists $j > t$ such that $x_t(v/x_j) \in I$.

It is clear from the definition that a polymatroidal ideal is weakly polymatroidal. The converse is not true in general, as the following example shows:

Let I be the polymatroidal ideal of $K[x_1, \ldots, x_6]$ which is generated by all squarefree monomials of degree 3. Let J denote the monomial ideal of $K[x_1, \ldots, x_6]$ generated by those monomials $u \in G(I)$ with $x_2 x_4 x_6 <_{\text{lex}} u$. Then the monomial ideal J is weakly polymatroidal, but not polymatroidal.

Theorem 12.7.2. *A weakly polymatroidal ideal I has linear quotients.*

Proof. Let $G(I) = u_1, \ldots, u_m$, where $u_1 > u_2 > \cdots > u_m$ in the pure lexicographical order with induced by $x_1 > x_2 > \cdots > x_n$. We show that I has linear quotients with respect to $u_1, \ldots, u_m$.

Fix a number j and let v be a monomial with $v \in (u_1, \ldots, u_{j-1}) : u_j$. Then $vu_j \in (u_i)$ for some $i < j$. Let $u_i = x_1^{a_1} \cdots x_n^{a_n}$ and $u_j = x_1^{b_1} \cdots x_n^{b_n}$. Then there exists an integer $t < n$ with $a_1 = b_1, \ldots, a_{t-1} = b_{t-1}$ and $a_t > b_t$. Therefore $x_t | v$, and in addition there exists $\ell > t$ such that $x_t(u_j/x_\ell) \in I$. Thus the set $A = \{u_k \colon x_t(u_j/x_\ell) \in (u_k)\}$ is nonempty. Let $u_s \in A$ be the unique element such that for any $u_k \in A$ $(k \neq s)$, we have either $\deg u_k > \deg u_s$ or $\deg u_k = \deg u_s$ and $u_k <_{\text{lex}} u_s$. One has $x_t(u_j/x_\ell) = u_s h$ for some $h \in S$. If $x_t | h$, then $u_j = u_s h'$ for some $h' \in S$, which is a contradiction since $u_j \in G(I)$. So we have $x_t^{b_t+1}$ divides u_s.

We claim that $u_s > u_j$ in the pure lexicographical order. On the contrary assume that assume that $u_s < u_j$. Let $u_s = x_1^{c_1} \cdots x_n^{c_n}$ with $c_1 = b_1, \ldots, c_{r-1} = b_{r-1}$ and $c_r < b_r$ for some $1 \leq r \leq n$. Since $x_t^{b_t+1} | u_s$, one has $r < t$. Then from the definition of weakly polymatroidal, one has $w = u_s x_r / x_k \in I$ for some $k > r$. Since $r < \ell$, $x_r | h$ and so $x_k h / x_r \in S$. From $w(x_k h / x_r) = x_t(u_j/x_\ell)$, $u_s <_{\text{lex}} w$ and $\deg w = \deg u_s$, we have $\notin G(I)$. Let $w = u_{s'} h'$ for some s' and $h' \in S$, $h' \neq 1$. Then $\deg u_{s'} < \deg w = \deg u_s$. This is a contradiction, since $u_{s'} \in A$. Therefore one has $u_s h \in (u_1, \ldots, u_{j-1})$, and hence $x_t u_j \in (u_1, \ldots, u_{j-1})$. Since x_t divides v, the proof is complete.   ⊔

Combining Theorem 12.7.2 with Theorem 8.2.15 we obtain

Corollary 12.7.3. *A weakly polymatroidal ideal is componentwise linear.*

Problems

12.1. Let $\mathcal{P} \subset \mathbb{R}^2$ denote the compact set consisting of those $(x, y) \in \mathbb{R}_+^2$ with $0 \leq x \leq 2$, $0 \leq y \leq 2$ and $x + y \leq 3$.
(a) Show that $\mathcal{P}$ is a polymatroid and find its ground set rank function.
(b) What is $\text{rank}(\mathcal{P})$?
(c) By using Theorem 12.1.4 find the vertices of $\mathcal{P} \vee \mathcal{P} \vee \mathcal{P}$.

12.2. Let $\mathcal{A} = (A_1, \ldots, A_d)$ be a family of nonempty subsets of $[n]$ and define the integer valued nondecreasing function $\rho_{\mathcal{A}} : 2^{[n]} \to \mathbb{R}_+$ by setting

$$\rho_{\mathcal{A}}(X) = |\{\, k \, : \, A_k \cap X \neq \emptyset \,\}|, \quad X \subset [n].$$

(a) Show that $\rho_{\mathcal{A}}$ is submodular. The polymatroid $\mathcal{P}$ with $\rho_{\mathcal{A}}$ its ground set rank function is called **transversal polymatroid** presented by $\mathcal{A}$.

(b) Let $P_{\mathcal{A}}$ denote the transversal discrete polymatroid presented by $\mathcal{A} = (A_1, \ldots, A_d)$. In other words, $P_{\mathcal{A}}$ is the set of all integer points belonging to the transversal polymatroid presented by $\mathcal{A} = (A_1, \ldots, A_d)$. Show that the set of bases of $P_{\mathcal{A}}$ is

$$B_{\mathcal{A}} = \{\, \epsilon_{i_1} + \cdots + \epsilon_{i_d} \, : \, i_k \in A_k, \, 1 \leq k \leq d \,\} \subset \mathbb{Z}_+^n.$$

(c) Let $P \subset \mathbb{Z}_+^4$ denote the discrete polymatroid of rank 3 consisting of those $\mathbf{u} = (u_1, u_2, u_3, u_4) \in \mathbb{Z}_+^4$ with each $u_i \leq 2$ and with $|\mathbf{u}| \leq 3$. Show that P cannot be transversal.

12.3. (a) Let $\mathbf{v}_1, \ldots, \mathbf{v}_n$ be vectors of a vector space. Let $\mathcal{M}$ denote the subset of $2^{[n]}$ consisting of those $F \subset [n]$ such that the vectors $\mathbf{v}_k$ with $k \in F$ are linearly independent. Show that $\mathcal{M}$ is a matroid.

(b) Let G be a finite graph with the edges $e_1, \ldots, e_n$. Let $\mathcal{M}$ denote the subset of $2^{[n]}$ consisting of those $F \subset [n]$ such that the subgraph of G whose edges are e_k with $k \in F$ is forest. Show that $\mathcal{M}$ is a matroid.

12.4. Let B the set of bases of a discrete polymatroid. We say that B satisfies the **strong exchange property** if, for all $\mathbf{u}, \mathbf{v} \in B$ with $\mathbf{u} \neq \mathbf{v}$ and for all i, j with $\mathbf{u}(i) > \mathbf{v}(i)$ and $\mathbf{u}(j) < \mathbf{v}(j)$, one has $\mathbf{u} - \epsilon_i + \epsilon_j \in B$.

(a) Show that the set of bases of a discrete polymatroid of Veronese type of Example 12.2.8 satisfies the strong exchange property.

(b) Find a discrete polymatroid whose set of bases does not satisfy the strong exchange property.

12.5. (a) Let $\mathcal{P} \subset \mathbb{R}^3$ denote the integral convex polytope which is the convex hull of $(0,0,0), (1,1,0), (1,0,1), (0,1,1)$. Does $\mathcal{P}$ possess the integer decomposition property?

(b) Let $\mathcal{P} \subset \mathbb{R}^3$ denote the integral convex polytope which is the convex hull of $(0,0,0), (1,1,0), (1,0,1), (0,1,1), (-1,-1,-1)$. Show that $\mathcal{P}$ possesses the integer decomposition property. Is the toric ring $K[\mathcal{P}]$ Gorenstein?

12.6. (a) The discrete polymatroid of Veronese type of Example 12.2.8 is called squarefree if each $d_i = 1$. Let B be the set of bases of a discrete polymatroid of squarefree Veronese type with $n = 2d$. Show that the base ring $K[B]$ is Gorenstein.

(b) Let B be the set of bases of a discrete polymatroid of Veronese type with each $d_i = d$. Show that the base ring $K[B]$ is Gorenstein if and only if n is divided by d.

12.7. Let $n = 3$ and let $a_1, a_2, a_3 > 0$ be integers. Let $P \subset \mathbb{Z}_+^3$ denote the discrete polymatroid of rank d consisting of those $\mathbf{u} = (u_1, u_2, u_3) \in \mathbb{Z}_+^3$ such that $u_1 \leq a_1$, $u_2 \leq a_2$, $u_3 \leq a_3$ and $u_1 + u_2 + u_3 \leq d$.

(a) When is P generic?

(b) When is the base ring $K[B]$ of P Gorenstein?

12.8. Let $\mathbf{a} = (a_1, \ldots, a_n) \in \mathbb{Z}_+^n$ with $a_n \geq 1$ and $\mathcal{P}_{\mathbf{a}}$ the integral polymatroid on the ground set $\{2, 3, \ldots, n\}$ whose ground set rank function ρ is given by $\rho(X) = \sum_{j=\min(X)}^{n} a_j$ for $\emptyset \neq X \subset \{2, 3, \ldots, n\}$. Let $K[\mathcal{P}_{\mathbf{a}}]$ denote the toric ring of $\mathcal{P}_{\mathbf{a}}$.

(a) If $\mathbf{a} = (0, 1, \ldots, 1, 2) \in \mathbb{Z}_+^n$, then show that $K[\mathcal{P}_{\mathbf{a}}]$ is Gorenstein.

(b) If $\mathbf{a} = (0, 1, 0, 2, 0, 3)$, then show that $K[\mathcal{P}_{\mathbf{a}}]$ is Gorenstein.

(c) If $\mathbf{a} = (0, \ldots, 0, a_n) \in \mathbb{Z}_+^n$, then show that $K[\mathcal{P}_{\mathbf{a}}]$ is Gorenstein if and only if a_n divides n.

12.9. Let $I \subset K[x_1, \ldots, x_6]$ be the monomial ideal generated by

$$x_1 x_3, x_1 x_4, x_1 x_5, x_1 x_6, x_2 x_3, x_2 x_4, x_2 x_5, x_2 x_6, x_3 x_5, x_3 x_6, x_4 x_5, x_4 x_6.$$

(a) Show that I is matroidal and unmixed.

(b) Show that I is not Cohen–Macaulay.

12.10. Let I be the polymatroidal ideal of $K[x_1, \ldots, x_6]$ which is generated by all squarefree monomials of degree 3. Let J denote the monomial ideal of $K[x_1, \ldots, x_6]$ generated by those monomials $u \in G(I)$ with $x_2 x_4 x_6 <_{\text{lex}} u$. Show that the monomial ideal J is weakly polymatroidal, but not polymatroidal.

Notes

A standard reference for the theory of matroids is the book by Welsh [Wel76]. Edmonds [Edm70] studied the polyhedral theory of polymatroids. Discrete polymatroids were introduced in [HH02]. One of the widely open outstanding conjectures due to White [Whi80] asserts that the defining ideal of the base ring of a matroid is quadratically generated. This conjecture can be extended to the base ring of a discrete polymatroid. Stronger conjectures even assert that the defining ideal of the base ring of a discrete polymatroid has a quadratic Gröbner basis, or at least is Koszul.

Polymatroidal ideals are ideals with linear quotients [CH03], and, as a consequence of a classical result due to Edmonds [Edm70], are closed under multiplication. In particular, all powers of a polymatroidal ideal have linear quotients. The notion of weakly polymatroidal ideals was introduced by [KH06], where only monomial ideals generated in one degree were studied.

Later, Fatemeh and Somayeh [FS10] generalized this notion to arbitrary monomial ideals.

A

Some homological algebra

A.1 The language of categories and functors

This is not an introduction to category theory but just a summary of some of the standard terminology used therein.

A **category** $\mathcal{C}$ is a class $\mathrm{Obj}(\mathcal{C})$ of **objects** together with a class $\mathrm{Mor}(\mathcal{C})$ of **morphisms**. Each morphism f has a unique source object $A \in \mathrm{Obj}(\mathcal{C})$ and a unique target object $B \in \mathrm{Obj}(\mathcal{C})$. If A is the source and B the target of f one writes $f\colon A \to B$ and says that f is a morphism from A to B. The class of morphisms from A to B is denoted by $\mathrm{Hom}(A, B)$.

For every three objects A, B and C a map

$$\mathrm{Hom}(A, B) \times \mathrm{Hom}(B, C) \to \mathrm{Hom}(A, C), \quad (f, g) \mapsto g \circ f,$$

called **composition of morphisms** is given such that the following axioms hold:

(A) (Associativity) If $f\colon A \to B$, $g\colon B \to C$ and $h\colon C \to D$ then

$$h \circ (g \circ f) = (h \circ g) \circ f.$$

(B) (Identity) For every object X, there exists a morphism $\mathrm{id}_X\colon X \to X$ called the **identity morphism** for X, such that for every morphism $f\colon A \to B$, we have $\mathrm{id}_B \circ f = f = f \circ \mathrm{id}_A$.

Examples of categories appear in all branches of mathematics. The simplest example of a category is the category of sets $\mathcal{S}$ whose objects are the sets and whose morphisms are the maps between sets. Other examples are for instance: the category $\mathcal{T}$ of topological spaces, whose morphisms are the continuous maps, or for a given ring R, the category $\mathcal{M}_R$ of R-modules, whose morphisms are the R-module homomorphisms. If R is graded we can also consider the category $\mathcal{G}_R$ of graded R-modules. The morphisms in this case are the homogeneous R-module homomorphisms of degree 0. As a special case of the last type of category we considered in Section 5.1 the category $\mathcal{G}$ of graded modules over the exterior algebra.

Let $\mathcal{A}$ and $\mathcal{B}$ be categories. A **covariant functor** $F\colon \mathcal{A} \to \mathcal{B}$ from $\mathcal{A}$ to $\mathcal{B}$ is a mapping that assigns to each object $A \in \mathcal{A}$ and object $F(A) \in \mathcal{B}$ and to each morphism $f\colon A \to B$ in $\mathcal{A}$ a morphism $F(f)\colon F(A) \to F(B)$ such that the following axioms hold:

(C) For all morphisms $f\colon B \to C$ and $g\colon A \to B$ in $\mathcal{A}$ one has

$$F(f \circ g) = F(f) \circ F(g).$$

(D) $F(\mathrm{id}_X) = \mathrm{id}_{F(X)}$ for all objects X in $\mathcal{A}$.

A **contravariant functor** $F\colon \mathcal{A} \to \mathcal{B}$ is defined similarly. The only difference is that it reverses the arrows of the maps. In other words to each morphism $f\colon A \to B$ in $\mathcal{A}$ the contravariant functor F assigns a morphism $F(f)\colon F(B) \to F(A)$, and for compositions of morphisms one has $F(f \circ g) = F(g) \circ F(f)$.

A typical example of this concept is the functor from the category of topological spaces to the category of abelian groups which assigns to each topological space X its ith singular homology group $H_i(X;\mathbb{Z})$. Indeed, a continuous map $f\colon X \to Y$ induces a group homomorphism $H_i(f;\mathbb{Z})\colon H_i(X;\mathbb{Z}) \to H_i(Y;\mathbb{Z})$ satisfying the axioms (C) and (D).

Other important examples are the functors Tor and Ext: let R be a ring, $\mathcal{M}_R$ the category of R-modules and $N \in \mathrm{Obj}(\mathcal{M}_R)$. Then for each integer $i \geq 0$, the assignments $\mathrm{Tor}_i^R(N, -)\colon \mathcal{M}_R \to \mathcal{M}_R$, $M \mapsto \mathrm{Tor}_i^R(N, M)$, and $\mathrm{Ext}_R^i(N, -)\colon \mathcal{M}_R \to \mathcal{M}_R$, $M \mapsto \mathrm{Ext}_R^i(N, M)$, are covariant functors, while $\mathrm{Ext}_R^i(-, N)\colon \mathcal{M}_R \to \mathcal{M}_R$, $M \mapsto \mathrm{Ext}_R^i(M, N)$, is a contravariant functor.

Special cases of these examples are the covariant functors $- \otimes_R N$ and $\mathrm{Hom}_R(N, -)$ and the contravariant functor $\mathrm{Hom}_R(-, N)$. The first of these functors is right exact, while the other two functors are left exact. Quite generally, if we have categories $\mathcal{A}$ and $\mathcal{B}$ where we can talk about exact sequences, for example in the categories $\mathcal{M}_R$, $\mathcal{G}_R$ or $\mathcal{G}$ mentioned above, we say that a functor $F\colon \mathcal{A} \to \mathcal{B}$ is **left exact** if for any exact sequence

$$0 \to A \to B \to C \to 0$$

in $\mathcal{A}$, the sequence $0 \to F(A) \to F(B) \to F(C)$ is exact for covariant F and $0 \to F(C) \to F(B) \to F(A)$ is exact for contravariant F. Similarly one defines right exactness. Finally F is called **exact** if F is left and right exact.

Let $\mathcal{A}$ be any one of the module categories $\mathcal{M}_R$, $\mathcal{G}_R$ or $\mathcal{G}$, and let $M \in \mathrm{Obj}(\mathcal{A})$. Then M is called **injective** if the functor $\mathrm{Hom}(-, M)$ is exact, it is called **projective** if $\mathrm{Hom}(M, -)$ is exact and it is called **flat** if $- \otimes M$ is exact. In Section 5.1 we have seen that the exterior algebra viewed as an object in $\mathcal{G}$ is injective.

Given two covariant functors $F, G\colon \mathcal{A} \to \mathcal{B}$. A family of morphisms $\eta_A\colon F(A) \to G(A)$ in $\mathcal{B}$ with $A \in \mathcal{A}$ is called a **natural transformation** from F to G, written $\eta\colon F \to G$, if for all $A, B \in \mathrm{Obj}(\mathcal{A})$ and all morphisms $f\colon A \to B$ the following diagram

$$
\begin{array}{ccc}
F(A) & \xrightarrow{\ \eta_A\ } & G(A) \\
{\scriptstyle F(f)}\downarrow & & \downarrow{\scriptstyle G(f)} \\
F(B) & \xrightarrow{\ \eta_B\ } & G(B).
\end{array}
$$

is commutative.

A natural transformation $\eta\colon F \to G$ is called a **functorial isomorphism** if there exists a natural transformation $\tau\colon G \to F$ such that $\tau_A \circ \eta_A = \mathrm{id}_{F(A)}$ and $\eta_A \circ \tau_A = \mathrm{id}_{G(A)}$ for all $A \in \mathrm{Obj}(\mathcal{A})$. It is customary to call an isomorphism $\alpha\colon F(A) \to G(A)$ functorial if there exists a functorial isomorphism $\eta\colon F \to G$ such that $\alpha = \eta_A$. An example of a functorial isomorphism is the isomorphism $M^\vee \to M^*$ given in Theorem 5.1.3.

A.2 Graded free resolutions

For this and the following sections of Appendix A we fix the following assumptions and notation. We let K be a field, $(R, \mathfrak{m})$ a Noetherian local ring with residue field K or a standard graded K-algebra with graded maximal ideal $\mathfrak{m}$. As usual we write S for the polynomial ring $K[x_1, \ldots, x_n]$. We let M be a finitely generated R-module, and will assume that M is graded if R is graded.

We let $\mathcal{M}(S)$ be the category of finitely generated graded S-modules, the morphisms being the homogeneous homomorphisms $M \to N$ of degree 0, simply called homogeneous homomorphisms. A **homogeneous homomorphism** $\varphi\colon M \to N$ of graded S-modules of degree d is an S-module homomorphism such that $\varphi(M_i) \subset N_{i+d}$ for all i. For example, if $f \in S$ is homogeneous of degree d, then the multiplication map $S(-d) \to S$ with $g \mapsto fg$ is a homogeneous homomorphism. Here, for a graded S-module W and an integer a, one denotes by $W(a)$ the graded S-module whose graded components are given by $W(a)_i = W_{a+i}$. One says that $W(a)$ arises from W by applying the **shift** a.

Now let M be a finitely generated graded S-module with homogeneous generators $m_1, \ldots, m_r$ and $\deg(m_i) = a_i$ for $i = 1, \ldots, r$. Then there exists a surjective S-module homomorphism $F_0 = \bigoplus_{i=1}^{r} Se_i \to M$ with $e_i \mapsto m_i$. Assigning to e_i the degree a_i for $i = 1, \ldots, r$, the map $F_0 \to M$ becomes a morphism in $\mathcal{M}(S)$ and F_0 becomes isomorphic to $\bigoplus_{i=1}^{r} S(-a_i)$. Thus we obtain the exact sequence

$$
0 \longrightarrow U \longrightarrow \bigoplus_j S(-j)^{\beta_{0j}} \longrightarrow M \longrightarrow 0,
$$

where $\beta_{0j} = |\{i\colon a_i = j\}|$, and where $U = \mathrm{Ker}(\bigoplus_j S(-j)^{\beta_{0j}} \to M)$.

The module U is a graded submodule of $F_0 = \bigoplus_j S(-j)^{\beta_{0j}}$. By Hilbert's basis theorem for modules we know that U is finitely generated, and hence we find again an epimorphism $\bigoplus_j S(-j)^{\beta_{1j}} \to U$. Composing this epimorphism with the inclusion map $U \to \bigoplus_j S(-j)^{\beta_{0j}}$ we obtain the exact sequence

$$\bigoplus_j S(-j)^{\beta_{1j}} \longrightarrow \bigoplus_j S(-j)^{\beta_{0j}} \longrightarrow M \longrightarrow 0.$$

of graded S-modules. Proceeding in this way we obtain a long exact sequence

$$\mathbb{F}: \cdots \longrightarrow F_2 \longrightarrow F_1 \longrightarrow F_0 \longrightarrow M \longrightarrow 0$$

of graded S-modules with $F_i = \bigoplus_j S(-j)^{\beta_{ij}}$. Such an exact sequence is called a **graded free S-resolution** of M.

It is clear from our construction that the resolution obtained is by no means unique. On the other hand, if we choose in each step of the resolution a minimal presentation, the resolution will be unique up to an isomorphism, as we shall see now.

A set of homogeneous generators $m_1, \ldots, m_r$ of M is called **minimal** if no proper subset of it generates M.

Lemma A.2.1. *Let $m_1, \ldots, m_r$ be a homogeneous set of generators of the graded S-module M. Let $F_0 = \bigoplus_{i=1}^r Se_i$ and let $\varepsilon \colon F_0 \to M$ be the epimorphism with $e_i \mapsto m_i$ for $i = 1, \ldots, r$. Then the following conditions are equivalent:*

(a) $m_1, \ldots, m_r$ *is a minimal system of generators of M;*
(b) $\mathrm{Ker}(\varepsilon) \subset \mathfrak{m}F_0$, *where $\mathfrak{m} = (x_1, x_2, \ldots, x_n)$.*

Proof. (a) $\Rightarrow$ (b): Suppose $\mathrm{Ker}(\varepsilon) \not\subset \mathfrak{m}F_0$. Then there exists a homogeneous element $f = \sum_{i=1}^r f_i e_i$ such that $f \notin \mathfrak{m}F_0$. This implies that at least one of the coefficients f_i is of degree 0, say $\deg f_1 = 0$. Therefore $f_1 \in K \setminus \{0\}$, and it follows that

$$m_1 = f_1^{-1} f_2 m_2 + \cdots + f_1^{-1} f_r m_r,$$

a contradiction.

(b) $\Rightarrow$ (a): Suppose m_1 can be omitted, so that $m_2, \ldots, m_r$ is a system of generators of M as well. Then we have $m_1 = \sum_{i=2}^r f_i m_i$ for suitable homogeneous elements $f_i \in S$. This yields the element $f = e_1 - \sum_{i=2}^r f_i e_i$ in $\mathrm{Ker}(\varepsilon)$ with $f \notin \mathfrak{m}F_0$, a contradiction. $\square$

Let M be a finitely generated graded S-module. A graded free S-resolution $\mathbb{F}$ of M is called **minimal** if for all i, the image of $F_{i+1} \to F_i$ is contained in $\mathfrak{m}F_i$. Lemma A.2.1 implies at once that each finitely generated graded S-module admits a minimal free resolution.

The next result shows that the numerical data given by a graded minimal free S-resolution of M depend only on M and not on the particular chosen resolution.

Proposition A.2.2. *Let M be a finitely generated graded S-module and*

$$\mathbb{F}: \ \cdots \longrightarrow F_2 \longrightarrow F_1 \longrightarrow F_0 \longrightarrow M \longrightarrow 0$$

a minimal graded free S-resolution of M with $F_i = \bigoplus_j S(-j)^{\beta_{ij}}$ for all i. Then

$$\beta_{ij} = \dim_K \operatorname{Tor}_i(K, M)_j \quad \text{for all } i \text{ and } j.$$

Proof. As a graded K-vector space $\operatorname{Tor}_i(K, M)$ is isomorphic to $H_i(\mathbb{F}/\mathfrak{m}\mathbb{F})$. However, since the resolution $\mathbb{F}$ is minimal all maps in the complex $\mathbb{F}/\mathfrak{m}\mathbb{F}$ are zero. Therefore $H_i(\mathbb{F}/\mathfrak{m}\mathbb{F}) = \mathbb{F}/\mathfrak{m}\mathbb{F} \cong \bigoplus_j K(-j)^{\beta_{ij}}$. $\qquad\Box$

The numbers $\beta_{ij} = \dim \operatorname{Tor}_i(K, M)_j$ are called the **graded Betti numbers** of M, and $\beta_i = \sum_j \beta_{ij} (= \operatorname{rank} F_i)$ is called the **ith Betti number** of M.

We conclude this section by showing that not only are the graded Betti numbers determined by a minimal graded free resolution but that in fact a minimal graded free resolution of M is unique up to isomorphisms.

Proposition A.2.3. *Let M be a finitely generated graded S-module and let $\mathbb{F}$ and $\mathbb{G}$ be two minimal graded free S-resolutions of M. Then the complexes $\mathbb{F}$ and $\mathbb{G}$ are isomorphic, that is, there exist isomorphisms of graded S-modules $\alpha_i \colon F_i \to G_i$ such that the diagram*

$$
\begin{array}{ccccccccccc}
\cdots & \longrightarrow & F_i & \longrightarrow & \cdots & \longrightarrow & F_1 & \longrightarrow & F_0 & \longrightarrow & M & \longrightarrow & 0 \\
& & \downarrow{\alpha_i} & & & & \downarrow{\alpha_1} & & \downarrow{\alpha_0} & & \downarrow{\mathrm{id}} & & \\
\cdots & \longrightarrow & G_i & \longrightarrow & \cdots & \longrightarrow & G_1 & \longrightarrow & G_0 & \longrightarrow & M & \longrightarrow & 0
\end{array}
$$

is commutative.

Proof. The existence of the isomorphism α_i will follow by induction on i once we have shown the following: let $\varphi \colon U \to V$ be an isomorphism of finitely generated graded S-modules, and let $\varepsilon \colon F \to U$ and $\eta \colon G \to V$ be homogeneous surjective homomorphisms with $\operatorname{Ker}(\varepsilon) \subset \mathfrak{m}F$ and $\operatorname{Ker}(\eta) \subset \mathfrak{m}G$. Then there exists a homogeneous isomorphism $\alpha \colon F \to G$ such that

$$
\begin{array}{ccc}
F & \xrightarrow{\ \varepsilon\ } & U \\
\downarrow{\alpha} & & \downarrow{\varphi} \\
G & \xrightarrow{\ \eta\ } & V
\end{array}
$$

is commutative. Indeed, let $f_1, \ldots, f_r$ be a homogeneous basis of F. Then

$$\varphi(\varepsilon(f_1)), \ldots, \varphi(\varepsilon(f_r))$$

is a homogeneous system of generators of V. Since η is a homogeneous surjective homomorphism, there exist homogeneous elements $g_1, \ldots, g_r \in G$ with

$\eta(g_i) = \varphi(\varepsilon(f_i))$ for $i = 1, \ldots, r$. Thus if we set $\alpha(f_i) = g_i$ for $i = 1, \ldots, r$, then $\alpha \colon F \to G$ is a homogeneous homomorphism which makes the above diagram commutative. Modulo $\mathfrak{m}$ we obtain the commutative diagram

$$
\begin{array}{ccc}
F/\mathfrak{m}F & \xrightarrow{\ \bar{\varepsilon}\ } & U/\mathfrak{m}U \\
{\scriptstyle\bar{\alpha}}\downarrow & & \downarrow{\scriptstyle\bar{\varphi}} \\
G/\mathfrak{m}G & \xrightarrow{\ \bar{\eta}\ } & V/\mathfrak{m}V.
\end{array}
$$

Since $\mathrm{Ker}(\varepsilon) \subset \mathfrak{m}F$, it follows that $\bar{\varepsilon} \colon F/\mathfrak{m}F \to U/\mathfrak{m}U$ is an isomorphism. Similarly, $\bar{\eta}$ and $\bar{\varphi}$ are isomorphisms. Thus $\bar{\alpha} = \bar{\eta}^{-1} \circ \bar{\varphi} \circ \bar{\varepsilon}$ is an isomorphism. Now by a homogeneous version of the Nakayama lemma it follows that α itself is an isomorphism. $\qquad\square$

A.3 The Koszul complex

We recall the basic properties of Koszul homology that are used in this book. Let R be any commutative ring (with unit) and $\mathbf{f} = f_1, \ldots, f_m$ a sequence of elements of R. The **Koszul complex** $K(\mathbf{f}; R)$ attached to the sequence $\mathbf{f}$ is defined as follows: let F be a free R-module with basis $e_1, \ldots, e_m$. We let $K_j(\mathbf{f}; R)$ be the jth exterior power of F, that is, $K_j(\mathbf{f}; R) = \bigwedge^j F$. A basis of the free R-module $K_j(\mathbf{f}; R)$ is given by the wedge products $e_F = e_{i_1} \wedge e_{i_2} \wedge \cdots \wedge e_{i_j}$ where $F = \{i_1 < i_2 < \cdots < i_j\}$. In particular, it follows that $\mathrm{rank}\, K_j(\mathbf{f}; R) = \binom{m}{j}$.

We define the differential $\partial \colon K_j(\mathbf{f}; R) \to K_{j-1}(\mathbf{f}; R)$ by the formula

$$
\partial(e_{i_1} \wedge e_{i_2} \wedge \cdots \wedge e_{i_j}) = \sum_{k=1}^{j} (-1)^{k+1} f_{i_k} e_{i_1} \wedge e_{i_2} \wedge \cdots \wedge e_{i_{k-1}} \wedge e_{i_{k+1}} \wedge \cdots \wedge e_{i_j}.
$$

One readily verifies that $\partial \circ \partial = 0$, so that $K_{\bullet}(\mathbf{f}; R)$ is indeed a complex.

Now let M be an R-module. We define the complexes

$$
K_{\bullet}(\mathbf{f}; M) = K_{\bullet}(\mathbf{f}; R) \otimes_R M \quad \text{and} \quad K^{\bullet}(\mathbf{f}; M) = \mathrm{Hom}_R(K_{\bullet}(\mathbf{f}; R), M).
$$

$H_i(\mathbf{f}; M) = H_i(K(\mathbf{f}; M))$ is the ith **Koszul homology module** of $\mathbf{f}$ with respect to M, and $H^i(\mathbf{f}; M) = H^i(\mathrm{Hom}_R(K_{\bullet}(\mathbf{f}; R), M))$ is the ith **Koszul cohomology module** of $\mathbf{f}$ with respect to M.

Let $I \subset R$ be the ideal generated by $f_1, \ldots, f_m$. Then

$$
H_0(\mathbf{f}; M) = M/IM \quad \text{and} \quad H_m(\mathbf{f}; M) \cong 0 :_M I = \{x \in M \colon Ix = 0\}.
$$

The Koszul complex $K_{\bullet}(\mathbf{f}; R)$ is a graded R-algebra, namely the exterior algebra of F, with multiplication the wedge product. We have the following rules whose verification we leave to the reader.

(i) $a \wedge b = (-1)^{\deg a \deg b} b \wedge a$ for homogeneous elements $a, b \in K(\mathbf{f}; R)$.

(ii) $\partial(a \wedge b) = \partial(a) \wedge b + (-1)^{\deg a} a \wedge \partial(b)$ for $a, b \in K(\mathbf{f}; R)$ and a homogeneous.

We denote by $Z_.(\mathbf{f}; R)$ the cycles of the Koszul complex and by $B_.(\mathbf{f}; R)$ its boundaries. Rule (ii) has an interesting consequence.

Proposition A.3.1. *The R-module $Z_.(\mathbf{f}; R)$ is a graded subalgebra of $K_.(\mathbf{f}; R)$ and $B_.(\mathbf{f}; R) \subset Z_.(\mathbf{f}; R)$ is a graded two-sided ideal in $Z_.(\mathbf{f}; R)$. In particular, $H_.(\mathbf{f}; R) = Z_.(\mathbf{f}; R)/B_.(\mathbf{f}; R)$ has a natural structure as graded $H_0(\mathbf{f}; R)$-algebra. Moreover, if I is the ideal generated by the sequence $\mathbf{f}$, then*

$$IH_.(\mathbf{f}; R) = 0.$$

Proof. Let z_1 and z_2 be two homogeneous cycles. Then $\partial(z_1 \wedge z_2) = \partial(z_1) \wedge z_2 + (-1)^{\deg z_1} z_1 \wedge \partial(z_2) = 0$, since $\partial(z_1) = \partial(z_2) = 0$. So $z_1 \wedge z_2$ is again a cycle, which shows that $Z(\mathbf{f}; R)$ is a subalgebra of $K(\mathbf{f}; R)$. Now let b be a homogeneous boundary and z a cycle. There exists $a \in K(\mathbf{f}; R)$ such that $\partial(a) = b$. It then follows that $\partial(a \wedge z) = \partial(a) \wedge z + (-1)^{\deg a} a \wedge \partial(z) = b \wedge z$, which shows that $b \wedge z \in B(\mathbf{f}; R)$. Similarly, we have $z \wedge b \in B(\mathbf{f}; R)$. This shows that $B(\mathbf{f}; R)$ is indeed a two-sided ideal in $Z(\mathbf{f}; R)$. Finally, since $H(\mathbf{f}; R)$ is a $H_0(\mathbf{f}; R)$-algebra, and since $H_0(\mathbf{f}; R) = R/I$, it follows that $IH(\mathbf{f}; R) = 0$.   ⊔

Corollary A.3.2. *If $(\mathbf{f}) = R$, then $H_.(\mathbf{f}; R) = 0$.*

Given an R-module M, then as in the preceding proof one shows that $Z_.(\mathbf{f}; R)Z_.(\mathbf{f}; M) \subset Z_.(\mathbf{f}; M)$ and that $B_.(\mathbf{f}; R)Z_.(\mathbf{f}; M) \subset B_.(\mathbf{f}; M)$. This then implies that $H_.(\mathbf{f}; M)$ is a graded $H_.(\mathbf{f}; R)$-module.

For computing the Koszul homology there are two fundamental long exact sequences of importance.

Theorem A.3.3. *Let $\mathbf{f} = f_1, \ldots, f_m$ be a sequence of elements in R, and denote by $\mathbf{g}$ the sequence $f_1, \ldots, f_{m-1}$. Furthermore, let M be an R-module and $0 \to U \to M \to N \to 0$ a short exact sequence of R-modules. Then we obtain the following long exact sequences:*

$$0 \to H_m(\mathbf{f}; U) \to H_m(\mathbf{f}; M) \to H_m(\mathbf{f}; N) \to H_{m-1}(\mathbf{f}; U) \to H_{m-1}(\mathbf{f}; M) \to \cdots$$
$$\cdots \to H_{i+1}(\mathbf{f}; N) \to H_i(\mathbf{f}; U) \to H_i(\mathbf{f}; M) \to H_i(\mathbf{f}; N) \to H_{i-1}(\mathbf{f}; U) \to \cdots$$
$$\cdots \to H_1(\mathbf{f}; N) \to H_0(\mathbf{f}; U) \to H_0(\mathbf{f}; M) \to H_0(\mathbf{f}; N) \to 0.$$

and

$$0 \to H_m(\mathbf{f}; M) \to H_{m-1}(\mathbf{g}; M) \to H_{m-1}(\mathbf{g}; M) \to H_{m-1}(\mathbf{f}; M) \to \cdots$$
$$\cdots \to H_{i+1}(\mathbf{f}; M) \to H_i(\mathbf{g}; M) \to H_i(\mathbf{g}; M) \to H_i(\mathbf{f}; M) \to \cdots$$
$$\cdots \to H_1(\mathbf{f}; M) \to H_0(\mathbf{g}; M) \to H_0(\mathbf{g}; M) \to H_0(\mathbf{f}; M) \to 0,$$

where for all i, the map $H_i(\mathbf{g}; M) \to H_i(\mathbf{g}; M)$ is multiplication by $\pm f_m$.

Proof. The short exact sequence $0 \to U \to M \to N \to 0$ induces the short exact sequence of complexes

$$0 \to K_{\bullet}(\mathbf{f}; U) \to K_{\bullet}(\mathbf{f}; M) \to K_{\bullet}(\mathbf{f}; N) \to 0$$

whose corresponding long exact sequence is the first fundamental long exact sequence.

As for the proof of the second fundamental long exact homology sequence we consider for each i the map

$$\alpha_i : K_i(\mathbf{f}; R) \longrightarrow K_{i-1}(\mathbf{g}; R)$$

defined as follows: let $a \in K_i(\mathbf{f}; R)$; then a can be uniquely written in the form $a = a_0 + a_1 \wedge e_m$ with $a_0 \in K_i(\mathbf{g}; R)$ and $a_1 \in K_{i-1}(\mathbf{g}; R)$. We then set $\alpha(a) = a_1$. Applying rule (ii) one immediately checks that $\partial \circ \alpha = \alpha \circ \partial$, so that

$$\alpha : K_{\bullet}(\mathbf{f}; R) \longrightarrow K_{\bullet}(\mathbf{g}; R)[-1],$$

where $[-1]$ denotes the shifting of the homological degree by -1.

Notice that $K(\mathbf{g}; R)$ is a subcomplex of $K(\mathbf{f}; R)$ and indeed is the kernel of α. Hence we obtain the short exact sequence of complexes

$$0 \longrightarrow K_{\bullet}(\mathbf{g}; M) \longrightarrow K_{\bullet}(\mathbf{f}; M) \longrightarrow K_{\bullet}(\mathbf{g}; M)[-1] \longrightarrow 0$$

whose corresponding long exact sequence homology sequence is the second fundamental long exact sequence.

It remains to be shown that the map $H_i(\mathbf{g}; M) \to H_i(\mathbf{g}; M)$ is multiplication by $\pm f_m$. In fact, the map is the connecting homomorphism. Thus for $[a] \in H_i(\mathbf{g}; M)$, we have to choose a preimage $b \in K_{i+1}(\mathbf{f}; M)$ under the map α for the cycle $a \in K_i(\mathbf{f}; M)$. Then the image of $[a]$ in $H_i(\mathbf{g}; M)$ is the homology class $\partial(b)$. In our case we may choose $b = a \wedge e_m$. Then $\partial(b) = \pm f_m a$, and $[a]$ maps to $\pm f_m[a]$, as desired. $\square$

The sequence $\mathbf{f} = f_1, \ldots, f_m$ is called **regular** on M, or an M**-sequence**, if the following two conditions hold: (i) the multiplication map

$$M/(f_1, \ldots, f_{i-1})M \xrightarrow{\ f_i\ } M/(f_1, \ldots, f_{i-1})M$$

is injective for all i, and (ii) $M/(\mathbf{f})M \neq 0$.

Regular sequences can be characterized by the Koszul complex.

Theorem A.3.4. *Let* $\mathbf{f} = f_1, \ldots, f_m$ *be a sequence of elements of R and M an R-module.*

(a) *If $\mathbf{f}$ is an M-sequence, then $H_i(\mathbf{f}; M) = 0$ for $i > 0$.*

(b) *Suppose in addition that M is a finitely generated R-module and that R is either* (i) *a Noetherian local ring with maximal ideal* $\mathfrak{m}$, *or* (ii) *a graded K-algebra with graded maximal ideal* $\mathfrak{m}$, *and that* $(\mathbf{f}) \subset \mathfrak{m}$. *In case* (ii) *we also assume that $\mathbf{f}$ is a sequence of homogeneous elements. Then we have: if $H_1(\mathbf{f}; M) = 0$, then the sequence $\mathbf{f}$ is an M-sequence,*

Proof. (a) We proceed by induction on m. Let $m = 1$. We have $H_i(f_1; M) = 0$ for $i > 1$, and the exact sequence

$$0 \longrightarrow H_1(f_1; M) \longrightarrow M \xrightarrow{f_1} M \qquad \text{(A.1)}$$

Since f_1 is regular on M, the kernel of the multiplication map $f_1 \colon M \to M$ is zero. Hence $H_1(f_1; M) = 0$, as well.

Now let $m > 1$, and set $\mathbf{g} = f_1, \ldots, f_{m-1}$. By induction hypothesis we have $H_i(\mathbf{g}; M) = 0$ for $i > 0$. Thus the second fundamental long exact sequence yields the exact sequence

$$0 \longrightarrow H_1(\mathbf{f}; M) \longrightarrow H_0(\mathbf{g}; M) \longrightarrow H_0(\mathbf{g}; M),$$

and for each $i > 1$ the exact sequence

$$0 = H_i(\mathbf{g}; M) \longrightarrow H_i(\mathbf{f}; M) \longrightarrow H_{i-1}(\mathbf{g}; M) = 0.$$

It follows that $H_i(\mathbf{f}; M) = 0$ for $i > 1$. Since $H_0(\mathbf{g}; M) = M/(\mathbf{g})M$, we see that $H_1(\mathbf{f}; M)$ is the kernel of the multiplication map $f_m \colon M/(\mathbf{g})M \to M/(\mathbf{g})M$. Thus $H_1(\mathbf{f}; M) = 0$ as well, since f_m is regular on $M/(\mathbf{g})M$.

(b) Again we proceed by induction on m. For $m = 1$ the assertion follows from the exact sequence (A.1). Now let $m > 1$. Since $H_1(\mathbf{f}; M) = 0$ by assumption, and since $H_0(\mathbf{g}; M) = M/(\mathbf{g})M$, we deduce from the exact sequence

$$H_1(\mathbf{g}; M) \to H_1(\mathbf{g}; M) \to H_1(\mathbf{f}; M) \to H_0(\mathbf{g}; M) \to H_0(\mathbf{g}; M)$$

that f_m is regular on $M/(\mathbf{g})M$ and that $H_1(\mathbf{g}; M)/(f_m)H_1(\mathbf{g}; M) = 0$. Since $f_m \in \mathfrak{m}$, Nakayama's lemma implies that $H_1(\mathbf{g}; M) = 0$. By our induction hypothesis we then know that $\mathbf{g}$ is an M-sequence, and since f_m is regular on $M/(\mathbf{g})M$, we conclude that $\mathbf{f}$ is an M-sequence. $\qquad \square$

Theorem A.3.4 has the following important consequence

Corollary A.3.5. *Let K be a field, $S = K[x_1, \ldots, x_n]$ the polynomial ring in n variables and M be a finitely generated graded S-module. Moreover, let $\mathbf{f} = f_1, \ldots, f_k$ be a homogeneous S-sequence. Then for each i there exists an isomorphism of graded $S/(\mathbf{f})$-modules*

$$\operatorname{Tor}_i^S(S/(\mathbf{f}), M) \cong H_i(\mathbf{f}; M).$$

In particular, for $\mathbf{x} = x_1, \ldots, x_n$ we have $\beta_{ij}(M) = \dim_K H_i(\mathbf{x}; M)_j$ and hence $\operatorname{proj} \dim M \le n$.

Proof. We compute $\operatorname{Tor}^S(S/(\mathbf{f}), M)$ by means of a free S-resolution of $S/(\mathbf{f})$. Since $\mathbf{f}$ is an S-sequence, Theorem A.3.4 implies that the Koszul complex $K_{\bullet}(\mathbf{f}; S)$ provides a minimal graded free S-resolution of $S/(\mathbf{f})$, so that

$$\operatorname{Tor}_i^S(S/(\mathbf{f}), M) \cong H_i(K_{\bullet}(\mathbf{f}; S) \otimes M) = H_i(\mathbf{f}; M).$$

Since this isomorphism respects the grading, the desired conclusion follows.

$\qquad \square$

A.4 Depth

The **depth** of M, denoted depth M, is the common length of a maximal M-sequence contained in $\mathfrak{m}$ (consisting of homogeneous elements if M is graded). In homological terms the depth of M is given by

$$\text{depth } M = \min\{i : \text{Ext}_R^i(R/\mathfrak{m}, M) \neq 0\} = \min\{i : H_{\mathfrak{m}}^i(M) \neq 0\}.$$

Here $H_{\mathfrak{m}}^i(M)$ is the ith local cohomology module of M; see A.7.

Proposition A.4.1. $\mathbf{f} = f_1, \ldots, f_m$ be an M-sequence contained in $\mathfrak{m}$ (consisting of homogeneous elements if M is graded). Then depth $M/(\mathbf{f})M =$ depth $M - m$.

Proof. We may assume $m = 1$. The general case follows by an easy induction argument. Thus let f by a regular element on M. The short exact sequence

$$0 \longrightarrow M \xrightarrow{f} M \longrightarrow M/fM \longrightarrow 0$$

gives rise to the long exact sequence

$$\cdots \rightarrow \text{Ext}^{i-1}(R/\mathfrak{m}, M) \xrightarrow{f} \text{Ext}^{i-1}(R/\mathfrak{m}, M) \rightarrow \text{Ext}^{i-1}(R/\mathfrak{m}, M/fM)$$

$$\rightarrow \text{Ext}^i(R/\mathfrak{m}, M) \longrightarrow \cdots$$

Since f is in the annihilator of each $\text{Ext}^i(R/\mathfrak{m}, M)$, this long exact sequence splits into the short exact sequences

$$0 \longrightarrow \text{Ext}^{i-1}(R/\mathfrak{m}, M) \longrightarrow \text{Ext}^{i-1}(R/\mathfrak{m}, M/fM) \longrightarrow \text{Ext}^i(R/\mathfrak{m}, M) \longrightarrow 0.$$

Let $t = \text{depth } M$. Then $\text{Ext}^i(R/\mathfrak{m}, M) = 0$ for $i < t$, and the short exact sequences imply that $\text{Ext}^i(R/\mathfrak{m}, M/fM) = 0$ for $i < t - 1$, while for $i = t$ they yield the isomorphism

$$\text{Ext}^{t-1}(R/\mathfrak{m}, M/fM) \cong \text{Ext}^t(R/\mathfrak{m}, M). \tag{A.2}$$

This shows that depth $M/fM = t - 1$, as desired. $\qquad\square$

The depth of a module can also be characterized by Koszul homology.

Proposition A.4.2. *Let* $\mathbf{x} = x_1, \ldots, x_n$ *be a minimal system of generators of* $\mathfrak{m}$. *Then*

$$\text{depth } M = n - \max\{i : H_i(\mathbf{x}; M) \neq 0\}.$$

Proof. We proceed by induction on the depth of M. If depth $M = 0$, then $\text{Hom}_R(R/\mathfrak{m}, M) \neq 0$. Hence there exists $x \in M$ such that $\mathfrak{m}x = 0$, and consequently $H_n(\mathbf{x}; M) \neq 0$.

Now let depth $M = t > 0$, and let $f \in \mathfrak{m}$ be a regular element on M. Since $fH_i(\mathbf{x}; M) = 0$ for all i, the long exact sequence of Koszul homology attached with the short exact sequence $0 \to M \to M \to M/fM \to 0$ splits into the short exact sequences

$$0 \longrightarrow H_i(\mathbf{x}; M) \longrightarrow H_i(\mathbf{x}; M/fM) \longrightarrow H_{i-1}(\mathbf{x}; M) \longrightarrow 0.$$

Since by Proposition A.4.1 we have $\operatorname{depth} M/fM = t - 1$, our induction hypothesis implies that $H_i(\mathbf{x}; M/fM) = 0$ for $i > n - t + 1$ and that $H_i(\mathbf{x}; M/fM) \neq 0$ for $i = n - t + 1$. Thus the short exact sequences of Koszul homology imply first that $H_i(\mathbf{x}; M) = 0$ for $i > n - t$, and then by choosing $i = n - t + 1$ that $H_{n-t}(\mathbf{x}; M) \cong H_{n-t+1}(\mathbf{x}; M/fM) \neq 0$. $\square$

Combining Proposition A.4.2 with Corollary A.3.5 we obtain

Corollary A.4.3 (Auslander–Buchsbaum). *Let M be a finitely generated graded $S = K[x_1, \ldots, x_n]$-module. Then*

$$\operatorname{proj\,dim} M + \operatorname{depth} M = n.$$

This is a special version of the Auslander–Buchsbaum theorem which is used several times in this book. More generally the Auslander–Buchsbaum theorem says that $\operatorname{proj\,dim} M + \operatorname{depth} M = \operatorname{depth} R$, if $\operatorname{proj\,dim} M < \infty$.

A.5 Cohen–Macaulay modules

Let M be an R-module. Since every M-sequence which is contained in $\mathfrak{m}$ is part of a system of parameters of M, it follows that $\operatorname{depth} M \leq \dim M$. The module M is said to be **Cohen–Macaulay** if $\operatorname{depth} M = \dim M$. The ring R is called a **Cohen–Macaulay ring** if R is a Cohen–Macaulay module viewed as a module over itself.

One important property of Cohen–Macaulay rings is that they are **unmixed**. In other words, $\dim R = \dim R/P$ for all $P \in \operatorname{Ass}(R)$. More generally, we have

$$\dim M = \dim R/P \quad \text{for all} \quad P \in \operatorname{Ass}(M), \tag{A.3}$$

if M is Cohen–Macaulay. This follows from the fact that $\operatorname{depth} M \leq \dim R/P$ for all $P \in \operatorname{Ass}(M)$. In particular, we see that a Cohen–Macaulay module has no embedded prime ideals, that is, all associated prime ideals of the module are minimal in its support.

An unmixed ring, however, need not be Cohen–Macaulay. For example, the ring

$$R = K[x_1, x_2, x_3, x_4]/(x_1, x_2) \cap (x_3, x_4)$$

is unmixed but not Cohen–Macaulay, since $\operatorname{depth} R = 1$, while $\dim R = 2$.

The Cohen–Macaulay property is preserved under two important module operations.

Proposition A.5.1. *Let M be a Cohen–Macaulay module, $\mathbf{f}$ an M-sequence with $(\mathbf{f}) \subset \mathfrak{m}$, and P a prime ideal in the support of M. Then $M/(\mathbf{f})M$ and M_P are again Cohen–Macaulay.*

Proof. Let $\mathbf{f} = f_1, \ldots, f_m$. Since $\mathbf{f}$ is part of a system of parameters of M, it follows that $\dim M/(\mathbf{f})M = \dim M - m$. Hence Proposition A.4.1 implies that $M/(\mathbf{f})M$ is Cohen–Macaulay.

In order to prove that M_P is Cohen–Macaulay, we use induction on depth M_P. If depth $M_P = 0$, then $P \in \operatorname{Ass}(M)$, and hence, according to (A.3), P is a minimal prime ideal of M. Thus $\dim M_P = 0$, so M_P is Cohen–Macaulay. If depth $M_P > 0$, then (A.3) implies that P is not contained in any associated prime ideal of M. Thus there exists $f \in P$ which is regular on M, and one may apply the induction hypothesis to $(M/fM)_P = M_P/fM_P$ to see that $\dim M_P - 1 = \dim M_P/fM_P = \operatorname{depth} M_P/fM_P = \operatorname{depth} M_P - 1$, from which the desired conclusion follows. $\square$

A.6 Gorenstein rings

Let M be an R-module. The **socle** of M, denoted $\operatorname{Soc}(M)$, is the submodule of M consisting of all elements $x \in M$ with $\mathfrak{m}x = 0$. Observe that $\operatorname{Soc}(M)$ has a natural structure as an $R/\mathfrak{m}$-module, and hence is a finite-dimensional K-vector space.

Proposition A.6.1. *Let M be a Cohen–Macaulay R-module of dimension d, and $\mathbf{f} = f_1, \ldots, f_d$ an M-sequence. Then*

$$\operatorname{Ext}_R^d(R/\mathfrak{m}, M) \cong \operatorname{Hom}_R(R/\mathfrak{m}, M/(\mathbf{f})M) \cong \operatorname{Soc}(M/(\mathbf{f})M).$$

In particular, $\operatorname{Ext}_R^d(R/\mathfrak{m}, M)$ is a finite-dimensional K-vector space.

Proof. We proceed by induction on d. If $d = 0$, we need only to observe that $\operatorname{Hom}_R(R/\mathfrak{m}, M) \cong \operatorname{Soc}(M)$. Now assume that $d > 0$. Then the isomorphism (A.2) yields $r_R(M) = r_R(M/f_1M)$. Applying our induction hypothesis to M/f_1M, the desired result follows. $\square$

Let M be a d-dimensional Cohen–Macaulay R-module. We set

$$r_R(M) = \dim_K \operatorname{Ext}_R^d(R/\mathfrak{m}, M).$$

The number $r_R(M)$ is called the **Cohen–Macaulay type** of M. A Cohen–Macaulay ring R is called a **Gorenstein ring** if the Cohen–Macaulay type of R is one.

H. Bass [Bas62] introduced Gorenstein rings as rings which have finite injective dimension, and showed that these are exactly the Cohen–Macaulay rings whose Cohen–Macaulay type is one.

In the proof of Proposition A.6.1 we have seen that if f is regular on M, then $r_R(M) = r_R(M/fM)$. Therefore induction on the length of an M-sequence yields

Proposition A.6.2. *Let M be a Cohen–Macaulay R-module and $\mathbf{f}$ an M-sequence. Then $r_R(M) = r_R(M/(\mathbf{f})M)$. In particular, if R is a Cohen–Macaulay ring and $\mathbf{f}$ is an R-sequence, then R is Gorenstein if and only if $R/(\mathbf{f})$ is Gorenstein.*

Corollary A.6.3. *Let $R = S/I$ where $I \subset S$ is a graded ideal, and let M be a graded Cohen–Macaulay R-module. Then $r_R(M) = r_S(M)$.*

Proof. Let $d = \dim M$ and $\mathbf{f}$ an M-sequence of length $n - d$. Then Proposition A.6.2 implies that

$$r_R(M) = \dim_K \operatorname{Hom}_R(R/\mathfrak{m}, M/(\mathbf{f})M)$$
$$= \dim_K \operatorname{Hom}_S(S/(x_1, \ldots, x_n), M/(\mathbf{f})M) = r_S(M).$$

$\square$

The sequence $\mathbf{x} = x_1, \ldots, x_n$ is an S-sequence and $S/(\mathbf{x}) \cong K$. Thus it follows from Proposition A.6.2 that S is a Gorenstein ring. A standard graded K-algebra R of the form $R = S/(\mathbf{f})$ with $\mathbf{f}$ is a homogeneous S-sequence, is called a **complete intersection**. As an immediate consequence of Proposition A.6.2 and Corollary A.6.3 we obtain

Corollary A.6.4. *Let R be a complete intersection. Then R is a Gorenstein ring.*

Not every Gorenstein ring needs to be a complete intersection. Indeed, the class of Gorenstein rings is much larger than that of complete intersections. A simple example of a Gorenstein ring which is not a complete intersection is the following: let $R = S/I$ where $S = K[x_1, x_2, x_3]$ and $I = (x_1^2 - x_2^2, x_1^2 - x_3^2, x_1 x_2, x_1 x_3, x_2 x_3)$. The ideal I is not generated by an S-sequence, because any S-sequence has length at most 3, while I is minimally generated by 5 elements. So R is not a complete intersection.

Next observe that $x_1^3 = x_1(x_1^2 - x_2^2) - x_2(x_1 x_2)$, and hence $x_1^3 \in I$. Similarly, we see that $x_2^3, x_3^3 \in I$. Obviously all other monomials of degree 3 belong to I, so that $(x_1, x_2, x_3)^3 \in I$. Since the generators of I generate a 5-dimensional K-subspace of S_2, we see that $H_R(t) = 1 + 3t + t^2$. The element $x_1^2 + I$ generates the 1-dimensional K-vector space R_2, and obviously belongs to the socle of R. In order to see that R is Gorenstein it suffices therefore to show that no nonzero element $f \in R_1$ belongs to the socle of R. In fact, let $ax_1 + bx_2 + cx_3$ be a nonzero linear form in S with $a, b, c, \in K$. We may assume that $a \neq 0$. Then $x_1(ax_1 + bx_2 + cx_3) \notin I$, because $x_1^2 \notin I$ and $x_1 x_2, x_1 x_3 \in I$.

In contrast to this example we have the following result.

Proposition A.6.5. *Let $I \subset S = K[x_1, \ldots, x_n]$ be a monomial ideal such that $\dim S/I = 0$. Then S/I is Gorenstein if and only if S/I is a complete intersection. If the equivalent conditions hold, then I is generated by pure powers of the variables.*

Proof. Let $I \subset S$ be graded ideal such that S/I is a zero-dimensional Gorenstein ring. We claim that I is an irreducible ideal. In fact, let $J \subset S$ be a graded ideal which properly contains I. There exists an integer k such that $\mathfrak{m}^k J \subset I$. Let k be the smallest such integer. Then $k > 0$ and $\mathfrak{m}^{k-1} J \not\subset I$. Let $x \in \mathfrak{m}^{k-1} J \setminus I$. Since $\operatorname{Soc}(S/I) \cong S/\mathfrak{m}$, and since $x + I$ is a nonzero element in $\operatorname{Soc}(S/I)$, it follows that $\operatorname{Soc}(S/I) \subset J/I$. Therefore, I can never be the intersection of two ideals properly containing I. In other words, I is irreducible, as asserted.

Now assume that I is a monomial ideal. Then I is irreducible if and only if I is generated by pure powers of the variables; see Corollary 1.3.2. Thus all assertions follow.   □

The Cohen–Macaulay type of a graded S-module has the following interesting interpretation.

Proposition A.6.6. *Let M be a Cohen–Macaulay graded S-module of dimension d. Then $r_S(M) = \beta_{n-d}(M)$. In particular, if $R = S/I$ is a Cohen–Macaulay ring of dimension d, then S/I is Gorenstein if and only if $\beta_{n-d}(R) = 1$.*

Proof. We proceed by induction on $\dim M$. If $\dim M = 0$, then $\operatorname{Soc}(M) \neq 0$. Let $\mathbf{x} = x_1, \ldots, x_n$ be the sequence of the variables of S, then $H_n(\mathbf{x}; M) \cong \operatorname{Soc}(M)$. Applying Corollary A.3.5 it follows that $r_S(M) = \dim_K H_n(\mathbf{x}; M) = \beta_n(M)$.

Now assume that $\dim M > 0$. After extending K, if necessary, we may assume that K is infinite. Then we find a linear form which is regular on M. After a change of coordinates we may assume that this linear form is x_n. Then $M/x_n M$ is a Cohen–Macaulay $S/x_n S$-module of dimension $d-1$. Applying the induction hypothesis and Proposition A.6.2, we see that $\beta_{n-d}^{S/x_n S}(M/x_n M) = \beta_{(n-1)-(d-1)}^{S/x_n S}(M/x_n M) = r_S(M/x_n M) = r_S(M)$. Thus it remains to be shown that $\beta_{n-d}^{S/x_n S}(M/x_n M) = \beta_{n-d}^S(M)$. Actually one has, $\beta_i^{S/x_n S}(M/x_n M) = \beta_i^S(M)$ for all i, because if $\mathbb{F}$ is a graded minimal free S-resolution of M, then $\mathbb{F}/x_n \mathbb{F}$ is a free $S/x_n S$-resolution of $M/x_n M$. Indeed, $H_i(\mathbb{F}/x_n \mathbb{F})$ is isomorphic to $\operatorname{Tor}_i(S/x_n S, M)$ for all i, and $\operatorname{Tor}_i(S/x_n S, M) = 0$ for $i > 0$, since x_n is regular on M.   □

Let $I \subset S$ be a graded ideal such that $R = S/I$ is a d-dimensional Cohen–Macaulay ring. Then the graded R-module

$$\omega_R = \operatorname{Ext}_S^{n-d}(R, S)$$

is called the **canonical module** of R. It can be shown that ω_R is a Cohen–Macaulay module of Cohen–Macaulay type 1; see [BH98, Chapter 3].

We denote by $\mu(N)$ the minimal number of homogeneous generators of a graded S-module. As a consequence of Proposition A.6.6 we obtain

Corollary A.6.7. *Let R be a standard graded Cohen–Macaulay ring. Then $\mu(\omega_R) = r_S(R)$. In particular, R is Gorenstein if and only if ω_R is a cyclic R-module.*

Proof. Let $\mathbb{F}$ be the minimal graded free S-resolution of R. Then

$$\omega_R = \mathrm{Coker}(F^*_{n-d} \to F^*_{n-d}).$$

Here N^* denotes the S-dual for a graded S-module N. Since F^*_{n-d} is a free S-module of the same rank as F_{n-d}, and since the image of the map $F^*_{n-d} \to F^*_{n-d}$ is contained in $\mathfrak{m}F^*_{n-d}$, it follows from Proposition A.6.6 that $\mu(\omega_R) = \mu(F^*_{n-d}) = \mu(F_{n-d}) = r_S(R)$, as desired. $\square$

The canonical module ω_R is a faithful R-module; see [BH98, Chapter 3]. Thus Corollary A.6.7 implies that R is Gorenstein if and only if $\omega_R \cong R(a)$ for some integer a.

The results stated in Proposition A.6.6 and in Corollary A.6.7 for the ring $R = S/I$ are equally valid if we replace S by a regular local ring and define ω_R in the same was as above.

A.7 Local cohomology

We maintain our assumptions on R and M from Section A.4. We set

$$\Gamma_{\mathfrak{m}}(M) = \{x \in M : \mathfrak{m}^k x = 0 \quad \text{for some} \quad k\}.$$

$\Gamma_{\mathfrak{m}}(M)$ is largest submodule of M with support $\{\mathfrak{m}\}$. It is easily checked that $\Gamma_{\mathfrak{m}}(-)$ is a left exact additive functor. The right derived functors $H^i_{\mathfrak{m}}(-)$ of $\Gamma_{\mathfrak{m}}(-)$ are called the **local cohomology functors**. Thus if $\mathbb{I}$ is an injective resolution of M it follows that

$$H^i_{\mathfrak{m}}(M) \cong H^i(\varinjlim \mathrm{Hom}_R(R/\mathfrak{m}^k, \mathbb{I})) \cong \varinjlim H^i(\mathrm{Hom}_r(R/\mathfrak{m}^k, \mathbb{I}))$$

$$\cong \varinjlim \mathrm{Ext}^i_R(R/\mathfrak{m}^k, M).$$

We quote the following fundamental vanishing theorem of Grothendieck:

Theorem A.7.1 (Grothendieck). *Let $t = \mathrm{depth}\, M$ and $d = \dim M$. Then $H^i_{\mathfrak{m}}(M) \neq 0$ for $i = t$ and $i = d$, and $H^i_{\mathfrak{m}}(M) = 0$ for $i < t$ and $i > d$.*

Corollary A.7.2. *M is Cohen–Macaulay if and only if $H^i_{\mathfrak{m}}(M) = 0$ for $i < \dim M$.*

In the graded case all local cohomology modules $H^i_{\mathfrak{m}}(M)$ are naturally graded R-modules and one calls the number

$$\mathrm{reg}(M) = \max\{j : H^i_{\mathfrak{m}}(M)_{j-i} \neq 0 \text{ for some } i\}$$

the **regularity** of M.

It has been shown by Eisenbud and Goto ([EG84] or [BH98, Theorem 4.3.1]) that in the case that M is a (finitely generated) graded S-module one has

$$\operatorname{reg}(M) = \max\{j \colon \operatorname{Tor}_i(K, M)_{i+j} \neq 0 \text{ for some } i\}.$$

Local cohomology can be computed by means of the **modified Čech complex**. We fix a system of elements $x_1, \ldots, x_n$ in $\mathfrak{m}$ which generates an $\mathfrak{m}$-primary ideal, and define the complex

$$\mathbb{C} \colon 0 \to C^0 \to C^1 \to \cdots \to C^n \to 0$$

with $C^k = \bigoplus_{1 \leq i_1 < i_2 < \cdots < i_k \leq n} R_{x_{i_1} x_{i_2} \cdots x_{i_k}}$. The differentiation $d^k \colon C^k \to C^{k+1}$ is defined on the component $R_{x_{i_1} x_{i_2} \cdots x_{i_k}} \to R_{x_{j_1} x_{j_2} \cdots x_{j_{k+1}}}$ to be $(-1)^{r-1}\alpha$. Here α is natural homomorphism $R_{x_{i_1} x_{i_2} \cdots x_{i_k}} \to (R_{x_{i_1} x_{i_2} \cdots x_{i_k}})_{x_{j_r}}$, if $\{i_1, i_2, \ldots, i_k\} \subset \{j_1, \ldots, \widehat{j_r}, \ldots, j_{k+1}\}$, and is the zero map otherwise.

For all i one has

$$H_{\mathfrak{m}}^i(M) = H^i(\mathbb{C} \otimes_R M). \tag{A.4}$$

We use (A.4) to compute the local cohomology of a Stanley–Reisner ring. Let Δ be a simplicial on the vertex set $[n]$ and let $R = K[x_1, \ldots, x_n]/I_\Delta$. In other words, $R = K[\Delta]$ is the Stanley–Reisner ring of Δ. We let $\mathbb{C}$ be the modified Čech complex of R with respect to the sequence $x_1, \ldots, x_n$. We note that $\mathbb{C}$ is a $\mathbb{Z}^n$-graded complex. The components of $\mathbb{C}$ are of the form R_x where x is homogeneous with respect to the $\mathbb{Z}^n$-grading. Let $\mathbf{a} \in \mathbb{Z}^n$; then we set

$$(R_x)_{\mathbf{a}} = \{\frac{r}{x^m} \colon r \text{ is homogeneous and } \deg r - m \deg x = \mathbf{a}\},$$

and extend this $\mathbb{Z}^n$-grading naturally to $\mathbb{C}$. This $\mathbb{Z}^n$-grading is compatible with the differentials of $\mathbb{C}$ and hence all local cohomology modules $H_{\mathfrak{m}}^i(R)$ are naturally $\mathbb{Z}^n$-graded.

Theorem A.7.3 (Hochster). *Let $\mathbb{Z}_-^n = \{\mathbf{a} \in \mathbb{Z}^n \colon a_i \leq 0 \text{ for } i = 1, \ldots, n\}$. Then*

$$H_{\mathfrak{m}}^i(K[\Delta])_{\mathbf{a}} = \begin{cases} \dim_K \tilde{H}_{i-|F|-1}(\operatorname{link}_\Delta F; K), & \text{if } \mathbf{a} \in \mathbb{Z}_-^n, \text{ where } F = \operatorname{supp}\mathbf{a}; \\ 0, & \text{if } \mathbf{a} \notin \mathbb{Z}_-^n. \end{cases}$$

Proof. Let F be a subset of the vertex set of Δ. The **star** of F is the set $\operatorname{star}_\Delta F = \{G \in \Delta \colon F \cup G \in \Delta\}$. Notice that $\operatorname{star}_\Delta F$ is a subcomplex of Δ. Let $\mathbf{a} \in \mathbb{Z}^n$; the $\mathbf{a}$-graded component $\mathbb{C}_{\mathbf{a}}$ of the modified Čech complex $\mathbb{C}$ of $K[\Delta]$ is a complex of finite-dimensional K-vector spaces, and there exists an isomorphism of complexes

$$\alpha \colon \mathbb{C}_{\mathbf{a}} \longrightarrow \operatorname{Hom}_{\mathbb{Z}}(\tilde{C}(\operatorname{link}_{\operatorname{star} H_{\mathbf{a}}} G_{\mathbf{a}}; K)[-j-1], K), \quad j = |G_{\mathbf{a}}|.$$

Here $G_{\mathbf{a}} = \{i \in [n] : a_i < 0\}$ and $H_{\mathbf{a}} = \{i \in [n] : a_i > 0\}$, and

$$\tilde{\mathcal{C}}(\text{link}_{\text{star}_{\mathbf{a}}} G_{\mathbf{a}}; K)[-j-1]$$

denotes the augmented oriented chain complex of $\text{link}_{\text{star }H_{\mathbf{a}}} G_{\mathbf{a}}$, homologically shifted by $-j-1$. Note that

$$\text{Hom}_{\mathbb{Z}}(\tilde{\mathcal{C}}(\text{link}_{\text{star }H_{\mathbf{a}}} G_{\mathbf{a}}; K)[-j-1], K) = (K\{\text{link}_{\text{star }H_{\mathbf{a}}} G_{\mathbf{a}}\}, e)[-j],$$

see Section 5.1.4.

The map α is defined as follows: let $x = x_{i_1} \cdots x_{i_k}$ with $i_1 < i_2 < \cdots < i_k$ and set $F = \{i_1, \ldots, i_k\}$. We first observe that

$$(R_x)_{\mathbf{a}} \cong \begin{cases} K, & \text{if } G_a \subset F \text{ and } F \cup H_a \in \Delta, \\ 0, & \text{otherwise.} \end{cases}$$

It follows that $(\mathbb{C}^i)_{\mathbf{a}}$ has a K-basis consisting of basis elements b_F indexed by $F \subset [n]$ with $|F| = i$, and such that $G_a \subset F$ and $F \cup H_a \in \Delta$. Now we let α^i be the K-linear map defined by

$$\alpha^i \colon (\mathbb{C}^i)_{\mathbf{a}} \to K\{\text{link}_{\text{star }H_{\mathbf{a}}} G_{\mathbf{a}}\}_{i-j}, \quad b_F \mapsto e_{F \setminus G_{\mathbf{a}}}.$$

Passing to homology, the map of complexes α yields the following isomorphism

$$H_{\mathfrak{m}}^i(K[\Delta])_{\mathbf{a}} \cong \tilde{H}^{i-|G_{\mathbf{a}}|-1}(\text{link}_{\text{star }H_{\mathbf{a}}} G_{\mathbf{a}}; K),$$

so that $\dim_K H_{\mathfrak{m}}^i(K[\Delta])_{\mathbf{a}} = \dim_K \tilde{H}_{i-|G_{\mathbf{a}}|-1}(\text{link}_{\text{star }H_{\mathbf{a}}} G_{\mathbf{a}}; K)$.

If $H_{\mathbf{a}} \neq \emptyset$, then $\text{link}_{\text{star }H_{\mathbf{a}}} G_{\mathbf{a}}$ is acyclic, and if $H_{\mathbf{a}} = \emptyset$, then $\text{star } H_{\mathbf{a}} = \Delta$, so that in this case $\text{link}_{\text{star }H_{\mathbf{a}}} G_{\mathbf{a}} = \text{link}_{\Delta} G_{\mathbf{a}}$. Thus the theorem follows from the fact that $H_{\mathbf{a}} = \emptyset$ if and only if $\mathbf{a} \in \mathbb{Z}_-^n$. $\qquad \square$

A.8 The Cartan complex

We give a short introduction to the Cartan complex which for the exterior algebra plays the role of the Koszul complex for the symmetric algebra.

Let K be a field, V a K-vector space with basis $e_1, \ldots, e_n$ and E the exterior algebra of V.

Let $\mathbf{v} = v_1, \ldots, v_m$ be a sequence of elements of degree 1 in E. The **Cartan complex** $C_{\boldsymbol{\cdot}}(\mathbf{v}; E)$ of the sequence $\mathbf{v}$ with values in E is defined as the complex whose i-chains $C_i(\mathbf{v}; E)$ are the elements of degree i of the free divided power algebra $E\langle x_1, \ldots, x_m \rangle$. Recall that $E\langle x_1, \ldots, x_m \rangle$ is the polynomial ring over E in the set of variables

$$x_i^{(j)}, \quad i = 1, \ldots, m, \quad j = 1, 2, \ldots$$

modulo the relations

$$x_i^{(j)} x_i^{(k)} = \binom{j+k}{j} x_i^{(j+k)}.$$

We set $x_i^{(0)} = 1$, $x_i^{(1)} = x_i$ for $i = 1, \ldots, m$ and $x_i^{(a)} = 0$ for $a < 0$.

The algebra $E\langle x_1, \ldots, x_m \rangle$ is a free E-module with basis

$$\mathbf{x}^{(\mathbf{a})} = x_1^{(a_1)} x_2^{(a_2)} \cdots x_m^{(a_m)}, \quad \mathbf{a} = (a_1, \ldots, a_m) \in \mathbb{Z}_+^m.$$

We say that $\mathbf{x}^{(\mathbf{a})}$ has degree i if $|\mathbf{a}| = i$ where $|\mathbf{a}| = a_1 + \cdots + a_m$. Thus $C_i(\mathbf{v}; E) = \bigoplus_{|\mathbf{a}|=i} E \mathbf{x}^{(\mathbf{a})}$.

The E-linear differential ∂ on $C_\bullet(\mathbf{v}; E)$ is defined a follows: for $\mathbf{x}^{(\mathbf{a})} = x_1^{(a_1)} \cdots x_m^{(a_m)}$ we set

$$\partial(\mathbf{x}^{(\mathbf{a})}) = \sum_{i=1}^{m} v_i x_1^{(a_1)} \cdots x_i^{(a_i - 1)} \cdots x_m^{(a_m)}.$$

One readily checks that $\partial \circ \partial = 0$, so that $(C_\bullet(\mathbf{v}; E), \partial)$ is indeed a complex. Moreover,

$$\partial(g_1 g_2) = g_1 \partial(g_2) + \partial(g_1) g_2 \tag{A.5}$$

for any two homogeneous elements g_1 and g_2 in $C_\bullet(\mathbf{v}; E)$.

Let $\mathcal{G}$ be the category of graded E-modules (in the sense of Definition 5.1.1), and let $M \in \mathcal{G}$. We define the complex

$$C_\bullet(\mathbf{v}; M) = M \otimes_E C_\bullet(\mathbf{v}; E),$$

and set $H_i(\mathbf{v}; M) = H_i(C_\bullet(\mathbf{v}; M))$. We call $H_i(\mathbf{v}; M)$ the ith **Cartan homology module** of $\mathbf{v}$ with respect to M. Note that each $H_i(\mathbf{v}; M)$ is a naturally graded E-module.

Proposition A.8.1. *Let $J \subset E$ be the ideal generated by the sequence $\mathbf{v} = v_1, \ldots, v_m$. Then $J H_\bullet(\mathbf{v}; M) = 0$.*

One good reason to consider the Cartan complex is the following result:

Theorem A.8.2. *For any graded E-module M and each $i \geq 0$ there is a natural isomorphism*

$$\mathrm{Tor}_i^E(M, K) \cong H_i(e_1, \ldots, e_n; M)$$

of graded E-modules.

For the proof of the theorem it suffices to show that $C_\bullet(e_1, \ldots, e_n; E)$ is acyclic with $H_0(e_1, \ldots, e_n; E) = K$. This will easily be implied by the next results.

Proposition A.8.3. *Let M be a graded E-module, $\mathbf{v} = v_1, \ldots, v_m$ a sequence of elements in E_1 and $\mathbf{v}'$ the sequence $v_1, \ldots, v_{m-1}$. Then there exists an exact sequence*

$$0 \longrightarrow C.(\mathbf{v}'; M) \overset{\iota}{\longrightarrow} C.(\mathbf{v}; M) \overset{\tau}{\longrightarrow} C.(\mathbf{v}; M)[-1] \longrightarrow 0$$

of complexes. Here ι is the natural inclusion map, while τ is defined by the formula

$$\tau(c_0 + c_1 x_m + \cdots + c_k x_m^{(k)}) = c_1 + c_2 x_m + \cdots + c_k x_m^{(k-1)},$$

where the c_i belong to $C_{k-i}(\mathbf{v}'; M)$.

The proof is straightforward and is left to the reader.

Corollary A.8.4. *There exists a long exact homology sequence*

$$\cdots \longrightarrow H_i(\mathbf{v}'; M) \overset{\alpha_i}{\longrightarrow} H_i(\mathbf{v}; M) \overset{\beta_i}{\longrightarrow} H_{i-1}(\mathbf{v}; M)(-1)$$
$$\overset{\delta_{i-1}}{\longrightarrow} H_{i-1}(\mathbf{v}'; M) \longrightarrow H_{i-1}(\mathbf{v}; M) \longrightarrow \cdots$$

of graded E-modules, where α_i is induced by the inclusion map ι, β_i by τ, and δ_{i-1} is the connecting homomorphism. If $z = c_0 + c_1 x_m + \cdots + c_{i-1} x_m^{(i-1)}$ is a cycle in $C_{i-1}(\mathbf{v}; M)$, then $\delta_{i-1}([z]) = [c_0 v_m]$.

We are now in a position to complete the proof of Theorem A.8.2 by showing that $C.(e_1, \ldots, e_n; E)$ is indeed acyclic: we show by induction on j that $H_i(e_1, \ldots, e_j; E) = 0$ for all $i > 0$. The assertion is clear for $j = 1$, since $C.(e_1; E)$ is the complex

$$\cdots \longrightarrow E x_1^{(2)} \overset{e_1}{\longrightarrow} E x_1 \overset{e_1}{\longrightarrow} E \longrightarrow 0,$$

and since the annihilator of e_1 in E is the ideal (e_1).

We now assume that the assertion is already proved for j, let $\mathbf{v} = e_1, \ldots, e_{j+1}$ and $\mathbf{v}' = e_1, \ldots, e_j$, and consider the long exact sequence

$$\cdots \longrightarrow H_i(\mathbf{v}'; E) \longrightarrow H_i(\mathbf{v}; E) \longrightarrow H_{i-1}(\mathbf{v}; E)(-1)$$
$$\longrightarrow H_{i-1}(\mathbf{v}'; E) \longrightarrow \cdots$$

We show by induction on i that $H_i(\mathbf{v}; E) = 0$ for $i > 0$. By our induction hypothesis (induction on j) we have $H_1(\mathbf{v}'; E) = 0$. Therefore we obtain the short exact sequence

$$0 \longrightarrow H_1(\mathbf{v}; E) \longrightarrow E/(\mathbf{v})(-1) \overset{\delta_0}{\longrightarrow} E/(\mathbf{v}')$$

Here δ maps the residue class of 1 in $E/(\mathbf{v})$ to the residue class of e_{j+1} in $E/(\mathbf{v}')$. Since the annihilator of e_{j+1} in $E/(\mathbf{v}')$ is generated by e_{j+1}, it follows from this sequence that $H_1(\mathbf{v}; E) = 0$.

Suppose now that $i > 1$. Our induction hypothesis (induction on j) and the above exact sequence yields

$$H_i(\mathbf{v}; E) \cong H_{i-1}(\mathbf{v}; E).$$

Applying the induction hypothesis (induction on i) we see that $H_i(\mathbf{v}; E) = 0$, as desired.

Let again $M \in \mathcal{G}$. The **Cartan cohomology** with respect to the sequence $\mathbf{v} = v_1, \ldots, v_m$ is defined to be the homology of the cocomplex $C^\bullet(\mathbf{v}; M) = {}^*\mathrm{Hom}_E(C_\bullet(\mathbf{v}; E), M)$. Explicitly, we have

$$C^\bullet(\mathbf{v}; M) : 0 \xrightarrow{\partial^0} C^0(\mathbf{v}; M) \xrightarrow{\partial^1} C^1(\mathbf{v}; M) \longrightarrow \cdots,$$

where the cochains $C^\bullet(\mathbf{v}; M)$ and the differential ∂ can be described as follows: the elements of $C^i(\mathbf{v}; M)$ may be identified with all homogeneous polynomials $\sum_a m_\mathbf{a} \mathbf{y}^\mathbf{a}$ of degree i in the variables $y_1, \ldots, y_m$ with coefficients $m_\mathbf{a} \in M$, where as usual for $\mathbf{a} \in \mathbb{Z}_+^n$, $\mathbf{y}^\mathbf{a}$ denotes the monomial $y_1^{a_1} y_2^{a_2} \cdots y_n^{a_n}$. The element $m_\mathbf{a} \mathbf{y}^\mathbf{a} \in C^\bullet(\mathbf{v}; M)$ is defined by the mapping property

$$m_\mathbf{a} \mathbf{y}^\mathbf{a}(\mathbf{x}^{(\mathbf{b})}) = \begin{cases} m_\mathbf{a} & \text{for} \quad \mathbf{b} = \mathbf{a}, \\ 0 & \text{for} \quad \mathbf{b} \neq \mathbf{a}. \end{cases}$$

After this identification ∂ is simply multiplication by the element $y_\mathbf{v} = \sum_{i=1}^n v_i y_i$. In other words, we have

$$\partial^i \colon C^i(\mathbf{v}; M) \longrightarrow C^{i+1}(\mathbf{v}; M), \quad f \mapsto y_\mathbf{v} f.$$

In particular we see that $C^\bullet(\mathbf{v}; E)$ may be identified with the polynomial ring $E[y_1, \ldots, y_m]$ over E, and that $C^\bullet(\mathbf{v}; M)$ is a finitely generated $C^\bullet(\mathbf{v}; E)$-module. It is obvious that cocycles and coboundaries of $C^\bullet(\mathbf{v}; M)$ are $E[y_1, \ldots, y_m]$-submodules of $C^\bullet(\mathbf{v}; M)$. As $E[y_1, \ldots, y_m]$ is Noetherian, it follows that the Cartan cohomology $H^\bullet(\mathbf{v}; M)$ of M is a finitely generated $E[y_1, \ldots, y_m]$-module.

Let $J \subset E$ be the ideal generated by $\mathbf{v}$. Then $J H^\bullet(\mathbf{v}; M) = 0$, and hence $H^\bullet(\mathbf{v}; M)$ is in fact an $(E/J)[y_1, \ldots, y_m]$-module. Viewing $(E/J)[y_1, \ldots, y_m]$ a standard graded E/J-algebra, the Cartan cohomology module $H^\bullet(\mathbf{v}; M)$ is a finitely generated graded $(E/J)[y_1, \ldots, y_m]$-module whose ith graded component is $H^i(\mathbf{v}; M)$ for $i \geq 0$, Notice that each $H^i(\mathbf{v}; M)$ itself is a graded E/J-module, so that $H^\bullet(\mathbf{v}; M)$ is a bigraded $(E/J)[y_1, \ldots, y_m]$-module with each y_i of bidegree $(-1, 1)$.

As in Chapter 5 we set $M^\vee = {}^*\mathrm{Hom}_E(M, E)$. Cartan homology and cohomology are related as follows:

Proposition A.8.5. *Let $M \in \mathcal{G}$. Then*

$$H_i(\mathbf{v}; M)^\vee \cong H^i(\mathbf{v}; M^\vee) \quad \text{for all} \quad i.$$

Proof. Since E is injective as shown in Corollary 5.1.4, the functor $(-)^\vee$ commutes with homology and we obtain

$$H_i(\mathbf{v}; M)^\vee \cong H^i({}^*\mathrm{Hom}_E(C_i(\mathbf{v}; M), E)) \cong$$
$$H^i({}^*\mathrm{Hom}_E(C_i(\mathbf{v}; E), M^\vee) \cong H^i(\mathbf{v}; M^\vee).$$

$\square$

By applying the functor ${}^*\mathrm{Hom}_E(-, M)$ to the short exact sequence of complexes in Proposition A.8.3 (with $M = E$) we obtain the short exact sequence of cocomplexes

$$0 \longrightarrow C^\bullet(\mathbf{v}; M)[-1] \longrightarrow C^\bullet(\mathbf{v}; M) \longrightarrow C^\bullet(\mathbf{v}'; M) \longrightarrow 0,$$

from which we deduce

Proposition A.8.6. *Let $M \in \mathcal{G}$. Then with $\mathbf{v}$ and $\mathbf{v}'$ as in A.8.3 there exists a long exact sequence of graded E-modules*

$$\cdots \longrightarrow H^{i-1}(\mathbf{v}; M) \longrightarrow H^{i-1}(\mathbf{v}'; M) \longrightarrow H^{i-1}(\mathbf{v}; M)$$
$$\xrightarrow{y_m} H^i(\mathbf{v}; M)(-1) \longrightarrow H^i(\mathbf{v}'; M) \longrightarrow \cdots$$

Proof. We show only that the map

$$H^{i-1}(\mathbf{v}; M) \to H^i(\mathbf{v}; M)(-1),$$

which is the dual of β_i, is indeed multiplication by y_m. We show this on the level of cochains. In order to simplify notation we set $C_i = C_i(v_1, \ldots, v_m; E)$ for all i, and let

$$\gamma: {}^*\mathrm{Hom}_E(C_{i-1}, M) \to {}^*\mathrm{Hom}_E(C_i, M)$$

be the map induced by $\tau: C_i \to C_{i-1}$, where

$$\tau(x^{(\mathbf{b})}) = \begin{cases} x_1^{(b_1)} \cdots x_m^{(b_m - 1)}, & \text{if } b_m > 0, \\ 0, & \text{otherwise.} \end{cases}$$

Our assertion is that γ is multiplication by y_m.

For all $x^{(\mathbf{b})} \in C_i$ and $ny^{\mathbf{a}} \in {}^*\mathrm{Hom}_E(C_{i-1}, M)$ with $n \in M$ we have $\gamma(ny^{\mathbf{a}})(x^{(\mathbf{b})}) = ny^{\mathbf{a}}(\tau(x^{(\mathbf{b})}))$. This implies that

$$\gamma(ny^{\mathbf{a}})(x^{(\mathbf{b})}) = \begin{cases} n, & \text{if } (b_1, \ldots, b_m) = (a_1, \ldots, a_m + 1), \\ 0, & \text{otherwise.} \end{cases}$$

Hence we see that $\gamma(ny^{\mathbf{a}}) = ny^{\mathbf{a}}y_m$, as desired. $\square$

The next proposition shows that for a generic basis $v_1, \ldots, v_n$ of E_1, the y_i act as generic linear forms on $H^\bullet(v_1, \ldots, v_n; M)$. We fix a basis $\mathbf{e} = e_1, \ldots, e_n$ of E_1. Then we have

Proposition A.8.7. *Let* $\mathbf{v} = v_1, \ldots, v_n$ *be a* K-*basis of* E_1 *with* $v_j = \sum_{i=1}^{n} a_{ij} e_i$ *for* $j = 1, \ldots, n$. *Then there exists an isomorphism of graded* K-*vector spaces*

$$\varphi \colon H^{\boldsymbol{\cdot}}(\mathbf{e}; M) \to H^{\boldsymbol{\cdot}}(\mathbf{v}; M)$$

such that

$$\varphi(fc) = \alpha(f)\varphi(c)$$

for all $f \in K[y_1, \ldots, y_n]$ *and all* $c \in H^{\boldsymbol{\cdot}}(\mathbf{v}; M)$. *Here* $\alpha : K[y_1, \ldots, y_n] \to K[y_1, \ldots, y_n]$ *is the* K-*algebra automorphism with* $\alpha(y_j) = \sum_{i=1}^{n} a_{ji} y_i$ *for* $j = 1, \ldots, n$.

Proof. Let $\beta \colon E[y_1, \ldots, y_n] \to E[y_1, \ldots, y_n]$ be the linear E-algebra automorphism deduced from α by the base ring extension E/K. Then $\beta(y_{\mathbf{e}}) = y_{\mathbf{v}}$, so β induces a complex isomorphism

$$C^{\boldsymbol{\cdot}}(\mathbf{e}; M) \longrightarrow C^{\boldsymbol{\cdot}}(\mathbf{v}; M), \quad g(y_1, \ldots, y_m) \mapsto g(\alpha(y_1), \ldots, \alpha(y_n)),$$

which induces the graded isomorphism $\varphi \colon H^{\boldsymbol{\cdot}}(\mathbf{e}; M) \to H^{\boldsymbol{\cdot}}(\mathbf{v}; M)$ with the desired properties.   □

The proposition shows that if we identify $H^{\boldsymbol{\cdot}}(\mathbf{v}; M)$ with $H^{\boldsymbol{\cdot}}(\mathbf{e}; M)$ via the isomorphism φ, then multiplication by y_i has to be identified with multiplication by $\alpha^{-1}(y_i)$.

B

Geometry

B.1 Convex polytopes

We briefly summarize fundamental facts on convex polytopes. All proofs will be omitted. We refer the reader to Grünbaum [Gru03] for detailed information about convex polytopes.

A nonempty subset X in $\mathbb{R}^n$ is called **convex** if for each $\mathbf{x}$ and for each $\mathbf{y}$ belonging to X the line segment

$$\{t\mathbf{x} + (1-t)\mathbf{y} : t \in \mathbb{R},\ 0 \leq t \leq 1\}$$

joining $\mathbf{x}$ and $\mathbf{y}$ is contained in X. If $X \subset \mathbb{R}^n$ is convex, then for each finite subset $\{\alpha_1, \ldots, \alpha_s\}$ of X its **convex combination** $\sum_{i=1}^s a_i \alpha_i$, where each $a_i \in \mathbb{R}$ with $0 \leq a_i \leq 1$ and where $\sum_{i=1}^s a_i = 1$, belongs to X.

Given a nonempty subset Y in $\mathbb{R}^n$, there exists a smallest convex set X in $\mathbb{R}^n$ with $Y \subset X$. To see why this is true, write $\mathcal{A} = \{X_\lambda\}_{\lambda \in \Lambda}$ for the family of all convex sets X_λ in $\mathbb{R}^n$ with $Y \subset X_\lambda$. Clearly $\mathcal{A}$ is nonempty since $\mathbb{R}^n \in \mathcal{A}$. Since each X_λ is convex with $Y \subset X_\lambda$, the intersection $X = \bigcap_{\lambda \in \Lambda} X_\lambda$ is again a convex set which contains Y. Since $X \in \mathcal{A}$ and since $X \subset X_\lambda$ for each $\lambda \in \Lambda$, it follows that X is a smallest convex set in $\mathbb{R}^n$ with $Y \subset X$, as desired.

The notation $\mathrm{Conv}(Y)$ stands for the smallest convex set which contains Y and is called the **convex hull** of Y.

It follows that the convex hull $\mathrm{Conv}(Y)$ of a subset $Y \subset \mathbb{R}^n$ consists of all convex combinations of finite subsets of Y. In other words,

$$\mathrm{Conv}(Y) = \{\sum_{i=1}^s a_i \alpha_i : \alpha_i \in Y,\ a_i \in \mathbb{R},\ 0 \leq a_i \leq 1,\ \sum_{i=1}^s a_i = 1,\ s \geq 1\}.$$

Definition B.1.1. A **convex polytope** in $\mathbb{R}^n$ is the convex hull of a finite set in $\mathbb{R}^n$.

Recall that a **hyperplane** in $\mathbb{R}^n$ is a subset $\mathcal{H} \subset \mathbb{R}^n$ of the form

$$\mathcal{H} = \{(x_1, \ldots, x_n) \in \mathbb{R}^n : \sum_{i=1}^{n} a_i x_i = b\},$$

where each $a_i \in \mathbb{R}$, $b \in \mathbb{R}$ and $(a_1, \ldots, a_n) \neq (0, \ldots, 0)$. Every hyperplane $\mathcal{H} \subset \mathbb{R}^n$ determines the following two closed half-spaces in $\mathbb{R}^n$:

$$\mathcal{H}^{(+)} = \{(x_1, \ldots, x_n) \in \mathbb{R}^n : \sum_{i=1}^{n} a_i x_i \geq b\};$$

$$\mathcal{H}^{(-)} = \{(x_1, \ldots, x_n) \in \mathbb{R}^n : \sum_{i=1}^{n} a_i x_i \leq b\}.$$

Let $\mathcal{P} \subset \mathbb{R}^n$ be a convex polytope. A hyperplane $\mathcal{H} \subset \mathbb{R}^n$ is called a **supporting hyperplane** of $\mathcal{P}$ if the following conditions are satisfied:

- Either $\mathcal{P} \subset \mathcal{H}^{(+)}$ or $\mathcal{P} \subset \mathcal{H}^{(-)}$;
- $\emptyset \neq \mathcal{P} \cap \mathcal{H} \neq \mathcal{P}$.

Definition B.1.2. A **face** of a convex polytope $\mathcal{P} \subset \mathbb{R}^n$ is a subset of $\mathcal{P}$ of the form $\mathcal{P} \cap \mathcal{H}$, where $\mathcal{H}$ is a supporting hyperplane of $\mathcal{P}$.

Theorem B.1.3. *A convex polytope $\mathcal{P} \subset \mathbb{R}^n$ has only a finite number of faces, and each face of $\mathcal{P}$ is again a convex polytope in $\mathbb{R}^n$.*

Theorem B.1.4. (a) *If $\mathcal{F}$ is a face of a convex polytope $\mathcal{P} \subset \mathbb{R}^n$ and if $\mathcal{F}'$ is a face of $\mathcal{F}$, then $\mathcal{F}'$ is a face of $\mathcal{P}$.*

(b) *If $\mathcal{F}$ and $\mathcal{F}'$ are faces of a convex polytope $\mathcal{P} \subset \mathbb{R}^n$ and if $\mathcal{F} \cap \mathcal{F}' \neq \emptyset$, then $\mathcal{F} \cap \mathcal{F}'$ is a face of $\mathcal{P}$.*

A **vertex** of a convex polytope $\mathcal{P} \subset \mathbb{R}^n$ is a point $\alpha \in \mathcal{P}$ for which the singleton $\{\alpha\}$ is a face of $\mathcal{P}$. Let $V = \{\alpha_1, \ldots, \alpha_s\}$ denote the set of vertices of $\mathcal{P}$. Write $\mathcal{P} - \alpha_i$ for the subset $\{\mathbf{x} - \alpha_i : \mathbf{x} \in \mathcal{P}\} \subset \mathbb{R}^n$. The **dimension** $\dim \mathcal{P}$ of $\mathcal{P}$ is the dimension of the vector subspace in $\mathbb{R}^n$ spanned by $\mathcal{P} - \alpha_i$, which is independent of the particular choice of α_i. The dimension of a face $\mathcal{F}$ of $\mathcal{P}$ is the dimension of $\mathcal{F}$ as a convex polytope in $\mathbb{R}^n$. An **edge** of $\mathcal{P}$ is a face of $\mathcal{P}$ of dimension 1. A **facet** $\mathcal{P}$ is a face of $\mathcal{P}$ of dimension $\dim \mathcal{P} - 1$.

Theorem B.1.5. *Let V denote the set of vertices of a convex polytope $\mathcal{P} \subset \mathbb{R}^n$. Then*

(i) $\mathcal{P} = \mathrm{Conv}(V)$;
(ii) *If $\mathcal{F}$ is a face of $\mathcal{P}$, then $\mathcal{F} = \mathrm{Conv}(V \cap \mathcal{F})$. In particular the vertex set of $\mathcal{F}$ is $V \cap \mathcal{F}$.*

Theorem B.1.6. *Let $\mathcal{F}_1, \ldots, \mathcal{F}_q$ denote the facets of a convex polytope $\mathcal{P} \subset \mathbb{R}^n$ and $\mathcal{H}_j \subset \mathbb{R}^n$ a supporting hyperplane of $\mathcal{P}$ with $\mathcal{F}_j = \mathcal{P} \cap \mathcal{H}_j$ and with $\mathcal{P} \subset \mathcal{H}_j^{(+)}$. Then*

$$\mathcal{P} = \bigcap_{j=1}^{q} \mathcal{H}_j^{(+)}.$$

Conversely,

Theorem B.1.7. *Let $\mathcal{H}_1, \ldots, \mathcal{H}_q$ be hyperplanes in $\mathbb{R}^n$ and suppose that $\mathcal{P} = \bigcap_{j=1}^q \mathcal{H}_j^{(+)}$ is a nonempty subset in $\mathbb{R}^n$. If $\mathcal{P}$ is bounded, then $\mathcal{P}$ is a convex polytope in $\mathbb{R}^n$. Moreover, if the decomposition $\bigcap_{j=1}^q \mathcal{H}_j^{(+)}$ is irredundant, then $\mathcal{P} \cap \mathcal{H}_1, \ldots, \mathcal{P} \cap \mathcal{H}_q$ are the facets of $\mathcal{P}$.*

B.2 Linear programming

Fix positive integers n and m and let $A = (a_{ij})_{\substack{1 \leq i \leq n \\ 1 \leq j \leq m}}$ be $n \times m$ matrix with each $a_{ij} \in \mathbb{R}$. The notation $A^\top$ stands for the transpose of A. Let $\mathbf{b} \in \mathbb{R}^n$ and $\mathbf{c} \in \mathbb{R}^m$. As in Chapter 11, for vectors $\mathbf{u} - (u_1, \ldots, u_n)$ and $\mathbf{v} - (v_1, \ldots, v_n)$ belonging to $\mathbb{R}^n$, we write $\mathbf{u} \leq \mathbf{v}$ if all component $v_i - u_i$ are nonnegative.

A **linear programming** is the problem stated as follows: Maximize the **objective function**

$$\mathbf{c}\,\mathbf{x}^\top$$

for $\mathbf{x} \in \mathbb{R}^m$ subject to the condition

$$A\mathbf{x}^\top \leq \mathbf{b}^\top, \quad \mathbf{x} \geq 0. \tag{B.1}$$

Its **dual linear programming** is the problem stated as follows: minimize the objective function

$$\mathbf{b}\,\mathbf{y}^\top$$

for $\mathbf{y} \in \mathbb{R}^n$ subject to the condition

$$A^\top \mathbf{y}^\top \geq \mathbf{c}^\top, \quad \mathbf{y} \geq 0. \tag{B.2}$$

A vector $\mathbf{x} \in \mathbb{R}^m$ satisfying (B.1) is called a **feasible solution**. Similarly, a vector $\mathbf{y} \in \mathbb{R}^m$ satisfying (B.2) is called a **feasible dual solution**. A feasible solution which maximizes $\mathbf{c}\mathbf{x}^\top$ is called an **optimal solution**, and a feasible dual solution which minimizes $\mathbf{b}\mathbf{y}^\top$ is called an **optimal dual solution**.

Theorem B.2.1 (Duality Theorem). *If $\mathbf{x}$ is a feasible solution and $\mathbf{y}$ is a feasible dual solution, then*

$$\mathbf{c}\,\mathbf{x}^\top \leq \mathbf{b}\,\mathbf{y}^\top.$$

We now come to the results which characterize vertices of convex polytopes in the language of linear programming.

Theorem B.2.2. *(a) Let $\mathcal{P} \subset \mathbb{R}^n$ be a convex polytope. Then for any $c \in \mathbb{R}^n$ there is a vertex α of $\mathcal{P}$ which maximizes $\mathbf{c}\,\mathbf{x}^\top$, where $\mathbf{x}$ runs over $\mathcal{P}$.*

(b) Let α be a vertex of a convex polytope $\mathcal{P} \subset \mathbb{R}^n$. Then there exists a vector $\mathbf{c} \in \mathbb{R}^n$ such that α is a unique member of $\mathcal{P}$ maximizing $\mathbf{c}\,\mathbf{x}^\top$, where x runs over $\mathcal{P}$.

A standard reference on linear programming and integer programming is Schrijver [Sch98].

B.3 Vertices of polymatroids

We now come to the problem of finding the vertices of a polymatroid. Let $\mathcal{P} \subset \mathbb{R}^n_+$ be a polymatroid on the ground set $[n]$ and ρ its ground set rank function. Recall from Theorem 12.1.3 (a) that

$$\mathcal{P} = \{\mathbf{x} \in \mathbb{R}^n_+ : \mathbf{x}(A) \leq \rho(A), A \subset [n]\}.$$

Thus in particular Theorem B.1.7 guarantees that $\mathcal{P}$ is a convex polytope in $\mathbb{R}^n$.

Given a permutation $\pi = (i_1, \ldots, i_n)$ of $[n]$, we set $A^1_\pi = \{i_1\}$, $A^2_\pi = \{i_1, i_2\}$, ..., $A^n_\pi = \{i_1, \ldots, i_n\}$. Let $\mathbf{v}(k, \pi) = (v_1, \ldots, v_n)$, where $k \in [n]$ and where

$$v_{i_1} = \rho(A^1_\pi),$$
$$v_{i_2} = \rho(A^2_\pi) - \rho(A^1_\pi),$$
$$v_{i_3} = \rho(A^3_\pi) - \rho(A^2_\pi),$$
$$\cdots$$
$$v_{i_k} = \rho(A^k_\pi) - \rho(A^{k-1}_\pi),$$
$$v_{i_{k+1}} = v_{i_{k+2}} = \cdots = v_{i_n} = 0.$$

Lemma B.3.1. *One has* $\mathbf{v}(k, \pi) \in \mathcal{P}$.

Proof. Let $\mathbf{v} = \mathbf{v}(k, \pi)$ and $A \subset [n]$. What we must prove is $\mathbf{v}(A) \leq \rho(A)$. Since ρ is nondecreasing, it may be assumed that $A \subset \{i_1, \ldots, i_k\}$. Let $j \in [n]$ denote the biggest integer for which $i_j \in A$. By using induction on $|A|$, one has $\mathbf{v}(A \setminus \{i_j\}) \leq \rho(A \setminus \{i_j\})$. Since

$$\mathbf{v}(A) = \mathbf{v}(A \setminus \{i_j\}) + v(i_j) \leq \rho(A \setminus \{i_j\}) + \rho(A^j_\pi) - \rho(A^{j-1}_\pi)$$

and since

$$\rho(A \setminus \{i_j\}) + \rho(A^j_\pi) \leq \rho(A) + \rho(A^{j-1}_\pi),$$

one has $\mathbf{v}(A) \leq \rho(A)$, as desired. $\square$

Lemma B.3.2. *Each point* $\mathbf{v}(k, \pi) \in \mathcal{P}$ *is a vertex of* $\mathcal{P}$.

Proof. Let $\mathcal{H}_j$ denote the hyperplane in $\mathbb{R}^n$ consisting of all points $(x_1, \ldots, x_n) \in \mathbb{R}^n$ with

$$x_{i_1} + x_{i_2} + \cdots + x_{i_j} = \rho(A^j_\pi),$$

where $1 \leq j \leq k$. Let $\mathcal{H}'_j$ denote the hyperplane in $\mathbb{R}^n$ consisting of all points $(x_1, \ldots, x_n) \in \mathbb{R}^n$ with

$$x_{i_j} = 0,$$

where $k + 1 \leq j \leq n$. One has $\mathcal{P} \subset \mathcal{H}^{(-)}_j$ and $\mathbf{v}(k, \pi) \in \mathcal{H}_j$ for all j. In other words, each hyperplane $\mathcal{H}_j$ is a supporting hyperplane of $\mathcal{P}$

with $\mathbf{v}(k, \pi) \in \mathcal{H}_j$. In addition, each hyperplane $\mathcal{H}'_j$ is a supporting hyperplane of $\mathcal{P}$ with $\mathbf{v}(k, \pi) \in \mathcal{H}'_j$. Hence Theorem B.1.4 (b) guarantees that $\mathcal{P} \cap (\bigcap_{j=1}^{k} \mathcal{H}_j) \cap (\bigcap_{j=k+1}^{n} \mathcal{H}'_j)$ is a face of $\mathcal{P}$. It is clear that $(\bigcap_{j=1}^{k} \mathcal{H}_j) \cap (\bigcap_{j=k+1}^{n} \mathcal{H}'_j) = \{\mathbf{v}(k, \pi)\}$. Hence $\mathbf{v}(k, \pi) \in \mathcal{P}$ is a vertex of $\mathcal{P}$, as required. □

We are now in the position to complete a proof of Theorem 12.1.4. For a vector $\mathbf{c} = (c_1, \ldots, c_n) \in \mathbb{R}^n$ we define a permutation $\pi = (i_1, \ldots, i_n)$ of $[n]$ such that

$$c_{i_1} \geq c_{i_2} \geq \cdots \geq c_{i_k} > 0 \geq c_{i_{k+1}} \geq \cdots \geq c_{i_n}$$

and consider the linear programming $(L_{\mathbf{c}})$ as follows:

$$\text{Maximize} \quad \mathbf{c}\,\mathbf{x}^\top$$

subject to

$$\mathbf{x} \in \mathcal{P}.$$

Lemma B.3.2 guarantees that $\mathbf{v}(k, \pi)$ is a feasible solution of $(\alpha_{\mathbf{c}})$. We prepare the $2^n - 1$ variables y_A, where $\emptyset \neq A \subset [n]$, and consider the linear programming $(L_{\mathbf{c}}^*)$ as follows:

$$\text{Minimize} \quad \sum_{\emptyset \neq A \subset [n]} \rho(A) y_A$$

subject to

$$\sum_{j \in A} y_A \geq c_j, \quad j = 1, \ldots, n$$

$$y_A > 0, \quad \emptyset \neq A \subset [n].$$

We then introduce the point $\mathbf{y}^* = (y_A^*)_{\emptyset \neq A \subset [n]}$ defined by setting

$$\mathbf{y}_{A_\pi^k}^* = c_{i_k};$$

$$\mathbf{y}_{A_\pi^j}^* = c_{i_j} - c_{i_{j+1}}, \quad j = 1, \ldots, k-1;$$

$$\mathbf{y}_A^* = 0 \quad \text{otherwise.}$$

Lemma B.3.3. *The linear programming $(L_{\mathbf{c}}^*)$ is the dual linear programming of $(L_{\mathbf{c}})$ with $\mathbf{y}^* = (\mathbf{y}_A^*)_{\emptyset \neq A \subset [n]}$ a feasible dual solution. Moreover, one has*

$$\mathbf{c}\,\mathbf{v}(k, \pi)^\top = \sum_{\emptyset \neq A \subset [n]} \rho(A) \mathbf{y}_A^*.$$

Proof. It is clear that $(L_{\mathbf{c}}^*)$ is the dual linear programming of $(L_{\mathbf{c}})$ with $\mathbf{y}^* = (\mathbf{y}_A^*)_{\emptyset \neq A \subset [n]}$ a feasible dual solution. Let $\mathbf{v}(k, \pi) = (v_1, \ldots, v_n)$. Then

$$\sum_{i=1}^{n} c_i v_i = \sum_{j=2}^{k} c_{i_j} (\rho(A_\pi^j) - \rho(_\pi^{j-1})) + c_{i_1} \rho(A_\pi^1)$$

$$= \sum_{j=1}^{k-1} (c_{i_j} - c_{i_{j+1}}) \rho(A_\pi^j) + c_{i_k} \rho(A_\pi^k)$$

$$= \sum_{\emptyset \neq A \subset [n]} \rho(A) \mathbf{y}_A^*,$$

as desired.   □

Theorem B.2.1 now guarantees that $\mathbf{v}(k, \pi)$ is an optimal solution of $(L_{\mathbf{c}})$. Thus in particular every vertex is of the form $\mathbf{v}(k, \pi)$. This fact, together with Lemma B.3.2 completes the proof of Theorem 12.1.4.

Example B.3.4. Let $n = 3$ and $\mathcal{P}$ the polymatroid given by the linear inequalities

$$
\begin{aligned}
x_1 &\quad &\quad &\leq 2; \\
&\quad x_2 &\quad &\leq 3; \\
&\quad &\quad x_3 &\leq 5; \\
x_1 + x_2 &\quad &\quad &\leq 4; \\
&\quad x_2 + x_3 &\leq 6; \\
x_1 &\quad + &\quad x_3 &\leq 6; \\
x_1 + x_2 + x_3 &\leq 7; \\
&\quad &\quad x_i &\geq 0.
\end{aligned}
$$

Let $\mathbf{c} = (7, 3, 1)$. Thus $k = 3$, $\pi = (1, 2, 3)$ and $\mathbf{v}(3, \pi) = (2, 2, 3)$. The dual linear programming $(L_{\mathbf{c}}^*)$ is to minimize the objective function

$$2y_{\{1\}} + 3y_{\{2\}} + 5y_{\{3\}} + 4y_{\{1,2\}} + 6y_{\{2,3\}} + 6y_{\{1,3\}} + 7y_{\{1,2,3\}}$$

subject to

$$
\begin{aligned}
y_{\{1\}} + y_{\{1,2\}} + y_{\{1,3\}} + y_{\{1,2,3\}} &\geq 7; \\
y_{\{2\}} + y_{\{1,2\}} + y_{\{2,3\}} + y_{\{1,2,3\}} &\geq 3; \\
y_{\{3\}} + y_{\{1,3\}} + y_{\{2,3\}} + y_{\{1,2,3\}} &\geq 1; \\
y_A &\geq 0.
\end{aligned}
$$

One has a dual feasible solution

$$\mathbf{y}^* = (\mathbf{y}_{\{1\}}^*, \mathbf{y}_{\{2\}}^*, \mathbf{y}_{\{3\}}^*, \mathbf{y}_{\{1,2\}}^*, \mathbf{y}_{\{2,3\}}^*, \mathbf{y}_{\{1,3\}}^*, \mathbf{y}_{\{1,2,3\}}^*) = (4, 0, 0, 2, 0, 0, 1)$$

with

$$\mathbf{c}\,\mathbf{v}(3, \pi)^{\top} = \sum_{\emptyset \neq A \subset [3]} \rho(A) \mathbf{y}_A^* = 23.$$

B.4 Intersection Theorem

The intersection theorem for polymatroids due to Edmonds [Edm70] has turned out to be one of the most powerful results in combinatorial optimizations. We refer the reader to Schrijver [Sch03] and Fujishige [Fuj05] for background on Edmonds' intersection theorem.

Theorem B.4.1 (Edmonds' Intersection Theorem). *Let $\mathcal{P}_1$ and $\mathcal{P}_2$ be polymatroids on the ground set $[n]$ and ρ_i the ground set rank function of $\mathcal{P}_i$ for $i = 1, 2$. Then*

$$\max\{\mathbf{u}([n]) : \mathbf{u} \in \mathcal{P}_1 \cap \mathcal{P}_2\} = \min\{\rho_1(X) + \rho_2([n] \setminus X) : X \subset [n]\}.$$

Moreover, if $\mathcal{P}_1$ and $\mathcal{P}_2$ are integral, then the maximum on the left-hand side is attained by an integer vector.

B.5 Polymatroidal Sums

Somewhat surprisingly, Theorem 12.1.5 is one of the direct consequences of Edmonds' intersection theorem. Let $\mathcal{P}_1, \ldots, \mathcal{P}_k$ be polymatroids on the ground set $[n]$ and ρ_i the ground set rank function of $\mathcal{P}_i$, $1 \leq i \leq k$. We introduce $\rho : 2^{[n]} \to \mathbb{R}_+$ by setting $\rho = \sum_{i=1}^{k} \rho_i$. It follows immediately that ρ is a nondecreasing and submodular function with $\rho(\emptyset) = 0$. We write $\mathcal{P}$ for the polymatroid on the ground set $[n]$ with ρ its ground set rank function.

Lemma B.5.1. *One has $\mathcal{P}_1 \vee \cdots \vee \mathcal{P}_k \subset \mathcal{P}$.*

Proof. Let $\mathbf{x} \in \mathcal{P}_1 \vee \cdots \vee \mathcal{P}_k$. Then $\mathbf{x} = \mathbf{x}_1 + \cdots + \mathbf{x}_k$ with each $\mathbf{x}_i \in \mathcal{P}_k$. Hence

$$\mathbf{x}(A) = \sum_{i=1}^{k} \mathbf{x}_i(A) \leq \sum_{i=1}^{k} \rho_i(A) = \rho(A).$$

for each $A \subset [n]$. In other words, $\mathbf{x} \in \mathcal{P}$. Thus $\mathcal{P}_1 \vee \cdots \vee \mathcal{P}_k \subset \mathcal{P}$. □

Lemma B.5.2. *One has $\mathcal{P} \subset \mathcal{P}_1 \vee \cdots \vee \mathcal{P}_k$*

Proof. Let $V_i = \{1^{(i)}, \ldots, n^{(i)}\}$ be a "copy" of $[n]$ and let V stand for the disjoint union $V_1 \cup \cdots \cup V_k$. We associate each subset $X \subset V$ with

$$\overline{X} = \{a \in [n] : a^{(i)} \in X \text{ for some } 1 \leq i \leq k\} \subset [n].$$

We introduce $\mu : 2^V \to \mathbb{R}_+$ by setting $\mu(X) = \sum_{i=1}^{k} \rho_i(\overline{X \cap V_i})$, where $X \subset V$. Let $\mathbf{x} \in \mathcal{P}$. We introduce $\xi : 2^V \to \mathbb{R}_+$ by setting $\xi(X) = \mathbf{x}(\overline{X})$, where $X \subset V$. It follows that both μ and ξ are ground set rank functions of polymatroids on the ground set V. Let $\mathcal{Q}_\mu$ (resp. $\mathcal{Q}_\xi$) be the polymatroid on V with μ (resp. ξ) its ground set rank function. Now, Theorem B.4.1 guarantees that

$$\max\{\mathbf{u}(V) : \mathbf{u} \in \mathcal{Q}_\mu \cap \mathcal{Q}_\xi\} = \min\{\mu(X) + \xi(V \setminus X) : X \subset V\}$$

$$= \min\{\sum_{i=1}^{k} \rho_i(\overline{X \cap V_i}) + \mathbf{x}(\overline{V \setminus X}) : X \subset V\}$$

$$= \min\{\sum_{i=1}^{k} \rho_i(X_j) + \mathbf{x}([n] \setminus \bigcap_{j=1}^{k} X_j) : X_j \subset [n]\}.$$

Since each ρ_i is nondecreasing, one has $\rho_i(X_j) \geq \rho(\bigcap_{j=1}^{k} X_j)$. Hence

$$\max\{\mathbf{u}(V) : \mathbf{u} \in \mathcal{Q}_\mu \cap \mathcal{Q}_\xi\} = \min\{\sum_{i=1}^{k} \rho_i(Y) + \mathbf{x}([n] \setminus Y) : Y \subset [n]\}.$$

Since $\mathbf{x} \in \mathcal{P}$, one has $\mathbf{x}(Y) \leq \rho(Y) = \sum_{i=1}^{k} \rho_i(Y)$ for all $Y \subset V$. Thus

$$\mathbf{x}([n]) = \mathbf{x}(Y) + \mathbf{x}([n] \setminus Y) \leq \sum_{i=1}^{k} \rho_i(Y) + \mathbf{x}([n] \setminus Y).$$

Consequently,

$$\min\{\sum_{i=1}^{k} \rho_i(X_j) + \mathbf{x}([n] \setminus \bigcap_{j=1}^{k} X_j) : X_j \subset [n]\} = \sum_{i=1}^{k} \rho_i(\emptyset) + \mathbf{x}([n]) = \mathbf{x}([n]).$$

In other words,

$$\max\{\mathbf{u}(V) : \mathbf{u} \in \mathcal{Q}_\mu \cap \mathcal{Q}_\xi\} = \mathbf{x}([n]).$$

Hence there is $\mathbf{u} \in \mathcal{Q}_\mu \cap \mathcal{Q}_\xi$ with $\mathbf{u}(V) = \mathbf{x}([n])$. Thus, in particular, since $\mathbf{u} \in \mathcal{Q}_\nu$, one has $\sum_{i=1}^{k} \mathbf{u}(a^{(i)}) \leq \mathbf{x}(a)$ for all $a \in [n]$. However, since $\mathbf{u}(V) = \mathbf{x}([n])$, it follows that $\sum_{i=1}^{k} \mathbf{u}(a^{(i)}) = \mathbf{x}(a)$ for all $a \in [n]$. We define $\mathbf{x}_i \in \mathbb{R}^n$, $1 \leq i \leq k$, by setting $\mathbf{x}_i(a) = \mathbf{u}(a^{(i)})$ for all $a \in [n]$. Then $\mathbf{x} = \mathbf{x}_1 + \cdots + \mathbf{x}_k$. Since $\mathbf{u} \in \mathcal{Q}_\mu$, one has $\mathbf{x}_i \in \mathcal{P}_i$, as desired. □

It follows from Lemmata B.5.1 and B.5.2 that the polymatroidal sum $\mathcal{P}_1 \vee \cdots \vee \mathcal{P}_k$ is a polymatroid on the ground set $[n]$ with $\rho = \sum_{i=1}^{k} \rho_i$ its ground set rank function. Moreover, if each ρ_i is integer valued, then $\rho = \sum_{i=1}^{k} \rho_i$ is integer valued. In other words, if each $\mathcal{P}_i$ is integral, then $\mathcal{P}_1 \vee \cdots \vee \mathcal{P}_k$ is integral. Finally, in the proof of Lemma B.5.2, if each $\mathcal{P}_i$ is integral and if $\mathbf{x} \in \mathcal{P}$ is an integer vector, then Theorem B.4.1 guarantees that $\mathbf{u} \in \mathcal{Q}_\mu \cap \mathcal{Q}_\xi$ can be chosen as an integer vector. Thus in particular each $\mathbf{x}_i \in \mathcal{P}_i$ is an integer vector. This completes the proof of Theorem 12.1.5.

B.6 Toric rings

Let $\mathcal{P} \subset \mathbb{R}^n_+$ denote an integral convex polytope of dimension d. If $\mathbf{u} = (u(1), \ldots, u(n)) \in \mathcal{P} \cap \mathbb{Z}^n$, then the notation $\mathbf{x}^{\mathbf{u}}$ stands for the monomial

$x_1^{u(1)} \cdots x_n^{u(n)}$. The **toric ring** $K[\mathcal{P}]$ is the subring of $K[x_1, \ldots, x_n, t]$ which is generated by those monomials $\mathbf{x}^{\mathbf{u}} t$ with $\mathbf{u} \in \mathcal{P} \cap \mathbb{Z}^n$. In general, we say that $\mathcal{P}$ possesses the **integer decomposition property** if, for each $\mathbf{w} \in \mathbb{Z}^n$ which belongs to $q\mathcal{P} = \{ q\mathbf{v} : \mathbf{v} \in \mathcal{P} \}$, there exists $\mathbf{u}_1, \ldots, \mathbf{u}_q$ belonging to $\mathcal{P} \cap \mathbb{Z}^n$ such that $\mathbf{w} = \mathbf{u}_1 + \cdots + \mathbf{u}_q$.

Lemma B.6.1. *If an integral convex polytope $\mathcal{P} \subset \mathbb{R}_+^n$ possesses the integer decomposition property, then its toric ring $K[\mathcal{P}]$ is normal.*

Proof. Since $\mathcal{P}$ possesses the integer decomposition property, it follows that the toric ring $K[\mathcal{P}]$ coincides with the Ehrhart ring [Hib92, pp. 97] of $\mathcal{P}$. Since the Ehrhart ring of an integral convex polytope is normal by Gordan's Lemma ([BH98, Proposition 6.1.2]), the toric ring $K[\mathcal{P}]$ is normal, as desired.   □

One of the most influential results on normal toric rings, due to Hochster [Hoc72], is the following:

Theorem B.6.2 (Hochster). *A normal toric ring is Cohen–Macaulay.*

Stanley [Sta78] and Danilov [Dan78] succeeded in describing the canonical module of a normal toric ring.

Theorem B.6.3 (Stanley, Danilov). *Let $\mathcal{P} \subset \mathbb{R}_+^n$ be an integral convex polytope and suppose that its toric ring $K[\mathcal{P}]$ is normal. Then the canonical module $\Omega(K[\mathcal{P}])$ of $K[\mathcal{P}]$ coincides with the ideal of $K[\mathcal{P}]$ which is generated by those monomials $x^u t^q$ with $u \in q(\mathcal{P} \setminus \partial\mathcal{P}) \cap \mathbb{Z}^n$.*

References

[AL94]   Adams, W.W., Loustaunau, P.: An Introduction to Gröbner bases. American Mathematical Society (1994)

[AHH97]   Aramova, A., Herzog, J., Hibi, T.: Gotzmann theorems for exterior algebras and combinatorics. J. Algebra, **191**, 174–211 (1997)

[AHH98]   Aramova, A., Herzog, J., Hibi, T.: Squarefree lexsegment ideals. Math. Z., **228**, 353–378 (1998)

[AHH00a] Aramova, A., Herzog, J., Hibi, T.: Shifting operations and graded Betti numbers. J. Algebraic Combin., **12**(3), 207–222 (2000)

[AHH00b] Aramova, A., Herzog, J., Hibi, T.: Ideals with stable Betti numbers. Adv. Math., **152**, 72–77 (2000)

[AAH00]   Aramova, A., Avramov, L.L., Herzog, J.: Resolutions of monomial ideals and cohomology over exterior algebras. Trans. Amer. Math. Soc., **352**, 579–594 (2000)

[AH00]   Aramova, A., Herzog, J.: Almost regular sequences and Betti numbers. Amer. J. Math., **122**, 689–719 (2000)

[BNT06]   Babson, E., Novik, I., Thomas, R.: Reverse lexicographic and lexicographic shifting. J. Algebraic Combin., **23**, 107–123 (2006)

[BF85]   Backelin, J., Fröberg, R.: Koszul algebras, Veronese subrings, and rings with linear resolutions, Rev. Roum. Math. Pures Appl., **30**, 549–565 (1985)

[Bas62]   Bass, H.: On the ubiguity of Gorenstein rings. Math. Z., **82**, 8–28 (1962)

[Bay82]   Bayer, D.: The division algorithm and the Hilbert scheme. PhD Thesis, Harvard University, Boston (1982)

[BS87a]   Bayer, D., Stillman, M.: A criterion for detecting m-regularity. Invent. Math., **87**, 1–11 (1987)

[BS87b]   Bayer, D., Stillman, M.: A theorem on refining division orders by the reverse lexicographical order. Duke Math. J., **55**, 321–328 (1987)

[BCP99]   Bayer, D., Charalambous, H., Popescu, S.: Extremal Betti numbers and applications to monomial ideals. J. Algebra, **221**, 497–512 (1999)

[BW93]   Becker, T., Weispfenning, W.: Gröbner Bases. Springer (1993)

[Big93]   Bigatti, A.M.: Upper bounds for the Betti numbers of a given Hilbert function. Comm. in Algebra, **21**, 2317–2334 (1993)

[BCR05]   Bigatti, A.M., Conca, A., Robbiano, L.: Generic initial ideals and distractions. Comm. Algebra, **33**, 1709–1732 (2005)

J. Herzog, T. Hibi, *Monomial Ideals*, Graduate Texts in Mathematics 260,   295
DOI 10.1007/978-0-85729-106-6, © Springer-Verlag London Limited 2011

296   References

[BK88]   Björner, A., Kalai, G.: An extended Euler-Poincaré theorem. Acta Math., **161**, 279–303 (1988)
[BW97]   Björner, A., Wachs, M.L.: Shellable nonpure complexes and posets. II. Trans. Amer. Math. Soc., **349**, 3945–3975 (1997)
[Bro79]   Brodmann, M.: The asymptotic nature of the analytic spread. Math. Proc. Cambridge Philos. Soc., **86**, 35–39 (1979)
[BC03]   Bruns, W., Conca, A.: Gröbner bases and determinantal ideals. In: Herzog, J., Vuletescu, V. (eds) Commutative algebra, singularities and computer algebra (Sinaia, 2002). Kluwer Acad. Publ. (2003)
[BC04]   Bruns, W., Conca, A.: Gröbner bases, initial ideals and initial algebras. In: L.L. Avramov et al. (eds) Homological methods in commutative algebra. IPM Proceedings, Teheran (2004), ArXiv: math.AC/0308102
[BG09]   Bruns, W., Gubeladze, J.: Polytopes, rings, and K-theory. Springer (2009)
[BH92]   Bruns, W., Herzog, J.: On the computation of a-invariants. Manuscripta Math., **77**, 201–213 (1992)
[BH98]   Bruns, W., Herzog, J.: Cohen–Macaulay rings. Revised edition. Cambridge University Press (1998)
[Bur68]   Burch, L.: On ideals of finite homological dimension in local rings. Proc. Camb. Philos. Soc., **64**, 941–948 (1968)
[CL69]   Clements, G.F., Lindström, B.: A generalization of a combinatorial theorem of Macaulay. J. Combin. Theory, **7**, 230–238 (1969)
[Con04]   Conca, A.: Koszul homology and extremal properties of Gin and Lex. Trans. Amer. Math. Soc., **356**, 2945–2961 (2004)
[CH94]   Conca, A., Herzog, J.: On the Hilbert function of determinantal rings and their canonical module. Proc. Amer. Math. Soc., **122**, 677–681 (1994)
[CH03]   Conca, A., Herzog, J.: Castelnuovo–Mumford regularity of products of ideals. Collect. Math., **54** 137–152 (2003)
[CHH04]   Conca, A., Herzog, J., Hibi, T.: Rigid resolutions and big Betti numbers. Comment. Math. Helv., **79**, 826–839 (2004)
[CLO92]   Cox, D., Little, J., O'Shea, D.: Ideals, varieties, and algorithms. Springer (1992)
[CLO98]   Cox, D., Little, J., O'Shea, D.: Using algebraic geometry. Springer (1992)
[CHT99]   Cutkosky, S.D., Herzog, J., Trung, N.V.: Asymptotic behaviour of the Castelnuovo–Mumford regularity. Compositio Math., **118**, 243–261 (1999)
[Dan78]   Danilov, V.I.: The geometry of toric varieties. Russian Math. Surveys, **33**, 97–154 (1978)
[Dir61]   Dirac, G.A.: On rigid circuit graphs. Abh. Math. Sem. Univ. Hamburg, **38**, 71–76 (1961)
[ER98]   Eagon, J.A., Reiner, V.: Resolutions of Stanley–Reisner rings and Alexander duality. J. of Pure and Appl. Algebra **130**, 265–275 (1998)
[Edm70]   Edmonds, J.: Submodular functions, matroids, and certain polyhedra. In: Guy, R., Hanani, H., Sauer, N., Schonheim, J. (eds.) Combinatorial Structures and Their Applications. Gordon and Breach, New York (1970)
[Eis95]   Eisenbud, D.: Commutative algebra with a view towards algebraic geometry. Springer (1995)
[EG84]   Eisenbud, D., Goto, S.: Linear free resolutions and minimal multiplicity. J. Algebra, **88**, 89–133 (1984)
[EK90]   Eliahou, S., Kervaire, M.: Minimal resolutions of some monomial ideals. J. Algebra, **129**, 1–25 (1990)

[EKR61] Erdös, P., Ko, P.C., Rado, R.: Intersection theorems for systems of finite sets. Quart. J. Math. Oxford Ser., **12**, 313–320 (1961)

[EVY06] Escobar, C.A., Villarreal, R.H., Yoshino, Y.: Torsion freeness and normality of blowup rings of monomial ideals. In: M. Fontana et al. (eds) Commutative algebra. Lect. Notes Pure Appl. Math., **244**, Chapman (2006)

[EV97] Estrada, M., Villarreal, R.H.: Cohen–Macaulay bipartite graphs. Arch. Math., **68**, 124–128 (1997)

[Far02] Faridi, S.: The facet ideal of a simplicial complex. Manuscripta Math., **109**, 159–174 (2002)

[Far04] Faridi, S.: Simplicial trees are sequentially Cohen–Macaulay. J. Pure Appl. Algebra, **190**, 121–136 (2004)

[FS10] Fatemeh, M., Somayeh M.: Weakly polymatroidal ideals with applications to vertex cover ideals. To appear in Osaka J. Math.

[FT07] Francisco, C.A., Van Tuyl, A.: Sequentially Cohen–Macaulay edge ideals. Proc. Amer. Math. Soc., **135**, 2327–2337 (2007)

[Fro90] Fröberg, R.: On Stanley–Reisner rings. In: Belcerzyk, L. et al. (eds) Topics in algebra. Polish Scientific Publishers (1990)

[Fuj05] Fujishige, S.: Submodular functions and optimization. Elsevier (2005)

[FHO74] Fulkerson, D.R., Hoffman, A.J., Oppenheim, R.: On balanced matrices. Math. Programming Stud., **1**, 120–132 (1974)

[Gal74] Galligo, A.: Apropos du théorème de préparation de Weierstrass. In: Fonctions de plusiers variables complexes. Lect. Notes Math., **409**. Springer (1974)

[Gar80] Garsia, A.: Combinatorial methods in the theory of Cohen–Macaulay rings. Adv. Math., **38**, 229–266 (1980)

[GVV07] Gitler, I., Valencia, C.E., Villarreal, R.H.: A note on Rees algebras and the MFMC property. Beiträge Algebra Geom., **48** 141–150 (2007)

[Gor00] Gordan, P.: Les invariantes des formes binaires. J. Math. Pures et Appliqués., **6**, 141–156 (1900)

[Gre98] Green, M.: Generic initial ideals. In: J. Elias, J.M. Giral, R.M. Miro-Roig, S. Zarzuela (eds) Proc. CRM -96, Six lectures on commutative algebra, Barcelona, Spain., **166**, Birkhäuser (1998)

[GP08] Greuel, G.M., Pfister, G.: A singular introduction to commutative algebra. Second, extended edition. With contributions by Olaf Bachmann, Christoph Lossen and Hans Schönemann. Springer (2008)

[Gru03] Grünbaum, B.: Convex polytopes, Second Ed. Springer (2003)

[GKS95] Grossman, J.W., Kulkarni, D.W., Schochetman, I.E.: On the minors of the incidence matrix and its Smith normal form. Linear Alg. and its Appl., **218**, 213–224 (1995)

[Har66] Hartshorne, R.: Connectedness of the Hilbert scheme. Publ. IHES, **29**, 5–48 (1966)

[Her01] Herzog, J.: Generic initial ideals and graded Betti numbers. In: Hibi, T. (ed.) Computational commutative algebra and combinatorics. Adv. Studies in Pure Math., **33**, Mathematical Society of Japan, Tokyo (2002)

[HH99] Herzog, J., Hibi, T.: Componentwise linear ideals. Nagoya Math. J., **153**, 141–153 (1999)

[HH02] Herzog, J., Hibi, T.: Discrete polymatroids. J. Algebraic Combin., **16**, 239–268 (2002)

[HH05] Herzog, J., Hibi, T.: Distributive lattices, bipartite graphs and Alexander duality. J. Algebraic Combin., **22**, 289–302 (2005)

298   References

[HH06]   Herzog, J., Hibi, T.: The depth of powers of an ideal. J. Algebra, **291**, 534–550 (2005)

[HHO09]   Herzog, J., Hibi, T., Ohsugi, H.: Unmixed bipartite graphs and sublattices of the Boolean lattices. J. Alg. Comb., **30**, 415–420 (2009)

[HHR00]   Herzog, J., Hibi, T., Restuccia, G.: Strongly Koszul algebras. Math. Scand., **86**, 161–178 (2000)

[HHT07]   Herzog, J., Hibi, T., Trung, N.V.: Symbolic powers of monomial ideals and vertex cover algebras. Adv. Math., **210**, 304–322 (2007)

[HHT09]   Herzog, J., Hibi, T., Trung, N.V.: Vertex cover algebras of unimodular hypergraphs. Proc. Amer. Math. Soc., **137**, 409–414 (2009)

[HHTZ08] Herzog, J., Hibi, T., Trung, N.V., Zheng, X.: Standard graded vertex cover algebras, cycles and leaves. Trans. Amer. Math. Soc., **360** 6231–6249 (2008)

[HHZ04a] Herzog, J., Hibi, T., Zheng, X.: Monomial ideals whose powers have a linear resolution. Math. Scand., **95**, 23–32 (2004)

[HHZ04b] Herzog, J., Hibi, T., Zheng, X.: Dirac's theorem on chordal graphs and Alexander duality. European J. Comb., **25**, 949–960 (2004)

[HHZ06]   Herzog, J., Hibi, T., Zheng, X.: Cohen–Macaulay chordal graphs. J. Combin. Theory Ser. A, **113**, 911–916 (2006)

[HRW99]   Herzog, J., Reiner, V., Welker, V.: Componentwise linear ideals and Golod rings. Michigan Math. J., **46**, 211–223 (1999)

[HSV91]   Herzog, J., Simis, A., Vasconcelos, V.W.: Arithmetic of normal Rees algebras. J. Algebra, **143**, 269–294 (1991)

[HT02]   Herzog, J., Takayama, Y.: Resolutions by mapping cones. Homology Homotopy Appl., **4**, 277–294 (2002)

[HT99]   Herzog, J., Terai, N.: Stable properties of algebraic shifting. Result in Math., **35**, 260–265 (1999)

[HTr92]   Herzog, J., Trung, N.V.: Gröbner bases and multiplicity of determinantal and Pfaffian ideals. Adv. Math., **96**, 1–37 (1992)

[Hib92]   Hibi, T.: Algebraic combinatorics on convex polytopes. Carslaw Publications, Glebe, N.S.W., Australia (1992)

[Hib87]   Hibi, T.: Distributive lattices, affine semigroup rings and algebras with straightening laws. In: Nagata, M., Matsumura, H. (eds.) Commutative Algebra and Combinatorics. Adv. Stud. Pure Math. **11**, North-Holland, Amsterdam (1987)

[Hoc72]   Hochster, M.: Rings of invariants of tori, Cohen–Macaulay rings generated by monomials, and polytopes. Ann. of Math., **96**, 318–337 (1972)

[Hoc77]   Hochster, M.: Cohen–Macaulay rings, combinatorics, and simplicial complexes. In: McDonald, B.R., Morris, R.A. (eds.) Ring theory II. Lect. Notes in Pure and Appl. Math., **26**. M. Dekker (1977)

[Hul93]   Hulett, H.A.: Maximum Betti numbers for a given Hilbert function. Comm. in Algebra, **21**, 2335–2350 (1993)

[Kal84]   Kalai, G.: A characterization of f-vectors of families of convex sets in $\mathbb{R}^n$. Israel J. Math., **48**, 175–195 (1984)

[Kal01]   Kalai, G.: Algebraic shifting. In: Hibi, T. (ed.) Computational commutative algebra and combinatorics. Adv. Studies in Pure Math., **33**, Mathematical Society of Japan, Tokyo (2002)

[Kat68]   Katona, G.: A theorem for finite sets. In: P. Erdös and G. Katona (eds.) Theory of Graphs. Academic Press (1968)

[KK79]   Kind, B., Kleinschmidt, P.: Schälbare Cohen–Macauley Komplexe und ihre Parametrisierung. Math. Z., **167**, 173–179 (1979)

[Kod00]   Kodiyalam, V.: Asymptotic behaviour of Castelnuovo–Mumford regularity. Proc. Amer. Math. Soc., **128**, 407–411 (2000)

[KH06]   Kokubo, M., Hibi, T.: Weakly polymatroidal ideals. Algebra Colloq., **13**, 711–720 (2006)

[KR00]   Kreuzer, M., Robbiano, L.: Computational commutative algebra 1. Springer (2000)

[KR05]   Kreuzer, M., Robbiano, L.: Computational commutative algebra 2. Springer (2005)

[Kru63]   Kruskal, J.: The number of simplices in a complex. In: Bellman, R. (ed.) Mathematical Optimization Techniques. University of California Press, Berkeley/Los Angeles (1963)

[Kun08]   Kunz, E.: Introduction to commutative algebra and algebraic geometry. New Edition. Birkhäuser (2008)

[Mac27]   Macaulay, F.S.: Some properties of enumeration in the theory of modular systems. Proc. London Math. Soc., **26**, 531–555 (1927)

[Mat80]   Matsumura, H.: Commutative Algebra. Second Edition. W.A. Benjamin (1980)

[Mat86]   Matsumura, H.: Commutative ring theory. Cambridge Unversity Press (1986)

[McM71]   McMullen, P.: The numbers of faces of simplicial polytopes. Israel J. Math., **9**, 559–570 (1971)

[Mil98]   Miller, E.: Alexander Duality for monomial ideals and their resolutions. ArXiv:math/9812095v1

[Mil00a]   Miller, E.: Resolutions and duality for monomial ideals. PhD Thesis, University of California at Berkeley (2000)

[Mil00b]   Miller, E.: The Alexander duality functors and local duality with monomial support. J. Algebra, **231**, 180–234 (2000)

[MH09]   Murai, S., Hibi, T.: Algebraic shifting and graded Betti numbers. Trans. Amer. Math. Soc., **361**, 1853–1865 (2009)

[MS04]   Miller, E., Sturmfels, B.: Combinatorial commutative algebra. Springer (2004)

[OH99]   Ohsugi, H., Hibi,T.: Toric ideals generated by quadratic binomials. J. Algebra, **218**, 509–527 (1999)

[Oxl82]   Oxley, J.G.: Matroid Theory. Oxford University Press (1992)

[Par94]   Pardue, K.: Nonstandard Borel-fixed ideals. PhD Thesis, Brandeis (1994)

[PS74]   Peskine, C., Szpiro, L.: Syzygies and multiplicities. C. R. Acad. Sci. Paris. Ser. A, **278**, 1421–1424 (1974)

[Rei76]   Reisner, G.A.: Cohen–Macaulay quotients of polynomial rings. Adv. in Math., **21**, 30–49 (1976)

[Roe01]   Römer, T.: Generalized Alexander duality and applications. Generalized Alexander duality. Osaka J. Math., **38**, 469–485 (2001)

[Roe01b]   Römer, T.: Homological properties of bigraded algebras. Ill. J. Math., **45**, 1361–1376 (2001)

[STT78]   Schenzel, P., Trung, N.V., Tu Coung, N.: Verallgemeinerte Cohen–Macaulay Moduln. Math. Nachr., **85**, 57–73 (1978)

[Sch98]   Schrijver, A.: Theory of linear and integer programming. Wiley (1998)

[Sch03]   Schrijver, A.: Combinatorial optimizations. Springer (2003)

[SV08]   Sharifan, L., Varbaro, M.: Graded Betti numbers of ideals with linear quotients. Le Matematiche, **63**, 257–265 (2008)

[SVV94]   Simis, A., Vasconcelos, W.V., Villarreal, R.H.: On the ideal theory of graphs. J. Algebra, **167**, 389–416 (1994)

[Sta75]   Stanley, R.P.: The upper bound conjecture and Cohen–Macaulay rings. Stud. Appl. Math., **54**, 135–142 (1975)

[Sta78]   Stanley, R.P.: Hilbert functions of a graded algebras. Adv. in Math., **28**, 57–83 (1978)

[Sta95]   Stanley, R.P.: Combinatorics and commutative algebra. Second Edition. Birkhäuser (1995)

[Stu90]   Sturmfels, B.: Gröbner bases and Stanley decompositions of determinantal rings. Math. Z., **205**, 137–144 (1990)

[Stu96]   Sturmfels, B.: Gröbner bases and convex polytopes. American Mathematical Society (1996)

[SH06]   Swanson, I., Huneke, C.: Integral Closure of Ideals, Rings, and Modules. Cambridge University Press (2006)

[Tay66]   Taylor, D.: Ideals generated by monomials in an R-sequence. PhD Thesis. University of Chicago (1966)

[TH96]   Terai, N., Hibi, T.: Alexander duality theorem and second Betti numbers of Stanley–Reisner rings. Adv. Math., **124**, 332–333 (1996)

[VV08]   Van Tuyl, A., Villarreal, R.H.: Shellable graphs and sequentially Cohen–Macaulay bipartite graphs. Journal of Combinatorial Theory Series A, **115**, 799–814 (2008)

[Vas98]   Vasconcelos, W.V.: Computational methods in commutative algebra and algebraic geometry, algorithms and computation in mathematics. With chapters by David Eisenbud, Daniel R. Grayson, Jürgen Herzog and Michael Stillman. Springer (1998)

[Vil90]   Villarreal, R.H.: Cohen–Macaulay graphs. Manuscripta Math., **66**, 277–293 (1990)

[Vil01]   Villarreal, R.H.: Monomial algebras. Marcel Dekker (2001)

[Vil07]   Villarreal, R.H.: Unmixed bipartite graphs. Rev. Colombiana Mat., **41**, 393–395 (2007)

[Wel76]   Welsh, D.J.A.: Matroid theory. Academic Press (1976)

[Whi80]   White, N.: A unique exchange property for bases. Linear Algebra Appl., **31**, 81–91 (1980)

[Yan00]   Yanagawa, K.: Alexander duality for Stanley–Reisner rings and squarefree $\mathbf{N}^n$-graded modules. J. Algebra, **225**, 630–645 (2000)

Index

$G(I)$, 5
M-sequence, 270
P-primary, 11
S-polynomial, 33
T-polynomial, 88
$\mathrm{supp}(f)$, 4
a-invariant, 99
d-linear resolution, 112
f-vector, 15, 234
h-vector, 15, 99, 101
ith skeleton, 16, 144
k-acceptable sequence, 201
x-condition, 187
x-regularity, 187
(strongly) stable monomial ideal, 103

Alexander dual, 17
almost
    regular element, 62
    regular sequence, 62
analytic spread, 198
annihilator
    numbers, 63, 92
    of an element, 10
antichain of a poset, 201
associated prime ideal, 10
augmented oriented chain complex, 82

base
    of a discrete polymatroid, 241
    of a polymatroid, 238
    ring, 250
binomial, 38, 184
    expansion, 106

ideal, 38, 184
bipartite graph, 154
boolean lattice, 157
Borel
    fixed, 57
    order, 58
    subgroup, 55
    type, 60
branch of a simplicial complex, 172
Buchberger's
    algorithm, 38
    algorithm in the exterior algebra, 91

canonical module, 276
Cartan
    cohomology, 282
    complex, 279
    homology module, 280
category, 263
chain of a poset, 156, 201
chord, 155
chordal graph, 155
clique complex, 155
clique of G, 155
closed walk, 185
Cohen–Macaulay
    module, 273
    over K, 132
    ring, 273
    type, 274
colon ideal, 6
combinatorial
    shifted complex, 212